THE SPACE TELESCOPE

The Space Telescope photographed in 1988, surrounded by work platforms in a clean room of the Lockheed Missiles and Space Company.

The Space Telescope

A study of NASA, science, technology, and politics

ROBERT W. SMITH
Smithsonian Institution
The Johns Hopkins University

with contributions by

Paul A. Hanle
Maryland Science Center

Robert H. Kargon
The Johns Hopkins University

Joseph N. Tatarewicz
Smithsonian Institution

The right of the
University of Cambridge
to print and sell
all manner of books
was granted by
Henry VIII in 1534.
The University has printed
and published continuously
since 1584.

CAMBRIDGE UNIVERSITY PRESS

CAMBRIDGE

NEW YORK   PORT CHESTER   MELBOURNE   SYDNEY

Published by the Press Syndicate of the University of Cambridge
The Pitt Building, Trumpington Street, Cambridge CB2 1RP
40 West 20th Street, New York, NY 10011, USA
10 Stamford Road, Oakleigh, Melbourne 3166, Australia

First published in 1989
Reprinted 1990 (twice)

Printed in the United States of America

Library of Congress Cataloging-in-Publication Data

Smith, Robert W. (Robert William), 1952–

The space telescope : A study of NASA, science, technology, and
politics / Robert W. Smith with contributions by Paul A. Hanle,
Robert H. Kargon, Joseph N. Tatarewicz.

p. cm.

Bibliography: p.

Includes index.

ISBN 0-521-26634-3 hard covers

1. Space Telescope History Project (U.S.)   I. Title.
QB500.267.S55   1989
522'.29–dc19   89–707
                                                        CIP

British Library Cataloging-in-Publication applied for

It should be emphasized, however, that the chief contribution of such a radically new and more powerful instrument would be, not to supplement our present ideas of the universe we live in, but rather to uncover new phenomena not yet imagined, and perhaps modify profoundly our basic concepts of space and time.

Lyman Spitzer, Jr., 1946

In our pluralistic society any major public undertaking requires for success a working consensus among diverse individuals, groups, and interests. A decision to do a large, complex job cannot simply be reached "at the top" and then carried through. Only through an intricate process can a major undertaking be gotten under way, and only through a continuation of that process can it be kept going.

James E. Webb, 1969

Time, not reason, separates real from absurd. Nothing is so certain as what is, nothing quite so unsure as what might be. What today is, is yesterday's possibility, a selection out of what-might-have-been's.

Edward W. Constant, 1980

James Beggs: The Space Telescope will be the eighth wonder of the world.

Edward P. Boland: It ought to be at that price.

1984

Contents

Contents

Preface

The Space Telescope Project is a historic venture, above all because it aims at placing a large astronomical observatory into orbit beyond the confines of the earth's atmosphere and promises to yield valuable new insights into the past, present, and future of the universe. But it also provides an example of what is increasingly important for the scientific enterprise: the mobilization of the resources of society for a scientific-technical goal. Today, a scientific instrument placed at the frontiers of knowledge represents a political and managerial achievement every bit as significant as the technical feat.

For the historian, the Space Telescope Project presents an invigorating challenge and a superb opportunity. The complexities of the enterprise demand approaches often far outside the ordinary training and experience of the historian. Moreover, never before have we had the chance to put historians in place at the beginning of a new scientific research program so potentially significant in its implications. This program is critically important both for its anticipated scientific results and for its impact on the scientific and technical communities it involves. All actors in this play – government, academe, industry – emerge more sophisticated for having participated. Similarly, the collaboration of the Smithsonian Institution and The Johns Hopkins University, with the cooperation of the National Aeronautics and Space Administration (NASA), has enabled historians of science and technology to be present at the origins and to look in a critical way at what may be termed "the new techno-politics."

The demands that such large-scale efforts place on the scientific and technical communities are enormous. Although critical and vivid issues, such as whether or not to pursue a manned space effort and whether or not to initiate a large, new high-energy physics facility in a period of rapidly improving technologies, are at the present time capturing the headlines in the news media, there are more fundamental long-range issues that beg for attention. Scientists, government officials, contractors, and policy analysts are fully aware that new instruments for scientific leadership — new institutions and new strategies — are urgently needed if there is to be successful advancement of science and high technology in the coming years. These changes directly concern not only scientists and engineers but also legislators, policymakers, and managers.

The problems of intersectoral policy have long been of major interest to both scholars and policymakers; it is only comparatively recently, however, that (apart from wartime situations) significant proportions of the scientific and technical communities have been drawn into this arena. Scientists are, at present, becoming accustomed to dealing with the several levels of government and with the "three-body problem" of government, academe, and private industry. Accordingly, the problem of the construction of a consensus within the scientific community has become far more complex than it once was. A united front concerning what to do and how to do it becomes more difficult to achieve and, at the same time, more important politically. The scientific community has had to learn to "sell" an instrument or facility — that is, has had to learn to build a broad political base in the larger community. It has become more sophisticated in employing the communications media to this end and has learned to use state and local political resources effectively.

An examination of such questions by historians and political scientists will illuminate science and technology policy formation at the national level and will also shed light on the interesting and underexplored impact of the new situation on the nature of contemporary research and the design of the scientific instruments themselves. As a contribution toward these ends, the Space Telescope History Project (STHP) was constituted in late 1982. It grew out of discussions among historians at Johns Hopkins and the National Air and Space Museum, successfully sought funds from NASA and later from the National Science Foundation, attracted historian Robert W. Smith to anchor the project, and began a multiphase effort relating to the history of the Space Telescope.

The immediate goals were (1) to produce a book, directed at a wide audience, on the origins and development of the instrument that would illuminate its complex scientific, technical, political, and managerial roots and (2) to establish a Resource Unit comprising guides to major historical sources, transcripts and tapes of oral histories taken in the course of the project, bibliographies, and organization charts. This Resource Unit is housed at the National Air and Space Museum.

Moreover, it was our intention to serve, in some fashion, as a pilot project for historical investigations of contemporary scientific and technical institutions, programs, and projects. We believe that such historical work, when its independence and freedom to investigate are jealously guarded, can make valuable contributions to the discipline of history, to policy formulation, and to the public understanding of modern science and technology.

Robert Kargon

Acknowledgments

This book is based on work partially supported by the National Aeronautics and Space Administration (NASA) under contract NASW-3691 and the National Science Foundation (NSF) under grant SES-8510336. The opinions, findings, and conclusions expressed herein are those of the author and do not necessarily reflect the views of NASA or the NSF.

The research and writing of this book have been greatly helped by numerous persons and institutions. Without the assistance of countless archivists, librarians, records managers, astronomers, engineers, and government officials, examination of many of the sources on which this study relies would have been impossible. Special thanks are due to staff members at the following institutions: NASA Headquarters, Washington, D.C. (in particular, the unfailingly helpful staff of the NASA History Office); the Marshall Space Flight Center, Huntsville, Alabama; the Goddard Space Flight Center, Greenbelt, Maryland; the Johnson Space Center, Houston, Texas, together with the staff of the Space Shuttle Chronology Project based at the center; the Space Telescope Data Center at Lockheed Missiles and Space Company, Sunnyvale, California; the Space Telescope Science Institute, Baltimore, Maryland; the Office of Management and Budget, Washington, D.C.; the Gerald R. Ford Presidential Library, Ann Arbor, Michigan; the Space Astronomy Laboratory, University of Wisconsin–Madison; the Milton S. Eisenhower Library of The Johns Hopkins University, Baltimore, Maryland; the Archives of the Rensselaer Polytechnic Institute, Troy, New York; and the Smithsonian Institution Archives and the Library of the National Air and Space Museum, Smithsonian Institution, Washington, D.C. A number of other people also spent considerable time in rummaging through private collections of papers in search of items, and to them, in particular to C. R. O'Dell and Ernst Stuhlinger for access to their daily notes, I extend my best thanks. In addition, John Bahcall and Lyman Spitzer, Jr., graciously allowed my colleague, Paul Hanle, to work systematically through their personal files during Paul's stay at the Institute for Advanced Study at Princeton. In the Astrophysics Division in NASA Headquarters, Robert Stencel and Edward Weiler were particularly helpful.

In addition to examining documentary sources, I, together with several colleagues, set out to document some of the sources on which this book rests by conducting oral history interviews. Eighty-seven people were interviewed on tape, and a list is included in Appendix 2. To all of these people, as well as those few who preferred that their interviews be off-tape, I am deeply grateful. The oral histories cited in the Appendix 2 references, unless specified otherwise, have been processed in the Department of Space History of the National Air and Space Museum, and I am appreciative of the efforts of Martin Collins and his staff, especially Jo-Ann Bailey and Deborah Hickle, for the superb support they have provided over several years. The National Air and Space Museum's wide-ranging collection of "Space Astronomy Oral History Interviews," secured under the direction of David DeVorkin, has also proved an invaluable resource.

In addition, I have also employed in the text a number of interviews in the extensive collection of the Center for History of Physics of the American Institute for Physics, and I am grateful to the center for permission to cite and quote from these interviews. The American Institute of Physics also awarded me a grant to travel to Houston to conduct several interviews with C. R. O'Dell. These interviews are now a part of the collection of the Center for History of Physics.

Hundreds of informal conversations with people involved in various ways with the Space Telescope program – discussions over lunch, chats during breaks in congressional testimony, arguments during plane rides to meetings, and so on – have also added much to my understanding of the Space Telescope's development. Joe Tatarewicz and I were, in addition, privileged to be allowed to observe many meetings on the Space Telescope. Thanks are especially due to the members of the Space Telescope Science Working Group for their tolerant acceptance of a historian at their meetings for so many years.

One of the themes of this book is the enormous size of the Space Telescope program. This has meant the participation of thousands of people in advocating, designing, building, and testing the Space Telescope and its associated ground system. Such a huge program presents the historian with severe problems of selection. To make the writing of this book possible, I have been compelled to be extremely selective and to focus on issues that I believe reflect broad themes in the telescope's history. One result of this approach is that many people who have played significant parts in the telescope's history do not appear in these pages. I hope these people will understand why this is so, even if they do not necessarily agree with my particular selection of topics.

That the account that follows is in many places less superficial

than it would otherwise have been is largely due to the efforts of *Acknowledgments*
many people who kindly consented to comment on chapters or drafts
of the book. My principal intellectual debts are to colleagues on the
Space Telescope History Project: Paul Hanle, Robert Kargon, and
Joe Tatarewicz. Paul Hanle, in addition, wrote the first draft for a
section of Chapter 4, and Joe Tatarewicz is chiefly responsible for
compiling the appendixes. Joe was a generous collaborator during
his one-year tenure as a Guggenheim Fellow, during which he examined links between the Space Telescope's history and the development of planetary science, and good friend throughout.

I was fortunate enough to receive helpful criticisms on a working
draft of the book from Robert C. Bless, Sylvia Fries, the late Leo
Goldberg, Richard Hirsh, and John Lankford. During his tenure in
the Martin-Marietta Chair of Space History at the National Air and
Space Museum (NASM), Leo Goldberg was also very supportive of
my efforts and gave freely of his time and advice on many topics.
During his year in the Chair of Space History, Herbert Friedman,
too, helped me to understand better the development of space astronomy. In addition, David DeVorkin, Riccardo Giacconi, Henry
Hitchcock, Allan Needell, and C. R. O'Dell each commented on
several chapters, and R. C. Henry commented on an early version
of Chapter 6. Sections of Chapters 4 and 5 were presented to a NASA/
NASM symposium in June 1987, and several participants, notably
W. Henry Lambright, who helped me to crystallize my ideas with
his notion of "ad-hocracy," made helpful comments. The colleagues
who gather at the Seminar in the History of Twentieth Century
Science, which Allan Needell has organized for several years at NASM,
have also provided a constant source of ideas. I alone am responsible
for any mistakes remaining in the book.

Research assistance was provided by a number of Johns Hopkins
undergraduates and graduate students. These included Bruce Hevly,
Erik Ledbetter, Lisa Rosner, Susan Siggelakis, and Mary Voss. I am
especially grateful to Ronald Brashear for his tenacious efforts over
several years. At the National Air and Space Museum, several volunteers worked under Joe Tatarewicz to help maintain sets of working files as well as a collection of slides and photographs. We were
particularly fortunate to obtain the services of Ben Pierce and Caro-
lyn Schmidt.

I would also like to acknowledge the extremely able assistance of
Patty Zegowitz in dealing with a host of administrative matters.
Gail Schley and Amanda Young aided too in the solution of numerous bureaucratic glitches. At Goddard, Dee Cartier somehow always
managed to find me an office to work in. The Space Telescope Science Institute generously provided an office for a period too. I also
very much appreciate the patience and help of the staff of the Cambridge University Press, particularly Helen Wheeler.

Acknowledgments   During the writing of this book and the research that preceded it, I was fortunate to be associated with two excellent and stimulating departments: the Department of Space History at the National Air and Space Museum, and the History of Science Department at The Johns Hopkins University. In fact, the staunch support of the museum and the History of Science Department at Hopkins proved essential to the completion of this book. Special thanks are due to Robert Kargon at Hopkins and to the National Air and Space Museum's directors and senior staff for their enthusiastic and resolute backing over several years. I owe a special debt to Paul Hanle for his unstinting encouragement. I am also very grateful for being given a completely free hand to define and pursue the research for, and later the writing of, this book as I thought fit.

Best thanks of all to Jennifer.

Robert W. Smith

Introduction

Thus Sr I have given you a short accompt of this small Instrument, which though in it selfe contemptible may yet be looked upon as an Epitome of what may be done according to this way.

Isaac Newton, 1669

THE YALE PROFESSOR'S IDEA

May 15, 1985. The atmosphere in the small, crowded meeting room at the Goddard Space Flight Center is tense. There are not enough seats to accommodate all those present, and people are jammed in the doorway. The project manager is patiently facing a barrage of probing questions and comments from astronomers anxious to learn the latest news about the testing and assembling of the Space Telescope, potentially the most powerful optical telescope ever built.

Everyone present knows that the long-awaited launch of the Space Telescope aboard the Space Shuttle *Atlantis* is no more than fifteen months away. Many of the astronomers, engineers, and administrators in attendance have devoted years to building the telescope, and they are pressing hard to ensure that there will be no blunders or cut corners at this crucial stage. They are fully aware that mistakes made now will be costly.

One of the astronomers present has attended hundreds of similar meetings on the Space Telescope. His involvement in arguing for, justifying, and then helping to develop the Space Telescope stretches back almost four decades, longer than the commitment of anyone else. He sits now, a tall, rather thin figure, intently watching and on occasion participating in the rapid ebb and flow of the discussion.

Lyman Spitzer, Jr., is one of the nation's most distinguished scientists. Born in Toledo, Ohio, almost seventy-two years earlier, he has won a host of scientific awards and prizes. One of a generation of U.S. scientists who learned the fundamentals of science politics during World War II, Spitzer's courteous and modest manner belies

I

The Space Shuttle *Atlantis* lifts off into the night sky from Cape Canaveral in November 1985. In early 1985, *Atlantis* was slated to launch the Space Telescope in 1986.

great determination to further those scientific projects in which he believes, and for such projects he is prepared to fight tenaciously. As a colleague who has been closely associated with him put it, "Lyman is a master of listening to everybody and doing in the end what he wants to do."[1]

The meeting at the Goddard Space Flight Center in 1985, however, is separated by a vast gulf from the events of 1946, a very important date in the history of the Space Telescope. Astronomers have long dreamed of operating telescopes above the obscuring layers of the earth's atmosphere. Writing in 1933, Spitzer's future teacher, H. N. Russell, looked forward to the time when astronomers would be able to make such observations. He even joked that when astronomers died, they might be "permitted to go . . . instruments and all, and set up an observatory on the moon."[2] But in 1946, Spitzer, then a thirty-three-year-old astronomy professor at Yale, became the first person to lay out in detail the enormous scientific benefits to be won by placing a large optical telescope in space, benefits that, in the face of numerous obstacles, have spurred on Spitzer and many others over so many years.

It was a difficult road that Spitzer and the builders of the Space Telescope traveled in transforming an idea into reality. The Space Telescope story is one of human undertakings involving far more

The Yale professor's idea

Two generations of Princeton astrophysicists: Henry Norris Russell (left) with Lyman Spitzer, Jr., in Spitzer's office at Princeton, ca. 1948. Their different career paths demonstrate some of the fundamental changes in the scientific enterprise since World War II. (Courtesy of Lyman Spitzer, Jr.)

than the launching of a spacecraft forty-three by fourteen feet. It involved speculative visions, brilliant insights, years of grinding work, triumphs and mistakes, and careers made and careers wrecked, as well as the tragedy of the explosion of the Space Shuttle *Challenger* in January 1986 and the lengthy delay that followed in the launching of the Space Telescope.

Moreover, if we are to understand the shaping of the Space Telescope and the forces that have sustained it over such a long period, as well as those that have fashioned its development, we must begin even further back in time than 1946. As we then move forward to the meeting room at Goddard Space Flight Center in 1985 and then toward the telescope's launch, we shall see that the form the Space Telescope has taken reflects much more than simply relentless and unyielding scientific and technical considerations.

It is, in fact, now a commonplace that the directions taken by large scientific and technological research projects of the last few decades have been heavily influenced by extrascientific pressures.[3] Indeed, one of this book's key arguments is that the Space Telescope has been a focus for activities resulting from the different and sometimes conflicting interests, the power struggles, and the contrasting objectives of widely varying institutions, groups, and individuals. The Space Telescope's design and the manner in which it has been built have therefore resulted from international collaboration and astonishingly complex interactions among the scientific community, the government, and industry.

But to better comprehend the forces that have shaped the telescope and to help the reader unfamiliar with telescopes and their evolution, this Introduction seeks first to define the place of the Space Telescope in the long history of telescope building. We then move on to describe some of the themes to be encountered in later chapters, particularly the Space Telescope as an example of the kind of enterprise known as Big Science. This book, indeed, is in many respects an in-depth account of one type of Big Science in the making, and the central thread that we shall follow is provided by the experiences of the astronomers who sought to win approval for, and then helped to build, the Space Telescope. One could find many ways to approach the telescope's history, and this is just one, but one that it is hoped will be a particularly fruitful approach. Certainly at this stage in the telescope's development, one cannot claim to have written a definitive account. Not only has the telescope's observing life yet to end, but its technology links to military spacecraft mean that a full analysis of its design heritage must await the lapse of many years.

The early chapters of this book focus on the ways in which various individuals, groups, and institutions were drawn into preliminary studies for a large optical telescope in space and how the scientific community, government, and industry were marshaled to develop ideas and support the building of the instrument. In Chapter 1, the focus is on attempts to interest astronomers in the telescope – how its advocates sought to persuade their colleagues of its worth. A central theme of this chapter, and of the first half of the book, is coalition-building, especially the coalition that would make the telescope feasible, not only in a technical sense but also in a political sense, by winning approval for it from the White House and Congress. In Chapter 2, the emphasis switches to NASA and the ways in which proponents of the Large Space Telescope (for that had become its name) sought to embed the telescope within NASA's own set of interests, including packaging it together with the Space Shuttle, and thereby win that agency's favor for its construction. Chapter 3 examines the early interactions among the astronomers, NASA, and industry, particularly how the agency's concern to cut the costs of the proposed spacecraft shaped important decisions on the telescope's design, as well as the structure of the program to build the telescope.

Chapters 4 and 5 trace the convoluted course of the three-year struggle between 1974 and 1977 to win approval for the telescope from the White House and Congress. That struggle led to a change in name, simply the Space Telescope, and it also had crucial consequences for the telescope's design, the nature of the program to build the telescope, and the telescope's future progress during the design and development phase. Should a large optical space telescope be built? – that is a question heard throughout the narrative in the first three chapters, but by Chapters 4 and 5 it has reached a crescendo, with large-scale lobbying campaigns by a broadly based coalition of supporters urging the White House and Congress to answer yes.

In Chapter 6 I take a step back to examine the history of the relationship between scientists and NASA, before analyzing the debate surrounding the various plans for the telescope's scientific operations. Much of this hard-fought debate centered on the issue of control: Who was to control its scientific operations, NASA or astronomers outside the agency? NASA's solution to this debate, however, also had much to do with the campaign to win approval for the telescope and the construction of a coalition to make the telescope politically feasible. Chapter 7 deals chiefly with the way in which NASA pieced together the program to design and build the Space Telescope, an international activity that required the coordination of numerous organizations, groups, and individuals in government, industry, and academe. The program design, I further argue in Chapters 8 and 9, was one of the main causes contributing to the difficulties encountered during the telescope's building. The main thrust of Chapters 8 and 9, nevertheless, is the development of the program between 1977 and the time in 1986 when the telescope seemed to be very close to launch – only to have the *Challenger* disaster intervene. In Chapter 10, I draw some conclusions from the history of the Space Telescope to date, particularly what that history has to say about NASA, about the influence of government patronage on the technology constituting a new scientific tool, and about Big Science and the nature of the scientific enterprise in the late twentieth century. But our starting point will be the Space Telescope's place in the history of telescope building, for, as will be seen, in some respects telescopes, because astronomers have long come to accept that the construction of ever more powerful telescopes is essential to the development of astronomy, seem to possess lives of their own.

A NEW TOOL

In 1669, Isaac Newton wrote to a friend to describe a new scientific instrument. The letter recounted how he had constructed it with his own hands and, with its aid, had seen moons orbiting Jupiter, as well as the crescent of Venus. This momentous device would soon propel its obscure maker to the very center of the scientific world.

Newton had, in fact, built a "reflecting" telescope, so called because it used reflections from two mirrors to produce focused images.[4] It was the first working example of a kind of telescope that would eventually become the primary research tool for astronomers. Newton, however, did not invent the telescope. Exactly who did and when is not clear.[5] What is certain is that a handful of natural

A drawing by Isaac Newton showing his first reflecting telescope. (Courtesy of the Royal Society.)

Types of telescopes: Newtonian reflector. Light enters the open front end of a Newtonian reflector. It travels to a concave primary mirror. From there it is reflected to a flat secondary mirror and directed to an eyepiece mounted in the side of the tube. The first working version of this type of telescope was built by Isaac Newton around 1669 (see also the drawing of Newton's telescope). It did not take engineers and scientists long to reject the Newtonian arrangement as a possible design for the Space Telescope. It would, for example, have compelled the designers to mount the scientific instruments on the side of the tube and so would have greatly complicated the Telescope's distribution of mass. (Courtesy of Stansbury, Ronsaville, Wood Inc.)

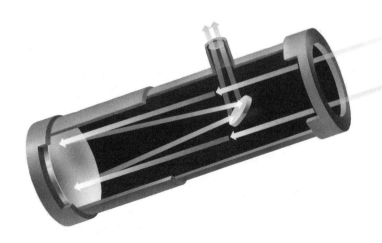

Types of telescopes: refractors. The first type of telescope to be invented was the refractor. Brought into prominence by Galileo early in the seventeenth century, the early refractor had a single objective lens. This lens, together with the eyepiece, brought the image to the eye. Large lenses, however, are difficult to make. In large part, this fact, in combination with the desire of astronomers for ever larger apertures to reach fainter and more distant objects, has meant that the major optical telescopes of the twentieth century have been reflectors.

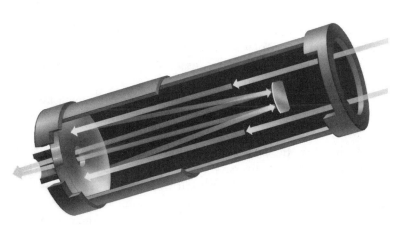

Types of telescopes: Cassegrain reflector. As in the Newtonian reflector, light enters the open front end of the Cassegrain reflector. It travels down the telescope tube to a concave primary mirror. From the primary, the light is reflected to a convex secondary mirror. Instead of being directed out of the side of the tube, as in the Newtonian reflector, it is then bounced back toward the primary mirror. A hole in the center of the primary permits the light to be collected by an eyepiece positioned behind the primary. In this way, the optical layout is "folded." This results in a more compact arrangement than for the Newtonian reflector. It also enables the instruments employed with a Cassegrain reflector to sit behind the primary mirror, a great advantage in balancing a large and massive telescope. The Space Telescope is itself a type of Cassegrain reflector. (Courtesy of Stansbury, Ronsaville, Wood Inc.)

philosophers were applying the new instrument to the study of the heavens sixty years before Newton devised his reflecting telescope. These earlier types of telescopes were "refracting telescopes." Refractors work on principles different from those of Newton's reflector. They contain systems of lenses and rely on refraction to produce focused images, not the mirrors and reflections Newton was to employ so adroitly.

Before the development of the refracting telescope, astronomers had been restricted in their studies to those objects visible to the naked eye. But from the early seventeenth century onward, the telescope gave them the chance to observe fainter astronomical bodies, as well as see familiar objects in novel and more detailed ways.

The most prominent of those to exploit the new refracting telescope was Galileo Galilei. Galileo was not the first to use the telescope for astronomy. He perhaps was not even the second,[6] but because of his writings advertising a string of spectacular findings, he was to be the leader in bringing the device to widespread attention.[7] Those discoveries included four moons of Jupiter, as well as the findings that our moon's surface is not smooth, as natural philosophers had long argued, but is pitted by craters and mountains, and that the Milky Way is composed of millions upon millions of stars. Indeed, "the telescope produced a vast change in kind, magnitude, and scope of the data base of astronomy, providing the observational materials on which a revolution would eventually become founded."[8] But the truly striking point is that, for the first time, a set of phenomena had been added to science that required the mediation of an instrument to be detectable. Such instruments today underpin the scientific enterprise and shape the research to be conducted; telescopes hold a central place among them.

BIG IS BETTER

Refracting telescopes continued to be used extensively for astronomical research until the late nineteenth century. They were nevertheless fated to fall victim to the insatiable desire of astronomers for light, and yet more light.

Other things being equal, the larger a telescope's main mirror or lens (and so the larger the light-collecting area), the dimmer the objects that can be observed. For practical purposes, refractors reached a limit in 1897. In that year a refractor was completed for the Yerkes Observatory of the University of Chicago that had a forty-inch objective lens. To build a refractor with a larger objective lens would be extremely taxing. First, the main lens would be so heavy that it would be a formidable task to hold it in place without it bending out of shape. Second, making the glass for such large lenses is very difficult. Third, lenses with diameters of more than forty inches would be so thick that much light would be lost in traversing them. By the late nineteenth century, reflectors were also undergoing rapid advances; they were cheaper than refractors and were better suited to the new kinds of knowledge astronomers would be pursuing early in the twentieth century.[9] In fact, although various bold astronomers had, usually on their own initiative and with their own money, built large reflecting telescopes in the eighteenth and nineteenth centuries, the first decades of the twentieth saw reflectors dominate and supplant refractors as the best tool for most types of astronomical research.

Foremost among the new generation of reflectors was the Hooker Telescope on Mount Wilson in California. First aimed toward the heavens in 1917, the Hooker Telescope's main, or primary, mirror is 100 inches across. In astronomical jargon it is referred to as the 100-inch Hooker Telescope, for reflectors are generally known by the diameter of the primary mirror, sometimes also referred to as the "aperture" of the telescope.

The Hooker Telescope was to play a very important role in the career of Edwin P. Hubble, the astronomer after whom the Space Telescope is named. Born in Marshfield, Missouri, in 1889, Hubble took a B.S. in mathematics and astronomy at the University of Chicago. He then went to Oxford as a Rhodes Scholar to read jurisprudence. After a brief period at a law office in Louisville, Kentucky, Hubble decided that astronomy was the career he really wanted to pursue, and he enrolled as a graduate student at Yerkes Observatory.

In 1917 he was offered a post at the Mount Wilson Observatory. This, however, was shortly after the United States had entered the Great War (World War I), and Hubble chose instead to enlist as a private in the U.S. Army and go to fight in Europe. The Mount Wilson position was nevertheless held open for him. At last, in 1919, at the age of thirty, Hubble "settled down to the work that was to bring him fame."[10]

At the start of the twentieth century, astronomers held what now appears as an incredibly restricted view of the physical universe. Almost all believed that there was only one galaxy of stars visible with even the largest telescopes, our own galaxy, the Milky Way. This star system, moreover, was estimated to be only a few thousand

Big is better

The Earl of Rosse's "Leviathan of Parsonstown." Completed in 1845, Rosse's giant reflecting telescope possessed a primary mirror seventy-two inches in diameter, making it the most powerful telescope built to that time. The telescope was mounted between piers of masonry. (Courtesy of the Director and Trustees of the Science Museum, London.)

Walter Adams (left), director of the Mount Wilson Observatory, James Jeans (center), and Edwin P. Hubble (right), after whom the Hubble Space Telescope is named, shown against the dome of the 100-inch Hooker Telescope in 1931. (Courtesy of the Archives of California Institute of Technology.)

light-years in diameter. By the late 1920s, such theories had been discredited, and to a large extent these changes in theory were driven by observations made with powerful telescopes on mountains in the western United States, chief among them the Hooker Telescope.

With the Hooker Telescope, Hubble, in part with his assistant Milton Humason, conducted researches in the 1920s and 1930s that convinced astronomers of the two central aspects of the current conception of the universe: first, that our Milky Way galaxy of stars is not alone, that space is populated by thousands of billions of galaxies; second, that these galaxies are not stationary, but are flying apart, and thereby exhibit the overall expansion of the universe.[11]

The Space Telescope's chief mission will be to try to observe the farthest objects in the universe. Hubble's profound discoveries, made by means of observations of extremely distant bodies, therefore explain why the Space Telescope has officially been named the Edwin P. Hubble Space Telescope, or Hubble Space Telescope for short.

The 100-inch Hooker Telescope that was to be so brilliantly used by Hubble owed much to the vision and dynamism of George Ellery Hale, the man who brought Hubble to Mount Wilson. Born in Chicago in 1868, Hale was a highly successful entrepreneur of science, as well as an accomplished astrophysicist. Even before he had reached the age of thirty, he had become the chief agent in bringing into existence the Yerkes Observatory, an institution based on what was then a novel principle: combining an astronomical establishment with a physical laboratory. Yerkes was also where Hubble was to receive his graduate training.

The visionary Hale was also the driving force behind the construction of what has been for many years the world's most powerful optical telescope, the 200-inch (five-meter) Hale Telescope perched atop Palomar Mountain in California.[12] Beside Hale's giant monument, Newton's reflecting telescope seems puny indeed. Whereas the primary mirror of Newton's instrument weighed a mere pound or two, that of the 200-inch Hale Telescope weighs some fourteen tons. The light-grasp of the 200-inch Hale also is vastly greater than

Big is Better

The 200-inch Hale Telescope on Palomar Mountain. An idea of the scale of the telescope can be gained from the fact that the observing cage, seen at the top of the telescope tube, is itself six feet in diameter. (Courtesy of the Archives of California Institute of Technology.)

Space Telescope stability and pointing accuracy. Distances on the sky are not given in our familiar linear measures (such as feet or miles), but in *angular measures*. Angular distances are measured in degrees. The Big Dipper extends for about 25 degrees (or 25° in the usual notation). This is about 5° larger than the span of one fully outstretched hand at arm's length. A one-foot ball that is placed about fifty-seven feet away covers an angle of 1°. This is twice the angular size of the sun or moon, each of which subtends an angle of about $\frac{1}{2}$°. When we consider the pointing and stabilization of the Space Telescope, it is necessary to talk in much smaller increments than degrees. In fact, a degree is subdivided into sixty minutes of arc, and each minute is subdivided into sixty seconds of arc. The moon, which extends for $\frac{1}{2}$° on the sky, therefore has an angular size of thirty minutes of arc, or 1,800 seconds of arc. The Space Telescope, however, has to be stabilized to 0.012 second of arc. This is equivalent to directing a laser in New York toward a dime in Washington, D.C., and keeping the laser beam trained on the dime.

that of Newton's telescope. Its huge collecting area and modern detectors means that an astronomer can use it to examine objects millions of times fainter than any Newton could ever observe.

THE SPACE TELESCOPE: AN OUTLINE

Despite the vast size difference between Newton's telescope and the Palomar Mountain 200-inch, they are linked by one of the most conspicuous features of reflecting telescopes: Their basic design has remained stable. Only in the last decade or so has there been much change in the design of reflectors, a change caused by the development of multi-mirror telescopes, in which light is collected by several mirrors rather than one mirror. Newton's telescope is, with only a little imagination, easily recognizable as the ancestor of the 200-inch. Indeed, many of the problems faced by Newton in making his crude instrument had to be addressed, although to incredibly greater degrees of accuracy, by the builders of the Palomar 200-inch and the Space Telescope. To summarize, there are three basic issues with which the builder of any reflecting telescope must be concerned. First, the mirrors must be accurately shaped. Second, there must be some means to maintain the alignment, as well as distance, between the primary and secondary mirrors. Third, the telescope must be capable of being pointed precisely to astronomical objects for long periods of time. If, for whatever reason, these three conditions are not met, be it with Newton's telescope or the Space Telescope, blurred and out-of-focus images will result.

Newton's telescope was essentially a tube six inches long, with mirrors at both ends. The front was open to admit light. At the back end was the primary mirror, the means of collecting the light from astronomical objects. In Newton's first instrument this was a small concave mirror about two inches in diameter, a mirror he had ground by hand. From the primary mirror the light was reflected back toward the open end of the tube to a much smaller "secondary mirror." Its task was to divert the reflected light to a lens system — the eyepiece — in the side of the tube. The eyepiece allowed an observer to inspect the image. When Newton placed his eye to the eyepiece and carefully adjusted the distance between the primary and secondary mirrors by use of a screw behind the primary, he could see in-focus images of those objects toward which he directed his telescope.

The most obvious feature of the twelve-ton Space Telescope is that it, too, is basically a tube with a mirror at each of its ends. Light from astronomical objects enters the open front end of the tube. It journeys to a concave primary mirror ninety-six inches in diameter, the surface of which has been ground and polished to exceptionally high accuracy. From there it is reflected back to a convex secondary mirror some twelve inches across. In order to keep the telescope's image in focus, the distance between the primary and secondary mirrors of the Space Telescope must be maintained within minute limits. Although they are 193 inches apart, the separation between the two mirrors must not change by more than a ten-thousandth of an inch. The light is not diverted to the side of the tube as in Newton's telescope. Instead, it is bounced back toward the primary mirror and through a hole in the middle of the primary. Such a telescope is termed a "Cassegrain reflector" after its French inventor. Once the light has passed through the hole, it is available for analysis by a battery of scientific instruments.

Modern-day astronomers rarely place an eye to their telescopes, nor will the users of the Space Telescope. The Space Telescope is a completely automated observatory, and the scientific data it receives must be radioed back to earth. All of the Space Telescope's instruments carry sophisticated electronic systems for detecting and examining the incoming light.

Photographic plates are still widely used in astronomy to record the images of large areas of sky. On board the Space Telescope, however, will be the modern electronic versions of a photographic plate, so-called Charge Coupled Devices. The Faint Object Camera, another electronic camera, at the heart of which is a kind of television tube, will also be carried by the telescope. The Space Telescope, in addition, will have a "High Speed Photometer," an instrument for measuring very rapid fluctuations in the light output of stars and galaxies. The telescope's guidance system can also be used to measure accurately the positions of astronomical bodies; it thereby will act as another scientific instrument. Two of the telescope's six instruments will be spectrographs, instruments that split the light from astronomical objects into its constituent wavelengths for analysis.

Isaac Newton not only built the first reflector but also was the first to realize that white light is a combination of different colors. Many of Newton's early optical experiments involved the division of sunlight into a spectrum by use of prisms.[13] Later researchers found that dark lines crossed the band of colors of the sun's light. These spectral lines, it would turn out, would change fundamentally the practice of astronomy and the goals of astronomers.

Until the middle of the nineteenth century, astronomers devoted themselves to saying *where* any astronomical body was, not *what* it was; the physical nature of distant stars was taken to be forever beyond human analysis. But in 1859 a German chemist discovered that a substance capable of emitting a certain spectral line would strongly absorb that same line. He then explained the appearance of dark lines in the spectrum of the sun by the absorption of the corresponding wavelengths in the atmosphere of the sun. Because studies in the laboratory could reveal which substances emitted or absorbed which lines, the composition of celestial bodies could then be investigated. By means of a comparison of the spectrum of the astronomical object with laboratory spectra, its composition could be probed. The spectral lines therefore provide a kind of celestial fingerprint that can distinguish between different substances. The analysis of starlight thus gave new and unprecedented powers to astronomers, and the instrument that made this possible was the spectroscope.

A spectroscope attached to a device to record the observed spectrum is known as a spectrograph. There will be two spectrographs aboard the Space Telescope: the Goddard High Resolution Spectrograph and the Faint Object Spectrograph. The former is for observing in detail brighter objects, whose light can be widely dispersed. The latter, in contrast, is principally for examining in somewhat less detail the light from fainter objects.

WHY SPACE?

The combination of large, ground-based optical telescopes (such as the Hale 200-inch) and their associated sensitive instrumentation makes for powerful astronomical tools. The $2 billion question (for such is the sum spent on the Space Telescope to ready it for launch) is, What added advantages are to be won by launching optical telescopes into space?

As astronomers have long been acutely aware, there are limits to what even the largest ground-based optical telescopes have been able to achieve. Central among these limits is that even though the largest telescopes usually are several thousand feet above sea level, they are still at the bottom of an ocean of air. Passage through the atmosphere refracts, or bends, starlight. A star's apparent position is rarely its actual position, and conditions in the atmosphere can drastically affect the quality of a telescope's images.

Nor is the atmosphere often steady or tranquil. It is also far from homogeneous; astronomers, in fact, think of it as composed of blobs called "seeing cells." Starlight entering the upper reaches of the atmosphere encounters many seeing cells on its passage to a telescope on the ground. As the cells change position, the incoming light is bent by various amounts. In consequence, the star's image shifts its apparent position and jiggles around the sky. This is familiar to anyone who has ever looked at a starry sky. With large telescopes, in which the diameter of the primary mirror usually is greater than the size of the seeing cells, the resulting image is composed of lots of small images or speckles. "The problem," wrote one astronomer, "is that there are many such speckle images, each faint and moving, so that time exposures average the speckles out." The result is an image, known by the term "seeing disk," much larger than any single speckle.[14] At any given telescope site, the size of the seeing disk varies with atmospheric conditions. However, from even the best sites on earth, it is unusual for the seeing disk of a star to be much smaller than one second of arc, and the worse the conditions, the larger the seeing disk.

The Space Telescope's images of stars, if all goes according to plan, will instead be far smaller than one second of arc, for by operating above the atmosphere, the telescope will avoid completely the problem of seeing. The wave nature of light also sets a limit to the quality of a telescope's images. If a telescope's mirrors are so accurately shaped that this limit is reached, then the optics are said to be "diffraction-limited." The Space Telescope's optics are very nearly diffraction-limited, and so the excellence of the telescope's images will be fixed only by the laws of optics, the quality of its mirrors, and how accurately and steadily the Space Telescope can be directed toward its targets, not by the state of the atmosphere.

The atmosphere presents still another major problem for astronomers. Most of the radiation arriving at the earth from outer space is blocked by the atmosphere from penetrating to the earth's surface. X-rays from high-temperature phenomena in stars and galaxies, for example, cannot reach the ground. Astronomers who wish to discover what the universe looks like in the "light" of x-rays must therefore fly their x-ray telescopes above the atmosphere.

Astronomers who use optical telescopes based on the ground can gain only a very restricted view of the universe. It is rather like tuning in to stations on a radio waveband. From the earth, only a few "stations" can be received. Moving above the atmosphere not only improves the reception but also makes it possible to tune in to all of the stations, or regions of the electromagnetic spectrum.

One portion of the electromagnetic spectrum that is almost com-

Introduction

The Spectrum. The earth's atmosphere keeps most of the electromagnetic radiation directed at the earth from ever reaching its surface. The atmosphere, for example, prevents the passage of x-rays from astronomical bodies from penetrating to the ground. To study the universe in x-rays, astronomers must therefore place their x-ray detectors aboard rockets or satellites. Also, ground-based optical astronomy is restricted almost entirely to the narrow spectral band corresponding to visible light. Radiation in the adjacent ultraviolet and infrared regions is almost totally blocked. However, the Space Telescope, because of its position almost 600 kilometers above the earth's surface can make observations in the optical, ultraviolet, and infrared regions of the spectrum. The upper edge of the shaded area marks the level at which the intensity of radiation at each wavelength is reduced to half of its original value.

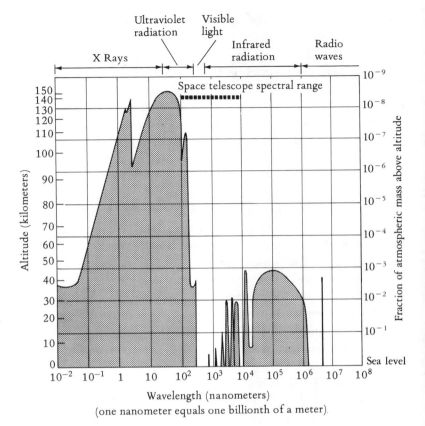

Wavelength (nanometers)
(one nanometer equals one billionth of a meter)

pletely blocked by the earth's upper atmosphere is the ultraviolet. The ultraviolet is also an important region of the spectrum for astronomers; it is the one in which the hottest stars radiate most of their energy. To pursue ultraviolet astronomy, it is essential to fly telescopes above the atmosphere.

The Space Telescope will orbit a few hundred miles above the earth. It is an optical *and* ultraviolet telescope, as its instruments are also capable of detecting ultraviolet light. The Space Telescope's instruments have been designed to be exchanged for new ones while the telescope remains in orbit. Plans for the telescope's future include flying scientific instruments sensitive in the infrared region of the spectrum, thereby increasing the telescope's wavelength coverage still further and enabling the Space Telescope to reach many more wavelengths than can an equivalent optical telescope on the ground.

There are still other advantages deriving from the telescope's position in space. First, the upper atmosphere glows faintly. This is a particular problem when astronomers wish to observe very dim objects, which can become lost in the so-called air glow. Even in space, there is still unwanted light from the "zodiacal light" (light due to sunlight scattered by particles of dust between the planets) and stray

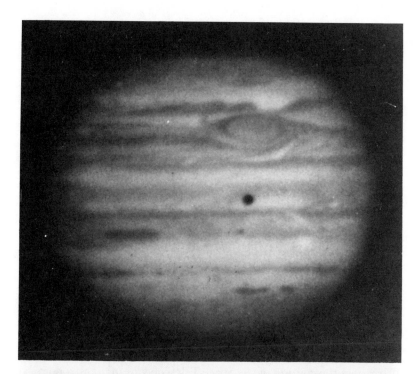

An idea of the advantages, in terms
of resolution, of a space-based telescope over a large ground-based telescope is provided by comparison of
two images of Jupiter. One was made
with a ground-based instrument, and
the lower came from a spacecraft
flying by the planet. If the Space
Telescope works to plan, it should
routinely provide images as good as
that from the flyby spacecraft, even
from its position in low earth orbit.
(Courtesy of NASA.)

17

starlight scattered by dust particles spread between the stars. The scattered sunlight and starlight nevertheless compose only a small fraction of that due to air glow. Hence, given its great resolving power, and by virtue of being above the air glow, the Space Telescope will reach much fainter astronomical objects than can ground-based telescopes that have far larger primary mirrors. Although the Space Telescope's mirror is not particularly large by the standards of ground-based telescopes, its position in space is designed to more than compensate for whatever it lacks in size. Without the imposition of the atmosphere, it should routinely provide images of stars whose seeing disks are around 0.1 second of arc in size, about ten times better than what is regarded as good performance for the best large ground-based telescopes.[15] The Space Telescope should detect stars about fifty times fainter than the faintest observable from the ground. So, roughly speaking, stars should be observed at distances about seven times greater than with ground-based telescopes, a significant jump.[16]

IMPLICATIONS FOR ASTRONOMICAL RESEARCH

What does the coming of the Space Telescope mean for astronomical research? First, if the Space Telescope works to plan for its anticipated fifteen to twenty years in orbit, many astronomers think that it will represent such an increase in observing capability that, much as Lyman Spitzer thought in 1946, perhaps the most significant discoveries to be made with its aid will be those they never expected. One of the chief lessons astronomers draw from the history of astronomy is that larger and more penetrating telescopes often prove crucial in making discoveries and in raising new questions, as well as in reshaping old questions about the universe. Thus, the building of the Space Telescope embodies for astronomers in part an act of faith, the belief that the construction of an optical telescope more powerful in many ways than any ever constructed will, as other pathbreaking telescopes have done in the past, make possible major new discoveries, and around these discoveries, new theories will be woven. As Laurence Fredrick, an astronomer who has been closely involved with the Space Telescope since the early 1960s, put it, if "it wasn't for that, I think we'd all quit!"[17]

In addition to what some participants in the building of the telescope have called the conscious expectation of the unexpected, astronomers have specific questions in mind for the telescope to tackle. Let us take just two examples. The first concerns cosmology, the study of the large-scale structure and evolution of the universe, generally accepted as the main scientific motivation for building the Space Telescope. One striking and frequently cited example of how the Space Telescope will be able to perform concerns the so-called

Cepheid stars in the Virgo cluster of galaxies. Astronomers expend prodigious amounts of effort in trying to measure the distances of astronomical bodies, but such are the errors in astronomers' measurements that even the distances to relatively nearby stars are known imprecisely. To reach remote objects, astronomers have to ascend increasingly rickety rungs of what is called the "distance ladder." Astronomers think that the cumulative errors in the various techniques are so big that by the time one is examining very dim galaxies, current estimates of distances may be somewhere between half as large and twice as large as the actual values. Such uncertainties are disturbing to astronomers because they mean, for example, that the universe's expansion cannot be gauged unambiguously. Will it expand forever, or will it eventually reverse its expansion and start to contract? Also, how much time has passed since the Big Bang? Without accurate distances to the remotest objects, it is difficult to reach firm conclusions on these basic questions, questions that have been at the heart of astronomy for over half a century and that astronomers armed with ground-based optical telescopes have threatened, but never managed, to solve to their satisfaction.

An unusual cluster of galaxies in the constellation of Hercules as imaged by the four-meter telescope on Kitt Peak in Arizona. The chief scientific motivation for the Space Telescope has always been cosmology and the study of such extremely distant objects. (Courtesy of the National Optical Astronomy Observatories.)

Introduction

Cepheids are fluctuating and very bright stars. But they vary their light output in a precise and regular manner. Moreover, the length of time between the peaks of a Cepheid's light output is directly related to its actual brightness. By comparing how bright the Cepheid appears to be and how bright it actually is, an astronomer can easily calculate its distance. Thus, astronomers regard Cepheid stars among their most useful tools to determine distances to galaxies. If all works to plan, the Space Telescope, unlike even the most powerful of ground-based telescopes, will possess sufficient light-grasp and resolving power that astronomers judge it will stand a good chance of detecting and measuring the light output of Cepheids in the Virgo cluster of galaxies, a huge collection of galaxies about sixty-five million light-years away. If so, the distance to the Virgo cluster of galaxies can be fixed to an unprecedented accuracy. Furthermore, the Virgo cluster of galaxies is itself a crucial rung in the distance ladder. Astronomers believe, therefore, that good estimates to the Virgo cluster will increase the precision with which the distances to even more remote galaxies and clusters of galaxies can be

The Crab nebula. One of the most studied objects in astronomy, the Crab nebula is the result of a supernova explosion that was observed from the earth in 1054 A.D. At the nebula's heart is a pulsar, an object high on the observing list of the Space Telescope's High Speed Photometer. (Courtesy of the National Optical Astronomy Observatories.)

20

measured, which in turn will mean that the universe's expansion and the time since the Big Bang can be better determined.

A second example of the possible uses of the Space Telescope focuses on astronomical objects that fluctuate very rapidly in brightness, numerous examples of which have been found in recent years. A well-known example is the so-called pulsar at the heart of the Crab nebula. So dense that a bit of the star the size of a grain of sand weighs thousands of tons, the pulsar beams out flashes of radiation thirty times per second. One of the Space Telescope's scientific instruments, the High Speed Photometer, has been explicitly designed to measure very rapid fluctuations and is sensitive enough to measure even those that occur in a hundred-thousandth of a second. To put it another way, the High Speed Photometer can make 100,000 measurements per second. The instrument's designers and builders hope that this will be valuable in examining many kinds of rapidly varying objects, including the pulsar in the Crab nebula, as well as potential "black holes." A black hole is a body whose gravitational pull is so great that not even light can escape its grip. Black holes cannot be seen directly, but as material falls into a black hole, it should exhibit rapid changes in light output. Among the programs planned for the High Speed Photometer will be the search for such in-falling matter and, by implication, the black hole into which it is falling.

BIG SCIENCE

In this Introduction we have already compared Newton's telescope and the Space Telescope. Such a comparison also vividly illustrates some of the changes in the scientific enterprise between the end of the seventeenth century and the end of the twentieth century.

Newton built his telescope in a short time, perhaps no more than a few days. He probably needed no more tools than he could readily obtain, in combination with his own wits and mechanical dexterity. Newton's construction and use of his telescope provide an example of "little science," science in which an individual largely conceives and executes a project, or in which a chief investigator leads or directs a small group or laboratory in conceiving and conducting a project. This sort of enterprise is very different from Big Science, the type of science usually associated with large multidisciplinary teams and big machines.

Optical astronomy, however, has traditionally been little science, and astronomers, even when employing big telescopes, have long been used to working alone or in small groups, perhaps with the aid of engineers and graduate students, building their own apparatus for use on telescopes and taking their own observations. It is also a field in which the United States has led the world for many decades.

Andrew Carnegie (left) and George
Ellery Hale beside the sixty-inch reflector on Mount Wilson in 1910. At
that time the sixty-inch was the most
powerful telescope in the world. By
late in the nineteenth century, large
telescopes usually were the products
of foundations and wealthy individuals such as Carnegie. That situation
was transformed after World War II,
when federal funds flowed into astronomy. (Courtesy of the Archives of
California Institute of Technology.)

In the late nineteenth and early twentieth centuries, the United
States became the chief power in optical astronomy, which until
recently accounted for almost all observational astronomy. Around
the turn of the century, the prosperity of American manufacturing
industries, along with the burgeoning banking and transportation
concerns, thrust enormous fortunes into the hands of a small number
of entrepreneurs and landowners. Some of these people donated part
of their wealth to the arts and sciences, and the money of this elite
often went to endow astronomical research: The observatories of Lick,
Yerkes, and Mount Wilson are all monuments to scientists, most
strikingly George Ellery Hale, who managed to wrestle successfully
with the wealthy and the generous. Astronomy, one of the least
immediately utilitarian of the sciences, thereby became the most
richly endowed of all the sciences in the United States.[18]

Despite this relative wealth, by the mid-1930s, as one writer
noted, "the United States could count fewer than 200 active astron-

omers, only a handful of whom were privileged to observe with the three 'modern' reflecting telescopes then located in favorable climates on the West Coast, the newest of these instruments being more than fifteen years old."[19] The inflow of federal funds to astronomy after 1945, one part of the postwar increase in federal support for science in general, dramatically altered this situation, providing money for students as well as for many modern and expensive facilities, some of them Big Science facilities.[20]

An example of what this transformation meant for astronomers is provided by the different career patterns of Spitzer and his predecessor as director of the Princeton University Observatory, H. N. Russell, perhaps the leading American astrophysicist of his generation. Between 1912 and 1947, Russell led a small department that was never particularly well-off in terms of funds, and its only real embellishment was Russell himself. Although Russell spent a small part of his time as an administrator and fund-raiser, most of his efforts were devoted to individual research. In contrast, Spitzer would head a generally expanding department and have responsibility for numerous multi-million-dollar projects, including the design and development of the major instrument aboard the $200 million *Copernicus* satellite, and he would assume leading roles in defining and advocating the Space Telescope program.

In contrast to the little science of Newton constructing his own reflecting telescope, the dominant features of the Space Telescope program are its elaborate organization and size and the remarkable range of institutions, groups, and individuals involved. The telescope is the product of the labors of many thousands of people in the United States and Europe of very diverse skills: astronomers, systems engineers, managers, mirror makers, electrical engineers, computer programmers, and numerous others. In the building of the telescope the emphasis has therefore been very much on a division of labor, organization, and control. The majority of astronomers who secure data with the aid of the telescope will be physically and managerially distant from direct control of the apparatus. A wide variety of specialists will be needed to supervise the scheduling of observations, the maintenance and calibration of the scientific data, and the operations and well-being of the Space Telescope. The telescope, to an extreme degree, thus represents the shift in the way astronomy has been funded and performed in the United States and in Europe, as well as the transformation of the astronomer's workplace. This is a kind of astronomy that surely even Edwin Hubble (who died in 1953) would not have recognized, let alone Newton.

The program to design, develop, and operate the Space Telescope in fact illustrates one kind of what is termed Big Science – but a kind of very Big Science that requires what some people have called a megaproject for its completion. In the last few decades, the sci-

23

entific enterprise has grown enormously and has become much more complicated than ever before.

Alvin Weinberg, the man who coined the term "Big Science," saw in its monuments (the huge rockets, the high-energy accelerators, the high-flux research reactors) the symbols of our time, just as surely as we find in the cathedral of Notre Dame a symbol of the Middle Ages. But Weinberg, along with many other scientists, was also deeply concerned by the rise of Big Science. He advocated that it be kept in what he saw as its proper place: in the national laboratories funded by the federal government. Weinberg warned that if it was not checked, Big Science would lead to the decline of science itself, for, among other evils, it tends to turn scientists into journalists in order to win funding for their projects, and it produces far too many science administrators. In the history of the Space Telescope we shall certainly see astronomers play a wide range of roles, in addition to that of scientific researchers: administrators, lobbyists, coalition builders, hardware managers, software managers, engineers, and institution builders, to name only some. In fact, the necessity for scientists in Big Science to be far more than just scientific researchers will be one of our themes.

Science has also grown big in other senses. Many of the activities of modern science rely on expensive and highly elaborate apparatus, as well as armies of professionals from many disciplines, not just scientists, to design, build, and operate these "Big Technologies."[21] And because the Big Technology – be it a physics accelerator, a radio telescope, or a spacecraft – is often a central feature of Big Science, this leads to the participation of industry, for its resources and expertise are essential to build the Big Technology.

Another characteristic of Big Science is its social organization. Big Science undoubtedly had its roots in the pre–World War II period, but the war taught many physical scientists new ways of pursuing their researches. As the provost of the Massachusetts Institute of Technology noted in 1953, there

was a time when scientific investigation was largely a matter of individual enterprise but the war taught scientists to work together in groups; they learned to think in terms of a common project, they were impressed by the progress to be made through unified action. A notable degree of this spirit has been transfused into the life of our larger universities. . . . The scale of research and the complexity of its techniques have grown beyond anything imagined a few decades ago. . . . It was the war that contributed principally to a major revolution in the method and spirit and scale of laboratory investigations.[22]

Big Science entails projects with common goals and involves large, multidisciplinary teams and hierarchical organizations. It is not uncommon in high-energy physics, for example, for a scientific paper to have more than a hundred coauthors.

Organization is crucial not only for execution of the scientific programs but also for the building of the technology. The kind of very Big Science represented by the Space Telescope is pursued and made possible by the coalition of a scientific community, government, and industry. Assembling such coalitions nevertheless takes a great deal of time. Organizing teams to develop engineering designs is one part of the preparation, but much time and effort also have to be spent to make a Big Science project politically feasible. Why is this? For a project such as the Space Telescope to go ahead, approval must be won from a wide variety of institutions, groups, and individuals. So not only was it necessary to assemble the telescope in an engineering sense, but the telescope's advocates among the astronomers, government, and industry also had to assemble a coalition for it in a political sense. As we shall see, this coalition eventually consisted of astronomers, planetary scientists, other interested scientists, NASA officials, the European Space Agency, members of the executive branch, and industrial contractors, congressmen and congressional staffers, and sympathetic journalists.

Weinberg also hit upon a central issue for the social and political relations of Big Science when he noted that by its nature it is usually extremely costly. As one writer argued more than a quarter of a century ago, "Without doubt, the most abnormal thing in this age of Big Science is money. The finances of science seem highly irregular and . . . they dominate most of the social and political implications."[23] Around $2 billion has been spent to prepare the telescope for orbit, making it, in part because of significant cost overruns and several lengthy schedule delays, the single most expensive scientific instrument ever built.

As historian of science Daniel Kevles has pointed out, the sheer bigness of Big Science is also enough to arouse questions.[24] The questions the Space Telescope's potential patrons raised were certainly demanding, and it was only after a long struggle that approval for it was won from astronomers, NASA, the White House, and Congress. The manner in which approval was secured will in fact tell us much about the nature of science policy-making in the United States. One commonly employed policy model, for example, is the leadership model, in which a powerful leader fashions a coalition in favor of a project. A case in which this model seems to have worked well is provided by President Kennedy's decision to commit the United States to go to the moon. However, with the Space Telescope, one is dealing with the kind of policy definition usual in post–World War II American science. That is, policy definition is started at low levels in the collection of executive agencies (such as NASA), whereas higher policymakers play only a coordinating role at the end of the process. In the decentralized, pluralistic structure of federal funding of post–World War II science, any overall centralized planning has generally proved to be impossible. As will be seen in the first five chapters, in the case of the Space Telescope one is dealing with the interplay of a range of interests in a policy process best characterized as ad hoc. Policy-making on the telescope was hammered out in a variety of arenas according to ever shifting rules and players; it was a process in which no one ever had a complete grasp of what was happening. The fundamental question of whether or not to approve the Space Telescope was constantly being reframed as the issues bearing on the decision were themselves reshuffled and repackaged, often because of the needs of coalition-building. More-over, the telescope's design and the associated program to build it were central elements in these changes. Issues of, say, congressional politics, which one perhaps might expect would not have any significant influence on the telescope's design and associated program, proved to be crucial.

The approval process therefore played a central role in shaping the telescope. Indeed, I shall argue that the telescope must be viewed as much more than a new astronomical instrument whose design was fixed solely by technical and astronomical requirements. Instead, I shall interpret the Space Telescope and the program to construct it as the products of a great range of forces: scientific, technical, social, institutional, economic, and political. From the complex interplay of these forces, there has emerged the Hubble Space Telescope.

I

Dreams of telescopes

We were just making it all up as we went along, addressing problems that
had never been addressed before.

C. R. O'Dell

Early in 1953, Gerard Kuiper, a leading astronomer, wrote to a
colleague on a proposal by Lyman Spitzer for an astronomical satellite. Kuiper was "greatly surprised that the project should be regarded as astronomical. I would place it with rocket research in a
class of very expensive technological developments that are bound
to benefit astronomy in the end, but which should be carried out
primarily for other reasons if at all."[1]

For many years the reaction of most astronomers to the idea of
orbiting a large optical telescope echoed that of Kuiper. Interest in
such an instrument grew slowly, but it was obvious to its advocates
that many astronomers would have to participate if it were to come
about. The central theme in this chapter is how those astronomers
such as Spitzer who championed a large telescope in space sought to
persuade their colleagues that this was an intensely exciting and
worthwhile endeavor, one that should be strongly backed by the
astronomical community.

The efforts of Spitzer and others must be seen in the context of
coalition-building. That is, a strong and broad-ranging coalition
would have to be assembled and mobilized in support of a large
space telescope before it could become *politically* feasible. And the
coalition-building process would in the end involve winning favor
from astronomers, NASA, industrial contractors, the White House,
Congress, the European Space Agency, other groups of scientists,
and even sympathetic journalists. But unless Spitzer and like-minded
astronomers could take the essential first step and convince a large
number of their astronomer colleagues of the telescope's worth, it
would never be built. Astronomers thereby acted as science policy

Rocket pioneer Hermann Oberth and his daughter view a model of the Space Telescope at the National Air and Space Museum, Washington D.C., in 1985. Oberth's writings in the 1920s contained the first serious speculations on orbiting telescopes.

entrepreneurs, advocates, and coalition builders for a new instrument, roles far outside those usually assumed by scientists in little science, but characteristic of Big Science.

FIRST THOUGHTS

In 1940, astronomer R. S. Richardson speculated on the possibilities of a 300-inch telescope placed on the moon's surface.[2] It was perhaps fitting that the idea was published in an issue of *Astounding Science Fiction.* At the very least, powerful rockets would be needed to carry telescopes aloft, and so before the end of World War II the idea of telescopes in orbit, let alone on the moon, seemed totally impractical. The extremely crude level of the current rocket technology implied that, at best, such instruments lay decades in the future. Even to those such as Lyman Spitzer who were fascinated by space, a telescope placed above the atmosphere was firmly in the realm of science fiction.

The war transformed this situation. The incentive to build weapons quickened the development of rocket technology, and the largest strides were taken by a team of German rocket engineers led by Walter Dornberger and a youthful engineer, the aristocratic Wernher von Braun, later to become a leader of the U.S. space program. Their efforts culminated in the A-4, later to be known as the V-2 guided missile ("Vergeltungswaffe-2").

Built largely by inmates of the Dora concentration camp under utterly barbaric conditions in a factory at Nordhausen, the V-2s may have taken a larger toll of lives in their construction than in their

A German A-4 rocket being readied for launch from test stand VII at Peenemünde (ca. 1943). (Courtesy of Smithsonian Institution, negative #77-14261.)

use. The V-2s were initially unleashed on the Allies in September 1944. The first two, both aimed at Paris, failed. But on September 8, a V-2 crashed to earth near the Port d'Italie in Paris, some 180 miles from its firing point, a distance far greater than any artillery shell could travel.[3] Thus began a lengthy campaign that saw the firing of well over two thousand V-2s at targets in England and on the continent of Europe.

The V-2s made terrifying if erratic weapons. They nevertheless provided an emphatic demonstration that the enormous and complex engineering problems posed by the construction of rockets powerful and reliable enough to lift astronomical instruments above the atmosphere had, to a large degree, been solved. The spectacular technical advances during the war, Spitzer recalls, "made it all seem possible."[4]

Although his own war work had been directed toward undersea weapons, Spitzer was soon acting as a consultant on the possibilities of orbiting telescopes. In January 1946, after many earlier discussions, industrialist Donald Douglas approached the U.S. Army Air Forces with a scheme for coordinating industry and government research and development with long-range strategic military planning.[5] His aim was what would now be called a think tank.

Douglas's plan met with an enthusiastic response from an air force keen to continue in some form the close relationship forged during the war between scientists and the military, the most spectacular result of which had been the development of atomic bombs. To further this aim, project RAND (a name coined by the combining of "research" and "development") was founded by Douglas Aircraft Company.[6] It was for RAND that Spitzer described the "Astronomical Advantages of an Extra-terrestrial Observatory," a seminal paper in the history of space telescopes.[7]

Spitzer's thinking had been influenced by several discussions in early 1946 with Leo Goldberg, soon to become director of the University of Michigan Observatory. Goldberg's interest in space research had been galvanized by the news of the V-2s, and in the spring and early summer of 1946 Spitzer explored with Goldberg and others the possibility of establishing a laboratory for high-altitude spectroscopy. On one such occasion, in July 1946, Goldberg vividly recalls sitting with Spitzer on a park bench in Washington and jotting down ideas for solar space experiments that such a laboratory might pursue.[8] Spitzer then incorporated these into his RAND report.[9]

In his report, Spitzer focused on the kinds of scientific programs, including studies of the sun, that could be conducted from space using a variety of sizes of space telescopes. He examined three options: What astronomical observations could be made from a satellite (1) without a telescope, (2) with a ten-inch telescope, and (3) with a monster telescope with an aperture of some 200–600 inches? The last option, Spitzer conceded, was "some years in the future," which is hardly surprising when we remember that in 1946 the great 200-inch Hale Telescope on Palomar Mountain had not even been completed, and the largest ground-based telescope had an aperture of 100 inches. Spitzer further suggested that the chief advantage of a 200–600-inch telescope in space would not be to supplement current ideas of the universe but rather to "uncover new phenomena not yet imagined, and perhaps to modify profoundly our basic concepts of space and time." This argument, in fact, would become for astronomers a central and widely used justification for a large space telescope.[10] It was a robust argument that, by separating the telescope's justification from particular astronomical questions, was timeless. Astronomical questions might be solved, and new ones pop up, but a large telescope in space would always be worthwhile, and Spitzer was able to deploy the argument in the 1980s as he had in 1946.

Spitzer's 1946 report was a small addition to a much larger RAND study ("Preliminary Design of an Experimental World-Orbiting Spaceship") on the feasibility of building, launching, and operating a satellite observatory for a variety of scientific purposes. Taking what the authors claimed was a "conservative and realistic" engineering approach, the study's conclusion was that it would cost around $150 million (1946 dollars) and take around five years to build and orbit such a 500-pound satellite.[11]

In the late 1940s, private foundations and the military provided the great bulk of the funds for scientific research in the United States. The huge cost of a scientific satellite apparently put its construction beyond the reach of a private foundation. If it were to be built, it would, it seemed, have to be funded by the military.[12] Although the technical resources and engineering expertise were available to embark on a scientific satellite project, in 1946 the political will to provide the funds was lacking, as no pressing near-term military uses for such a satellite were foreseen once it was in orbit.[13] Plans for U.S. satellites would, for the time being at least, remain on the drawing board.

Spitzer's own report on the "Astronomical Advantages of an Extra-Terrestrial Observatory" was classified and took years to emerge into public sight. It thus had no direct impact on the thinking of astronomers. Even his public efforts to arouse his colleagues met with little success. For a long time they "were all quite startled to have a serious astronomer talking about what one could do" with large telescopes in space.[14] In 1953, for instance, the prominent astronomer Gerard Kuiper told a colleague: "I had not heard of the Spitzer satellite proposal. . . . Unless there should be some advantage in getting government services used to the notion that astronomers are capable of asking for a lot of money, I would regard the inclusion of this project hazardous and probably undesirable."[15]

SCIENCE ON ROCKETS

The late 1940s and early 1950s nevertheless saw much pioneering research in space science in the United States. To a considerable extent, it was set in motion by the existence of research groups that had flourished during the war, now seeking new lines of research, as well as by the availability of military funding and large numbers of captured V-2 components that could be assembled into completed rockets. The V-2s had been seized at the end of the war in Europe and transported to the White Sands missile range in New Mexico. Von Braun and about four hundred of the German engineers and

A V-2 rocket lifts off from White Sands in New Mexico in late 1946. It was loaded with numerous items of scientific equipment, including trailing wire antennas (seen near the rocket's fins) designed to investigate the properties of the ionosphere. (Courtesy of Smithsonian Institution, negative #80-3827.)

scientists who had built the V-2s had surrendered to U.S. forces near the end of the war. Many of them were soon back at work on the V-2s, this time under the direction of the U.S. Army.[16]

The army undertook a series of launchings of the V-2s to acquire experience with large missiles, but it also allowed scientists to place experiments aboard some of the rockets. These experiments were of use to the army because of its interest in the effects of the atmosphere on radio communications, as well as the possibility of using the sun's light in guidance systems for missiles.[17]

The V-2 program lasted until 1952, and the sixty-three rockets fired provided considerable data on conditions in the earth's atmosphere. Those researchers also clearly demonstrated to those involved that the pursuit of space science could be a risky undertaking. For example, on June 28, 1946, staff of the Naval Research Laboratory attempted to observe the ultraviolet spectrum of the sun by flying a spectrograph aboard a V-2. The flight ended in calamity as the rocket

V-2s made large craters when they hit the New Mexico desert. In order to prevent obliteration of the scientific and engineering payloads, schemes were devised to break apart the rockets while they were airborne. The various parts would then fall to earth more slowly, and the equipment could be recovered. (Courtesy of Smithsonian Institution, negative #80-4095.)

and spectrograph smashed into the New Mexico desert, leaving a crater eighty feet in diameter and thirty feet deep. Only a bucketful of identifiable debris was recovered.[18]

Following that disastrous start, the Naval Research Laboratory's next flight was a technical and scientific triumph. Before the rocket reentered the atmosphere, the tail unit, containing the new spectrograph, separated successfully from the main body of the rocket. It fell to earth, and the researchers recovered the exposed photographic film, containing the record of the spectrum, undamaged. The film showed clearly that the farther the rocket traveled through the ozone layer, the more of the sun's ultraviolet spectrum was visible.[19]

Rockets other than the V-2s – such as the American-developed Aerobees and Vikings – also came into service. These, however, were still what are known as "sounding rockets." Such rockets lift instruments to make measurements at high altitudes and are fired along a vertical, or nearly vertical, path, falling back to earth after reaching a peak altitude. A V-2, for example, could reach an altitude of about one hundred miles. Using rockets in such a way, astronomical observations can be obtained for only several minutes before the instruments arch back into the atmosphere under the pull of gravity. Most of an instrument's flight time is therefore spent journeying up, and then down, into the atmosphere.[20]

The chief concern of nearly all of the early space researchers was the upper atmosphere, not observations of astronomical bodies. Although the sun was often studied, the space scientists generally undertook these observations as a means of gauging changes in the conditions and behavior of the earth's atmosphere, not because of fascination with our nearest star. Space scientists also found it far easier to observe the bright sun than the other, vastly fainter, celestial bodies. It was not until 1955, in fact, that astronomical objects other than the sun (or "nighttime objects," as they are called) were observed by detectors flown aboard a rocket.[21]

Sometimes experimenters made use of balloons instead of rockets to carry instruments to high altitudes.[22] For example, a group at the Princeton University Observatory flew a twelve-inch telescope aboard the Stratoscope I balloon to heights of about fifteen miles to take detailed photographs of the sun.[23]

SPUTNIK

Despite what could be achieved with their current rockets and balloons, many space scientists eagerly looked forward to a time when they would be able to fly instruments above the atmosphere for periods far longer than those they had already achieved. That stage in the development of space science would be ushered in by experiments in orbiting satellites.

At the time of Spitzer's RAND report, satellites had been deemed by the military to be too expensive for the rewards they would offer. Given the escalation of the cold war, such an attitude did not persist long. U.S. companies made advances in guidance and control techniques and in the miniaturization of some hardware while working on other rocket programs, and the military services decided that these advances made a satellite more feasible technically.[24] The navy, the air force, and other groups therefore began studying the design and use of artificial satellites to be launched into orbit by large, powerful rockets. In early 1949, for example, the air force directed RAND to begin extensive studies on the potential military uses of earth satellites, including the possible advantages of using such spacecraft for reconnaissance.[25]

During the early 1950s, the feeling developed among U.S. space scientists, many of whom knew of at least some of the military developments, that the era of scientific satellites would soon dawn.[26] In January 1956 there was even a conference on the scientific uses of earth satellites. Over thirty papers were delivered, and the optimistic mood was captured by one speaker who proclaimed that the "speculation and conjecture of a few years ago have given way to specific proposals for achieving successful establishment of small vehicles in earth-circling orbits. The technical achievements of the last few years leave but little doubt that a satellite can be successfully launched within the next few years."[27]

In response to growing pressure from scientists for a satellite program, the U.S. National Academy of Sciences and the National Science Foundation had jointly agreed in 1955 to seek approval to orbit a scientific satellite during the forthcoming International Geophysical Year (IGY). The IGY, designed to promote an internationally concerted study of the earth's atmosphere and oceans, was to begin officially in July 1957 and run until December 1958. The United States planned to launch its first IGY satellite aboard a navy Vanguard rocket. Although the goals were ostensibly scientific, there were also distinct military overtones. In physicist Philip Morrison's apt expression, the "bulk in the jacket pocket of the IGY showed up very well to many who knew. But who knew? It was not made public to those who run while they read."[28]

The Soviet Union also made plain its intention to orbit a satellite as part of the IGY. Russian delegates therefore appeared in Washington, D.C., in October 1957 for a conference on IGY rockets and satellites.

"The subdued sense of anticipation that pervaded the sessions," one leading participant remembered, "stemmed from the awareness that preparations had been under way for some time, that the IGY was already in full swing, and that the first artificial satellite must soon appear over the horizon. But those expectations did not diminish the surprise and dismay felt by U.S. scientists when the launching of Sputnik I was announced on the evening of 4 October 1957."[29]

For some, there were mixed emotions. James Van Allen, already a leading space scientist, could not suppress his excitement at the technological feat, marking it down in his notebook as a "Brilliant achievement!"[30] He and many other people decided straightaway that a new age had dawned, the "space age."

The 184-pound Russian satellite orbited the earth every ninety minutes. Marked by its persistent "beep, beep" radio signals, mocking the United States as Secretary of State John Foster Dulles put it,[31] *Sputnik I*'s journey had enormous repercussions. Perhaps its most far-reaching effect was that it provoked a wave of near hysteria in the United States, for as Van Allen had realized immediately, here was a "tremendous propaganda coup for the USSR."[32]

To a nation with a profound distrust of Soviet intentions, that unexpected demonstration of Russian technological prowess was startling and deeply disturbing. Underlying this response was the realization that a Soviet missile powerful enough to launch a satellite into space could also be used to carry atomic weapons from Russia to obliterate American cities. Science and technology figured centrally in America's dreams of abundance and world leadership, but now they appeared in America's nightmares of a dangerous world with uncertain prospects of security. *Sputnik I* was nine times heavier than the first planned U.S. satellite. That weight disparity implied a big Russian lead in large rockets, a point reinforced when the even more massive *Sputnik II* was launched soon after *Sputnik I*. To make

35

matters worse, on December 6, 1957, in full view of the nation's television cameras, the first attempt to launch a U.S. satellite ended in disaster. The Vanguard rocket rose a few feet from its launch pad and faltered; as it fell back, the fuel tanks ruptured, and the rocket folded and crashed back to the ground surrounded by flames and billowing smoke.

The reactions to *Sputnik* were to a large degree genuine expressions of shock and unease. They were also to a significant extent controlled, guided, and fashioned by certain politicians and a section of the news media anxious to use *Sputnik* as a stick with which to beat the Eisenhower administration.[33]

Eisenhower's previous approach to missiles and satellites had been, according to one of his biographers, to let each military service "develop its own program and hope that one of them would score a breakthrough. The result had been failure. The generals and admirals squabbled with one another, made slighting remarks about their fellow services' efforts, and ignored the Secretary of Defense." Given the buffeting he had to endure from the political storm whipped up by *Sputnik*, Eisenhower could not hold his ground. Facing the overwhelming political pressure of nearly all Democrats, most Republicans, and a majority of columnists and scientists, on April 2, 1958, the president requested Congress to establish a civilian space agency.[34] It was to be termed the National Aeronautics and Space Administration (NASA). The bill establishing the agency was swiftly passed through the Congress in the spring and summer of 1958, and NASA came into being on October 1.

Here we should note an important point. Although NASA was created as a civilian agency, the new technologies that allowed men and machines to move around in space always had national security implications and applications. The "relationship between civilian and national security efforts in space," political scientist John Logsdon maintains, "is defined much more by policy than by differences in technology. The technological reality is that almost any space capacity – earth observation, environmental monitoring, communications, transportation, in-orbit construction, and so forth – can be used for either civilian or national security purposes."[35] The separation of national security from civilian research and development of space technology would therefore be almost meaningless.[36] NASA might be a civilian agency, but as, for example, secret Space Shuttle flights in the 1980s would evince, it would do many things of military importance. Nor would this be all one-way traffic, for techniques and technology developed for national security purposes would also be exploited by NASA. Indeed, the Space Telescope would itself, as we shall see, be a consequence in part of this transfer of know-how between the national security and civilian worlds.

Not only had *Sputnik* helped to bring NASA into existence; it also
had reshaped the thinking of politicians to such an extent that many
old assumptions about the federal funding of space activities were
promptly swept away. The usual rules on Capitol Hill lapsed for a
time as the legislators could hardly dispense money to NASA fast
enough. As NASA's first administrator, T. Keith Glennan, recalled
in 1982, "The Congressional attitude was: Why can't we do this
tomorrow? . . . I don't remember in those first years ever going up
on the Hill where I wasn't asked: What did you ask for? Don't you
need more money?" "Only a blundering fool" could have come back
from Capitol Hill with a result detrimental to the agency.[37]

At first, however, the precise role of NASA was somewhat obscure. The agency had therefore to carve out a niche for itself while
several determined rivals eagerly attempted to place themselves at
the center of the space stage, including most noticeably the U.S.
Air Force.

Another key question arose: Should NASA compete with the Soviets in a race to achieve space spectaculars, or should it pursue a
more scientifically based program of research? The political pressure
to counter, and surpass, the achievements of the Soviets meant that
the American riposte to *Sputnik* would lead to a space race; in effect,
it was the pursuit of the cold war by another means. Science would
nevertheless play a far from insignificant part in the chase, and budgets for space science would reach levels undreamed of in pre-*Sputnik*
days, when money for space science had come largely from the military, with small amounts from the Atomic Energy Commission and
the National Science Foundation.

There were certainly numerous scientists eager to pursue the new
opportunities to perform space research. Many responded to a call
in 1958 by Lloyd V. Berkner, chairman of the newly established
Space Science Board of the prestigious and influential National
Academy of Sciences. Berkner wanted suggestions for projects to
follow those of the International Geophysical Year. Approximately
two hundred replies arrived at the National Academy. These were
sent to NASA, and a number of them, together with suggestions
dispatched directly to the agency, helped to provide a focus for a
NASA Space Sciences Working Group on "Orbiting Astronomical
Observatories (OAOs)."[38] Such plans had support at the highest levels of the government. When President Eisenhower's science advisor
spoke to him "about the potential gains in national prestige if we
establish the first astro-observatory on a satellite [the] President was
very much interested and said he would certainly be in favor of
proceeding vigorously."[39]

OBSERVATORIES IN SPACE

NASA and several astronomers were keen to move quickly to devise space observatories. But here NASA and its advisory committees faced one of the central issues in the history of modern astronomy: How was the agency to decide the priorities for funding the various branches of astronomy? In particular, was it better to pursue, say, x-ray astronomy, about which astronomers knew very little, and the pursuit of which was therefore something of a shot in the dark, or should preference be given to ultraviolet astronomy, from which many astronomers were confident of exciting results?[40] Very hot stars, in contrast to the sun, radiate most of their energy in the ultraviolet, and many of the spectral lines due to matter strewn between the stars were also calculated by astronomers to fall into this region, for example. As NASA's head of space science claimed in 1966, "for the initial exploration in virgin fields [such as, for example, x-ray astronomy], relatively simple instrumentation and limited scope research programs often suffice to permit rapid exploitation of new technologies and breakthroughs of understanding."[41] The OAOs were being funded by NASA for reasons of prestige as well as scientific results, and simple instrumentation held out to the agency's managers little prospect of short-term benefits in terms of prestige. If, instead, the OAOs were to concentrate on ultraviolet astronomy, the spacecraft would have to be technically sophisticated, for they would have to build on centuries of observations in the optical region of the spectrum, a region next to the ultraviolet. Ultraviolet astronomy, in consequence, promised notable results with the aid of complex spacecraft. Together with its advisory committees, NASA therefore opted to put its early emphasis squarely on ultraviolet astronomy.

In addition to the choice of the wavelength range on which to concentrate, there were other pressing issues: How should the OAOs be constructed? What scientific objectives should drive the engineering designs of the orbiting telescopes? How should the development of the telescope program be organized? NASA had assembled a working group to provide advice on some of these questions.

Two of the members had sat together in a Washington, D.C., park some thirteen years earlier and had debated scientific programs for a satellite telescope. Leo Goldberg and Lyman Spitzer now had the chance to turn their speculations into reality, Spitzer as the leader (or, in the term NASA employed, "Principal Investigator") of a Princeton group to build a telescope to observe the ultraviolet light of stars and material between the stars, and Goldberg as the principal investigator for an experiment to study the sun.

The working group began its meetings in March 1959, and from its discussions and negotiations there evolved a common spacecraft design. In all the OAOs the astronomical instruments were to be located in the spacecraft's central tube, which was to be forty inches in diameter and ten feet long. The diameter of the tube also determined the maximum size of an OAO telescope's primary mirror.[42] Plans for an OAO to observe the sun, however, were dropped by NASA in favor of a series of simpler orbiting solar observatories. OAOs would therefore focus on observations of stars and galaxies.

GETTING STARTED

In 1960 and 1961 NASA issued "requests for proposals" (RFPs) for the OAO spacecraft and the astronomical instruments to be flown aboard them. These were calls for interested companies to describe how they would meet the technical requirements for the observatories. The companies' responses were then reviewed by NASA.

From these deliberations, Grumman Aircraft Corporation was picked by NASA as the chief contractor for the spacecraft. In addition, the Goddard Space Flight Center at Greenbelt, Maryland, newly founded, with many staff volunteers transferred from the nearby Naval Research Laboratory, was to provide the overall project management.[43]

The first OAO contained experiments devised by research groups at the University of Wisconsin, the Lockheed Missiles and Space Division, the Goddard Space Flight Center, and the Massachusetts Institute of Technology.[44] NASA, the contractors, and the astronomers were entering largely unexplored territory, and they were having to start almost from scratch in all sorts of areas. An astronomer who would later become a central figure in the Space Telescope Project recalled his brief involvement as a graduate student at Wisconsin with the early planning for the first OAO:

We were sitting down and figuring out how to make a spacecraft point. Things like that. . . . You know, we hadn't built an astronomical spacecraft at that point. In today's parlance, NASA-ese, you would call it, it was systems engineering. But it was systems engineering *ab initio*. We were just making it all up as we went along, addressing problems that had never been addressed before.[45]

The technical demands not only were often new but also were taxing. Perhaps chief among these was how to point the observatory at one small area of the heavens for long periods while it orbited hundreds of miles above the earth at some 17,000 miles per hour. It was one thing to develop pointing and control systems to direct 100–200-pound rocket payloads at the sun, but quite another to build pointing systems for satellite observatories weighing several thousand pounds, which needed to be aimed at stars, planets, and galaxies, all immensely fainter than the sun. In the early 1960s, not even

39

rockets, which space scientists had worked with for over a decade, could be targeted at nighttime objects. Rather, once a rocket was free of the drag of the atmosphere, it would start to wobble like a top. These movements were unplanned and uncontrolled. Hence, a major problem for space scientists was to determine just where an experiment had been pointed at each moment during its flight, a task that required determining the rocket's path.[46]

In addition, once an OAO was in orbit, it would not be possible to make repairs. A simple fault that might be remedied by a few turns of a screwdriver if the observatory were on the ground could wreck the mission and destroy years of effort, as well as waste many tens of millions of dollars.[47] The spacecraft builders, in consequence, emphasized making components reliable. They also sought to ensure that a great deal of redundancy could be built into the observatories. In this way, the failure of one component would not necessarily cripple the entire spacecraft.[48]

To add to the obstacles, spacecraft engineering was a relatively new field not only for astronomers and NASA managers and engineers but also for the companies and their engineers who were involved in designing and developing the spacecraft. Grumman constructed the OAOs, and like most of its major competitors in the field, it had vast experience and expertise in building aircraft, but the design and development of satellites presented fresh problems. The tough challenges were not all technical; there were managerial and organizational issues to be thrashed out as well. One major recurrent issue concerned the sorts of working relationships that would be forged between NASA and groups of space researchers. Physical scientists were used to taking charge of their research projects. NASA, however, would do business in a very different manner and, as an organization, would strongly resist any loss of control to outside groups, as we shall see.[49]

THE FIRST ORBITING ASTRONOMICAL OBSERVATORY

Despite these institutional difficulties and the novelty of many of the technical problems, the OAOs were planned to be NASA's largest, as well as most elaborate, scientific satellites. The size and complexity of the spacecraft worried the leaders of the groups building the scientific experiments. In fact, the astronomers had pressed for some smaller and simpler telescopes to be placed in orbit. They contended that to make the best use of the OAOs, "it would be very helpful to have preliminary experience with smaller astronomical payloads."[50]

NASA managers ruled this out, claiming that the astronomy program had access to only a restricted number of launch vehicles, as well as emphasizing that it would not be much quicker to build a smaller observatory. If a smaller observatory were built, the work on the full-size OAOs would be delayed.[51]

The claim about the restricted number of launch vehicles, however, is difficult to accept, particularly as twenty years later one of the managers centrally involved with this decision wrote that

exposed directly to the outside pressures to match or surpass the Soviet achievement in space, NASA moved more rapidly with the development of observatory-class satellites and the larger deep-space probes than the scientists would have required . . . some of the most intense conflicts between NASA and the scientific community arose later over the issue of the small and less costly projects versus the large and expensive ones — conflict that NASA's vigorous development of manned spaceflight exacerbated.[52]

Prestige was an important consideration for NASA managers in fixing the speed of the orbiting observatory program and therefore the technical priorities. Against the clamor to compete with the Soviets, the voice of the astronomers calling for a more conservative program was lost.

Perhaps not surprisingly, the first OAO took longer to complete than was anticipated. The launch was originally planned for 1963, but it had to be postponed. The 3,900-pound *OAO-1* (containing more than 440,000 parts and over thirty miles of electrical wire) also underwent rigorous qualification and testing procedures that had not originally been included in the project plans.[53]

Despite the delays, by early 1966 *OAO-1* was at Cape Kennedy ready for launch.[54] The time before launch was nerve-wracking for the scientists as, first for one reason, and then another, lift-off was deferred. Twice the rocket engines were ignited, and twice they shut down. A tornado forced another delay. Each time the launch was scrubbed, Arthur D. Code and his University of Wisconsin team had to rework the satellite's commands, often working almost twenty-four hours straight.[55] Finally, on April 8, the Atlas-Agena rocket raised itself slowly from its launch pad and accelerated into orbit.

The observatory was deposited perfectly in space. However, when the system used to point the satellite to its target stars was turned on by ground controllers, there were electrical discharges. The discharges knocked out some control circuitry, including the control on the batteries. As the batteries kept charging, they overheated. By April 10, contact with the satellite had been lost. When tracking radar picked up several pieces of the observatory, it seemed likely that the batteries had exploded.[56] Over the objections of the scientists, NASA had taken too large a step in moving straight to the complex OAOs and had fallen flat on its face.

For the Wisconsin group, it was, one team member recalls, a "big mess, total disaster." After years of labor on the satellite, they were

The ill-fated *OAO-I* undergoing its final check before launch at Cape Kennedy. (Courtesy of NASA.)

devastated: "We never even got a chance to turn on our instruments, so the most we could say at that time, after nearly five years of effort was, well, we launched with the instruments off and they stayed off. They did what they were supposed to do."[57]

A special review board was established by NASA to scrutinize the entire OAO program.[58] The management and technical approaches were overhauled, and by December 1968 the Wisconsin group's experiment, this time paired with a package of ultraviolet telescopes developed by the Smithsonian Astrophysical Observatory, was ready to fly once more. Lifted into orbit aboard an Atlas-Centaur rocket, *OAO-II* functioned well. It returned vast quantities of data for four and a half years and did much to open up new areas of astronomy as it was employed to measure ultraviolet emissions of galaxies, stars, planets, and comets.[59]

The sense of loss deriving from the catastrophic failure of *OAO-I* was matched following the third launch of an OAO. That satellite, also borne aloft on an Atlas-Centaur rocket, carried a thirty-six-inch ultraviolet telescope designed by a group at the Goddard Space Flight Center. The Goddard team never found out how the satellite would have performed in space, for a bolt holding the protective shroud over the observatory failed to detach after launch. The Atlas-Centaur rocket thus had to carry the observatory and its shroud into orbit, far more weight than had been planned. It was an impossible task. The observatory plunged earthward without even reaching orbit.[60]

For Albert Boggess III, the leader of the Goddard science team and one of the major figures in the development of ultraviolet as-

The launch of *OAO-II* aboard an Atlas-Centaur from Cape Kennedy in December 1968. (Courtesy of NASA.)

tronomy, "it was certainly a bleak day when the thing went into the ocean. In fact it was a difficult time for all of us, and a difficult time for me to think about even in retrospect, because I think there were some pretty obvious mistakes that had been made."[61]

The OAO program's mixed fortunes continued with the launch in August 1972 of the fourth OAO, later to become known as *Copernicus* after the Polish astronomer Nicholas Copernicus (1473–1543). The principal experiment for *Copernicus* was Princeton University's ultraviolet telescope. Lyman Spitzer and his group had devised it to observe the spectra of bright stars and of matter spread between the stars.[62] A thirty-two-inch primary mirror was at the heart of the largest astronomical telescope yet sent into space, and it was housed in the heaviest unmanned satellite orbited by NASA (some 4,900 pounds). *Copernicus* was also the most technically sophisticated of the OAO series. It could point to a star with the unprecedented accuracy of 0.1 second of arc and maintain a stability of 0.02 second of arc

43

for lengthy periods. This was equivalent to aiming *Copernicus* at, and locking onto, an object the size of a basketball from a distance of 400 miles.[63]

Copernicus and the Wisconsin experiments on *OAO-II* were widely judged to be technical and scientific successes, and that bolstered NASA's confidence that a space telescope much larger than any in the OAO series could be built.[64] But the story of the OAOs also illustrates the heartbreaking disappointment sometimes to be endured in space astronomy, particularly in NASA's first decade or so. Two of the four missions ended disastrously; yet the other two sent back to earth high-quality science data for periods longer than planned, *Copernicus* for nine years, and *OAO-II* for over four years.

Some groups in the space science program fared much better than others, but unexpected and sometimes overwhelming problems and the complete loss of missions were certainly not peculiar to the OAO program. Reflecting on the struggles as well as ill-luck borne by his research group at Harvard in the Orbiting Solar Observatory program, Leo Goldberg wryly noted that at one point "it was beginning to look as if the primary purpose of space research was to build character. But as trying as these setbacks were at the time, they made us appreciate the successes all the more when they came."[65]

The OAOs also illustrate a fundamental point that would become central in the debate on large telescopes in space: Was it better to build toward a particular astronomical spacecraft by small increments or by taking big strides? The OAOs certainly represented large increases in capability over what had gone before. Despite the two failures, it is nevertheless questionable that it would have been cheaper or quicker to have worked toward them via an intermediate stage. An intermediate spacecraft, however, probably would have been more productive scientifically. As far as the astronomers were concerned, the smaller spacecraft would have produced scientific data and provided a test bed for instrumentation to be used in the larger observatory. Political forces, however, had prevailed over scientific reasons as NASA managers had dictated the jump to the OAOs.

CRITICS

Even in the late 1950s and early 1960s, when the OAO program was still in the conceptual stage, a few astronomers were eager to race ahead and build a space telescope whose primary mirror would be much larger than the thirty-two inches of *Copernicus*. Nor was it clear in the late 1950s that NASA would soon have sole control of civilian space astronomy activities and that such a large telescope would have to be NASA's responsibility. Thus, a consortium of universities (the Association of Universities for Research in Astronomy, AURA) for some years received funds from the National Science

Foundation for studies of a fifty-inch orbiting telescope.[66] AURA even established a "Satellite Telescope Sub-Committee" chaired by Lyman Spitzer, with Arthur D. Code, C. D. Shane, and Aden Meinel as members. Spitzer and Code were, of course, principal investigators on the OAO series, but it was Meinel, himself a strong and vocal advocate of space telescopes, who became the main driving force behind the AURA plans. Meinel, in fact, was the first director of the AURA-run Kitt Peak National Observatory, and he was also in charge of the observatory's Space Division from 1959 until early 1961. Although AURA did build and operate a fifty-inch, remotely controlled ground-based telescope, on Meinel's departure from Kitt Peak for another position the plans for the AURA orbiting telescope faded.[67]

The AURA space telescope was just one of the many ideas floated in the early 1960s. Although the scientific advantages of a large telescope in space were obvious to astronomers, by no means all thought that such an instrument should be developed. The issue was not the scientific results to be achieved, but rather the timing and priorities: When should it be built, and how much money and effort should be expended on such a telescope? The space age was only a few years old, and launch-vehicle failures were common; compared with the conditions for operating telescopes on the ground,

Artist's conception of an orbiting telescope, prepared for a visit by AURA representatives to the U.S. Army Ballistic Missile Agency in Huntsville, Alabama, in 1959. Note the astronauts tending the telescope from a nearby space station. (Courtesy of Frank K. Edmondson.)

LARGE ORBITAL TELESCOPE by AURA

optical space astronomy was a risky business. Thus, many influential astronomers contended, why not take one's chances with cloudy weather and do ground-based astronomy and produce good science, in preference to chasing the chimera of a large telescope in space? Why not wait until the technical basis for such instruments was more solidly founded? Optical space astronomy, by the standards of ground-based astronomy, was extremely expensive, and so some astronomers argued that the money would be better spent on ground-based telescopes. Many ground-based 200-inch telescopes, one argument ran, could be purchased for the $200 million plus cost of three OAO spacecraft.[68]

For some astronomers, space astronomy was in bad odor because of the enormous, and, to them, largely wasted, expense of space programs. For Fred Hoyle, a controversial but brilliant astronomer,

the cart has been put before the horse. Instead of presenting the orbiting observatory as an ancillary to the space program, the orbiting observatory has been offered as one of the important reasons for implementing the space program. As part of the campaign for selling the space program, the advantages of observation from space have been promulgated almost with Madison Avenue techniques. The key points in the logic of this sales technique run like this: You all know the wonderful discoveries that have been made with big telescopes, like those at Mt. Wilson and Palomar. O.K.? Now we have something that will make those big telescopes seem like a kid's spy glass. So although it's going to cost plenty, it'll be worth it.[69]

Hoyle derided this argument, and he urged instead the construction of more ground-based telescopes.[70]

The dominant goals of the space program were hardly those of space scientists. The U.S. civilian space program, and in particular the moon landings, was being carried out largely to enhance the prestige and demonstrate in a very public arena the technological muscle of the United States, not to secure scientific results.[71] At times there was substantial pressure on the agency to curtail its space science program in the interest of an all-out effort to reach the moon. NASA administrator James Webb had resisted these forces, arguing that it would eventually be necessary to look beyond Apollo. The United States should therefore "pursue an adequate well-balanced space program in all areas, including those not directly related to the manned lunar landing."[72]

The organization and politics of science funding also meant that the money provided for ground-based and space-based astronomy could not be switched in any simple manner between the two. That this was so did not prevent some astronomers thinking it was otherwise. This mistaken belief led to resentment. Writing to the NASA administrator in 1970, Leo Goldberg, then director of the Harvard College Observatory, summed up matters when he commented that space astronomy had been "carried out by a relatively tiny number of scientists while most astronomers looked on from outside, sometimes with envy at the amount of money being spent and often with suspicion and even hostility."[73]

X-ray, and most gamma-ray, ultraviolet and infrared astronomy can be pursued only from space. Optical astronomers, however, had a choice between space and ground-based researches. This choice led to a divide between those who wanted to put optical astronomy into space and those who remained convinced of the priority of ground-based astronomy, who were familiar with its techniques, and who were not inclined to enter new territory. In his *Scientific Autobiography,* the great German physicist Max Planck contended that a "new scientific truth does not triumph by convincing its opponents and making them see the light, but rather because its opponents eventually die, and a new generation grows up that is familiar with it."[74] With the introduction of the new technology of space astronomy, the issue was somewhat different. Certainly there was no simple generational split between advocates and critics of optical space astronomy, between the young and the old. Several of the leading proponents — Arthur D. Code, Leo Goldberg, Lyman Spitzer, and Fred Whipple among them — were senior, well-established astronomers. In fact, their seniority meant that they were better placed to take risks and to break out into new fields than were younger colleagues who had yet to find secure positions.

The criticisms of space astronomy, however, were not simply due to conservatism. What astronomers regarded as enormous strides had been taken with a variety of powerful ground-based telescopes throughout the 1960s. These included, for example, the exciting discovery of the spectacularly energetic and enigmatic quasars, objects that seemed to be much smaller than the known galaxies but that appeared to emit vastly more energy. Plenty of opportunities therefore remained to pursue forefront astronomical research with earth-bound optical instruments.

In the United States, the divide between space astronomers and ground-based astronomers was to a considerable degree the product of, and was reinforced by, history and geography. To simplify somewhat, for West Coast astronomers with ready access to large telescopes at good sites, optical space astronomy was not nearly so appealing as for East Coast or Midwest astronomers. The East Coast and Midwest astronomers had to contend with much poorer observing conditions and had access to less powerful optical telescopes than their colleagues in the West.

In 1969, University of Wisconsin astronomer Robert C. ("Bob") Bless took a break from his work on *OAO-II.* At the request of NASA, he toured various observatories and astronomy departments to discuss optical and ultraviolet space astronomy, particularly with younger astronomers. Bless found a general recognition that space astronomy

47

did have something useful to contribute to astronomy. "The few who were not interested," Bless told a leading science manager at Goddard Space Flight Center,

expressed doubts that NASA would pay any attention at all to their opinions. Lack of responsiveness to outside astronomical opinion was also the primary reason given by a very small minority of astronomers for not being willing, under the present circumstances, to devote any time to the follow-on OAO program if it were funded. These astronomers were mainly on the West Coast and certainly one of the reasons for this feeling is their access to large telescopes. They can thus pursue profitable and exciting ground-based work more easily than Midwest or East Coast astronomers and so are less willing to gamble with the uncertainties of space astronomy. However, this feeling was occasionally expressed, though much less vigorously, elsewhere. Clearly NASA will have a problem in convincing many astronomers that even after the program is under way it will continue to receive NASA support and that outside input will be welcomed and acted upon.[75]

ADVOCATES

In striking contrast to the critics of space astronomy, those astronomer advocates of a large telescope in space considered the scientific potential of such an instrument to be so impressive that they deemed it essential to lay plans as soon as possible. The early 1960s, we should remember, were years of almost unbounded optimism about what could be achieved in space. It was a time of rapidly rising space budgets, but also a time when the wide-ranging and complex challenges posed by the pursuit of optical and ultraviolet astronomy from satellites perhaps had not been fully or widely grasped. For the astronomers, there were, as the OAOs were to evince, some hard lessons ahead.

The first launch of an OAO, however, was four years in the future when in the summer of 1962 the Space Science Board organized a major review of the scientific research to be pursued in space. Over two hundred people gathered in Iowa City with representatives from organizations that included NASA, a variety of government agencies, the National Academy of Sciences, and the Space Science Board. A substantial portion of the proceedings was devoted to space astronomy. The Working Group on Astronomy also proclaimed that the recent orbiting of two satellites to observe the sun meant that "the narrow bounds which the terrestrial atmosphere has always imposed on the exploration of the full astronomical spectrum are breached still further. The spectrum opens out, not just for the few precious seconds when a sounding rocket or the X-15 [rocket plane] climbs the apex of its quick trajectory, but for weeks and even months at a time."[76]

As might be expected of a group composed to a considerable ex-

Science under the wing of national security. Space scientists pursued all sorts of means of securing data from high altitudes, including mounting spectrographs on X-15 rocket planes launched from B-52s. (Courtesy of NASA.)

tent of space enthusiasts several of whom were active on the OAOs, the Working Group on Astronomy saw a large telescope in space as the natural outgrowth of the OAO program. The first three observatories were estimated by NASA to cost at least $200 million, and so it was apparent to the working group that a much larger instrument, with a primary mirror of, say, 100 inches or more, would represent a huge investment for astronomy. Certainly it would be of such scope that it could not be achieved by NASA and its contractors solely with the aid of the members of one or two university astronomy departments or a government or industrial laboratory. A Large Space Telescope (for such it was then termed, often by its acronym LST) would surely become a truly national enterprise that would require the cooperation of government, industry, and academe. "For this reason," the final report argued, "it is vital that [the telescope's] scientific justification receive the most careful and comprehensive consideration by the astronomical and related scien-

49

tific communities." Not only would the project be very costly, but also it would be technically demanding. Hence, "the thinking time necessary for its conception and initiation cannot start too soon."[77]

The Working Group on Astronomy also touched on what they saw as the major technical questions that were raised by the Large Space Telescope. First, what size primary mirror should it have? Second, should it be located on the moon or in space? Third, should astronauts maintain the telescope, by, for example, changing its failed equipment? Fourth, what new techniques needed to be developed for the Large Space Telescope to be successful? All of these questions were to be posed many times in the telescope debate throughout the 1960s and (except for the possible moon basing) the 1970s.

In the working group's opinion, the next optical/ultraviolet telescope to be launched into space after the OAOs should represent significant increases in light-grasp and resolution. On one of the evenings of that conference, Spitzer had even given a visionary lecture on the advantages of a 400-inch space telescope.[78] The working group nevertheless thought that to opt for a size larger than about one hundred inches might be too big a leap, pushing the technology too far and risking failure.

The group also favored, for several reasons, placing a telescope in earth orbit rather than on the moon. It would be considerably more expensive, the group reckoned, to put a telescope on the moon because of the cost of transferring equipment over such a large distance. It would also be easier to reach the telescope for repairs and maintenance if it were in earth orbit. For these tasks, astronauts might use a nearby space station as a base (in 1962, as we shall see in the next chapter, the construction of a space station was viewed by many people, and not only those in NASA, as a natural and likely near-term development in the U.S. space program). As the authors of the final report from the Iowa meeting maintained,

any very large telescope is so complex that it seems certain to require regular maintenance. Thus, the very first thinking about the large-aperture space telescope must include the concept of manned access not only to permit initial adjustments, if necessary, and later maintenance and repair operations, but also to permit the even more important activities of installation of new and different auxiliary equipment, and the recovery of data or material that cannot be easily telemetered [back to earth].[79]

The working group also spent some time on the kinds of astronomical problems that might be tackled with such a telescope, but, as with Spitzer's 1946 paper, the chief claim was that a space telescope with an aperture of around one hundred inches would represent such a vast increase in observing power that it could not fail to produce novel and far-reaching scientific results. The group even suggested that the National Academy of Sciences establish a study team to examine the telescope idea further. This was not a unanimous view; a minority report cautioned that at "a time when not a single image of a celestial body has been obtained in a satellite, it is premature to convene a group to study a space telescope larger than the 38-inch telescope of the OAO."[80]

SEEKING A COALITION

The proposal for a study to be conducted by the National Academy of Sciences failed to attract enough backing to become reality. The arguments of the Large Space Telescope's supporters nevertheless bore some fruit as NASA funded a series of feasibility studies of the instrument. Such studies are essential tools in helping NASA decide what projects it should ultimately fund to completion, and the first extensive study on the Large Space Telescope was executed by Boeing in the mid-1960s for NASA's Langley Research Center.

In keeping with most of the early NASA-sponsored studies, the subject was, in the NASA terminology, a *manned* astronomical observatory. Again, as was to be the case with most of the studies performed during the 1960s, Boeing assumed that the telescope would be operated in conjunction with a manned space station. Indeed, Boeing envisaged that the telescope would even be linked directly to a space station by use of a spring system.[81] Boeing also focused on a telescope with a primary mirror of 120 inches.[82] This size was chosen because the Langley group had earlier decided that a telescope with a 120-inch mirror was the largest that could be diffraction-limited and carried aloft by the Saturn series of launch vehicles (then seen as the likely means of launching a large space telescope). Boeing and other contractors were to agree on and employ this size in their studies, and so for over a decade the central feature of the planned Large Space Telescope's design was a 120-inch primary mirror.[83]

The construction of a large space telescope seemed to a number of people in NASA to be an obvious goal. Indeed, for some, the issue was not "Should this project be pursued?" but rather "How and when should it be pursued?" In 1965, in fact, NASA administrator James E. Webb wrote that such a telescope was a prospective major program for the agency. He even authorized a study group to report on how large-scale scientific programs might be managed by the agency, with a large space telescope program taken as a specific example.[84]

There was also discussion within NASA of what was called a "National Astronomical Space Observatory." The observatory was in effect a group of spacecraft that would carry instruments to observe a wide variety of wavelengths, not just the optical, ultraviolet, and

Boeing's 1965 conception of a manned orbiting telescope. An idea of the scale of the telescope can be gathered from the relative sizes of the astronaut and the spacecraft.

infrared regions of the spectrum that would be accessible to a large space telescope.[85] Space telescopes, one might say, were very much in the air in the mid-1960s.

Part of the reason for this increased interest was another study organized by NASA and the Space Science Board. Around one hundred NASA staff members, astronomers, and consultants assembled in 1965 at Woods Hole, Massachusetts, to consider once again plans for some principal areas of space research. The charge to the Working Group on Optical Astronomy was to "examine the future needs of optical astronomy for large aperture orbiting telescopes of a generation beyond the orbiting astronomical instruments which are now being readied for launching."[86] In contrast to the Iowa meetings of 1962, there was in 1965 an added confidence among the participants that large space telescopes could be built. In the intervening years, powerful rockets had placed tons of equipment into orbit; the series of Gemini space flights, which in June 1965 had included the

Astronaut Ed White performs the
first U.S. space walk in June 1965.
(Courtesy of NASA.)

first space walk by a U.S. astronaut[87] had for many people shown
the long-term potential for repair and maintenance of spacecraft in
orbit. Industrial contractors had performed several engineering studies
on large space telescopes, and Grumman's progress in building and
testing the OAOs promised speedy progress in ways of pointing
scientific spacecraft, perhaps the biggest engineering challenge confronting the advocates of a large space telescope. Their feet planted
on what they argued was a solid technical base, the members of the
Working Group on Optical Astronomy therefore pushed for the
construction of what they called a "Large Orbital Telescope" (LOT).
Such an instrument, the working group judged, would be "uniquely
important to the solution of the central astronomical problems of
our era."[88]

The group's strong backing of the LOT is not surprising, as it
was composed largely of "believers." The chairman was Lyman Spitzer,
and among the other eight members were such strong advocates as
Arthur D. Code and Aden Meinel. Fred Whipple – another principal investigator on the OAO series, director of the Smithsonian Institution's Astrophysical Observatory, and long-time proponent of
space astronomy – was also a member.[89] Laurence W. Fredrick of
the University of Virginia and William G. Tifft of the University

Aden Meinel (left) and Arthur D. Code, two champions of large optical telescopes in space, seen inspecting a honeycomb-type mirror in the late 1950s. (Courtesy of Space Astronomy Laboratory, University of Wisconsin–Madison.)

of Arizona also sat on the group. Fredrick was a consultant to NASA's Langley Research Center on manned orbiting telescopes and had already written at length on such instruments, and Tifft was solidly in favor of manned space telescopes.

The working group's firm advocacy was nevertheless tempered by the concern that because the LOT would surely be very costly (certainly of the order of several hundreds of millions of dollars) and would require such a major effort by so many astronomers, it should not be undertaken until a significant fraction of the astronomical community favored the program. If a start were attempted before a sturdy base of support had been formed, the enterprise might well founder.

NASA also emphasized how essential it was to win over astronomers to the telescope's cause. Homer Newell, NASA's head of space science, addressed the American Astronomical Society in 1966:

In the past we have built our program around interested key scientists, with the concurrence and endorsement of their peers. However, for anything so large as a telescope comparable in size to our largest ground-based instruments . . . the cost to the Nation would be so great that the enthusiastic support and willing participation of an interested scientific community will be essential. An enlarged participation is essential not merely in the design, development, and operation of space facilities, but also in the pursuit of theoretical and laboratory research and of ground-based astronomical observations necessary to support the space program.[90]

How, then, to counter criticisms of space astronomy and begin to assemble such a coalition for a telescope? The Working Group on Optical Astronomy recommended that to help convert the Large Orbital Telescope "agnostics" and "atheists," to lobby for the telescope, and to begin an orderly examination of the technical problems such a telescope would present, some part of the group should continue in existence as an ad hoc panel under the National Academy of Sciences.

A similar proposal had not been acted on after the Iowa meetings in 1962, but in 1965 it was accepted. Seven of the astronomers in the working group joined what was titled the "Ad Hoc Committee on the Large Space Telescope," with Spitzer again chairman.[91] The three invited participants from NASA included the chief of astronomy at NASA Headquarters, Nancy Roman.

Born in 1925, Roman had developed a keen interest in astronomy during junior high school. After studying the subject as an undergraduate at Swarthmore, she received her Ph.D. from the University of Chicago in 1949. She had joined the Chicago faculty before moving in 1955 to the Naval Research Laboratory. In 1959, she had been approached by a NASA staffer after a lecture: "By the way, do you know anyone who would like to come and work for NASA and set up a program in space astronomy?" Before that time she had deliberately steered away from space science, "but the idea of my coming in with an absolutely clean slate to set up a program that I thought was likely to influence astronomy for 50 years was just a challenge that I couldn't turn down."[92] Roman joined NASA and began the tough job of helping to build NASA's astronomy program, often in the teeth of opposition from astronomers who were convinced that the money would be more profitably spent on ground-based telescopes. In so doing, she showed herself to be a hard in-fighter who was not overly concerned with bruising egos in her support for programs.

By 1965, then six years immersed in space astronomy, Roman was intimately aware of its difficulties and challenges. She thought that the promoters of the Large Orbital Telescope tended to underestimate the technical problems ahead, although she judged that even at this early stage, more feasibility studies were worthwhile, "because after all, we were never going to get there if we didn't start somewhere, and this was as good a place as any to start. I just wasn't at the point where I thought we were going to be doing it in a few years."[93] Roman, as we shall see, was also to play an important role in winning approval for the telescope by developing arguments in its favor, briefing managers and engineers at NASA and executive-branch officials, responding to questions from Congress, and often acting as the NASA Headquarters link to astronomers.

Nancy Roman, seen here in the early 1980s, was a leading advocate of the Large Space Telescope and then the Space Telescope in NASA Headquarters from the mid-1960s to 1980. (Courtesy of Nancy Roman.)

"CHAIRMAN SPITZER'S LITTLE BLACK BOOK"

The first meeting of the National Academy of Sciences "Ad Hoc Committee on the Large Space Telescope" was held in April 1966. The next three meetings were expanded to include as guests and observers a number of leading astronomers. At the California Institute of Technology in July 1966, for example, the focus was on the uses of an LST for studies of galaxies, observational cosmology, and interstellar matter. The visitors included astronomers from observatories on the West Coast of the United States. As noted earlier, the astronomers at those institutions had access to powerful telescopes at good observing sites and had thus far tended to be the least interested in space astronomy. Spitzer and his committee were carrying their message to where, from their perspective, it was most needed.

By June 1967, the committee's work had progressed to the stage where the members were together in Washington, D.C., for a three-day session. Their draft report was refined over the rest of that year.[94] Spitzer wrote a summary of the committee's findings the next year,[95] but the final report was not published until late 1969. *Scientific Uses of the Large Space Telescope* was, nevertheless, a crucial document in crystallizing ideas about what a large space telescope could do and why such an instrument should be built. But perhaps its most important role was to demonstrate that the prestige of the National Academy of Sciences had been put behind the telescope idea.[96] Published at the height of the Cultural Revolution in China, the most vivid image of which was militant Red Guards brandishing Chairman Mao's "little red book," the black-cover report of the Spitzer committee swiftly became known as "Chairman Spitzer's little black book."

The first point to note about the little black book is that the name "Large Orbital Telescope," used at Woods Hole in 1965, had been changed back to "Large Space Telescope." In considerable part this was due to the membership on the committee of William G. Tifft, a junior but enthusiastic proponent of employing astronauts to perform all sorts of tasks in space astronomy.[97] When the Spitzer report was drafted, astronauts had yet to set foot on the moon, and to the committee's members it seemed impossible that a telescope could be sited at a lunar base in the near future. Even for Tifft, it appeared "inevitable that the first instrument would go into orbit and I even agreed with that. But I said let's not commit; let's not freeze it to that. Let's keep the basic term as general as possible."[98] Rather than preempt matters, the "Large Orbital Telescope," clearly destined for earth orbit, became the neutrally named "Large Space Telescope."[99] Fred Whipple quipped that it should be called the "Great Optical

Device," giving it the appealing acronym "GOD." This suggestion,
however, was not pursued.[100]

The Spitzer committee had also arrived at four major conclusions on the use of the Large Space Telescope. First, it would make a "dominant contribution to our knowledge of cosmology," that is, to our understanding of the structure, scale, and evolution of the universe. Stars one hundred times dimmer than the faintest detected from the ground could be observed because of the telescope's very high resolution and light-gathering power. The distances to galaxies could then be measured more accurately, and such distant galaxies could be observed that the Large Space Telescope might be able to reveal the very fate of the universe. In other words, the LST might determine if the universe is to expand forever, the galaxies dispersing into eternity, or if it will eventually start to contract and race toward what would later become known as the "Big Crunch," with all of its material falling back on itself. Second, "in many other fields of astronomy . . . the LST would give important and decisive information."[101] Third, "an efficient space astronomy program cannot be carried out by the LST alone. A continuing series of smaller telescopes is also required." It was desirable to gain engineering experience by building telescopes smaller than the 120-inch Large Space Telescope, and these smaller telescopes, as well as sounding rockets, could be exploited to conduct research programs to be followed up by the more powerful Large Space Telescope. Fourth, "the most effective utilization of powerful space telescopes requires a substantial increase in the number of ground-based instruments."[102] Careful to address the concerns and anxieties of ground-based astronomers about space astronomy gobbling up all funds for astronomy, and seeking to accommodate the aspirations of such astronomers in the LST plans, the committee members contended that the number of facilities for ground-based astronomy should be at least doubled. Even this increase would be but a small percentage of the cost of the LST and would, they argued, greatly "increase the value of the space astronomy program both by improving scientific insight and by avoiding space observations that could have been made from the ground."

The Spitzer committee had raised the flag of the Large Space Telescope to a prominent position. As a committee of the National Academy of Sciences composed largely of leading astronomers, it carried considerable influence. Who would rally to the cause?

2

Building a program

That's the project I would like to see Marshall do. The Large Space Telescope – I'd support it.

Wernher von Braun

However influential the deliberations of the Spitzer committee were among astronomers, far more than the backing of the National Academy of Sciences would be required to initiate the construction of the Large Space Telescope. By the late 1960s, the days had long since passed when astronomer-entrepreneurs could imitate George Ellery Hale and secure funds to construct giant telescopes by persuading wealthy philanthropists to part with their money, at least not the several hundred million dollars the Large Space Telescope promised to cost. If the telescope were to be built, it could be done only with strong commitments of money, technical resources, and manpower from the federal government. And given the structure of U.S. science funding, such support would have to be won from NASA, far and away the chief sponsor of space astronomy in the United States following its founding as an agency in 1958.

It is indeed characteristic of Big Science that many of the key decisions to accept or reject such enterprises are made by bureaucratic organizations, not professional groups of scientists. In other words, the idea of the Large Space Telescope, which had originated with astronomers, had to be integrated within the bureaucratic aims of NASA if the telescope were to be built, for it was NASA that could provide its funding. The agency would therefore have to be prepared to commit political capital, manpower, and resources to the telescope, its own advisory groups would have to back the telescope enthusiastically, and NASA would have to be ready to press the telescope's cause before numerous institutions and groups, including the White House and the Congress. For it to come into being, the design and development of such a telescope would have to be embedded within NASA's changing set of goals, policies, and

ambitions, as well as the agency's technical and management practices. The plans to build the telescope would thus be subjected to —
partly shaped by, and partly themselves shaping — the shifting sands
of NASA's institutional and political circumstances. It is to NASA,
then, that we now turn.

NASA

NASA is a central player, although by no means the only player, in
the U.S. civilian space program. The private sector and the Department of Commerce, to take two examples, are also engaged in civilian space efforts. And although it is a civilian agency, NASA has
nevertheless been involved in various national security activities. Until
the early 1980s, NASA also received the bulk of funds going to
U.S. space endeavors, but from that time on the Department of
Defense began to spend more on space than NASA. Nor are civilian
space policies neatly segregated from military policies; in fact, the
two sets of policies often intersect. Almost any space capability can
be applied for both civilian and military purposes, and military space
policies therefore provide some of the background against which to
set civilian space policies, including, as will be seen in Chapter 5,
the policy-making on the Large Space Telescope.

NASA often appears from the outside to be a monolithic organization. It is in fact a loose coalition of quite disparate groups. NASA
is headquartered in Washington, D.C., but there are various "field
centers" around the United States, including, for example, the Marshall Space Flight Center in Huntsville, Alabama. The role of NASA
Headquarters is essentially to develop agency policy and to provide
overall direction and guidance for the agency, as well as to act as a
"buffer" between the field centers and the Congress and the White
House. Headed by an administrator and deputy administrator, NASA
Headquarters contains several major divisions. The most important
ones for the history of the Space Telescope have been the Office of
Space Science and Applications, the Office of Manned Space Flight,
and the Office of Advanced Research and Technology. The Office of
Space Science and Applications, for example, oversees all NASA's
space science projects, but the feasibility studies and the design and
development work for a scientific spacecraft are largely the field centers' responsibility. Indeed, the detailed engineering and management work on NASA projects is generally conducted at the field
centers.[1] NASA is in fact a highly decentralized agency, one result
of which is that there is an inherent tension between the field centers
and NASA Headquarters. The field centers thus sometimes view
Headquarters as a kind of rival institution, a point that will be of
importance later in our story.

One of the central issues in NASA's history, and in the history of government-purchased technology in the United States generally, is how to buy, maintain, and operate research equipment, in particular, whether to use a system of arsenals or employ independent contractors.[2] The U.S. Army has made particular use of arsenals, for instance. NASA has instead generally preferred to follow the course charted by the U.S. Air Force in exploiting and fostering to the fullest the resources of private industry for most of the products and services it uses.

By shying away from creating its own manufacturing facilities, NASA has consequently been in large degree a technical management agency. NASA engineers and scientists generally have not built flight hardware. Rather, as Arnold Levine writes,

> they have planned the program, drafted the guidelines, and established the parameters within which the product is to be developed. Viewed in this light, the rationale for an in-house staff has largely been to enable NASA to perform those functions that no government agency has the right to contract out, functions enumerated by a former Director of the Bureau of the Budget [now the Office of Management and Budget] as "the decisions on what work is to be done, what objectives are to be set for the work, what time period and what costs are to be associated with the work, what the results are expected to be . . . the evaluation and the responsibilities for knowing whether the work has gone as it was supposed to go, and if it has not, what went wrong and how it can be corrected on subsequent occasions." Without the experience of actually building a spacecraft or performing experiments, [field] center personnel could not effectively select industrial contractors or supervise flight projects running to hundreds of millions of dollars.

The in-house work in advanced research and technology at centers such as Marshall and the Goddard Space Flight Center in Maryland has been done not only for its own sake but also because the agency decided that it would help keep NASA personnel abreast of the latest developments in their fields. They would thus be far better placed to judge the work of contractors and to ensure that, for example, cost estimates were realistic. Also, field centers would be more likely to attract and keep the best people if they had the opportunity to continue to work with the hardware needed for space programs. The manner in which NASA has structured its activities, then, has had other significant effects. "By separating evaluation and production," Levine argues, "NASA acquisition policy has had three especially important long-range consequences: the delegation of technical direction and monitoring to the centers, the refusal to set up a production capacity already existing in the private sector, and the refusal to create operating divisions intermediate between public and private sectors."[3] To take an example from Chapter 1, the Orbiting Astronomical Observatories were built by Grumman and its subcontractors, but a field center, Goddard, was charged with the overall management of their development.

STYLES

In thinking about particular machines or spacecraft there is a great temptation to focus on hardware and software, rather than on human activity. A consequence is that the final machine is then seen as somehow inevitable, its design fixed and propelled by the inexorable demands of purely technical considerations. In fact, the history of technology — including, as we shall see, the history of the Space Telescope — provides numerous cases in which technological innovation has had to be interpreted in a manner that has emphasized many other factors too.[4] Certainly technical factors alone cannot explain how two NASA centers, given the same set of requirements for a spacecraft, can produce their own distinctive designs for that spacecraft, a circumstance that was to occur in the Large Space Telescope program. Instead, what does become clear from NASA's history is that delegation of technical direction and monitoring to the agency's field centers has meant that each NASA center has taken on a distinct management and engineering style that has had to be carefully considered in analyzing the development of hardware with which the various centers have been involved.

A center's style is a complex product of its history. Through their own particular sets of institutional interests, their participation in different projects, and their work with various contractors, NASA centers acquire their own unique funds of experience and develop their own ways of tackling certain types of problems. These resources, both physical and intellectual, then fashion a center's approach to later projects.

In the 1960s, for example, Goddard's focus was on unmanned scientific and applications satellites. The engineers, managers, and scientists at Goddard did not have to worry about crew safety, a fundamental concern in designing spacecraft in which astronauts are to fly. Goddard had also become extremely knowledgeable about, and adept at carrying out, projects that were of relatively small scale. The Goddard center therefore viewed its activities in terms of series of spacecraft, not one-of-a-kind vehicles devised for a single flight.

Although several NASA centers were involved with space science, Goddard was very much the agency's lead center for astronomy. So when Goddard looked toward possible designs for the Large Space Telescope, the center's engineers and scientists naturally viewed it as a program that might succeed, and be closely based on, the technology developed for the Orbiting Astronomical Observatories (OAOs). Goddard engineers also proposed that the systems incorporated into the OAOs could be gradually scaled up to meet the more stringent needs of the Large Space Telescope.[5] Goddard, indeed, fashioned plans for "follow-on" OAOs that would serve to bridge the gap between a three-meter telescope and the OAOs.[6] Unlike Marshall, Goddard stressed the telescope as an unmanned and automated satellite, although one capable of having various parts and subsystems changed in orbit by astronauts.

Grumman, Goddard's prime contractor for the OAOs, was also pushing hard to play the same role for the Large Space Telescope. By 1969, Grumman had completed a study of a three-meter telescope in which they had identified a variety of approaches to its construction, but all of which took the OAOs as their foundation.[7]

Goddard's management and engineering styles were certainly different from the more centralized and generally more conservative approaches pursued at the Marshall Space Flight Center in Huntsville, Alabama. One writer has characterized those NASA centers reporting to the Office of Manned Space Flight at NASA Headquarters as "semi-autonomous, almost baronies, within the [Office of Manned Space Flight] framework."[8] Marshall was one such barony. Created in 1960 by transferring to NASA the core of the U.S. Army Ballistic Missile Agency's Development Operations Division, Marshall's intimate links with the army, and the army manner of working, meant that the center was inclined to go against the NASA grain to some extent and on occasion build hardware. Such an approach was also close to the approach that had been used by von Braun and his team at Peenemünde, several of whom had become leaders at Marshall.[9] The Marshall center also relished large-scale projects and had managed the hugely complicated and demanding program to construct the rocket engines to power the Apollo spacecraft to the moon. Marshall's history and orientation were therefore very much those of a center aimed toward placing, and using, astronauts in space.

Goddard and Marshall were not the only NASA centers active in early studies of telescopes in space. As noted in Chapter 1, Langley Research Center in Hampton, Virginia, had become engaged in such planning in the early 1960s. Studies were conducted for the center by a variety of contractors, but all bore on Langley's investigations of a possible manned orbiting telescope that might become part of a space station.[10]

In the early 1960s, NASA planners expected that if orbiting telescopes were to be visited regularly by astronauts, then a space station would have to be in orbit nearby. So for Langley, astronomical observatories were valued less for the astronomy they could perform than as logical adjuncts of, as well as solid justifications for, a space station program. From the very inception of NASA, in fact, a space station program has been widely seen within the agency as an essen-

Wernher von Braun (second from
right) in 1956. Others in the photograph include (going counterclock-
wise from the back left), Major General H. N. Toftoy, Ernst Stuhlinger,
Hermann Oberth, and Eberhard Rees
(later to be director of the Marshall
Space Flight Center). (Courtesy of
NASA.)

tial objective, and its proponents often have advocated space telescopes to help legitimate their arguments for a station.

One of those advocates was Wernher von Braun. Born in 1912 in
Wirsitz, Germany, von Braun had been fascinated by rocketry since
the late 1920s. Although he first indulged his passion as a pastime,
by 1932, he, as well as other German rocket pioneers, were at work
on rockets for the German Army. His technical expertise, allied to
a charismatic and persuasive manner, led to rapid advancement, and
he became one of the leaders of the Nazi rocket program in World
War II (see Chapter 1). At the end of the war, he came to the United
States to work on guided missiles for the U.S. Army.

In many ways, one can regard von Braun as the principal guide
to the paths of space exploration that NASA has followed and has
attempted to follow. Von Braun, in effect, articulated the agency's
basic goals even before the agency had been created. Certainly space

63

stations had long been in von Braun's thoughts and played a pivotal role in his ideas on space travel. In 1952, for example, he had publicly urged the construction of an enormous wheel-shaped, triple-deck space station.[11] Von Braun envisaged that the station would be accompanied by an astronomical observatory that would sometimes be visited by astronauts. "This observatory," von Braun contended, "will be used mainly to record the outer reaches of the universe, from the neighboring planets to the distant galaxies of stars. This mapping of the heavens will produce results which no observatory on earth could possibly duplicate."[12]

When the Marshall Space Flight Center was formed in 1960, von Braun became its first director. Although Marshall's overriding concern in its first decade was to develop the rockets that would provide the thrust to carry astronauts to the moon, various studies of space stations and related projects were nevertheless conducted at the center. It was these that had led some Marshall staff members to develop an interest in space telescopes, particularly in questions of how astronauts might service various kinds of astronomy "modules" attached to space stations or located nearby.

During the 1960s, several Marshall groups also examined the problems of placing mirrors and large optical telescopes in space.[13] From 1965 to 1967, for example, the Perkin-Elmer Corporation, in combination with the Lockheed Missiles and Space Company, investigated for Marshall a series of large orbiting telescopes whose mirror sizes would range from one-half to three meters.[14]

TO LEAP OR WALK?

Despite all the studies of telescopes that were being performed, NASA still had to decide if and when it would seek to build a large space telescope. The ambitions of space scientists always exceed NASA's capacity to support their plans, and usually many worthy possible programs are jostling for scarce funds. One of the agency's most difficult but most important tasks is therefore its selection of which projects to approve and which to reject. The decision to fund a project depends to a large degree on the relative power of its supporters, particularly those within NASA. In 1969 the National Academy of Sciences approved the telescope project, a crucial first step in bringing it into being. The next step for the telescope's advocates was to gain the patronage of NASA as an agency, not just the interest of various people within the system. But before the agency's favor could be won, the idea of building such a telescope would have to endure several NASA reviews. An important part of this process is played by the advisory groups that NASA itself sets up using outside scientists. A project that does not win their enthusiastic backing has little chance of being funded.

One such advisory group was the "Astronomy Missions Board." Established in 1967, the board's task was to advise NASA in the planning and conduct of all agency missions to create and operate astronomical experiments in space. The main board received recommendations from a number of panels, and the panel on "Optical Space Astronomy" was chaired by Lyman Spitzer. Spitzer's panel included four astronomers from our small cast of characters already familiar with and favorably inclined toward the Large Space Telescope: Arthur D. Code, Nancy Roman, Harlan Smith, and Fred Whipple. All had participated in the Woods Hole meeting in 1965, and all had later served on Spitzer's National Academy of Sciences committee. It is hardly surprising, then, that the Optical Space Astronomy panel proposed a large space telescope as a central goal for NASA. For them, the telescope was *the* tool for tackling many astronomical problems.[15]

The panel also had to face another knotty and pivotal question: how to make the passage from the first few OAOs, with maximum possible mirror sizes of forty inches, to a large space telescope with a far larger primary mirror of 120 inches. Would it be possible, for example, to scale up the OAOs until they reached the appropriate size, as Goddard and Grumman argued, or would wholly new technical approaches be needed to build the 120-inch telescope? Here we should note a significant point: When the astronomers spoke of a 120-inch large space telescope, they were in fact referring to a 120-inch *diffraction-limited telescope,* one whose performance would be determined by the laws of physics and by the accuracy with which the telescope could be pointed, not by the flaws in its optical system. If the industrial contractors could not make mirrors 120 inches in diameter that were diffraction-limited, then the astronomers would press for the primary to be larger than 120 inches to compensate for the reduced optical performance, say 150 or 180 inches.

Even if the telescope's mirror really was to be 120 inches in diameter, the transition to the Large Space Telescope obviously implied large increases in both size and complexity as compared with existing astronomical spacecraft. For most of the 1960s, those interested in the 120-inch telescope regarded the gap as too big to vault. So underpinning the debates of Spitzer's Optical Space Astronomy panel, as well as those of the Astronomy Missions Board itself, was the belief that the final design of the 120-inch telescope should be reached by an evolutionary process, not a quantum jump. The assumption among NASA, the contractors, and astronomers was that this should be achieved by building one or more intermediate space telescopes. The experience gained in that exercise could then be employed in designing and fabricating the Large Space Telescope itself. A series of spacecraft would also have the advantage, from the perspective of space astronomers, of ensuring a steady flow of data.

Existing research groups could be kept occupied, new research groups could be formed, and astronomers could be kept continuously involved in the field of optical space astronomy. Such an approach would help to fashion a stronger base of support for the building of the final 120-inch telescope.

The issue for NASA, the contractors, and the astronomers, therefore, was to determine what kind of intermediate telescope to construct. Until that could be decided, and one concept selected, the question of building the Large Space Telescope inevitably had to remain in abeyance. But NASA would find it immensely difficult to pick the road down which to travel to the telescope, and for a few years the debate on the telescope became hopelessly bogged down as competing interests vied to promote their own versions of the intermediate telescope.

STEPPING-STONES

Studies of telescopes to bridge the gap between the OAOs and the Large Space Telescope had begun in the mid-1960s. One idea under review at NASA was an orbiting telescope called "ASTRA."[16] NASA's initial plans featured ASTRA spacecraft with forty-inch primary mirrors, although that was later increased to sixty inches.[17]

The Astronomy Missions Board maintained that at least two ASTRA observatories should be flown. In what the board described as its "optimum" program, it was even bolder, proposing four ASTRA launches, the first targeted for 1976, the last for 1983. Because the Large Space Telescope was listed for a 1981 launch in the optimum program, the first three ASTRAs could serve as engineering and scientific test beds for the bigger telescope.

A significant point in the planning of both the Astronomy Missions Board and NASA's ASTRA group was that both assumed that the ASTRA spacecraft would be "national facilities."[18] In other words, although a small group of astronomers would develop the scientific instruments to be carried on the ASTRA observatories, any astronomer could propose to make observations with those instruments once they were in orbit.

Such a plan was in decided contrast to existing practice. To that time, the experiments flown on space science missions had been the products of teams, each led by a scientist designated the principal investigator. The reward for the labors of the principal investigator and his or her team on a space astronomy project was that they chose which astronomical objects to observe and, for a certain period, had exclusive rights to all of the scientific data collected. This arrangement tended to cause resentment among astronomers who were not part of the system and so felt excluded from the "charmed circle." But even on the very first OAO there was a small "Guest Observer"

program to allow astronomers who had not worked on the experiments to make observations. For *OAO-C, Copernicus,* the program
was extended further. So whereas to a limited extent the OAOs were
themselves national facilities, NASA intended to make the process
even more open for ASTRA.

In 1968, NASA estimated that an ASTRA observatory would
cost between $200 and $450 million, large sums that correspond
roughly to the range $635 to $1,430 million in 1982 dollars.[19] If
executed as planned, ASTRA would clearly be an expensive project,
and because the Large Space Telescope would be much more complex and would stay in orbit for a decade or more, it would likely
be considerably more costly. Given the critical attitude with which
many ground-based astronomers regarded optical astronomy from
space, restricting the telescope's observations to only a limited group
(the principal investigators and their teams) would be no way for
NASA to win friends for space astronomy. It would be difficult for
the telescope's advocates to justify it at that price if it were for the
benefit of only a small number of astronomers. Also, the body of
data the telescope would return during its long life would be so vast
that it would surely overwhelm a handful of principal investigators
and their teams, and much of the data might never be examined. In
consequence, NASA decided early that for a combination of political
and scientific reasons, the Large Space Telescope would be a "national facility," to be launched into space for the benefit of all astronomers – in the political arena, to win wider support for optical
space astronomy, and in the scientific realm, to ensure the participation of many qualified scientists.

PHASED STUDIES

Several times in this chapter there have been references to studies
conducted for NASA by contractors. Such studies are a fundamental
part of NASA's planning process and were certainly essential in the
development of the Large Space Telescope program. But to better
understand these studies, we must delve into a little "NASA-ese."

During the mid-1960s, NASA introduced into its projects a system of "Phased Project Planning." Although it has been argued that
Phased Project Planning was nothing more than a normal management sequence for research and development, it was one of NASA
administrator James Webb's tools for strengthening management
control over all new projects. If nothing else, it introduced a new
set of terms into NASA's vocabulary.

By the late 1960s, the NASA terminology had evolved and become essentially set: A project's "Phase A" constitutes preliminary
analysis, analysis aimed at answering the basic question whether or
not a spacecraft can be built to the desired specifications.[20] Cost is

not a major concern in Phase A. In the next phase, Phase B, more refined designs are established, and costs are better defined. NASA and its contractors examine various approaches, but the goal is for NASA to decide the technical and scientific requirements to be used in writing the contracts for Phase C/D, the detailed design and development phase in which final designs are selected, metal is cut, components are assembled, and the spacecraft is built. Phase C/D also marks the "great divide," the distinction between conception and implementation. This is because before Phase C/D can begin, the White House and the Congress must approve funding for the project.[21]

During the 1960s, when NASA and its contractors were conducting feasibility studies on large space telescopes, the agency was in a low-key Phase A, even if it was not officially so designated. At that stage, the studies tended to be done piecemeal as money became available. "We'd find [$50,000] here, another [$100,000] here, and we would look at those [studies] and see what else we needed," recalls Jesse L. Mitchell, head of the Physics and Astronomy Division in NASA Headquarters. In this manner, Mitchell and others kept the telescope visible within the agency, and contractors were being encouraged to become interested in and informed about the telescope.[22] The next step for the telescope advocates within NASA would be to gear up the agency's planning and move into an official Phase A.

A turf fight within NASA Headquarters prevented the agency's managers from reaching a consensus on how this was to be done. At stake was the direction of any large space telescope program and the sort of telescope to be built. The Office of Advanced Research and Technology had for several years funded studies of space telescopes and was contending with the Office of Space Science for control of the early stages of any large space telescope program. The issue was whether the first flight of the telescope should (1) be aimed toward simply proving the technology and demonstrating that a diffraction-limited 120-inch large space telescope could indeed be built or (2) be geared to performing science. If the former, the Office of Advanced Research and Technology would hold sway; if the latter, the program would be in the hands of the Office of Space Science.

Even if they could not agree on the type of telescope to build, the two offices could agree on the need for more feasibility studies.[23] Both offices therefore sponsored a special workshop on the technology needed to build optical telescopes. In that workshop, held at Marshall in the spring of 1969,[24] there was one substantive question at issue: Was it technically feasible at that time to build a large space telescope?

Most of the eighty-three papers delivered by NASA, industrial, and university engineers, as well as astronomers, were devoted to

68

summaries and reviews of what were seen by NASA and its contractors as key technical problems.[25] There were sessions, for example, on mirrors and optical materials. Mirrors for space telescopes would be subjected to powerful vibrations on launch, as well as widely varying temperatures once in orbit. Ideally, the mirror they needed should not expand or contract, or lose its accurate figure, with the vibrations or the alterations in temperature. What material, then, was best for telescope mirrors that had to endure the exceptional rigors of long flights in space? Optical engineers knew that fused silica retains its shape for long periods, but that fused silica changes size with temperature shifts much more than does so-called ultra-low-expansion glass. Ultra-low-expansion glass was, however, more of an unknown quantity. If it should prove unstable and slowly alter its size during the many years the telescope was planned to be in space, the telescope's performance might gradually and inexorably worsen.

The material for the mirror and the accuracy to which the mirror could be polished were therefore central concerns for NASA and its contractors. The mirror was considered by many to be what NASA engineers sometimes refer to as a project's "tall pole." That is, shaping the mirror to the precision astronomers wanted might entail such difficulties that it would hold up the project, or "tent," and other problems would tend to get lost beneath the canvas — problems that would be present, certainly, but not so pressing or so visible. Early in a project, NASA and its contractors search for ways to "collapse" such a tall pole.

In consequence, in 1969, NASA stepped up its research on mirrors for a large space telescope. Not only was the Optical Telescope Technology Workshop held, but also there was a review by NASA of what facilities there were within the agency for building and testing optical systems, together with further studies of mirrors by university and industrial contractors.[26]

TEMPTING VON BRAUN

The choice of material for the primary mirror was only one of the matters facing the planners of the Large Space Telescope program. Addressing the Optical Telescope Technology Workshop on "NASA Goals and Objectives," Nancy Roman, the chief of astronomy in NASA Headquarters, frankly conceded that "our plans are still somewhat vague and somewhat flexible." NASA hoped that the workshop might suggest "some hints of the more promising approaches" to reach the goal of a large space telescope. Roman explained:

We want to go on to the intermediate sized telescope. We can go one of two ways, and, in fact, we will probably use aspects of both ways. We can go through the [Apollo Telescope Mount], which is essentially a manned mission making full use of man and the versatility and maintenance that he can provide, or we can bypass the major experiences of man and rely primarily on the techniques developed to handle automated spacecraft. In either case, the chances are that the intermediate spacecraft will make some use of the fact that man can work in space and, at the same time, will be planned so that it can operate in an automated mode for long periods of time without requiring the presence of man.[27]

Roman thus posed the crucial question: Was it better to evolve toward the Large Space Telescope via ASTRA, or was it preferable to build a 120-inch "precursor" telescope as soon as possible – in effect, fly a kind of prototype 120-inch – even if it would not have diffraction-limited performance?

Space telescopes were also high on the agenda of the "Astronomy Planning Panel," part of an agency-wide study in 1969 of possible post-Apollo programs. The planning panel was composed of NASA astronomers and engineers interested in space astronomy. They, too, examined the question of a 120-inch "precursor" or prototype Large Space Telescope as a stepping-stone to the final Large Space Telescope. But the planning panel challenged the assumption that the diffraction-limited telescope must follow the flight of an intermediate telescope. Instead, the panel proposed a more aggressive, and so more risky, approach in which the concepts to be used on the telescope in space would be tested and verified on the ground, not by the use of smaller intermediate telescopes or the flight of a precursor telescope. Such "all-up testing," as it is called, had been introduced into the development of the Titan II missile,[28] and NASA had successfully employed all-up testing in 1967 in the first flight of the enormous thirty-six-story-high Saturn V moon rocket. In so doing, four planned flights had, in effect, been compressed into one, and hundreds of millions of dollars had been saved.[29] Employing an all-up testing approach for the Large Space Telescope would be, the Astronomy Planning Panel concluded, "a relatively high risk program and requires that major program decisions related to optical concept selection, operating, orbit and the role of man be made by 1975. However, it is potentially the quickest and cheapest way of achieving an LST."[30]

There were numerous ideas among NASA and its contractors on designs for intermediate telescopes and the final Large Space Telescope, even a variety of technical approaches and program philosophies, but until NASA could decide which sort of intermediate telescope(s) to build, the 120-inch telescope would be delayed. And to one well-placed, but decidedly partial, NASA observer, matters looked badly confused. Henry L. Anderton, a leading manager in the Office of Advanced Research and Technology, appealed to von Braun at

Marshall to step in and end what he saw as a muddle. In September 1969, Anderton told von Braun the following:

I am surprised and chagrined that Marshall hasn't picked up the pieces of the space telescope fiasco and come forward with a solution to NASA's problem. Marshall in general, and you in particular, are always ready for a good exercise in innovation, so I assume that either you haven't noticed the problem or someone there feels you shouldn't get involved in it. Since I don't know which it is, it seems best to mention the problem to you personally. Then if you don't want to play the white knight, I will at least have tried my best.

NASA has had a very difficult time crystallizing its plans for a large optical space telescope. [The Office of Space Science and Applications] doesn't really know what it wants – some like a follow-on OAO and others want ASTRA. George Mueller [head of the Office of Manned Space Flight] wants a two-meter stellar telescope on the second [orbital] workshop – and would like it diffraction limited. [The Office of Advanced Research and Technology] has been working on a technology telescope for some time to give us the basic knowledge for the three-meter instrument of the eighties. Goddard had a proposal for follow-on OAO's which simply says, "we have experience – give us the job," and doesn't address any of the real problems of the job. In short everyone is on top dead center and won't talk to one another in a meaningful way.

Marshall, on the other hand, has everything available to come up with a proposal that at least nearly satisfies everyone. You worked on ASTRA, you have George Mueller's 72-inch telescope (or whatever goes in [Apollo Telescope Mount-B]), and you have the [Office of Advanced Research and Technology] telescope which your Astronics Laboratory people have been working on for four years. You are almost our only hope. Wouldn't you like to step in and solve the problem? In our case we came to Marshall with our technology telescope because you are innovative, knowledgeable, and willing to take a risk. This seems like one of those times when such characteristics could pay off for us all.[31]

Von Braun conceded that the obstacle to driving the telescope program forward was still the lack of accord in NASA Headquarters between the Office of Space Science and the Office of Advanced Research and Technology. However, he did not rise to Anderton's bait to try to assume control of the Large Space Telescope program by some kind of bureaucratic "end run." Instead, von Braun diplomatically advised selecting scientific groups for the first Large Space Telescope and "planning our missions as part of the Space Station and Space Shuttle programs."[32] Von Braun, as we shall see, did nevertheless very much want to bring the Large Space Telescope program to Marshall.

NASA'S CHANGING CIRCUMSTANCES

On July 20, 1969, Neil Armstrong descended from the lunar module, announced that "it's one small step for man, one giant leap for mankind," and walked on the moon. Yet this spectacular triumph for NASA was achieved at a time of deep uncertainty within the agency. Addressing a joint session of Congress in May 1961, President Kennedy had charged NASA with "achieving the goal, before this decade is out, of landing a man on the moon and returning him safely to earth." What was to follow now that that goal had been achieved? What new programs should NASA seek to pursue, and, more important, in what directions would the agency be allowed to move – or be driven – by Congress and the White House?

NASA's disarray over the matter of the Large Space Telescope was one element of a more widespread uncertainty over the agency's future. Much hinged on whether or not NASA would be permitted to build a space station. For example, if a space station were not built, astronauts could not move regularly from a space station to tend a nearby telescope. The plan to repair and maintain the telescope in orbit would of necessity then become a plan that did not require astronauts on a regular basis.

Although NASA was trying to muster support for a space station, it was by no means clear that the agency's campaign would succeed. NASA's budget had been reduced precipitately from the level reached in the mid-1960s, when monies were lavished on the Apollo program. The political support once enjoyed by the space program was fading, and during the years of the Johnson administration the high costs of the Great Society programs and the escalating war in Vietnam were taking funds out of other areas of the federal budget. NASA was just one agency that was being squeezed.

The changed circumstances did, however, lead to an important decision for NASA, even if it was by default. In late 1969, the agency's declining budget and pressure from the White House prompted NASA to close the Electronics Research Center in Cambridge, Massachusetts, the center that had run the Office of Advanced Research and Technology's telescope studies, and in 1970 the center was transferred to the Department of Transportation. Its demise left the Office of Space Science securely in command of the Large Space Telescope program and disposed of one bureaucratic roadblock to the telescope's progress.

The Electronics Research Center's optical equipment was transferred to Goddard. Among these items was a machine for grinding and polishing mirrors up to two meters in diameter.[33] In fact, the Office of Space Science had begun looking toward a two-meter space telescope as a likely precursor to the three-meter telescope, thereby skipping altogether more modest programs such as ASTRA.

One of the NASA people promoting such a jump was Marcel J. ("Marc") Aucremanne, who by 1970 was spending a considerable part of his time on the Large Space Telescope. After training as an electrical engineer, in 1959 he had joined the fledgling NASA

Headquarters in Washington, D.C., as a spacecraft program manager.[34] As program manager for advanced programs and technology, and later as the telescope's program manager, much of the day-to-day legwork on the telescope within NASA Headquarters – providing briefings, drafting memoranda, as liaison to the field centers, answering inquiries from inside and outside NASA, and so on – would fall to him.

The two-meter telescope, Aucremanne argued, had several advantages as the bridge to the three-meter. Equipment for making and testing two-meter mirrors was available. The one-meter Stratoscope II mirror, carried aloft by a balloon, had already been built and flown. It was of a special lightweight construction, such as would be used for the Large Space Telescope's mirror, and had also been polished to a diffraction-limited level. A jump to a mirror size of two meters would therefore not be unreasonable and would allow many of the subsystems to be used in the three-meter telescope to be tested in orbit.[35] In its studies for Marshall on a "Large Telescope Experiment Package," Perkin-Elmer had also selected a two-meter primary mirror. Two meters, moreover, was the suggested size for the mirror in the proposed stellar Apollo Telescope Mount, a telescope that was under study for flight aboard possible space stations and that was being backed by the Office of Manned Space Flight.

In June 1970, the Large Space Telescope and the plans for a two-meter precursor telescope were given a prominent position in briefings for NASA's senior managers. Nancy Roman contended that the telescope would "contribute substantially to every problem in astronomy,"[36] and Marc Aucremanne described how much work had already been done to develop the technology required. He reckoned that a diffraction-limited, two-meter precursor telescope could be flown by 1976, and he anticipated that the three-meter Large Space Telescope could be in orbit by 1980.

ENTER THE SPACE SHUTTLE

In June 1970, at the time of those briefings by Roman and Aucremanne, NASA was working hard to win approval to build some kind of Space Shuttle. But at that date the Space Shuttle design was far from fixed, and the program to construct it would not win presidential approval until 1972.[37] Notwithstanding these points, NASA rapidly incorporated the possible use of the Space Shuttle into its planning for the telescope. In his briefing, Aucremanne had emphasized that the telescope's existing designs had been fashioned on the assumption that it would be launched by a Titan III rocket, with certain constraints that that would impose in terms of weight, volume, and strength necessary to survive lift-off. However, some subsystems for the telescope could be built in the form of modules, and

73

Grumman's 1970 conception of what eventually became the Space Shuttle. This artist's drawing illustrates its possible use to maintain and refurbish the Large Space Telescope. (Courtesy of Pete Simmons.)

so when the Space Shuttle became operational, astronauts could rendezvous with the telescope "for the very important functions of instrument update, maintenance, or replacement of malfunctioning modules."[38] At this stage of NASA's planning, Aucremanne was advancing only aspirations for the shuttle, not facts that had been demonstrated to everyone's satisfaction. But, as we shall see again and again, the packaging together of the telescope and the shuttle would become a major means by which NASA managers sought to move the telescope from a marginal to a central position in the agency.

NASA and its contractors had also begun by 1969 to argue that the use of the Space Shuttle would significantly lower the costs of the telescope. In that year, Grumman had completed an "OAO/LST Shuttle Economics Study." It had two explicit goals: first, to determine what the program would cost if the launch vehicle were (a) the Titan III rocket or (b) the Space Shuttle; second, to examine how the availability of the Space Shuttle might alter the mission objectives and ways of operating the telescope.

With the Space Shuttle, Grumman calculated the price of the

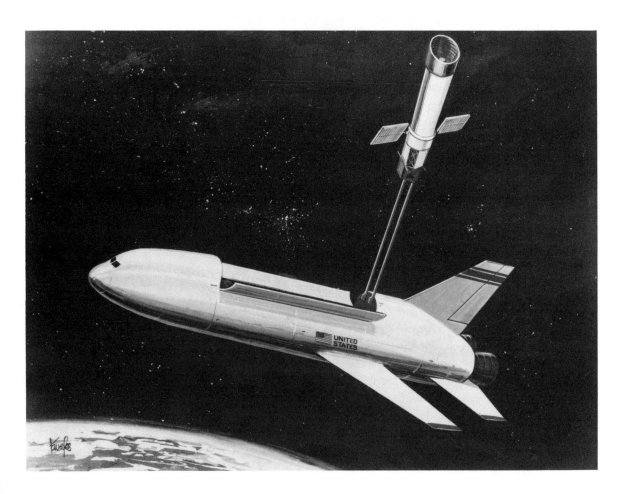

Large Space Telescope program to be $465 million, but without the shuttle, and using instead a series of Titan III launchers, $638 million. Underlying both approaches, however, was the assumption that flights of a precursor telescope would be needed before a diffraction-limited three-meter telescope could be flown.

Grumman, too, stressed the flexibility that a Space Shuttle would provide. If there was a major failure, the telescope could even be plucked out of orbit, placed in the orbiter's payload bay, and returned to earth for repair. More advanced scientific instruments could be exchanged every few years for those on board the telescope. In that way, the instruments would be kept close to the state of the art. Grumman also claimed that the current (1970) spacecraft technology was adequate to build the telescope, and the shuttle would be deployed to repair or preempt failures in orbit by replacing components. This approach, Grumman estimated, would be less expensive than developing components to last for the telescope's entire lifetime.

At the heart of Grumman's study, there was nevertheless another issue: to persuade NASA that the Grumman and Goddard approach to building a Large Space Telescope by upgrading the Orbiting Astronomical Observatories was the best one. Even the choice of the title "OAO/LST Shuttle Economics Study" was directed to this end, as it explicitly linked the Orbiting Astronomical Observatories to the Large Space Telescope.[39]

LOW'S APPROVAL

The Grumman shuttle study was completed in September 1970. By that time, the rounds of briefings and reviews had led the NASA administrator to ask the Office of Space Science to prepare plans on how the agency might begin formal Phase A/B studies of the Large Space Telescope.[40] On October 29, 1970, the Office of Space Science's ideas were presented in a crucial briefing to NASA's acting administrator, George Low. The emphasis was on how the telescope's scientific objectives had evolved over the previous seven years, the amount of preparatory engineering and scientific work that had already been done, and the strong interest in the program throughout the agency. In addition, the Office of Space Science wanted to advance the program in NASA by forming management groups. These groups would then act as a focus for the agency's overall efforts on the telescope.

In 1966, a senior advisory group of scientists from outside the agency had urged NASA to create a consortium of universities for the purpose of providing the project leadership for the Large Space Telescope program. As we shall see in Chapter 6, NASA had rejected this, not wishing to lose control to an outside body. Now, in

75

1970, the Office of Space Science proposed that Goddard and Marshall each complete Phase A studies. In so doing, the two NASA centers would compete for the role of lead center, the center to provide the overall project management for the telescope.

This issue, however, was not straightforward. NASA Headquarters was not concerned solely with which might be the "best" or most appropriate center in terms of management or technical expertise and resources. There also were questions for NASA Headquarters about how much manpower would be available at each of the centers to work on the telescope, as well as the wider issue of NASA's own institutional interests, and whether these were to be furthered most by choosing Goddard or Marshall as lead center. No definite management scheme was thus proposed at the October 1970 briefing to George Low. Nevertheless, one possibility that was raised by the Office of Space Science was for Marshall to assume the role of lead center up to and including the time of the launch of the telescope, after which responsibility would be handed to Goddard. Certainly by mid-1970 some in NASA Headquarters were already strongly inclined to adopt this management structure, even *before* the centers had embarked on the planned competition.[41]

Design conceptions of the 120-inch Large Space Telescope presented to NASA's acting administrator George Low in the course of a crucial briefing in October 1970. (Courtesy of NASA.)

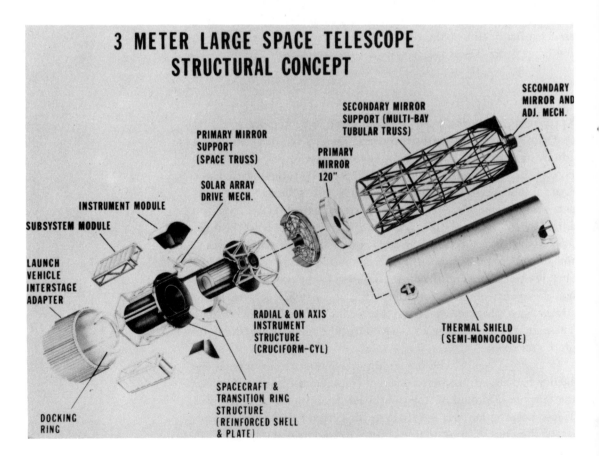

76

The final point broached in the briefing to George Low was how quickly NASA should proceed with the program. That was extremely important for NASA, because as the telescope's cost was reckoned to be several hundred million dollars, its price and the schedule on which it would be built would, as we shall see in Chapters 3, 4, and 5, have to be weighed by its potential patrons extremely carefully against other agency, national, international, and scientific goals. For example, if astronomers strongly opposed the telescope and pressed for other programs they regarded as more attractive, the telescope's justification would be undercut, or if Congress and the White House chose radically new aims for NASA, these might conflict with building the expensive Large Space Telescope. In neither case would the telescope be politically feasible.

Low nevertheless approved the Office of Space Science's recommendations to form management groups: an LST Task Team (to be chaired by Aucremanne) and a Scientific Advisory Group (to be chaired by Roman).[42] The Large Space Telescope program had now become solidly enmeshed within the NASA bureaucracy and had started to

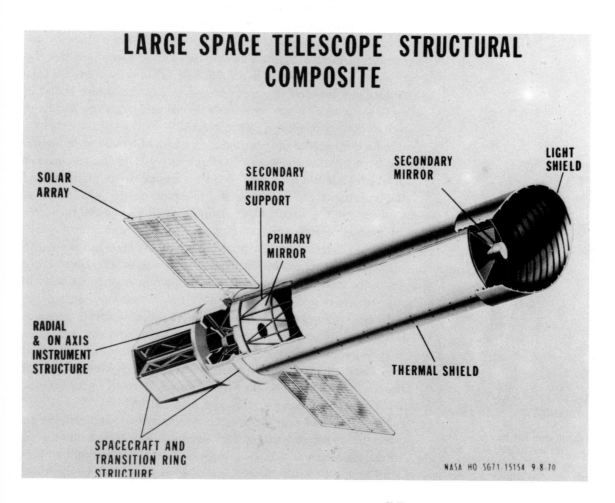

LARGE SPACE TELESCOPE STRUCTURAL COMPOSITE

SOLAR ARRAY

SECONDARY MIRROR SUPPORT

SECONDARY MIRROR

LIGHT SHIELD

PRIMARY MIRROR

RADIAL & ON AXIS INSTRUMENT STRUCTURE

THERMAL SHIELD

SPACECRAFT AND TRANSITION RING STRUCTURE

NASA HQ SG71-15154 9-8-70

occupy a significant place in the set of interests of the Office of Space Science, as well as those of Goddard and Marshall. In fact, by late 1970 the various pieces that its advocates reckoned to be necessary for the Large Space Telescope program were slowly coming together: There was a small but influential group of astronomer advocates of the telescope, there were active and interested engineering groups at Goddard and Marshall, as well as at several contractors, there was firm support for the telescope within the Office of Space Science at NASA Headquarters, and there was an agency-level commitment to at least study the telescope in detail.

IN HALE'S FOOTSTEPS

Once Low had given approval to form management groups, the pace of NASA's Large Space Telescope efforts picked up. By early November, Jesse L. Mitchell, director of physics and astronomy programs at NASA Headquarters, was writing to Marshall that "with this encouragement from Dr. Low I propose to proceed as rapidly as our limited resources allow with the LST planning."[43]

A graduate of the Alabama Polytechnic Institute, with a B.S. in aeronautical engineering, Mitchell had been a pilot and crew commander in Europe during the war. In 1947 he had joined the National Advisory Committee on Aeronautics' Langley Research Laboratory as a research engineer. After a year in the White House in 1959 as an aide to the president's special assistant for science and technology, Mitchell had joined NASA.[44]

As a boy in Alabama during the 1930s, Mitchell's imagination had been fired by the story of George Ellery Hale and the manner in which Hale had been the prime schemer and driving force behind the building of the most powerful telescopes in the world. During a vacation in the 1950s, Mitchell and his wife had visited the Corning glassworks to see the prototype blank of the mirror for the 200-inch Hale Telescope on Palomar Mountain. As Hale had been with his giant telescopes, Mitchell was fascinated by the Large Space Telescope's scientific potential, and he sought eagerly for ways to move the telescope program ahead.

In late 1970, Mitchell fastened on the idea of adapting one of Hale's favorite stratagems: Get a mirror blank, and then somehow or other conjure up the money for the rest of the telescope. Mitchell thus asked the Marshall center to transfer various kinds of optical material in its stores to the Goddard and Marshall optical shops. These items included a seventy-two-inch mirror blank.[45]

Mitchell found bureaucratic niceties distasteful, and early in his career he had found that a demonstration was "worth more than 10,000 calculations." If the seventy-two-inch mirror could be polished to a diffraction-limited level, he reasoned that would show

Marshall engineer Jean Olivier in 1987. He was to play an important role in the telescope project from the early 1970s. (Courtesy of NASA.)

conclusively to the doubters within NASA and outside that mirrors accurate enough for the Large Space Telescope could indeed be fashioned. In so doing, the project's "tall pole" could be partly collapsed, and additional support won for the telescope.[46]

Meanwhile, the wheels of management put into place by the Office of Space Science began to turn. On February 25, 1971, the LST Task Team met in Washington, D.C. Its ten members represented a wide variety of interests within NASA, with each member being responsible for a particular area of expertise likely to be needed in designing the telescope.[47]

Marshall and Goddard, too, increased their efforts on the Large Space Telescope. Marshall, for example, had formed a special Phase A team headed by Jean Olivier (pronounced as Oliver). Olivier would come to play a lead role in the design and building of the Space Telescope, eventually as chief engineer, and would win a reputation as one whose views carried great weight throughout the program.[48] By 1971, Marshall had already worked on a variety of space tele-

In Hale's footsteps

During the 1960s and early 1970s, Marshall worked on many ways of exploiting a space station. These included employing a space station to service modules containing astronomical apparatus. (Courtesy of NASA.)

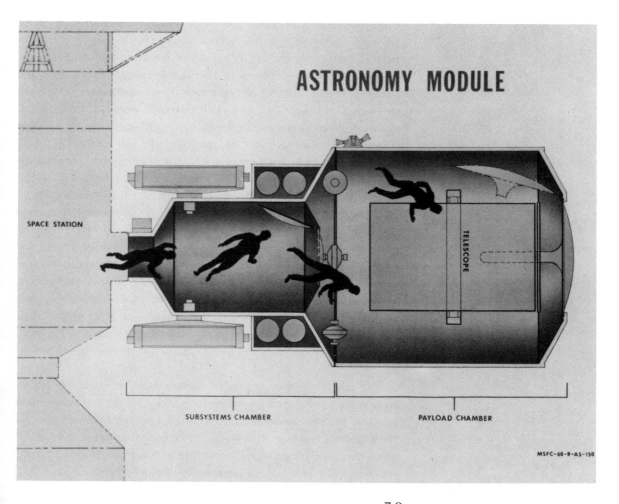

scope concepts, particularly as they related to possible space station designs, but Olivier and other Marshall engineers now expanded their own series of studies, both at the center and through contracts with industry and universities.

In July 1971, the "Large Space Telescope Science Steering Group," chaired by Nancy Roman, assembled for its first meeting. This group's ostensible assignment was to define the scientific objectives for the telescope and advise on the ways NASA might meet them. During the two days of discussions, the talk frequently turned to the question of the size of the Large Space Telescope's primary mirror, a fundamental issue, because the mirror size would be the most influential factor in determining the dimensions of the rest of the telescope.

John Naugle, head of the Office of Space Science at NASA Headquarters, told the members of the steering group that the telescope was one of the agency's principal goals in space astronomy in the 1970s. He nevertheless warned that it might be necessary to drop to, say, a 100-inch space telescope (which would doubtless be cheaper than a 120-inch instrument) so as not to draw too much money from, and so disrupt, the rest of NASA's astronomy programs. With these cautionary words, Naugle sounded a theme that would reverberate throughout the entire period of planning and building the telescope: the trade-off between cost and scientific performance, and the appropriate split between the two. There was never, as we shall see, a clear or simple answer on this matter to which all the different parties engaged in the telescope program could agree, and so it would give rise to numerous disputes, battles, and negotiations.

Technical and scientific issues were not the only ones debated by the Science Steering Group. As for the National Academy of Sciences committee chaired by Lyman Spitzer a few years earlier, a central question was the political issue: how to gain the astronomical community's enthusiastic endorsement of the telescope. One astronomer suggested that NASA should emphasize that the telescope was part of a "balanced program." By that time, "balanced program" was a heavily loaded term. It carried the implication that optical space astronomy should not dominate ground-based astronomy, and so it was used by ground-based astronomers eager to ensure their funds would not be siphoned off to support space activities, as well as by space astronomers anxious not to alarm ground-based astronomers.[49]

The relationship of ground-based astronomy to optical space astronomy was also a major worry for Mitchell as he and other managers in NASA Headquarters searched for ways to win support for the telescope, support that would be indispensable if it were to become politically feasible. Even before the Science Steering Group meeting, Mitchell had suggested that the agency settle on a 2.5-

meter Large Space Telescope. He proposed building a mobile ground-based telescope of that size. Such a telescope would allow, among other things, the "earliest possible start on an LST, based on the available technology," as well as permit experience to be gained with hardware similar to that to be used with the Large Space Telescope, "as opposed to more and more paper." It might even "yield major innovations in design and operation of ground-based telescopes," as well as be an "effective means of bringing together the ground-based and space-based astronomical community."[50] Although the idea of a mobile telescope went nowhere, the concern in NASA that had inspired it – the need to link more tightly the interests of ground-based astronomers to those of space astronomers – remained.

EARLY DESIGNS

By the time of the early work in Phase A, the design of the Large Space Telescope had already been divided by NASA and its contractors into three main sections: (1) the "Optical Telescope Assembly" (OTA), (2) the "Support Systems Module" (SSM), and (3) the "Scientific Instruments" (SIs). These terms will become very familiar, as they would continue to underpin the telescope's design in the following years.

The Optical Telescope Assembly at this stage of the design process was essentially a reflecting telescope. At its heart was a three-meter mirror that would collect and reflect light onto a convex secondary mirror. The secondary mirror would then redirect the light back through a hole in the primary mirror to a variety of scientific instruments. In other words, the Optical Telescope Assembly would be a kind of Cassegrain telescope (see the Introduction).

The Support Systems Module would provide the Optical Telescope Assembly and the scientific instruments with power, radio communications with earth, pointing and control, and other support systems necessary to ensure that the observatory would function successfully. At that stage of the planning, though all the prospective designs were decidedly different, they shared the common feature that the Support Systems Module was located behind the Optical Telescope Assembly and in effect was bolted onto the back of the OTA.

During 1971 and early 1972, Goddard and Marshall developed their different telescope concepts. These were intended by the centers to provide a base of knowledge for further, more comprehensive, design studies, and these concepts were arrived at, and fashioned by, a variety of in-house and contractor studies. The Itek Corporation designed a possible Optical Telescope Assembly for Marshall, for example.[51] Marshall engineers themselves worked on a concept for the Support Systems Module, although Lockheed, Martin, and General Dynamics/Convair also devised their own Support Systems Module designs.[52] All of the various conceptions of the Support Systems Module were compatible with the Itek-developed Optical Telescope Assembly and with a proposed set of scientific instruments, the design of which was the charge of a contractor to Goddard.[53]

That Goddard should manage the development of designs for possible scientific instruments led to an awkward institutional arrangement. It meant that Marshall and Goddard were linked at the same time as they were vying with one another to become the lead center for the telescope program. That there often is competition among different divisions within American companies is hardly a novel observation, and in like manner, NASA centers, by their nature, often tend to adopt a highly competitive attitude toward other centers. Yet Goddard, because of its responsibility for the instruments, was providing information to Marshall on possible scientific instruments for the telescope to aid Marshall in the construction of its telescope design. Communication between the two centers, perhaps not surprisingly, was far from good.[54] The Marshall–Goddard relationship was not to be an easy one in future years, and sometimes it would be remarkably bad. But the seeds of discord and confusion were sown in the fertile soil of a contest to be the lead center for the Large Space Telescope.

PROGRAM DESIGN

In early 1972, the Office of Space Science announced that the studies at Goddard and Marshall had progressed sufficiently for Mitchell to tell the two center directors that "it is apparent to me that management decisions should be made as soon as possible if we are to continue to make good progress toward the launch and operations of the LST." Headquarters wanted an intensive effort within the next few months so that the lead center could be chosen by March 1972.[55] As both centers recognized, here would be an important step in the construction of what can be termed the "program design," the way in which the various organizations that would be involved in designing and building the telescope would themselves be assembled and ordered.

The decision on the lead center would be made at NASA Headquarters. The issue, as already noted, was not simply that of picking the "best" Phase B proposal according to some set of agreed on and unambiguous technical and management criteria. A key point for NASA Headquarters managers was how much manpower would be available at each center to work on the telescope, an issue that was particularly acute for Marshall. In 1972, agency budget cuts meant

that the numbers of NASA staffers and contractors were still falling. Program design
Marshall had been hit particularly hard, and the center's personnel,
together with support contractors, had tumbled from 15,435 in fiscal 1967 to 9,654 in fiscal 1972.[56] Without new projects, Marshall's future looked bleak, and the center was often raised in discussions between the Office of Management and Budget and NASA
as a target for closure.

By 1972, NASA's ambitious plans to follow Apollo with a program aimed to land astronauts on Mars had been sunk by loss of
political support and a reduced agency budget. From the wreckage
of those schemes, NASA had finally won approval from the White
House in January 1972 to build the Space Shuttle.[57] Although it
was large, the shuttle program was by no means comparable in size
to the Apollo program, and so NASA would shrink even further in
size.

"We still had the [Orbiting Astronomical Observatories and the
Orbiting Geophysical Observatory]," Mitchell recalls, "so that
Goddard just was committed up to the gills as far as manpower
went, and it was obvious, to me at least and to most people, that as
we came out of Apollo, that the manpower was going to be available
at the Manned Spacecraft Centers."[58] Also, the attitudes of the two
center directors toward the Large Space Telescope were distinctly
different. John Clark, the director of Goddard, was leading a center
that had done well in securing new programs. Its future, as NASA
headquarters had recognized, was assured. Securing the Large Space
Telescope, for Goddard, was far from a life-or-death issue. Pete Simmons, an official in Grumman's Orbiting Astronomical Observatory
project, recalls battling on this very point with Clark "innumerable
times." Try as he might, he could not persuade Clark to fight hard
to bring the Large Space Telescope to Goddard. Perhaps this was
because in early 1972 Goddard was having serious problems in managing a number of existing projects. As NASA's deputy administrator lamented in March of that year at a senior management review,
"almost every one" of Goddard's major projects "was shown with a
red flag."[59]

Marshall, in comparison with Goddard, was very anxious to secure the program to build the telescope and pursued it vigorously.
In 1965, a well-placed member of the Space Sciences Laboratory at
Marshall had assumed that "of course, [Marshall] cannot compete
(nor should it try to compete) with [Goddard] in the field of astronomy."[60] By 1971, matters were quite different. An astronomer remembers that "the Marshall people, my goodness, they practically
had a brass band and red carpet any time NASA Headquarters people would come down to Marshall to see what they were doing with
what became [the Space Telescope]."[61]

Marshall's aggressive attitude came from the very top. Jesse Mitchell

83

vividly recalls attending a meeting at Marshall in the late 1960s. After dinner, he strolled with the center director, Wernher von Braun. They discussed Marshall's future, "and I asked [von Braun] about the Large Space Telescope. He expressed to me there 'Jesse, that's the project I would like to see Marshall do. The Large Space Telescope — I'd support it.' "[62]

But Goddard was the permanent NASA center for space astronomy and, in sharp contrast to Marshall, had a large number of resident astronomers. In the late 1960s, Marshall did not have a single staff member with a Ph.D. in astronomy. On the other hand, it had never devoted itself solely to building large rocket engines. As noted earlier, throughout the 1960s there were several small groups active at Marshall in mirror and telescope studies. The center had also been involved in space science since the late 1950s, although never to anything like the same extent as Goddard. Marshall had managed the Princeton Stratoscope II balloon-borne telescope program and was the lead center for the complex Apollo Telescope Mount to be flown aboard *Skylab,* for example. In addition, NASA Headquarters had appointed Marshall lead center for the series of "High Energy Astronomy Observatories" (HEAOs) that were planned to be launched during the 1970s. Through the center's involvement with Apollo there was also a rich fund of experience at Marshall on how to manage large-scale programs, such as the Large Space Telescope would surely prove to be. Goddard's proven expertise, instead, was in the management of many programs of relatively small scale.

It was, however, the manpower argument that was decisive. Writing in 1977 at the end of his period as NASA administrator, James Fletcher told his successor that "closing Marshall has been on [the Office of Management and Budget's] agenda ever since I came to NASA [in 1971], although from time-to-time they have also suggested [the Jet Propulsion Laboratory], Ames and Lewis. We have always resisted this very strongly on the basis that (a) the initial cost of replacing the facility would be very high and (b) we couldn't afford to risk the Space Shuttle program." Marshall had unrivaled skills and knowledge in building rocket engines, and those would be essential to the shuttle's development. NASA, Fletcher wrote, had fended off the thrusts of the Office of Management and Budget by putting new programs at Marshall, "such as space telescope, HEAO, etc.," as well as allowing Marshall "to do a considerable amount of work on the Shuttle to make good use of their personnel."[63]

Although Goddard's recent management performance had not impressed senior managers in NASA Headquarters, it is in the light of what these managers regarded as vital agency interests that we should see NASA's decision on the lead center. The Office of Space Science had, in fact, favored making Marshall the lead center for

84

about two years, and the competition for that prize was almost irrelevant.

On April 25, 1972, Naugle wrote to NASA deputy administrator George Low that he had thoroughly reviewed the studies and advanced technical development work on the Large Space Telescope, and he had opted for Marshall as the lead center. Naugle also maintained that "it is technically feasible to develop this three-meter optical telescope, and . . . it can be placed in operation in the decade of the 1980s as an essentially permanent observatory in space through the marriage of automated spacecraft technology and the unique capabilities of the shuttle transportation and maintenance systems."[64]

On May 5, Naugle broke the news to Eberhard Rees, director of Marshall. Naugle told Rees that he had already discussed with Goddard's director the possibility that Goddard would still participate in building the Large Space Telescope. Goddard, Naugle hoped, would work with Marshall in a manner similar to that already worked out for construction of the High Energy Astronomy Observatories.[65]

When Rees and Clark discussed the choice of Marshall as lead center on May 11, Clark reported that when "John Naugle told me the decision, I told him I was not surprised. I couldn't imagine him coming to any other conclusion." Clark was convinced that it was best to do the mission and data operations from Goddard, as any other decision would be "extremely costly." He nevertheless hedged when Rees asked if Goddard would manage the scientific instruments.[66] The extent of Goddard's involvement in the telescope program would provoke numerous discussions and squabbles between Marshall and Goddard and would be a running theme in the telescope program.

CONCLUSION

The choice of Marshall as the lead center meant that the way in which NASA would structure its program to design and build the telescope was crystallizing. The selection of Marshall had another meaning, too, for following the endorsement of the telescope by the National Academy of Sciences and the Spitzer committee, more essential steps on the road to negotiating approval for the telescope had been taken: The telescope had become solidly embedded in the aims of both the Office of Space Science in NASA Headquarters and the Marshall Space Flight Center, and by packaging the telescope together with the Space Shuttle, the telescope program was gaining wider interest in the space agency. NASA as an organization now had a firm commitment to at least study the telescope in some detail, and the coalition advocating the telescope's construction was growing in strength.

3

Astronomers, industry, and money

I knew that the competition would be far behind us, and this was by far the most critical concern of Marshall and ourselves.

Domenick Tenerelli

Your man is over there.

Lyman Spitzer, Jr.

In 1972, NASA was pondering if, when, and how to start the Large Space Telescope's Phase B, the phase in which the agency's contractors would produce a variety of telescope designs, and the program's cost would be better defined. NASA Headquarters, however, would have to approve the transition from Phase A to Phase B. Such a shift is by no means automatic, and many projects have foundered at this point. Sometimes a scheme will be deemed technically unrealistic, but in the deliberations on the telescope's move to Phase B it was to be its expense that would emerge as the central issue for NASA Headquarters managers. Costs, as well as detailed engineering studies, would force major changes in the way the telescope would be designed and in how NASA would plan the program to build it. Thus the drive intensified to produce a telescope that not only would be feasible in an engineering sense but also could win support from astronomers, NASA, Congress, and the White House, and thereby be politically feasible. That drive will provide the main theme in this chapter, but it will also entail examining how astronomers and industry were engaged in the program, the roles they played, and the shifting power relationships among NASA, the astronomers, and the contractors, in particular the extent to which the telescope's designs were the creations of industry.

COSTS

In March 1972, Eberhard Rees, a Peenemünde veteran who had succeeded von Braun as director of Marshall, sent his center's esti-

86

mate of the cost of the Large Space Telescope program to NASA
Headquarters. That estimate, between $570 and $715 million (1972
dollars), covered design and development, building an engineering
model of the telescope, the "precursor LST," and the final LST.[1]
These numbers alarmed senior managers at NASA Headquarters.
For George Low, NASA's deputy administrator, they were impos-
sibly high. Low judged that if the telescope's price could not be
substantially reduced, it would never be approved by the Office of
Management and Budget and by Congress. If so, the telescope would
never be approved. Marshall's task, as the center was told repeatedly
by NASA Headquarters, was to emphasize ways of reducing costs.[2]

In November 1972, NASA administrator James Fletcher was
briefed on the telescope. Marshall had begun stressing "low-cost
systems" and potential ways of saving money. Even so, Marshall
calculated that if the telescope stayed in orbit for its planned fifteen
years, the cost would be about $700 million ($2 billion in 1987
dollars). That was a great deal of money, but World War II, notes
one historian, "vastly expanded perceptions of how much money
might productively be spent creating advanced tools for scientific
research. The War also dramatically changed expectations about how
much of this money might be forthcoming from the federal government."[3] Certainly, since the war, many groups of professional scientists have developed expensive research ambitions. However, the
government agencies that channel most of the funds to them cannot
meet all of their desires, and the costs of the numerous potential and
worthy programs always outstrip the available money. There is, then,
a natural tension between the clients (the scientists) and their patrons (the government agencies), a tension that would be quite visible in the case of the Large Space Telescope.

The problem was that at a cost of $700 million, the telescope was
projected to be the most costly scientific instrument ever built.
Headquarters viewed the figure of $700 million as too extravagant,
and that sparked an effort by the agency and its contractors to design
a cheaper, and thus politically more feasible, telescope. In planning
for the Large Space Telescope in 1972, Marshall and Goddard were
therefore compelled to mediate between the astronomers' aspirations
and NASA Headquarters' aim to produce a less expensive telescope.

In reality, this meant that Marshall and Goddard and their contractors were embarked on a search to produce a sound engineering
design for the telescope while cutting the cost of its design and
development, and even perhaps the telescope's planned scientific
performance. But reshaping the telescope's design was a difficult and
delicate task. It also entailed more than engineering analyses for
NASA managers, because it would have to be done while still maintaining and even extending the support of the astronomers: Cut the
telescope's performance too much, and the astronomers might lose

interest, but produce too costly a design, and it might never be built.

CONTRACTORS

In addition to design changes, NASA Headquarters sought other ways to reduce costs. Two ways in which NASA did this were to attempt to stimulate competition between contractors and to seek to interest a number of major contractors. Much turned for NASA on the important issue of whether the telescope would be built using a "prime contractor" or "associate contractors."

In NASA's way of doing business, the telescope would be designed and built almost entirely by industrial contractors, not by NASA or the astronomers. The choice between prime contractors and associate contractors would then structure how the agency was to work with its contractors and how those contractors would interact with each other. To make this point a little clearer, consider the construction of a house. With a prime contractor, one builder has overall responsibility. The builder lets various subcontracts − for plumbing, for instance − but the builder has overall responsibility for the job. On the other hand, if the house is built by associate contractors, then each associate contractor will erect its own section of the house, and let its own subcontracts. But in this case, a third party has to confirm, for example, that walls and floors meet where they should and that the house is not jerry-built. If NASA were to use associate contractors for the Large Space Telescope, there would be one Optical Telescope Assembly associate contractor and one Support Systems Module associate contractor. In that case, Marshall or another contractor would have to act as overseer and ensure that all the parts of the telescope fit together correctly. Associate contractors would also mean a more elaborate management structure than with a prime. Despite that drawback, both Marshall and Goddard had favored associate contractors in their Phase A proposals. Marshall, for example, had claimed that if it acted as the overseer of the associate contractors, it would be cheaper, as the agency would not have to pay a fee to the prime contractor to manage the other contractors.[4]

But possible cost savings were not the only benefits Marshall foresaw from the use of associate contractors. Marshall had checked which companies had the capability to build the Support Systems Module and the Optical Telescope Assembly. In Marshall's opinion, only large aerospace companies of the size of, say, Lockheed and Boeing had the expertise and resources to manage the design and development of the overall telescope system. If a prime contractor were to be used, this role would, because of NASA's management approach, inevitably have to be played by an aerospace company. If so, an

optical company would be subcontracted to the aerospace company
to construct the Optical Telescope Assembly. Yet Marshall knew of
only two optical companies (Itek and Perkin-Elmer) that had the
skill to do the job and also wanted to participate in building the
telescope.[5] Using a prime contractor would also mean that in the
bids for the Phase C/D contracts, two aerospace companies would
be paired with the two optical houses. The other aerospace companies that had not been so paired would then be cut out of the competition. But if associate contractors were employed, NASA would
be able to pick what it reckoned to be the best Optical Telescope
Assembly proposal and the best Support Systems Module proposal,
and then perform the pairing itself. So NASA reasoned that opting
for a prime contractor might well produce a less competitive, and
ultimately more costly, selection of companies.

Marshall also preferred associate contractors because that system
would enable the center to manage directly both the Optical Telescope Assembly and Support Systems Module, rather than working
through a prime to manage the Optical Telescope Assembly. As
Marshall's project manager was to contend, these sections of the
spacecraft, as well as the scientific instruments, should all be managed with a tight rein and in close combination, so "that scientific
and technical requirements could be properly balanced, appropriate
trade offs made to reduce costs, interfaces well defined, and so costs
kept to a minimum."[6]

CHANGES

At a meeting with NASA administrator Fletcher on December 21,
1972, Marshall agreed that two optical contractors would independently produce designs for the Optical Telescope Assembly and that
plans for these contracts should soon be put "in train."[7] However,
toward the end of the meeting, Fletcher set the telescope program
in a new direction. He struck an ominous note, one that was to be
heard again and again for the next few years, warning Marshall that
NASA's financial position "was not conducive to [the] initiation of
large projects," and certainly not a billion-dollar Large Space Telescope program. Fletcher instead suggested that $300 million was a
reasonable cost target for the design and development of a "Mark I"
telescope, which might later be upgraded.[8]

The follow-up meetings at NASA Headquarters immediately after
Fletcher's remarks were perhaps even more significant. There, Marshall, NASA Headquarters, and Goddard managers discussed the
engineering approach to building the telescope, as well as ways of
saving money. In particular, Marshall consented to adjust its planning so that the program's cost to the end of the telescope's first year
in space would, as Fletcher had proposed, not exceed $300 million.

Other decisions were reached too. In the first of several flip-flops on this issue, Marshall proposed to consider returning the telescope to earth for maintenance during the early years of its operation, instead of following an approach that would stress repair in orbit. The argument ran as follows: To develop instruments and subsystems that could be replaced in orbit would require innovative, and doubtless expensive, designs. Exchanging systems on the ground would simplify the design and also lower the costs. In particular, it would lower the design and development costs to a level closer to that wanted by Fletcher, even if this approach would mean higher operational costs. Marshall agreed that plans for a percursor telescope should be abandoned.[9] In so doing, a long-held assumption was swept unceremoniously aside.

For years, a precursor telescope, or perhaps precursor telescopes – be it a fourth Orbiting Astronomical Observatory, ASTRA observatories, a precursor 120-inch telescope, or a Stellar Apollo Telescope Mount – had been widely viewed in NASA and outside as an essential stepping-stone to the Large Space Telescope. Such a concept was now dead. Although everyone had recognized and accepted that the precursor was to lead to the 120-inch Large Space Telescope, there was never widespread agreement on exactly what the precursor itself was to be. No strong base of support had developed for any one concept among the astronomers, the contractors, or NASA. The precursor was in consequence politically vulnerable and to a large degree had become a project in search of a reason. Once Marshall had decided that there was no compelling technical need to build one, and that by canceling it money could be saved, the precursor's fate was sealed.

Also, by 1975, several of what were, in certain respects, later claimed to be a kind of precursor to the Large Space Telescope, the "Big Bird" photoreconnaissance satellites, had already been orbited and established as operational. The link between astronomical space telescopes and earthward-looking telescopes is discussed in Chapter 5. Here it is sufficient to note that the Big Bird satellites have been reported to have carried large mirrors, together with sophisticated instrumentation. Perhaps the Big Bird flights meant that for those NASA managers, contractors, and astronomers in the know, a kind of precursor to the Large Space Telescope had already been launched, and there was no need for NASA to construct another precursor. Also, the *Copernicus* satellite, the second successful Orbiting Astronomical Observatory, had been launched in 1972, and although its main mirror was only thirty-two inches in diameter, the working of its guidance and control system had helped to assure NASA, and enabled the agency to claim in congressional testimony, that the precise pointing required of the Large Space Telescope could be achieved.

Dropping the precursor telescope idea was not the only crucial

change the managers of the Large Space Telescope program were considering in early 1973. Another involved the way the telescope would be built: a shift to the so-called protoflight concept. Until the early 1970s, almost all major U.S. space hardware had been built only after constructing a prototype. The prototype was then used for tests, and the results from the tests guided the final design. The need to cut the cost of the proposed Large Space Telescope program, however, drove Marshall to look to an alternative scheme, one in which the prototype would also be the flight spacecraft. By adopting that approach, it could save the time and cost of fabricating a complete set of hardware simply for testing.[10]

Although the protoflight concept has certain decided advantages over the traditional method of building a spacecraft, the Marshall engineers recognized that it carried its own drawbacks. In fact, there are two main extra risks that it brings to a program. First, tests still must be carried out to investigate the spacecraft's performance, and these, in the absence of a prototype, must be run on the flight hardware. Such tests must be meticulously documented and planned, in addition to being performed extremely carefully so as to ensure that nothing is done to the flight hardware that might jeopardize the spacecraft. The second major problem with the protoflight concept is that the results of these tests may indicate that design changes are needed. Such alterations then have to be made on the hardware that is to be flown. Certainly spacecraft engineers would much prefer flight hardware to be designed in one piece, not subjected to modifications and changes. Also, some kind of engineering model or "breadboard" (a board on which electrical or electronic circuit diagrams may be laid out and experimental circuits constructed) is needed even when using the protoflight concept. Such a model, though it is by no means an exact replica of flight hardware, does enable engineers to develop some analytical and experimental tests that will provide guidance in building the protoflight spacecraft.[11]

ENTER THE ASTRONOMERS

While NASA was considering the ways its contractors might build the telescope, the agency also was examining what sort of role the astronomers should play in this process. To begin this part of our story, we switch to a hot June evening in 1972 on Montesano, a beautiful wooded hill overlooking Huntsville, Alabama, a town not too far removed from its past of dirt roads, farming, and cotton mills. By 1972 it was most famous for its association with the Marshall Space Flight Center and the U.S. moon program.

Many of the German rocket engineers who had joined Wernher von Braun in surrendering to U.S. forces at the end of the war, and who had subsequently played such an important part in the activities of Marshall, had made their homes on Montesano. One such was Ernst Stuhlinger. When *Explorer I,* the very first U.S. satellite and the riposte to *Sputnik,* had been launched in January 1958, it had been Stuhlinger who had pushed the button in the control room that put *Explorer* into a circular orbit.[12]

But now Stuhlinger was acting in his capacity as Marshall's associate director for science. He was hosting a reception at his home for the members of the Large Space Telescope Science Steering Group, the chief body of scientists investigating the telescope's scientific goals and design. During the group's talks earlier in the day, the members had vigorously debated the qualities needed by the person chosen to be "project scientist" for the telescope. This would be an important position. That person not only would have to defend the scientific integrity of the telescope within NASA and be the chief conduit for scientific advice to the project's engineers but also would have to be senior enough to carry clout with his or her astronomer colleagues. The last was a significant point if traditional ground-based astronomers were to be assured that the Large Space Telescope was worthwhile and were to provide their support – support that would be essential for the program's initiation. We therefore see, once again, an example of the wide variety of roles imposed on scientists by the demands of Big Science, for the project scientist would have to be at different times, among other things, an advocate, a coalition builder, a publicist, a scientist-engineer, and a manager.

Stuhlinger was keenly interested in the appointment of the project scientist, and he was quizzing Lyman Spitzer on who could do the job. Spitzer himself was an obvious candidate, but he was not prepared to leave Princeton to work full time at Huntsville.[13] Over a dozen names had been bandied about during the day's discussion. Spitzer, however, pointed to one of the Science Steering Group and declared, "Your man is over there."[14]

Spitzer had indicated C. R. ("Bob") O'Dell. Born in Illinois, O'Dell had, from an early age, wanted to be an astronomer. After a brief and uneasy flirtation, as a graduate student, with the Wisconsin work on the first Orbiting Astronomical Observatory, he had shifted to ground-based astronomy. In so doing, he entered a very fast track indeed. By 1966 he had become director of the famous Yerkes Observatory of the University of Chicago, at the remarkably young age of twenty-nine.[15]

When he had joined Yerkes, he had been to all intents and purposes a conventional ground-based astronomer, and like many such astronomers he was chary of space astronomy. But O'Dell's exposure to a variety of current and future space programs as a member of NASA's Astronomy Missions Board from 1967 to 1969 and as a member of President Nixon's transition team in 1968 persuaded

92

C. Robert ("Bob") O'Dell (second from left) was appointed the Space Telescope's project scientist in 1972, a position he would hold for over a decade. The others in the 1980 photograph are program manager Don Burrowbridge (left), project manager William Keathley (second from right), and Lockheed manager William Wright. (Courtesy of C. R. O'Dell.)

him of space astronomy's worth and potential. Passionately fond of flying – he was to write a textbook on aerobatics in 1980 – O'Dell, like Spitzer, felt an emotional pull toward space activities and space travel. Moreover, by the end of the Astronomy Missions Board's activities in 1969, O'Dell had decided that the "Space Telescope is going to be the big thing," that it would be "the most important instrument in astronomy during my lifetime."[16]

Once the project scientist job had been offered to him, he did not dwell long on the decision. O'Dell felt that the timing was right for him to move from Yerkes, although he left initially on a three-year leave of absence. On September 5, 1972, O'Dell was in Huntsville as the Large Space Telescope's project scientist.[17]

O'Dell had held a prominent position as director of Yerkes and had a reputation for being politically canny. That he would isolate himself from his astronomer colleagues by moving to Marshall, an astronomical backwater, despite being a top-rank NASA center, was noteworthy for several of them. O'Dell's move gave the program added credibility and helped to persuade these astronomers that NASA was truly serious about funding the Large Space Telescope.[18] During Phase B he would become a leading figure on the project, and in large part this was because of his central role in providing a focus for the aspirations and goals of non-NASA astronomers.

When O'Dell took up his new position, his first aim was to fashion an institutional means to secure high-quality scientific advice while the telescope was being designed and built: How to arrange this, and how to attract the most talented astronomers to participate in the design process?

Drawing on the discussions of the Phase A Science Steering Group, of which he had been a member, O'Dell soon devised a "Science

Management Program for the LST." As an advocate of the astronomical community within NASA, he brought to his position a belief, widespread among astronomers outside of NASA, that civil service astronomers generally were not of the same caliber as university-based astronomers and that the telescope's development and operations should not be entrusted solely to them. As O'Dell recalls, "I was one of those people guilty of a very real prejudice about the Goddard . . . science organization of astronomers – again, this was before the great success of [the International Ultraviolet Explorer[19]]. And also there was a certain natural rivalry that was built into the appointment, that is between the two centers, and since I was [at Marshall], and [Goddard] was not really happy with that arrangement, still wanted to have [the Space Telescope] there, it seemed a natural thing to take an aggressive attitude toward them."[20] O'Dell wanted to draw the outside astronomical community into the telescope program as much as possible, to establish institutional mechanisms to ensure that it would be *their* telescope, not Goddard's telescope.

The main tool by which O'Dell sought to accomplish this was a new scientific advisory group. He wanted such a group to have some members whose job it would be to maintain a "broad" view of the telescope and its capabilities, but also to include the leaders of the "Instrument Definition Teams." These teams would define the number and kinds of scientific instruments to be flown and what the specific scientific instruments were to do, identify the scientific and technical factors affecting their design, and provide preliminary instrument designs for use by industrial contractors in their own planning.[21] Industry would then produce the detailed designs of the scientific instruments.

Up to that time, NASA had planned a "universal instrument" for the telescope. The design of the universal instrument incorporated a variety of cameras, spectrographs, and infrared instruments. Several mirrors would then be placed in the path of the light entering the telescope to divert parts of the beam into the separate instruments. But the Phase A Science Steering Group had judged this arrangement to be too complicated. Indeed, it was sometimes known, rather unaffectionately, as the "kludge."[22] Rather than press on with designing the universal instrument, O'Dell wanted to rethink the entire approach to the scientific instruments.

George Pieper, Goddard's director of earth and planetary sciences, thought O'Dell's scheme for involving the astronomical community in designing the telescope "in general" acceptable, but he bristled at some of its aspects. Writing to NASA Headquarters, Pieper protested three major issues on which, he warned, "it should be understood that [Goddard's] participation is predicated." The main requirement was that Goddard be responsible for the full collection of scientific instruments and that the set of instruments be delivered to Marshall as an integrated, tested, and calibrated unit.[23] In other words, Goddard insisted on a free hand in managing the development of the scientific instruments, which, at that point, Goddard argued, should follow the concept of the universal instrument. Nor did Goddard want to be directed on this issue by Marshall.

Despite the wide differences between the Marshall and Goddard positions, an agreement was thrashed out at a meeting involving O'Dell, Pieper, and another of Goddard's leading science managers. Most of Pieper's points were conceded, and Goddard acceded to O'Dell's proposal for involving astronomers outside of NASA in the Large Space Telescope program.[24] More important, perhaps, than those results was that the meeting involved some early exchanges in what was to be a generally difficult and sometimes egregious relationship between Marshall and Goddard, a relationship that would lead the Goddard study manager to write as early as 1976 that the

Enter the astronomers

The "kludge." Early concept for the telescope's collection of scientific instruments. (Courtesy of NASA.)

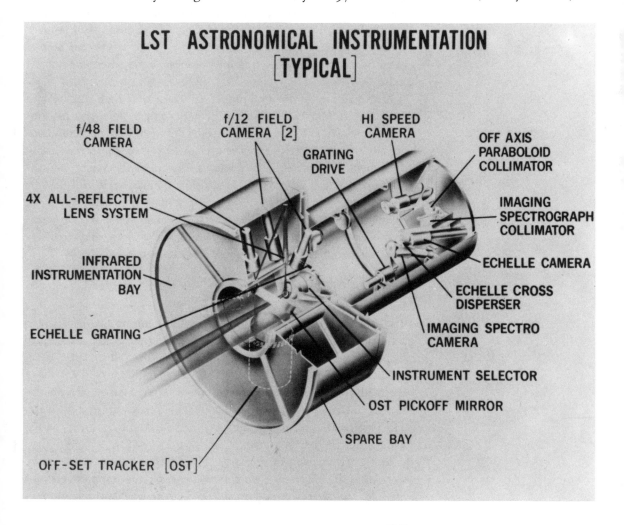

LST ASTRONOMICAL INSTRUMENTATION [TYPICAL]

f/48 FIELD CAMERA

f/12 FIELD CAMERA [2]

HI SPEED CAMERA

GRATING DRIVE

OFF AXIS PARABOLOID COLLIMATOR

4X ALL-REFLECTIVE LENS SYSTEM

IMAGING SPECTROGRAPH COLLIMATOR

INFRARED INSTRUMENTATION BAY

ECHELLE CAMERA

ECHELLE CROSS DISPERSER

IMAGING SPECTRO CAMERA

ECHELLE GRATING

INSTRUMENT SELECTOR

OST PICKOFF MIRROR

SPARE BAY

OFF-SET TRACKER [OST]

problems had led to "a tremendous amount of wasted effort and dollars."[25]

CREATING TEAMS

With Goddard's approval, NASA moved to assemble the group of astronomers for which O'Dell had pressed, the Science Working Group, together with the associated Instrument Definition Teams.[26] This process began in December 1972 with a public notice inviting astronomers to submit proposals to join in the preliminary design of the scientific instruments, as well as various aspects of the telescope itself (such as optical systems).

To increase interest in, and awareness of, the telescope, O'Dell decided to hold briefings at various points around the United States as part of his general policy of "going out and drumming up business, beating the bushes." The "dog and pony show," as he jokingly referred to it, opened at the California Institute of Technology on January 16, 1973, with presentations by representatives of Marshall and Goddard. Next day it moved to Chicago, ending up on January 18 at the Harvard College Observatory.[27] There was even a special Large Space Telescope briefing for European astronomers at Frascati, Italy.

NASA received some 118 proposals. These were then checked and evaluated at Marshall by several panels composed largely of non-NASA reviewers. From these scrutinies emerged a set of recommendations on the Instrument Definition Teams to be formed, who their members were to be and who was to sit on the Science Working Group. After reviews of these recommendations within NASA Headquarters, the agency announced publicly that there were to be six Instrument Definition Teams.[28] Together with the five "at large" members of the Science Working Group, there was a total of thirty-eight astronomers selected, almost all from outside of NASA.[29]

ENTER INDUSTRY

The astronomers were not the only ones gearing up for the Phase B activities. As already noted, NASA is to a large degree a technical management agency. It would therefore be the industrial contractors, not NASA or the astronomers, who would carry the bulk of the load of the design and development work on the telescope, as well as construct almost all of the telescope's hardware and software. The participation of many industrial firms in the telescope program would thus be indispensable.

With the division of the Large Space Telescope into three main sections – the Optical Telescope Assembly, the Support Systems Module, and the scientific instruments – there were two main groups of contractors preparing to bid for Phase B contracts: aerospace companies bidding for the Support Systems Module contracts and corporations with expertise in building large optical systems, or "optical houses," as they are known, bidding for the Optical Telescope Assembly.

Lockheed Missiles and Space Company, for example, was busy preparing its bid for the Support Systems Module. An aerospace giant, it had relatively little experience with satellites for space science, but possessed a vast wealth of knowledge about building satellites for national security uses − including photoreconnaissance satellites − as well as other purposes. Lockheed, moreover, had already investigated astronomical space telescopes for several years.[30] From 1965 to 1967, for instance, Lockheed had been paired with the Perkin-Elmer Corporation in a study for Marshall of various space telescopes, and from 1968 to 1970 it had engaged in the Marshall-managed Large Space Telescope Experiment Program (see Chapter 2).[31]

How should Lockheed prepare to win a Phase B contract, a definite step toward being awarded the lucrative Phase C/D contract? Most of Lockheed's work for NASA to that time had been based on the incredibly versatile Agena spacecraft. Winning a Large Space Telescope contract therefore had the added incentive for Lockheed of perhaps being the key that would open the door to more agency contracts.

Domenick Tenerelli, a member of the spacecraft structures organization at Lockheed, was one of a small group of Lockheed employees called to a meeting in mid-1972 to discuss the company's plans. He had eight years of experience in spacecraft engineering and while working full time for Lockheed at Sunnyvale, California, was also doing research for a Ph.D. in aeronautics and astronautics at nearby Stanford University. He had heard that Lockheed would bid for the Large Space Telescope and so had spent some time reading up on the telescope. Tenerelli had then quickly become heavily engaged in analyzing various engineering trade-offs as Lockheed assembled its "core" LST team.[32]

Lockheed engineer Domenick Tenerelli.

Lockheed staff members reviewed their early findings in October 1972.[33] To win a contract, Lockheed knew it would have to pursue a well-defined marketing strategy, as well as engineering strategy. Nor was there any simple division between the two. Lockheed's "master strategy," in fact, had several elements. First, the company intended to base its designs for the Support Systems Module on the so-called Satellite Control Section it had already developed for other satellites. The Satellite Control Section, ten feet in diameter and seventy-eight inches long, was in effect an already existing version of the Support Systems Module required by the Large Space Telescope. Lockheed would adapt its Satellite Control Section to provide the thermal control, electrical power, communications, data management, and pointing control on which the telescope itself would rely.[34] Second, Lockheed would look to the Support Systems Module to make a "major penetration" of the market arising from the need to build payloads for the Space Shuttle. In mid-1972 it certainly appeared to many people in the aerospace industry that the shuttle would create a bonanza for builders of spacecraft. After all, why plan to launch so many shuttle flights if there were few payloads to lift into orbit? The third part of Lockheed's master strategy was to understand the customer thoroughly, that is, learn as much as possible about how Marshall went about its work, and thereby improve Lockheed's chances of submitting a winning bid for Phase C/D.

For Lockheed, the Large Space Telescope was seen as a vehicle to develop, by use of its Satellite Control Section, a generalized capability. In fact, the contractors were not starting from scratch and seeking the best way to build a one-of-a-kind spacecraft, the Large Space Telescope, an approach that surely would have proved prohibitively expensive. Rather, Lockheed's telescope design, like the designs of the other contractors, would involve the company's own commercial and institutional interests, the company's history, and detailed engineering analyses.

In November 1972, at a special briefing for industry representatives, Marshall announced that the Phase A studies were to end the following month and that Phase B contracts would be let the next year.[35] Of particular concern to Lockheed and the other potential builders of the Support Systems Module was the design that Marshall, together with its contractors, had fashioned in Phase A.

The Support Systems Module would sit behind the three-meter primary mirror. Also behind the mirror would be the scientific instruments located inside a pressurized compartment. Marshall had included a pressurized compartment to allow astronauts to move around the instruments, to replace failed components, for example, without having to wear bulky and cumbersome space suits. To ease the transfer of astronauts between the Space Shuttle and the telescope, Marshall proposed that the two be directly linked by a "docking adapter." That telescope design bore some striking resemblances to designs already generated at the center for the astronomy modules and observatories intended for use with a space station. Marshall had thus drawn on the concepts and resources it had already developed in its space station investigations, a point that is hardly surprising, particularly as some of the Marshall engineers who were now hammering out Large Space Telescope designs had worked on these earlier studies.

There were, then, many studies and planning efforts under way on the telescope. However, formal approval from NASA Headquarters to start Phase B had yet to be granted, nor would it be until

Marshall had drastically revised its planned program to design and build the telescope.

REVISED PLANS

In December 1972, NASA administrator Fletcher had suggested that the Large Space Telescope program cost no more than $300 million. Marshall judged this a "reasonable goal."[36] Reasonable it might be, but exactly how was Marshall to meet the limit? It is important to remember that in all of their cost-cutting exercises, the Marshall managers and engineers were treading a narrow line. In particular, Marshall could save money by lowering the telescope's scientific requirements and thus the quality of the science the telescope would ultimately perform. Yet, if Marshall went too far in reducing the telescope's planned performance, it would lose the support of the scientists, as well as defeat the very purpose of building the telescope.

But the need to cut costs forced Marshall to make some severe changes to the program, changes that were reported to NASA Headquarters in a major meeting in April 1973. In the original plan, presented to headquarters as recently as December 1972, Marshall had included an engineering model, a precursor LST, and an extensive program of building spare subsystems so that failed components could be rapidly replaced, as well as the Large Space Telescope itself. In December 1972, Marshall had even proposed to test the telescope's systems by use of the engineering model, in addition to test-flying the precursor LST and actual LST. However, by April 1973, Marshall had opted for the protoflight concept. Indeed, by April, all that was left of the December program was the Large Space Telescope and a shortened list of spares.

The method of maintaining the telescope, too, had been radically changed. In the original December plan, Marshall had called for astronauts to perform extensive maintenance and repairs in orbit. By April, Marshall had decided that it preferred to make only simple repairs in orbit, with the chief means of maintaining the telescope being to return it to earth, a scheme that Marshall had earlier regarded as a backup.[37]

By adopting these and other measures, Marshall reckoned that the program's cost could be brought down into the range of $290 to $345 million in fiscal year 1973 dollars (through the first year of operation, but excluding the cost of launching the telescope into orbit with the shuttle and the costs of civil service and support contractors). The pressure from headquarters to produce a less expensive telescope program (pressure largely deriving from the judgment of NASA top management regarding how much funding for the telescope program they would be able to win from Congress and the

White House) had driven Marshall to revise drastically its approach to how the Large Space Telescope was to be constructed.

In part we can interpret these shifts in the program as an effort by NASA to push expenditures well into the future, to transfer costs from the design and development phase to the time when the telescope would be in orbit. The focus for NASA became largely the cost of the telescope's design and development, *not* the total cost of building, operating, and maintaining the telescope in orbit. It would, in fact, be the telescope's design and development costs that would become the crux of a long-running battle between NASA and the Congress, not the overall cost. But a number of the changes adopted were to be far from permanent. For instance, once into the building of the telescope, the idea of regular returns of the telescope to the ground would be dropped. To a degree, the cost-cutting measures would prove to be only paper exercises.

One interpretation of this process is that some of the items were cut by Marshall because they sounded plausible and their removal would generate a lower cost, thereby making the telescope more feasible politically. Headquarters then agreed to the changes, because that is what it wanted to hear, but both parties suspected that ultimately they would need to expand the program, to get more spares later, for example. By then, approval for the program would have been gained, and enough dollars would have been spent to prevent the telescope being canceled. Another interpretation is that the program's managers were endeavoring to produce a low-cost program through some innovative approaches. If that was so, the attempt did not really work, as technical concerns forced the agency to reintroduce many items that had earlier been deleted.

But the important point for this part of our story is that Marshall's revised plan satisfied NASA Headquarters that the issues raised at the December 21 meeting had been dealt with. Headquarters thus gave Marshall the green light to press ahead into Phase B and issue "Requests for Proposals" to industry for the Optical Telescope Assembly.[38] Marshall and Goddard could begin their own detailed definition studies of the telescope, as well as start the Phase B efforts of the contractors.

PLANNING THE SCIENCE

The move into Phase B also meant that the Science Working Group that O'Dell had maneuvered to bring into existence could get down to business. The first meeting was held in June 1973, and there Jesse Mitchell underscored two issues that were to recur again and again in the next few years. First, he cautioned the group to remember that the importance of the Large Space Telescope would have to be justified repeatedly to astronomers and laymen alike. Second, the

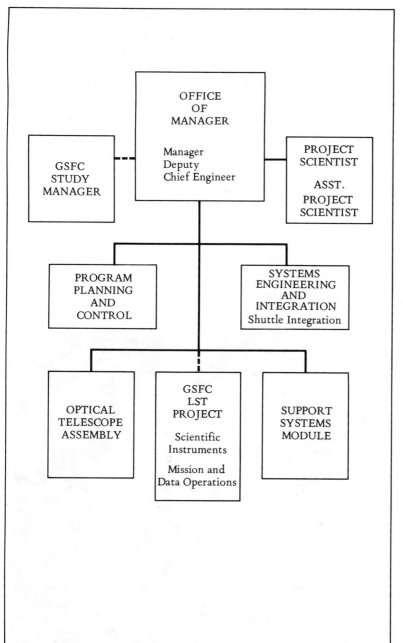

Large Space Telescope organization chart. (Courtesy of Stansbury, Ronsaville, Wood Inc.)

astronomers had to help NASA decide how to build the telescope at the lowest cost. The telescope's design would emerge through engineering and scientific analyses as well as tests of hardware, and the advice of the astronomers would be needed in the negotiations that would be central to the design process.[39]

Without enthusiastic backing from other astronomers and laymen, Mitchell was telling the group's members, there could be no telescope. Moreover, if the astronomers could not aid the engineers to produce a low-cost telescope, then the program would never reach the hardware phase. This gathering set the tone for the other Phase B Science Working Group meetings. Discussions at which one might suppose that only scientific matters would be probed involved much more, for, as is characteristic of the kind of Big Science represented by the telescope, technical and scientific issues were inextricably interwoven with other factors, in this case political, economic, and social factors, and the Science Working Group provided the main forum in which astronomers could debate them.

The astronomers also started their own detailed studies of possible scientific instruments after the first meeting of the Science Working Group. In late 1972, NASA had identified three major instrument possibilities: a diffraction-limited camera, a low-dispersion spectrograph, and a high-dispersion spectrograph. In addition to the Instrument Definition Teams to examine these instruments, NASA had formed three other definition teams.[40] The principal scientific

An early meeting of the Large Space Telescope Science Working Group, probably 1973. Those shown include members as well as observers. Back row, left to right: Gerry Neugebauer, Ivan King, Harlan Smith, John Bahcall, Allan Sandage, Nancy Roman, and James A. Downey III (project manager). Front row, left to right: C. R. O'Dell, Robert Danielson, Ernst Stuhlinger, and Pete Simmons (Martin-Marietta). Not all the members are shown in the photograph. (Courtesy of Ernst Stuhlinger.)

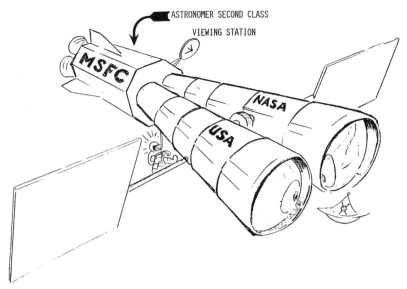

A cartoon that made the rounds of the telescope program in the mid-1970s showing the "Large Space Binoculars," or LSB.

task of these teams was to turn broad scientific objectives and aspirations into detailed scientific and technical requirements, requirements that could be used in eventually building the instruments for the telescope.

As Nancy Roman has described, the aim was for the designs of the scientific instruments and the design of the telescope to advance together, because "the Telescope wasn't going to be very effective without them. And secondly, you wanted to influence the Telescope design so that it would in fact be appropriate for the instruments."[41] It would be pointless, for example, to design an instrument that was too large to fit into the available space. But NASA's lack of planning funds meant that the astronomers' own studies were not as sophisticated as they might have liked. As a member of one of the instrument teams recalls, they did not have at their disposal "any complicated ray tracing programs that might be used to really test out a scheme, and things were pretty schematic in their layouts. No detailed studies were being done."[42] It was, moreover, the industrial contractors who were to identify and investigate in detail the trade-offs to be made in the instrument designs, trade-offs such as, for example, scientific performance versus engineering feasibility.

One major problem with which both the astronomers and contractors wrestled was how to position the scientific instruments around the "focal plane," the section of the telescope where the light from astronomical bodies was brought to a focus. Should there be some kind of arrangement that would swing each instrument in turn to the field of view, or should all of the scientific instruments share a common field of view, with each being allotted a segment?[43]

At a meeting of one of the Instrument Definition Teams in late

1973, astronomer Ivan King proposed that "LST should never stop looking." In King's opinion, the shutter of one of the telescope's cameras should be open every time the telescope was pointed and stabilized. This manner of operating the instrument became known as the "serendipity mode," for by collecting images in such a fashion, discoveries might be made serendipitously.[44] So even if observations were being taken with, say, a spectrograph, a camera would still be securing data. Here was a strong argument for the astronomers in favor of positioning the scientific instruments so that they would all share a common field of view, rather than placing them so that one instrument would take observations at any one time.

A ground rule established by NASA for the telescope planning phase was that instruments would be designed so that they could be replaced, either in orbit or while the telescope was being refurbished on the ground. The point was that the telescope was being designed for a fifteen-year life in space. In an ideal situation, the telescope would have at all times in orbit a fully working complement of instruments. New instruments would thus have to be substituted for those that failed, and if NASA chose to develop later generations of scientific instruments beyond those initially flown, they would have to be switched with the first generation of instruments. The Science Working Group also pressed for the instruments to share some common characteristics. In particular, the group urged that the so-called axial instruments – those arranged axially around the focal plane – be designed so that each would fit into a module that would remain the same size for all instruments.[45] Changing the instruments would therefore entail swapping one module for another. For the astronomers, that was another persuasive reason for the instruments to share a common field of view.[46]

DETECTORS

For the astronomers, the choice of individual scientific instruments also was intimately related to the choice of detectors to be incorporated into them to catch and reveal the light from the bodies at which the instruments would be directed. The telescope's scientific performance would in fact be determined to a large degree by the detectors. Because no ideal detector for all purposes existed, the choice of those for the telescope's instruments would necessarily be a compromise among many factors, involving a variety of complex trade-offs in terms of, for example, cost, schedule, engineering feasibility, and scientific requirements. A possible scientific instrument might promise excellent science, but if, for example, its detectors could not be developed for use in space by the time of the telescope's launch, it would hardly be prudent for NASA to plan to fly such an instrument.

It is one thing to build detectors for use on ground-based telescopes, where adjustments and repairs can be made easily; it is quite another to devise detectors to function remotely for many years in the hostile environment of space. In space, even one minor failure might bring a detector's life, and therefore the life of the instrument containing it, to an end.

From the vantage point of the Woods Hole discussions on the Large Space Telescope in 1965, there had appeared to be two possibilities for the main detector to produce images: fine-grain film or, looking to the future, a type of electronic detector known as an image tube. However, those astronomers pushing a Large Space Telescope very much took the attitude that they would select the best detectors available for flight and adapt them for use.[47]

Photographic plates have been used in astronomy for well over a century. Although plates can cover a large area of the sky in a single exposure, they have major drawbacks. As one astronomer has pointed out, "photographic plates are inefficient and insensitive; even the best of them throw away 98 percent of the incident blue light and virtually all of the red light. They are uneven, chemically dirty, and worst of all they have such a limited storage capacity per unit area that, despite their appalling insensitivity, they become saturated by the light from the very bluest part of the sky in less than an hour."[48] Film also has the disadvantage that within a month or so of being in the environment of space, the occasional exposure to penetrating radiation causes the photographic emulsion to fog. Given that NASA planned that in its orbits around the earth the telescope would pass through the "South Atlantic Anomaly," a region of trapped energetic particles, film would have to be heavily shielded. Also, there would have to be some system of collecting the film before it became fogged, or of processing it on board the telescope. In the latter case the telescope would have to carry electronic devices to "read" the film and beam back to earth signals telling what the film revealed.

During the 1960s, Marshall had studied how astronauts might retrieve film from space telescopes, particularly in conjunction with possible space stations. Here, in fact, is one of the pivotal themes in the history of space science: To what extent should astronauts be involved in the execution of experiments? Whereas NASA's Office of Manned Space Flight was ever anxious to press the use of astronauts, the great majority of astronomers, both within and outside the agency, tended to be much more wary.

One group of generally skeptical astronomers had constituted the ASTRA Working Group, which in the late 1960s, as we saw in Chapter 2, had been planning a sixty-inch space telescope. The ASTRA Working Group wanted to ensure that if astronauts were to be engaged in the ASTRA program, the astronauts would be used to replace scientific instruments and effect repairs, not run the observatory by operating the instruments, a role the working group feared might be forced upon them as a justification for manned space flight. The astronomers planning the Large Space Telescope took a similar line. Roman also recalls that "we didn't really want anybody around the Space Telescope very frequently, because every time you did, there was a chance of damage, and there certainly was a chance of contamination. We wanted to keep people away from there as much as we could. So I don't think we would ever have gone to film and frequent rendezvous unless we were absolutely forced to it."[49]

Film, though it has been used successfully for flights of relatively short duration by photoreconnaissance satellites, poses decided difficulties for a flight planned to last many years, perhaps ten to twenty years. But by 1973, the problem of retrieving the information collected by scientific instruments aboard spacecraft had been faced by NASA for over a decade, particularly in the unmanned missions launched to the moon and planets. The need for an alternative to film for planetary science and space astronomy had led NASA to supply funds to researchers with the aim of improving television-type detectors. These detectors had the essential ability to gather and relay images back to earth. On the first day of the very first meeting of the Science Steering Group back in the telescope's Phase A in 1971, the astronomers had quickly decided that they would assume the use of electronic detectors in place of film. The astrono-

A 1969 illustration prepared for Marshall to illustrate the use of an astronaut to collect and replace film canisters from a large telescope in space. (Courtesy of NASA.)

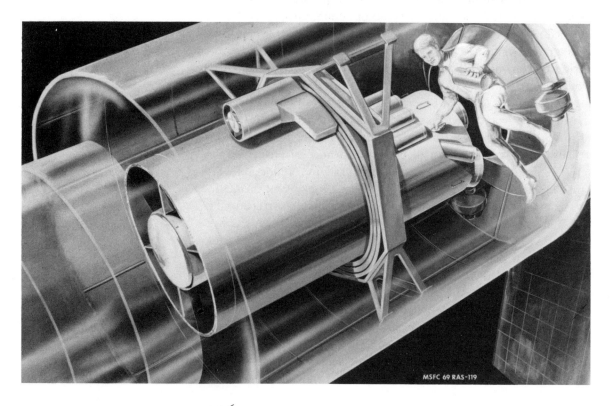

MSFC 69 RAS-119

mers deemed film "impractical and more importantly subject to rapid deterioration in the space environment."[50]

The Science Steering Group had returned to the use of detectors the next day. John Lowrance of the Princeton University Observatory reported on a comparative study he had made of various detectors, with particular emphasis on the "Secondary Electron Conduction Vidicon," generally referred to as the SEC Vidicon. The Princeton group's development of the SEC Vidicon, based on a Westinghouse television-type tube, had been supported since 1964 through a series of grants and contracts from NASA. As has often been the case for astronomical detectors, a mature "off-the-shelf" product was being adapted to astronomical ends. One reason why this is common is that astronomers generally want to observe such faint objects that they place exceptional demands on the detectors they employ. But astronomers do not compose a sufficiently large and lucrative market to entice suppliers of detectors to adapt their devices to cater to astronomy alone. Astronomers and astronomy-minded engineers therefore usually have to convert commercial products into astronomically useful detectors. Almost invariably this is a demanding task, and certainly it was so for the SEC Vidicon.

The development of the SEC Vidicon was initially directed toward the planned "Princeton Advanced Satellite" and, later, the proposed ASTRA orbiting telescopes. Although those missions were not flown, the studies of the SEC Vidicon had brought together at Princeton what was seen in NASA as a good team that was well placed to continue to improve the detector, a detector that had been explicitly fashioned to be used aboard astronomical satellites and seemed to many people to be far ahead of its rivals.

There was also a strong institutional link between Princeton and the Large Space Telescope, a link that went well beyond the participation of Lyman Spitzer. Many earlier telescope studies had also been conducted at Princeton. The extensive Princeton programs of rocket, balloon, and satellite space astronomy, as well as detector development, were all parts of a drive by Princeton toward advancing the cause of, and their own stake in, the Large Space Telescope.[51] So, as one well-placed observer has pointed out, "small wonder that the [SEC Vidicon] was identified as *the* detector, and the way to go."[52] The issue of which detectors to fly on the Large Space Telescope was nevertheless far from settled, and in fact it would become the central technical problem for the Phase B Science Working Group.

Although the SEC Vidicon was widely expected among NASA personnel and the astronomers to be the detector for most of the telescope's instruments, other detectors also were under study. Charge Coupled Devices (CCDs) were principal among these for producing images of areas of the sky.[53] CCDs are examples of so-called solid-state devices that work on a principle different from that of the SEC Vidicon. To be brief, the CCD is a type of silicon chip. The light pattern falling on a CCD produces a charge replica of itself, with more charge produced and collected where the light is the brightest. The light therefore produces a type of electrical photograph. The image is, in effect, developed in the silicon chip and can be amplified and then displayed on a television screen. By the mid-1980s, CCDs were in widespread use, perhaps most commonly in home video cameras, but the concept on which the CCD is based had been discovered and verified experimentally by researchers at the Bell Laboratories only in 1970.[54]

By 1973 and the start of Phase B, CCDs, although there had been little time to adapt them for astronomical uses, nevertheless already appeared to some astronomers to have long-term potential. CCDs were attractive because they were small and lightweight and consumed little power, all highly desirable features for a detector to be orbited in a spacecraft. In addition, they were more sensitive than other electronic detectors, were durable, and could produce images of objects that varied enormously in brightness, an important point for use in astronomy.

On the debit side, there were some serious problems with the CCDs. First, they could cover but a fraction of the area of sky available to an SEC Vidicon. Second, imperfections in the silicon chips meant that there would be all sorts of flaws in their images. Third, and perhaps most important, the available CCDs were poor at detecting ultraviolet light. Bright sources of ultraviolet light produced only a negligible response; yet the study of astronomical objects in this region of the spectrum was one of the major attractions for those astronomers who advocated the Large Space Telescope.

In an attempt to improve the CCD's ultraviolet response, a group at Goddard was studying what happened when CCDs were placed inside a television tube. The idea was that the incoming photons could be converted into electrons, and the CCDs could then be used to detect the electrons, not photons. By, in effect, exchanging the ultraviolet photons for electrons, the device's ability to record such photons' light could be much enhanced. These hybrid devices were known as "Intensified Charge Coupled Devices" (ICCDs).[55]

NASA managers impressed on the astronomers that they were also anxious to avoid having the scientific instruments slow down the pace of the project. This problem had often occurred in the past, and so the choices of detectors would have to be made in a timely fashion.[56]

NASA funds for developing detectors were nevertheless extremely limited. Hence, at a meeting of the Science Working Group in February 1974, the members debated the cost of adapting detectors

for space flight and how many different types it was practical to
consider, and how the money might be distributed among them.
The working group was concerned that not enough was being spent
on detectors. Project scientist O'Dell agreed, but pointed out that
within the overall budget, detectors had a relatively good priority,
as they accounted for about a fourth of the total. The problem was
that the overall program budget was low: The original request of
$10 million had been chopped to $6.2 million. The result was that
most of the money for detectors was being spent on the SEC Vidicon, although there was a smaller effort on CCDs.[57]

Detectors remained a central concern for the Science Working
Group. At the May 1974 meeting, the mood was one of austerity,
as it had begun with a call by the project manager for economies
and a request that the existing telescope design be made less extrav-
agant. In response, the working group settled on two detectors that,
with only certain exceptions, would be employed in *all* instruments.
Again, funding, schedule pressures, and NASA's way of doing business were fundamentally shaping scientific and technological choices.

The two detectors the astronomers helped to select were the SEC
Vidicon and the ICCD. NASA estimated the cost for adapting both
for use in space aboard the telescope at $12 to $17 million, a substantial fraction of the estimated $40 million that was expected to
be available for scientific instruments for Phase C/D.[58]

Whichever one of these two detectors was flown would be manufactured by industry. There were, however, important differences
between the development of the SEC Vidicon and that of the CCDs.
The SEC Vidicon was based on a television tube built by Westinghouse that was attracting little interest outside of the telescope program. The tubes were being custom-made for Princeton and seemed
to have no long-term future. In contrast, the CCD was a focus of
industry attention. This was in large part because of a drive to convert these devices into replacements for Vidicon television tubes in
commercial and home video cameras. Companies had therefore geared
the image sizes of the CCDs under development toward the video
market. Here the CCDs scored over the Vidicons in terms of, for
example, size, weight, power consumption, and durability. The
challenge for the industrial developers of the CCDs was to match
the Vidicon's advantages in such areas as spectral range, field of
view, and cost.

The Goddard group pursuing ICCDs was also consciously "pig-
gybacking" onto a commercial device that Texas Instruments was
investigating. The need to maintain the interest of Texas Instruments in long-term commercial benefits, the Goddard group knew,
was essential if that company was to continue to provide its own
development funds.[59] In addition to the attention from industry,

the military was extremely interested in advancing the state of the art in CCDs. Again, this led to investments in industry research and development programs.[60]

By the mid-1970s, then, there was widespread research on CCDs because of what can be described as a strong mix of "market pull" and "invention push." That is, there were clearly defined markets for the devices that were pulling along their development, and NASA, the military, and industry itself were providing funds to push the improvement of the invention.[61] Both of these factors were to assist in transforming the CCDs into detectors for astronomical instruments. But that would be no simple matter, for it would take the efforts of some highly gifted astronomers and engineers to "tame" the commercial products to the special demands of astronomical sensing.[62] Nor did it seem very likely to many people that it could be done in time for the telescope's launch.

INDUSTRY AND PHASE B

While the astronomers were examining instruments and detectors, the industrial contractors were simultaneously undertaking studies of potential spacecraft designs. The first two major design contracts let by NASA were for the Optical Telescope Assembly. They were worth $800,000 each to Itek and Perkin-Elmer and were to last seventeen months.[63]

Itek and Perkin-Elmer were to focus on the design of the Optical Telescope Assembly; however, that assembly could not be studied in isolation from the rest of the spacecraft. The power to be consumed by the heaters in the Optical Telescope Assembly, for example, would have to be provided by energy from the spacecraft's solar arrays and controlled in part by the Support Systems Module. Hence, the Optical Telescope Assembly and Support Systems Module designs had to advance together to some degree. Both optical houses, Itek and Perkin-Elmer, therefore joined with other contractors who could provide them with expertise and knowledge in areas outside their own spheres of competence.

Despite working to the same guidelines, Itek and Perkin-Elmer developed somewhat different designs; certainly Marshall had no intention of directing them to examine the same concepts. Indeed, NASA's own aim in a Phase B study is to investigate a variety of potential designs and to avoid being locked into any one design too early.[64] But despite the distinct approaches, the contractors and NASA managers and engineers had identified at the start of Phase B several key issues they would each have to address. Two of those for the Optical Telescope Assembly were the choice of material for the primary mirror and how close to the diffraction limit this mirror could be polished. Itek had already begun work on a seventy-two-inch

Itek craftsman at work on the seventy-two-inch test mirror. (Courtesy of NASA.)

mirror blank. The rough grinding was finished in September 1972. By November, the mirror was being meticulously polished and tested, and by July 1973 the mirror had been coated and verified to be extremely accurate.[65]

The Itek test mirror was of a type known as an "egg crate." This term derives from the resemblance of these kinds of mirrors to old-fashioned egg crates formed of interleaving vertical slats. The mirror's core is not solid, but is composed of a lattice of glass plates. The weight is thereby much reduced over that of a solid mirror, a highly significant point for space applications, where one of the chief design considerations is to save weight. Moreover, Itek had demonstrated publicly and convincingly that such a precisely shaped egg-crate mirror could be fashioned with relatively little difficulty, even if at seventy-two inches it was considerably smaller than the 120 inches planned for the Large Space Telescope's mirror.

Another major problem for NASA and its contractors was the means to guide and stabilize the telescope. If the completed telescope was to perform to the negotiated requirements, it would have to be capable of being aimed at an astronomical target with a pointing stability of 0.005 second of arc, an angle on the sky about 360,000 times smaller than the angle that is subtended by the diameter of the full moon. So taxing was this requirement that it was widely viewed in NASA and outside as the most difficult technical challenge the designers and builders would have to overcome. To use a term introduced in Chapter 2, for these people, the pointing and control system had become the project's tall pole. So not only did the telescope have to be pointed extremely accurately, means also had to be devised to keep it locked on its astronomical targets. This task was crucial because there would inevitably be tiny disturbances that would act to move the spacecraft away from its targets, disturbances known as "jitter." Jitter might arise from the motions of the gyroscopes used in pointing, for example. Should the entire spacecraft be moved if small corrections in its position were needed (a method known as body pointing)? Or should the secondary mirror of the Large Space Telescope be shifted slightly to compensate for the spacecraft's minor motions (a method known as image motion compensation)? During Phase A, Bendix had performed studies for Marshall that argued that body pointing alone was sufficient. Marshall, however, was not convinced. Hence, the center's Phase A design concept also incorporated a movable secondary mirror. But more studies persuaded Marshall that control moment gyroscopes and reaction wheels alone could point and stabilize the telescope. If so, a moving secondary would not be essential, even though Perkin-Elmer argued it promised to give the best performance.

Marshall's basic engineering approach was to use the simplest available systems where possible, and for pointing and control that would mean using either control moment gyroscopes or reaction wheels alone, but preferably not the two in combination. Both control moment gyroscopes and reaction wheels rely on spinning motions to move the spacecraft they are directing. Control moment gyroscopes are large enough to apply a "controlling torque," or turning force, to the spacecraft. They often are referred to as momentum-exchange or momentum-storage devices by spacecraft engineers, as the turning motions produced by outside forces acting on the spacecraft can be transferred to the control moment gyroscopes, rather than shifting the spacecraft itself.[66] Reaction wheels, in contrast, rotate at different speeds, and varying the speeds alters the turning forces they apply to the spacecraft. The principle is essentially that if the reaction wheels spin in one direction, then, in response, the spacecraft spins the other way.

Marshall was concerned that a movable secondary mirror would be much more complex than a system of body pointing. But the choice of the simpler scheme also had to be weighed by designers against other factors, such as its performance, cost and how risky it would be. Through its Phase B investigations, Perkin-Elmer judged that it would be extremely awkward, if not impossible, to test the body pointing technique fully while the telescope was being fabricated on the ground.[67] For the contractors to invent ways to do ground testing might then be very expensive, perhaps not even feasible. If that were the case, body pointing would involve some uncertainty.[68]

Another essential part of the pointing and control system was the design of the fine-guidance sensors. In Phase B, these sensors probably gave more people more headaches than any other part of the spacecraft. It would, moreover, be the fine-guidance sensors that would perform a crucial role in keeping the telescope locked on its targets. The use of star images was at the root of the concept for the fine-guidance sensor under study at both Itek and Perkin-Elmer. The basic idea was that light from a star would enter a sensor. If the star started to drift out of the sensor's field of view, an electrical signal would be sent to the control moment gyroscopes or reaction wheels. These would then act to shift the position of the spacecraft slightly, the aim being to once more place the star centrally in the sensor's field of view. With two such fine-guidance sensors locked on two separate stars, the spacecraft's motions could be controlled.

While Itek and Perkin-Elmer were working on the design of the fine-guidance sensors and the rest of the Optical Telescope Assembly, the aerospace companies that hoped to win Phase B Support Systems Module contracts – due to be released in early 1974 – were assembling their teams. Each aerospace company was also performing studies it hoped would improve its position in the forthcoming competition, as well as deepen its understanding of possible designs for the Large Space Telescope.

Lockheed, for example, had already conducted some subcontract work for Perkin-Elmer in Phase B. But to strengthen its chances of winning one of the Support Systems Module contracts, Lockheed sought additional help and teamed with Bendix and TRW – Bendix to provide additional expertise in the design of pointing and control systems, and TRW in communications and data management.

In 1974, Lockheed judged that the pointing and control systems composed "the most important single technical challenge of the program."[69] Here Bendix's participation was attractive because of its corporate knowledge in the design of systems for spacecraft attitude control and stabilization. Bendix, in addition, had already performed pointing and control studies for Marshall in Phase A and the early part of Phase B.

Domenick Tenerelli was one of the engineers at Lockheed who was particularly anxious to run jitter tests for the Large Space Telescope, that is, to take a test vehicle (a scale model of the actual telescope), operate the control moment gyroscopes (then viewed by Lockheed as the likely means of controlling the telescope's slews across the sky), and check how the test vehicle responded. Lockheed also decided to apply external forces to the test vehicle to see how it would react – in effect, to shake it and measure what happened. "This was a test program that I pushed hard to get," Tenerelli recalls, "and I will say one thing, though. It was quickly accepted, because I knew that the [other companies] would be far behind us, and this was by far the most critical concern of Marshall and ourselves."[70] Lockheed thus took an engineering model from another

Lockheed's structural dynamic test vehicle – in effect, an engineering model of the planned telescope – being readied for vibration tests. (Courtesy of Lockheed.)

program, in effect, a three-quarter-scale model of the Large Space Telescope, with a simulated Optical Telescope Assembly, scientific instruments, Support Systems Module, and primary and secondary mirrors, and ran tests.[71] Lockheed could then exploit the engineering model and test data to demonstrate to Marshall that it had the pointing problems well in hand.

The other Support Systems Module contractors pursued related areas and sought to gain competitive advantages in their own ways. Martin-Marietta, for example, built a machine to simulate the telescope's fine pointing, and Boeing developed a simulator to investigate the Space Telescope's pointing and control system.[72]

The replies to NASA's request for proposals, coming from the five hopeful companies who were bidding to win the Support Sys-

tems Module contracts, were due July 8, 1974. In late June they
were therefore heavily engaged in preparing their responses.

On Capitol Hill, a very different kind of drama was being played
out. The presidency of Richard Nixon was drawing to a close as the
congressional hearings on the Watergate matter inexorably eroded
the president's support. The other business of Congress nevertheless
went on. Imagine the consternation that swept through the telescope project and the ranks of its actual and prospective contractors
when on June 20, 1974, the news broke that Congress had denied
funding for the telescope. The entire project was in danger of being
canceled, and the project's participants faced the depressing prospect
that their labors had been wasted. Whether or not those advocating
and working on the telescope could bring the Congress within the
assembly of institutions, groups, and individuals that favored the
telescope would now be put to the test.

4

Selling the Large
Space Telescope

It was a common problem. We had to get the funds or nobody would win.

Pete Simmons

I think it was just the greatness of the program, the potentialities of the program . . . which attracted me.

John Bahcall

In 1967, NASA's planetary science program was in disorder. Budget pressure due to the escalating cost of the war in Vietnam, President Johnson's Great Society programs, and the racial strife in American cities was forcing Congress to take a hard look at a planned mission called "Voyager" that would entail sending unmanned probes to Mars.[1] However, when the Manned Spacecraft Center in Houston chose the first week in August 1967 to send twenty-eight potential contractors a request for proposals to study a manned mission to Venus and Mars, "Voyager" was doomed. It had come to be viewed by Congress as an intended first step toward a hugely expensive journey by astronauts to Mars, and so it was swiftly killed.[2]

Big Science projects, because of their high costs, if nothing else, attract great attention from Congress and, as in the case of the ill-fated "Voyager," can be defeated on Capitol Hill. The process by which such projects are approved or rejected features complex interactions among the White House, the Congress, the proposing government agency, and a variety of other interested parties. Before returning explicitly to the Large Space Telescope, we shall therefore examine the political process to which it was subjected, one that threatened it with the same fate as "Voyager." In so doing, we shall see in this chapter and the next that astronomers assumed the roles of telescope advocates, publicists, coalition builders, and lobbyists – all roles fashioned by the demands of the kind of Big Science that the telescope represented, a pivotal demand among which was to win approval for the telescope program.

A major space project's eventual users often are involved in winning approval for it via various lobbying activities, as are the companies that hope to win the contracts to build the hardware, software, and associated facilities. Nor is the lobbying simply a matter of convincing congressmen of the value of a particular project. In addition, the executive branch, including the president, the Office of Management and Budget, the Office of Science and Technology Policy, and the sponsoring agency, must favor a project if it is even to get to the Congress. Hence, a large-scale project must be introduced and "sold" to many different groups and constituencies.

During this selling process, the proposed project may come under pressure to be altered in scope and content, sometimes subtly, on occasion radically. At one extreme, it can even be canceled if the interests and goals it represents do not mesh with those of its potential patrons, or if, like the planned "Voyager" mission to Mars, it is deemed by the Congress or the White House to be ill-thought-out or premature.

Even if a project does gain the approval of the executive and legislative branches, the checks and balances within the system can greatly change the final form of the project, as, for example, by reducing its overall funding or by granting approval with the understanding that international partners be sought. Thus, before NASA will subject a project to the wider political process, it must be extensively studied and reviewed by one or more of the agency's field centers and by contractors, as well as by NASA Headquarters.

It is also NASA Headquarters that decides a project's priority. NASA investigates the feasibility of many more potential missions than ever reach the advanced planning stage. Of those that do progress that far, many fail to be approved for Phase C/D, the design and development phase in which detailed designs are produced and the hardware built. In fact, the course that a project must negotiate to be approved for a so-called new start is long and littered with obstacles, and we have seen in the last three chapters what a tortuous process it can be to reach even Phase B.

First, a project has to compete with many others for approval within NASA and gain substantial support inside and outside the agency before NASA managers will opt for a new start. Next, the space agency has to bring the project to the Office of Management and Budget (OMB) and convince OMB (and perhaps the Office of Science and Technology Policy, too) that federal funding is appropriate. The various agencies and departments of the executive branch (NASA, the Department of Housing and Urban Development, the Department of Transportation, and so on) send competing claims for shares in the federal budget to the OMB. OMB's role in prepar-

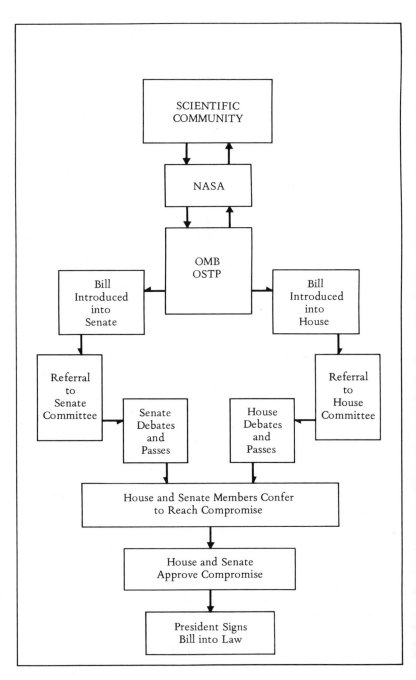

A political process: the route for transforming an idea for a major scientific facility into reality. Although this illustration shows the major hurdles to be negotiated in bringing a major new scientific facility into being, it is necessarily a simplification. In actual practice, the process is much more complex, as it also involves many feedback loops. Also, the process can take many years to complete. (Courtesy of Stansbury, Ronsaville, Wood Inc.)

ing and administering the annual budget thus gives it enormous power in the government, for if OMB decides that the submitted project should receive funding, it will be included in the president's budget message. This message is submitted to Congress in January each year. If the project reaches Congress, it is then scrutinized by four subcommittees (two in the House of Representatives and two in the Senate) that deal with NASA funding. The entire legislative process must therefore be gone through *twice,* a point that will be of great importance in the story of the telescope. An authorization bill must be passed by the House and the Senate, and both bodies must also pass an appropriations bill to grant the money to undertake the proposed activity. The final law, containing the funds for the project as an appropriation, is then sent to the president. If the president signs the bill into law (and that is not certain, for the president may disagree with Congress and veto it), OMB notifies NASA that funds are available and the project can begin.[3]

The issue of when to bring a scientific project before OMB and the Congress for new-start approval is therefore crucial and is one that NASA examines at length. Agency managers weigh many factors. Certainly it is not simply a question of the quality of the science promised by the project. Among the other points that NASA managers consider are, for example, the following:

1 The state of the technology required for the spacecraft and experiments: Is it already in hand? Will it be necessary to advance the state of the art?
2 Political receptivity to the project: What are the attitudes likely to be in OMB, in the Office of Science and Technology Policy, in the administration, and in Congress?
3 Does the project fit within NASA's budget constraints?
4 Is the project seen as a potential competitor for funds within NASA, that is, from an agency standpoint, draining funds away from more desirable activities?
5 Does NASA have the appropriate numbers of people available to manage and provide technical backup for the project?
6 In the case of an astronomical project, how much support for it is there within the astronomical community?[4]

By 1973, NASA had been conducting feasibility studies for the Large Space Telescope for a decade, and the Phase B studies were well under way. Around fifty astronomers were involved in defining the scientific instruments and the scientific guidelines, and numerous major contractors were eager to participate. Hence, for Jesse Mitchell and others, the time was swiftly approaching when the agency should try for a new start:

I thought the scientific requirements were beautiful. The technology was just around the corner. We had a commitment from Marshall Space Flight Center and enough manpower at Goddard to support them in the scientific area. . . . We had the support of the outside scientific community, and most important support in NASA, not just in the Office of Space Science but also in the [Office of Manned Space Flight]. They needed us and we needed them. The timing was right.[5]

The Space Shuttle and Large Space Telescope had already been presented together to the Congress on several occasions. In 1971, for example, the head of NASA's Office of Manned Space Flight had emphasized to a House committee that for major payloads, such as the Large Space Telescope, the Space Shuttle allowed the planning to be more flexible. Shuttle payloads could be checked out by astronauts in orbit; they could be maintained and refurbished in orbit, and even returned to earth if necessary.[6] During hearings in 1972, John Naugle had also claimed that "total program costs can be substantially reduced with the Shuttle since it will permit the reuse of the spacecraft."[7] NASA also contended that because the Space Shuttle would allow repairs to be performed in orbit, it would be unnecessary to do much testing of spacecraft on the ground. Costs would therefore be reduced. As we shall see, that set of arguments, founded on the planned capabilities of the Space Shuttle and none of which had yet been demonstrated, would be at the heart of NASA's planned approach to building the telescope, its claim that it could be done at relatively low cost, and so the agency's approach to selling the telescope to the White House and Congress.

SELLING STRATEGIES

Small projects usually do not appear before Congress for funding until they are poised to enter Phase C/D. For the costly Large Space Telescope, however, the strategy evolved by Mitchell and others in NASA Headquarters was to bring it before the Congress as a line item in the president's budget even while it was still in Phase B. The project could be discussed, and the reasons for pursuing it explained by NASA.[8] In that way, a base of support could be built on Capitol Hill for the time, a year or so later, when the telescope would be submitted to the lawmakers as a new start. Moreover, the Office of Manned Space Flight in NASA would use the Space Shuttle to justify the Large Space Telescope, and the Office of Space Science would use the telescope to justify the shuttle. The telescope and shuttle, NASA hoped, would therefore continue to march forward together.

Mitchell retired from the agency in mid-1973, but it was the strategy he had helped to formulate that was adopted. The costs of the Phase B studies themselves were sufficiently high (around $10 million) that they could not be met out of NASA's "Supporting Research and Technology" program without having a major effect on all the other items in the program. NASA, agency managers would later argue, had little choice but to include the telescope as a line item in Phase B, strategy or no strategy.[9]

Some of what might be termed the selling of the Large Space Telescope had taken place even before the telescope appeared in the president's budget for fiscal year 1975 (which reached Congress in January 1974). For example, in 1972, even Superman had enthused over the telescope in a story in which, among other stirring events, Superman had cleaned the telescope's optics with his "heat vision." As was to be commonplace during the next few years, the story proclaiming the virtues of the telescope, in this case with Superman's aid, had been planted by an advocate of the telescope. On this occasion it was Pete Simmons, a space astronomy manager at Grumman and a great LST fan. Via Grumman's military space work, he had joined the Orbiting Astronomical Observatory program in 1969. There he had become fascinated by astronomy and received numerous informal tutorials from the OAO astronomers, as well as one of the staff members of the Space Science Board. Simmons became deeply attached to the Large Space Telescope, and he energetically sought ways to advance its cause.[10] Superman's aid was just one method.

Simmons was also the driving force behind a symposium in Washington, D.C., in early 1974: "The Large Space Telescope — A New Tool for Science." It brought together engineers, astronomers, and managers to document in a public forum how much work had already been done in preparing for the telescope and to proclaim, to an audience that would include congressional staffers, among others, that the telescope could indeed be built and that the scientific rewards would be enormous.

After a series of generally upbeat scientific and technical papers, the concluding talk was delivered by John Naugle. He struck a cautious note when he warned that the buildup of the Large Space Telescope program had to take place at the right pace: too fast, and the astronomers and public would not back the program; too slow, and those involved would lose interest. Naugle also stressed that

scientists must recognize that where they are dependent upon public support for their endeavors, they must communicate the importance of their endeavors to the public — the knowledge they have gained and its significance. This enables the public to participate, in many cases vicariously, in these activities. If scientists devote perhaps one-tenth of the creative energy devoted to understanding the universe to explaining to the public the reasons for and importance of what they are doing, then I think the problems that we have in obtaining support for basic research will disappear.[11]

Naugle would soon have a highly dramatic vindication of his words.

ON CAPITOL HILL

To help us understand the selling of the telescope, let us first recall the state of funding for the space program in the mid-1970s. Following President John F. Kennedy's decision in 1961 to send astronauts to the moon, annual budgets had risen rapidly to accommodate the Apollo program. Space science funds rose with them and remained generous through about 1966, but the combined effects

Even Superman was enlisted to help
sell the Large Space Telescope.
(Courtesy of D.C. Comics.)

of growing inflation and budget decreases began to cut deeply into space science activities in the second half of the 1960s, although there was a rise in the early 1970s due to the Viking program to Mars.[12] After NASA failed to muster support for two major new programs – a manned flight to Mars and an orbiting space station – it finally won approval from President Nixon for the Space Shuttle in 1972. The agency's budget leveled off at around $3 billion in fiscal year 1974, which was only about one-third of the real amount provided in the most lavishly funded years of Apollo. Money for space science, and any multi-hundred-million-dollar project within it such as the Large Space Telescope, was being examined closely in that skeptical and austere time. Add to this picture congressional pressure for NASA to begin attacking the energy problems caused by the first oil crisis of 1973 and we can understand why Naugle was cautious.[13]

At first, however, all went well for the telescope's advocates. At hearings before the House Authorizations Subcommittee, the telescope was well received.[14] But that was the lull before the storm, the first rumblings of which were heard on March 26, 1974, when senior NASA staff members testified before the House subcommittee responsible for NASA appropriations, chaired by Edward P. ("Eddie") Boland. When the discussion turned to the Large Space Telescope, NASA managers were asked only one substantive question: How much would it cost?

That was the overriding issue for Congress and the issue that would shape the debate on the telescope for the next three years. That it was the most important issue is not surprising. Few congressmen have a scientific or engineering background. Despite the technical expertise of many congressional staffers, Congress is hardly able to engage the proponents of scientific projects on equal technical and scientific footing, particularly as those scientists who oppose any given project often will keep their objections private. So, unlike most of the policy issues that Congress deals with, the initiation of a major new scientific project engages the advocates of the project, but usually no organized opposition. A serious and detailed discussion on the scientific merit of the project is therefore generally preempted, and the only question left to fight over is the project's cost.

It was certainly the telescope's cost that in March 1974 John Naugle had to defend before the House Appropriations Subcommittee. Naugle, principal proponent of the space science portion of the NASA budget, responded that a "ballpark" figure for the price of the telescope would be $400 to $500 million. NASA administrator James Fletcher added that the money would be distributed over a period of about six years. Although no one objected, Naugle sensed that further justification was in order. He emphasized that the telescope was to be launched by the vaunted Space Shuttle and operated over a period of ten to fifteen years.

The exchanges on the Large Space Telescope lasted less than five minutes, but the subcommittee returned to the telescope the next day. There the questioning became much more aggressive, for the congressmen were concerned about the apparent proliferation of costly telescopes appearing before them. The subcommittee had appropriated new-start funds for the "Very Large Array" (VLA) radio telescope in 1972, estimated to cost some $80 million, and in the fiscal 1975 budget NASA was again asking for over $6 million for the construction of a 120-inch telescope in Hawaii. There seemed to be no "rational plan" for major astronomical facilities, and one congressman even exclaimed that "we are exploding with telescopes."[15]

Congressman George E. Shipley of Illinois also pressed Fletcher on why the telescope could not be deferred. When Naugle argued that the Large Space Telescope was strongly supported by astronomers, Shipley fired back that "I don't think you could come up with any type of telescope that the astronomers wouldn't support strongly. That is a weak argument."[16] Shipley was also concerned that although the current request was for planning funds, that was, in effect, a toe in the door. It would, he feared, then lead the subcommittee inexorably on to funding the building of the telescope so as not to waste the money already spent on studies. The subcommittee had given NASA an easy ride the previous year, and senior NASA managers had expected the same sort of treatment in 1974. They were wrong.

The Large Space Telescope had few friends on the House Appropriations Subcommittee. Hence, when on June 21 the subcommittee submitted its report, the recommendation was that all of the $6.2 million requested for planning funds for the telescope be denied. The subcommittee also pointed out that "the LST is not among the top four priority telescope projects selected by the National Academy of Sciences, and suggests that a less expensive and less ambitious project be considered as a possible alternative."[17] When the full Committee on Appropriations, reflecting the subcommittee's views, sent the fiscal 1975 appropriations bill to the House of Representatives, the House passed it by a roll-call vote of 407 to 7. All funds for the Large Space Telescope had thereby been deleted.[18] The project was therefore in danger of being deferred for several years; perhaps the idea might even be dropped by NASA because of the opposition in Congress.

REACTIONS

To the astronomers following the proceedings in Washington, D.C., from the sidelines but generally naive of the workings of Congress, the news came as a thunderclap from a clear sky.[19] Their thoughts nevertheless turned quickly to what could be done to salvage the situation. Although the project was in deep trouble, it still was conceivable to NASA, the astronomers, and contractors that if senators and representatives could be persuaded in a very short time that the telescope was a worthwhile enterprise, the Senate might pass a bill to fund planning studies. The disagreement in the House and Senate appropriations bills would then have to be resolved through contacts between the two bodies, probably a formal House-Senate conference.[20] If the Senate view prevailed, the telescope would be restored to the budget. The telescope might be down, but because the entire legislative process had to be gone through twice, it was not out.

The Senate Appropriations Subcommittee had already held its hearings, and the questioning had not been friendly. Plenty of work thus needed to be done to turn congressional sentiment around. The Senate would report (i.e., vote on) its appropriations in several weeks. In the meantime, astronomers and contractors would mount vigorous lobbying campaigns to revitalize the project.

Itek and Perkin-Elmer were leading the Phase B studies on the Optical Telescope Assembly, and several major aerospace companies were soon to submit proposals for study contracts for the Support Systems Module. These contractors, who had already invested heavily in terms of time and resources in putting together teams and in performing studies, and who hoped to win the lucrative contracts for the design and development phase, embarked on lobbying campaigns to convince the congressmen to back the telescope. To increase their impact, the contractors worked together in a loose coalition.

Pete Simmons was well versed in the ways of Capitol Hill, as well as political campaigns.[21] In 1974 he had become Large Space Telescope manager at McDonnell Douglas, and as Simmons recalls,

It was a common problem. We had to get the funds or nobody would win, right? And so I told [the other contractors] that I was doing this and they could do something comparable and they could go down and make these presentations [to staffers and congressmen], which they did do. I know that Martin and TRW and Lockheed all went and did things like this. Also the science community, I was personally calling guys. . . . Every time I talked with scientists I would call Bob O'Dell and tell him what I had said and what the scientist had said, so that he knew. There were a lot of people coming from everywhere to help get this restored. A lot of people.[22]

Certainly McDonnell Douglas was very active in defending the Large Space Telescope. On June 22, the day after the telescope had been knocked out of the budget, Congressman James Symington, with two of his personal staff, Simmons, and a member of McDonnell

Douglas's marketing team, met to discuss the telescope's status.
Symington was chairman of the House subcommittee dealing with
NASA authorizations, and his district included the McDonnell
Douglas plant in St. Louis. He was angry that the House subcommittee had denied planning funds for the telescope, and he gave his
energetic backing and assistance to McDonnell Douglas's plans for
reinstating the telescope into the budget. Those plans included briefing both Missouri senators and their staff members on the telescope,
meeting (with Symington's aid) with other congressmen and staffers, contacting scientists to press them to speak out, contacting the
other members of their own LST team to take action (these included
Honeywell, the Radiation Division of Harris Intertype, and Math-
ematica, Inc.), and contacting other members of industry who might
be able to help. A member of McDonnell Douglas's marketing staff
also issued regular progress reports.[23]

Such activities were common for the contractors. Lockheed, for
example, had a full-time representative working in Washington on
behalf of the telescope, and on July 5 he reported to Marshall that
on that date two senior members of Lockheed had seen representatives of ten congressmen.[24]

An important point here is that the astronomers and contractors
were not fettered in the same manner as NASA staff members were.
As civil service employees, the amount of contact the latter could
have with the lawmakers and their aides was limited, and the exchanges usually were arranged through the agency's congressional
liaison office. NASA employees often had to rely on their clients, in
this case the potential telescope contractors and astronomers, to promote the agency's programs. The arguments that the two client
groups were making about the telescope were also distinct, as the
astronomers not only had to advocate the Large Space Telescope but
also had to address the issue of why the telescope had not been the
number-one priority for astronomers – one of the reasons the House
Appropriations Subcommittee had denied funding.

BAHCALL AND SPITZER

Two Princeton astronomers, John Bahcall (based at the Institute for
Advanced Study) and Lyman Spitzer (of the Princeton University
Observatory), were to become the principal astronomer lobbyists for
the telescope in 1974. Both were members of the Science Working
Group and thus were intimately aware of what the Large Space Telescope was and what it was intended to do. But why did Bahcall and
Spitzer become the leaders of the selling campaign for the astronomers? There was not then, as there is now, a lobbying role for the
professional society of astronomers, the American Astronomical Society. Although the society's executive officers could speak out as

John Bahcall in 1985. (Courtesy of Pearl Gerstel.)

individuals, the society itself was unable to champion the telescope, lest that threaten its tax-exempt status.[25] Nor was it clear, given the feelings of many astronomers toward space astronomy, that all of the executive officers of the society would want to recommend the telescope. Also, some astronomers shunned such political activities as base, beneath the dignity of science, but the two Princeton astronomers thought otherwise. They fully accepted that one cannot win at politics without playing, and for the supporters of the Large Space Telescope, that would entail negotiations, bargaining, coalition-building, and compromises of their original plans.

Spitzer had experience in lobbying Congress because of his involvement in the nuclear-fusion program, as well as the Orbiting Astronomical Observatories program. Indeed, it was after conversations among Bahcall, Martin Schwarzschild (another Princeton astronomy professor), and Spitzer that it occurred to Bahcall and Spitzer that "we were the logical people" to lobby for the telescope.[26] Not only was the Large Space Telescope widely seen as the "Lyman Spitzer Telescope,"[27] but also Spitzer was familiar with the workings of Capitol Hill and aware of the techniques and strategies needed to convince Congress to fund space astronomy. Beyond that, he enjoyed extremely high standing among astronomers, both ground-

based and space astronomers, and Bahcall, the younger partner and
himself a respected astronomer, possessed great physical energy and
(it was to turn out) political acumen, as well as relish for the fight.
The attributes of one complemented those of the other excellently.
Spitzer, despite his habit of often wearing sandals with no socks,
even to formal meetings, was every inch a distinguished Princeton
professor and statesman of science whose style was to present "facts
and clear, cool logic," and Bahcall was his charming, though sometimes brash and feisty, foil.

Born in Shreveport, Louisiana, in 1934, Bahcall entered the University of California at Berkeley with the expectation that he would
become a rabbi. He started his undergraduate career as a philosophy
major,[28] but after more or less stumbling into physics, he switched
fields. Bahcall received his Ph.D. from Harvard in 1961, and after
a brief stay at Indiana and nine years at Caltech, in 1971 he joined
the Institute for Advanced Study at Princeton to pursue theoretical
astronomy.

At the time the telescope was in Phase A, Bahcall "knew nothing
about space astronomy and was not in any way involved."[29] When
he did learn something of the telescope, Bahcall was immediately
attracted because "it was a program which led to unpredictable
breakthroughs, and I found that terribly exciting."[30] After joining
the Phase B Science Working Group, Bahcall felt himself "more
and more mesmerized by the greatness of the project."[31] So when
the telescope was threatened with cancellation, he became angry and
determined to campaign to prevent that. As O'Dell recalls, "no one
had more conviction of the rightness and value of Space Telescope
than John. Just no one involved anywhere in the project. And when
John says, 'What can I do about this?' he means it. . . . I was
feeding him ammunition and he was just blasting, blasting away,
charging off in the right directions, and just all on his own. No one
delegated him. No one asked him. He was just there and willing to
do it."[32]

There were strong ties among Bahcall, Spitzer, and O'Dell, the
project scientist; in fact, there was what might be called a
Princeton–Huntsville axis in play at that time. In late April and
early May 1974, for example, O'Dell had spent a few weeks on leave
at the Institute for Advanced Study at Princeton, and during that
stay he had conferred at length with both Bahcall and Spitzer.[33] The
ties, moreover, went back considerably further; O'Dell and Spitzer,
for instance, had been mountaineering together in the late 1960s.
It was thus a natural development – given the relationship of trust
and friendship among O'Dell, Bahcall, and Spitzer – that the astronomers' lobbying effort should have been led by the two Princeton enthusiasts.

Bahcall and Spitzer twice visited Washington to meet with congressmen and staffers, and throughout July and August they dispatched letters to Capitol Hill, telephoned, and encouraged their astronomer colleagues to organize lobbying campaigns of their own. Links were also established at the highest levels in NASA as Bahcall and Spitzer reported to NASA administrator James Fletcher and others in NASA Headquarters by letter and telephone. They also stayed in touch with O'Dell in Huntsville, relaying to him the results of their activities.

At first, O'Dell had had little idea of the workings of Congress, but he had soon been given a crash course in the ways and functions of Capitol Hill through conversations with contractors, particularly Max Hunter of Lockheed and Simmons of McDonnell Douglas.[34] O'Dell therefore acted as a conduit for information between the astronomers and NASA. As he watched the proceedings from ringside, he offered coaching on whom to contact. O'Dell was a civil servant and so was unable to participate openly, but he pressed Bahcall and Spitzer, as well as other leading members of the astronomical community, to maintain the pressure. That the scientists would take their case directly to Congress was far from unprecedented. For example, in 1946 the Federation of Atomic Scientists was founded with the goal of lobbying for civilian control of the atomic energy program, and it worked successfully for defeat of the May–Johnson bill, which proposed to give the military a central role in the development of nuclear power.[35] But as Bahcall recounted in 1983, it was also a distinct advantage for them that they were astronomers, for as oddities on Capitol Hill – a testimony in itself of how unusual it was for astronomers to take a case directly to the lawmakers – it was relatively easy for them to gain access to congressmen and staffers.[36]

There was another bonus for the telescope's supporters. Although the ostensible function of the members of the Phase B Science Working Group and Instrument Definition Teams was to define the scientific requirements and suggest potential scientific instruments, their central position in the Large Space Telescope Project meant that there also existed a sizable group of vocal advocates widely spread in the astronomical community. Once again we have a vivid example of the interpenetration of science, technology, and politics in the kind of Big Science represented by the telescope, for the working group became the core group of those astronomers lobbying Congress for the telescope. After all, if the astronomers working on the telescope were not prepared to fight for it, who would? Before the deletion of the telescope from the budget, the astronomers on the working group had been helping to sell the telescope to their colleagues; now they would help to sell the telescope to Congress.

PRIORITIES

The House Appropriations Subcommittee had denied funding for the telescope in part because of the relatively low priority astronomers had assigned to the telescope. The subcommittee had cited as its authority *Astronomy and Astrophysics for the 1970s*, a report by a prestigious astronomy survey committee of the National Academy of Sciences. Chaired by the distinguished Caltech astronomer Jesse L. Greenstein, in 1972 the committee had responded to strong political pressures to set priorities rather than simply lay out a shopping list of desirable projects. Greenstein's committee recommended four first-priority projects for the coming decade:[37] a large array of radio telescopes (ultimately sited in New Mexico), a program to develop new electronic detectors and large ground-based telescopes, the development of infrared astronomy, and a high-energy astronomical observatory for x-ray and gamma-ray studies.

The Large Space Telescope was listed among the second-priority projects, with the statement that it "has extraordinary potential for a wide variety of astronomical uses," and the committee "believes that it should be a major goal in any well-planned program of ground and space-based astronomy."[38] In fact, the Bureau of the Budget (later the Office of Management and Budget) had recommended to Greenstein that his committee not consider the Large Space Telescope for the 1970s.[39] Nor did NASA press the telescope before the committee, a decision perhaps prompted by Greenstein's criticisms of optical space astronomy and known uneasiness about the potential cost of the telescope. Perhaps NASA reasoned that if it had urged the inclusion of the telescope for the 1970s, Greenstein might have shot the proposal down, leaving the agency in an even worse position.[40] The result was that the telescope had barely squeaked into the Greenstein committee's report as an afterthought. During the final round of meetings on the committee's conclusions, George Field, who was impressed by the telescope's potential, argued for its inclusion. Together with Princeton astronomer Donald Morton, he wrote a draft section calling for a sequence of spacecraft leading to the Large Space Telescope. Despite Greenstein's initial objection that the telescope would not be ready until the 1980s and that the committee's charge was the astronomy of the 1970s, Field and Morton's recommendation was included.[41] By early 1974, however, the National Academy of Sciences' Space Science Board had decided that the impetus of the Greenstein study had been spent. Some of the findings were obsolete because of new scientific discoveries. President Nixon's go-ahead to NASA in 1972 to build the Space Shuttle had also radically altered the ground rules of space astronomy: As soon as the shuttle became available, space scientists would no longer rely (or, as many might have said, be allowed to rely) on expendable launch vehicles, and it would be possible to visit spacecraft in earth orbit to replace a faulty scientific instrument, for example. Hence, in February 1974 the Space Science Board had decided to embark on a new study of priorities for space science. That study was to elevate the telescope to the number-one priority for space astronomy, but it was still some way from completion when the House subcommittee deleted the telescope's planning funds.[42]

The Greenstein committee was composed of influential astronomers, and their arguments had to be taken seriously. Moreover, in the same set of hearings in which it argued for the 120-inch Large Space Telescope, NASA had exploited the Greenstein committee's report in its defense of a 120-inch infrared telescope to be sited in Hawaii. That the House subcommittee had turned the Greenstein report on the Large Space Telescope is, then, not surprising.[43]

Bahcall and Spitzer, however, both contended that Boland's House subcommittee had misinterpreted the Greenstein committee's comments on the LST. The two Princeton astronomers sought to persuade the House's counterpart, the Senate subcommittee, chaired by the Democrat William Proxmire of Wisconsin, that their reading was the correct one.[44] On July 1, 1974, Bahcall wrote to Proxmire and told him that when in 1970 the Greenstein committee had begun its meetings, "the engineering and scientific feasibility of the LST had not been fully established and it was felt that a two-step approach beginning with a smaller telescope would be desirable." Bahcall repeated NASA's claim that it was feasible without the intermediate step, and this technological leap was indeed the most economical way to build the telescope. For Proxmire, Bahcall also marshaled arguments of fiscal and managerial economy: (1) It would be wasteful of money and effort to halt the project in midstream. (2) The Large Space Telescope and the Space Shuttle designs were related, so the programs should be carried out together. (3) The Large Space Telescope was needed to balance the space science program (balance, that is, between planetary exploration, which had thus far received the bulk of NASA's space science money, and the astronomy of objects beyond the solar system, the central reason for proposing the telescope). (4) There was European interest in supplying some of the funds for the project, and so the costs to the United States could be reduced.[45]

Although everything that Bahcall had written had the ring of truth, taken in sum it could prove dangerous by promising too much. To call the telescope "the ultimate tool of optical astronomy," as the Greenstein committee had done and Bahcall and others had reinforced, was to tempt Congress to consider the instrument as the only telescope needed for a long time to come. To some extent, the VLA radio telescope, funded by the National Science Foundation, had

been sold in that same manner just two years earlier. Congressman
George Shipley of Illinois protested to Spitzer that "we just approved the VLA telescope which was going to solve the riddles of
the universe. Now you want another telescope. What's wrong with
you fellows?"[46]

The astronomer lobbyists tried to counter such objections by dis-
tinguishing clearly between previous instruments and the Large Space
Telescope, and they had to underline the Large Space Telescope's
priority. To do so, Bahcall and Spitzer decided to act on a suggestion from Congressman Talcott and to enlist the aid of the Greenstein committee. Because of the science advisory structure NASA
had established, this was not as hard as it might have been. O'Dell
and Spitzer had themselves been members of the committee, and of
the other twenty-one members, three were members of the telescope's instrument definition teams or sat on the Science Working
Group, six were members of university departments with instrument definition team or working group members, and another three
were in universities with instrument definition team or working group
members.

Nevertheless, "I think everybody was concerned," recalls Bahcall,
"that we not cut other projects out while we advocated Space Telescope. That was a really tough thing."[47] At that time there was still
considerable opposition to the telescope. As we saw in Chapters 1
and 2, there had always been criticism from those optical astronomers concerned that the telescope's cost might starve ground-based
astronomy of funds, as well as some of those who supported the
pursuit of space astronomy in other wavelengths. But as Bahcall
points out, the opponents "were not organized." The critics also
expressed arguments such as this:

"There's a finite amount of money in Washington to support astronomy.
If they dump so much into the Space Telescope, there won't be enough to
support ground-based astronomy." Or another sentiment which was often
implicit or explicit was, "Space Telescope or space astronomy is East Coast
astronomy. West Coast astronomy, i.e. ground based astronomy, will suffer if space astronomy is promoted at its expense." There was [also] a real
sentiment for not getting involved in politics. That is, astronomers and
scientists worked in their laboratories, and they didn't confuse politics with
science.[48]

Despite such objections and whatever private concerns they may have
had, all twenty-three members of the Greenstein committee were
solicited and agreed to Bahcall's draft statement justifying the pre-
cedence of the telescope: *"In our view, Large Space Telescope has the
leading priority among future space astronomy instruments."*[49]

Bahcall and Spitzer sent the results of their poll of the Greenstein
committee to both Boland and Republican Burt L. Talcott of Cali-

fornia, the ranking minority member of Boland's subcommittee. They nevertheless cautioned that the findings had to be regarded as unofficial until the appearance of a full review of priorities by the Space Science Board.

Greenstein was privately concerned that some of the selling of the telescope was potentially harmful to ground-based astronomy. Following a suggestion from Congressman Symington, he had nevertheless been moved to have his own reinterpretation of his committee's report entered into the *Congressional Record*. "I am confident," Greenstein emphasized,

> that had we not had in mind budget limitations, [and] the at that time unsolved technological problems, and had we fully realized the wide range of discovery that we have had even in the last three years, we would not have taken quite so "conservative" an attitude. Astronomers felt then and feel now that the LST is the ultimate optical telescope and that together with a well-balanced, ground based program, it will open up new vistas for the human mind to contemplate.[50]

The Space Science Board, too, would shortly provide strong public support for the Large Space Telescope as the first choice for space astronomy.[51] The astronomers now presented a solid front, in public at least. The congressional attacks had in fact helped to bring the remaining dissenting astronomers into line.

TOWARD A CONFERENCE

With the Greenstein committee and the Space Science Board behind the telescope, the issue of the Large Space Telescope's priority dissolved swiftly. The telescope lobbyists still nevertheless faced the problem of how to persuade the Senate Appropriations Subcommittee to fund the Large Space Telescope studies. If it did so, that would force the House and Senate to resolve their differences on the LST's funding, probably as part of a House-Senate conference on the NASA budget.

The prospects did not look bright at the end of June. William Proxmire, the subcommittee's chairman, had taken an aggressive line with NASA at the Senate hearings in April.[52] However, Bahcall and Spitzer's lobbying, aided by a visit from James Fletcher, had won support in the person of Republican senator Charles Mathias of Maryland, the ranking minority member of Proxmire's subcommittee. After an introduction arranged by the head of Princeton's Institute for Advanced Study, Mathias, a strong supporter of basic science, became a convert to the cause, and his advocacy would prove crucial both then and in future funding battles.

It was, in fact, following an amendment moved by Mathias that the Senate Appropriations Subcommittee voted to restore all of the deleted $6.2 million for planning studies in fiscal 1975. On August 16, the full Senate approved this move.[53] Next, as the supporters of the telescope had hoped, the discrepancy between the House and Senate allocations of funds for the telescope would go to conference.

On August 9, President Richard Nixon, embroiled in the Watergate scandal, had resigned the presidency. On taking office, President Gerald Ford called for frugality in government spending. Addressing a joint session of Congress a few days later, Ford urged that "if we want to restore confidence in ourselves as working politicians, the first thing we all have to do is learn how to say 'no.' "[54]

That was not good news for the Large Space Telescope advocates, for it meant that efforts to constrain the budget would dominate the House–Senate conference when it met August 20. An agreement nevertheless emerged between the two sides based on NASA's suggestion that it should be allowed to spend up to $3 million to continue planning studies on the telescope. Here, then, was a partial victory for the telescope's supporters in NASA and for the contractor and astronomer lobbyists. The Large Space Telescope had squeaked through, and the appropriation was a modest endorsement of the idea for the telescope.

There was, however, more to the compromise than simply a loss of planning funds. The conferees had also decided that NASA had to consider "substantial participation of other nations in a less expensive project to be launched at a later date."[55]

The $3 million appropriation, a big cut from the $6.2 million originally requested, meant that NASA saw no realistic hope of completing all the Phase B studies in fiscal 1975. A new start for fiscal 1976 was out of the question, particularly as NASA administrator Fletcher was "pretty sure there's no way we can sell a new $300 to $500 million program for FY 1976."[56] The project members would have to regroup and follow Congress's direction to pursue international partners, as well as produce a less expensive Large Space Telescope.

A MATING DANCE

From the time of its inception in 1958, NASA has pursued international participation in its programs. Indeed, one of the goals of the National Aeronautics and Space Act of 1958, the charter for the civilian space program that established NASA, was "cooperation by the United States with other nations and groups of nations in work done pursuant to the Act and in the peaceful applications thereof."[57] Even before the telescope had run into problems with Congress, there had been talks on participation between NASA and the European Space Research Organization (ESRO), as well as the British Science Research Council.

ESRO had been established in 1964, and its first big project, and one of the reasons for the organization's birth, was the so-called Large Astronomical Satellite (LAS), roughly comparable to an Orbiting Astronomical Observatory. In 1966, three detailed designs for the LAS were submitted and evaluated, and a British design was selected. Launch was planned for around 1970, leaving a mere three years for the detailed design and development, a short period indeed. As H. C. van de Hulst, one of ESRO's founding fathers, has written, "in those years our optimism had not yet clashed with the facts." Adequate funds were extremely difficult for ESRO to come by. When ESRO decided that the LAS project would require about 40 percent of the organization's available resources for four or more years beyond 1968, it was dropped.[58]

The Europeans' LAS was nevertheless to be reborn, though in more modest form, as the International Ultraviolet Explorer (IUE). After the demise of the LAS, the group of British astronomers involved in the design of its scientific payload managed to interest NASA in a small ultraviolet astronomy satellite. NASA had plans of its own for a similar spacecraft, and in 1970 the IUE came into being as a joint project among ESRO, NASA, and the United Kingdom, with NASA providing the bulk of the funding.[59]

Consequently, by the early 1970s, links had been established between optical space astronomers in the United States and Europe, and in 1973 three Europeans had been selected as members of the NASA Phase B Instrument Definition Teams. ESRO, too, was still interested in a large telescope in space. Because it did not possess sufficient resources to carry out the entire project itself, ESRO turned to collaboration with NASA. Thus, when in February 1974 there was a NASA/ESRO review of science programs, the issue of European cooperation on the Large Space Telescope was raised, and ESRO agreed to study the possibility of providing one of the scientific instruments.[60] The discussed cooperation therefore matched the usual way in which the United States had been involved in international space science endeavors to the mid-1970s, that is, as the dominant partner who defines the program and who then invites other nations to participate by proposing experiments. NASA usually had shied away from joining with other nations as equal partners in space science enterprises that were jointly planned, funded, and implemented.

From NASA's perspective there were, nevertheless, distinct advantages in European collaboration in the telescope program. First, the funds NASA had available for the scientific instruments were limited; here might be a chance to enlarge that budget and increase the number of instruments under study. Second, the telescope program would have a broader base of political support, thereby making it easier to initiate and more difficult to cancel — an important

Technicians at the Goddard Space Flight Center in 1977 prepare the International Ultraviolet Explorer for launch. (Courtesy of NASA.)

consideration for the agency's managers. Third, it would mesh well with the agency's overall policy of encouraging international cooperation.

In April 1974, ESRO sent an observer to a meeting of the Science Working Group, and in May, ESRO's Astrophysics Working Group recommended that analyses be carried out in Europe regarding three potential instruments, as well as associated detectors. An ESRO Large Space Telescope Group, including the three European members of the NASA Instrument Definition Teams, was established to perform the studies.[61]

By the early summer of 1974, then, there was a definite move toward European involvement in the telescope program via provision of a scientific instrument. The House-Senate conference decision of August 21 made the issue even more germane, as the conference had told NASA to consider "substantial participation of other nations." But securing "substantial participation" was proving difficult. Although ESRO's contribution of one scientific instrument might cover around one-fifth of the instrument budget, that was hardly "substantial" compared with the total costs to design and develop the telescope, perhaps $10 million out of a total of some $400 million. NASA Headquarters therefore asked Marshall to identify additional practical ways in which other nations might cooperate in building the telescope. At the same time, NASA wanted to be wary: The Large Space Telescope would be pushing the state of the art in almost all respects, and the U.S. government had invested heavily over the years to develop this technology, some of which was classified.[62] Agency managers did not want to give away the store. Marshall's task was thus to find "clean" interfaces, to pick out sections of the telescope that could be built by the Europeans without their requiring such data on the rest of the telescope that U.S. companies would lose proprietary and classified information.

The negotiations on the possible provision of scientific instruments by ESRO or the United Kingdom ran into a problem when ESRO and the United Kingdom took the position that if they were to provide instruments, they could not do so in an open competition. Rather, NASA would have to agree in advance that ESRO would have one or more instrument slots on the telescope. Neither ESRO nor the United Kingdom wanted to expend funds on studies if they were not guaranteed to bear fruit.

Some of the members of the Science Working Group considered that unfair, because U.S. astronomers would have to compete against each other to get their scientific instruments aboard the telescope. Preselection of an instrument before the competition also seemed to the working group members unlikely to ensure the best possible complement of instruments.[63]

The reward expected by the Europeans for providing an instrument would be a certain guaranteed amount of observing time for European astronomers. This observing time would be blocked off from U.S. astronomers, a fact that also drew criticism from the working group.[64] NASA had suggested the alternative that the Europeans might consider the cash purchase of observing time on the telescope, rather than participation with an instrument. Although that might have solved the agency's problems with U.S. astronomers, the European response, as a NASA official conceded, was "lukewarm." Certainly that flew in the face of ESRO's fundamental policy to spend as much of its program funds as possible in member states. In any case, as the telescope was intended by NASA to be an international observatory, it might well be that even without any financial contribution, European astronomers would win 10 to 15 percent of the observing time in open competition.[65]

In October 1974, three NASA staffers flew to Europe to discuss

European participation further. On October 7 they were in London for a meeting with the United Kingdom's Science Research Council. Two days later they were in Paris for talks at ESRO Headquarters. Both meetings were described by a NASA participant as "cordial," but again the clear message was that neither the United Kingdom nor ESRO would be prepared to compete for an instrument.

At the Paris meeting, an ESRO engineer discussed other ways his agency might cooperate with NASA. He identified two candidate areas: the solar array and the electrical power systems.[66] ESRO was already building the solar array for the International Ultraviolet Explorer, and so the possibility of supplying such a subsystem for the telescope was a natural choice to be investigated further.

Despite the political advantages of NASA/European collaboration, the Science Working Group still resisted noncompetitive selection of any flight instrument. As political scientist John Logsdon argued, there "has been an undercurrent of ambivalence among U.S. space scientists and NASA managers about European involvement in NASA missions, whatever the stated policy."[67] A NASA project manager will generally try to minimize the number of interfaces in a program, for with each additional interface the management of the program becomes more complex. NASA managers sometimes liken interfaces to intersections on a highway, in that they are not the only places where accidents occur, but they are accident-prone. European involvement would add another interface to an already managerially complicated project, one over which NASA's managers would not have their usual degree of control.

Another reason for the ambivalence of U.S. space scientists − an intensely competitive community in any case − toward international participation was the suspicion that in some cases a foreign experiment had been chosen by NASA over a rival U.S. experiment not on the basis of merit but because it would be provided to NASA essentially free of charge. Such a selection, of course, means one less U.S. experiment aboard the spacecraft in question. When, in 1974, for example, a leading U.S. space scientist had protested vigorously to the NASA administrator about a possible European contribution to another proposed NASA program, Naugle told NASA administrator Fletcher that

we are encountering increasing resistance from U.S. scientists and increasing interest from Congress. Congress views such co-operation as a reduction in funding requirements, whereas the U.S. scientists regard such missions which will carry U.S. and foreign experiments as a reduction in their opportunities to do research.[68]

Thus, the space science community is sometimes inclined, as one of NASA's leading science managers put it, to speak with a "forked tongue." "On the surface you'll hear nice words about, science knows no political boundaries, it's universal, good of all mankind, free interchange of ideas, open, public. Then you come to the other side. It's competitive as hell."[69] No one objected to a competitive selection of a European instrument, but some did object trenchantly to selection by management fiat.

In January 1975, O'Dell polled all the members of the Science Working Group to get their views on European cooperation. He found that "on one extreme were those who felt that LST would sell without compromising the principle of fair competitive selection, while at the other extreme were those who felt without some ESRO participation the program was dead – as such, a compromise was necessary." Given the perilous political state of the project, the working group, taken as a whole, was prepared to back down and "tolerate" the noncompetitive selection of a scientific instrument.[70] And given the project's relative shortage of funds, it was unlikely that NASA would have enough money for a fourth, let alone a fifth, scientific instrument. So, as was openly discussed, the astronomers were getting a scientific instrument "for free" by European participation.[71] ESRO, in turn, was getting the opportunity to participate in an important space astronomy project, perhaps the major one of the late twentieth century, one ESRO itself could not fund in the foreseeable future.

By early 1975, Marshall had concluded that ESRO should not provide a major subsystem for the telescope – such as, for example, the power system – because of the complexity of the interfaces that would result and the attendant management problems. Nor was ESRO particularly interested in such an option. ESRO was, however, still examining whether or not it might provide the solar cells or other portions of the solar array. Informal talks had also begun with NASA on ways in which ESRO might support the telescope during its life in orbit, a less sensitive way of providing funds than building sections of the spacecraft.[72]

CONGRESSIONAL PRESSURE

Early 1975 saw the start of another round of congressional testimony on the NASA budget, with NASA managers again asking for planning funds for the Large Space Telescope. It was also a time of U.S.–Soviet détente. Perhaps its most spectacular image was that provided by the Apollo–Soyuz mission that was to lead to the dramatic linking in space of U.S. and Russian spacecraft. Taking his cue from these events, Edward P. Boland, the House Appropriations Subcommittee chairman, asked whether or not NASA had considered the Russians for a partnership in the telescope project. NASA administrator Fletcher replied that it might be worth discussing, but given the danger of technology transfer that it would have entailed, there is no evidence that NASA ever took this seriously. In fact, NASA's lack of success in finding an international partner came under strong fire. George Shipley was especially hard on what he saw as NASA's lack of progress. Why, he quizzed the NASA administrator, could not a partner meet one-half or one-third of the total telescope cost? When told that ESRO might provide a solar-cell array costing $5 to $10 million, Shipley derided the amount as "peanuts in a program like this." Fletcher explained that it would take a lot of "peanuts" to get up to the one-third to one-half level, and they were striving to do so. By NASA standards, however, ESRO had only a limited science budget. If the European contribution was to reach, say, half of the total LST costs, special permission would be required from the ESRO member countries to provide the funding. Such extra funding, moreover, was already being provided for the costly Spacelab project, which involved ESRO constructing a large module to fit into the Space Shuttle's payload bay.

Another NASA manager cautioned that the agency faced problems of technology transfer. "The LST does use a fair amount of previously developed technology," he testified, "but has some new technology in it, technology which is not available at this time to the European Community. Therefore, they would not be able to work with it." Shipley was not impressed. He could not see how, given the urban and housing problems in the country, the Large Space Telescope could be a priority item at $400 to $500 million, and he warned the NASA managers that "unless you do come in with a lesser figure and you do come in with some signed contracts so to speak, from participating countries, I for one will do everything I can to oppose this."[73]

This buffeting could only have increased NASA's eagerness to secure international participation. By mid-1975, the discussions between the two space agencies had progressed to the state where some of the preliminary terms of a possible agreement had been thrashed out. By early summer, NASA and ESRO (soon to be incorporated with the European Launcher Development Organization into ESA, the European Space Agency) saw the need for a clear mandate if the negotiations were to progress further.[74] To decide on such a mandate, a working group was established at a NASA/ESA cooperation meeting in June. The group's charge was to prepare a detailed proposal on ways in which ESA might cooperate with NASA.[75]

CONCLUSIONS

The political problems the telescope program had encountered in 1974 had underscored how marginal it was to the interests of some congressmen. For the telescope's advocates, the problems had provided two main lessons. First, the telescope's cost had to be lowered;

second, a stronger and more widely spread base of support was essential. Spurred on by the actions of Congress, NASA had searched for an international partner and had sought a basis for cooperation. The political problems had also helped to bring the remaining dissenting astronomers into line and had led to at least the start of the forging of links between the astronomers and contractors. The result was that by mid-1975, the coalition in favor of the telescope was considerably stronger than it had been a year earlier. It was, however, still not strong enough to win approval for the telescope, as we shall see in Chapter 5.

5

Saving the Space Telescope

Socrates: Shall we make astronomy the next study? What do you say?
Glaucon: Certainly. A working knowledge of the seasons, months, and years
is beneficial to everyone, to commanders as well as to farmers and sailors.
Socrates: You make me smile, Glaucon. You are so afraid that the public
will accuse you of recommending unprofitable studies.

Plato, *Republic*[1]

You know, someone said, what good is it going to do when you get it out
there? What does it do? What does it find? And somebody says, "More of
the same." Is that correct?

Congressman Edward P. Boland, 1976

. . . we are talking about the salvation of the world.

James C. Fletcher, 1975

In August 1974, Congress had set two conditions on providing
planning funds for the Large Space Telescope: that NASA investigate (1) international collaboration on the telescope program and (2)
the building of a less costly LST. There then followed a subtle and
drawn-out process in which NASA, the contractors, and astronomers sought to repackage and adjust the technology and scientific
objectives to the agency's view of what the "market" – in this case,
Congress – would bear, and in so doing accommodate a different
and wider set of interests in the telescope's design. NASA's plan to
include a European-built scientific instrument was one example of
this. The redesigned telescope must therefore be interpreted not as
a product of technical and scientific considerations alone but as the
result of a complicated interaction and interpenetration of technical,
scientific, political, economic, institutional, and social forces. How
these forces shaped the policy-making on the telescope, as well as
the negotiations and debates on the telescope's revised design and
the program to build it, will be the focus of this chapter.

143

BACK TO THE DRAWING BOARD

At the Science Working Group meeting of May 1974, Marshall project manager James Downey spoke on the program's finances. He reckoned that the telescope they were planning would cost $325 million (1973 dollars), but, he warned, the existing design was "perhaps too extravagant."[2] In the midst of the campaign in the summer of 1974 to secure Phase B funds, Lyman Spitzer sided with Downey.

Spitzer was intimately aware of sentiment on Capitol Hill through his lobbying. He had spoken, too, with NASA administrator James Fletcher, who had emphasized that "there's no way we can sell a new $300 to $500 Million program" for the following year. Spitzer therefore expected *"real problems"* when the telescope was defended in the next round of congressional hearings.[3]

What conclusions were to be drawn? Spitzer reluctantly decided that the telescope, as then designed, was perhaps too expensive to be sold. After talking the matter over with O'Dell,[4] he explained to all the other members of the working group that the scientific arguments for the Large Space Telescope were "extremely strong ones to most people." However, the "most serious question that has been raised . . . both by Congressmen and by astronomers, concerns the estimated cost of the program." Spitzer related how one prominent member of the Greenstein committee had expressed a fear that the cost of the telescope might be so high that it would draw funds away from ground-based astronomy. If that view persisted, support for the telescope would be eroded. Spitzer suggested that the contractors and the Science Working Group expend much more effort examining the possibility of a smaller, lower-performance telescope.

He argued that one obvious way to cut costs was to reduce the size of the primary mirror and thereby scale down the rest of the spacecraft. Although Spitzer proposed that the aperture range of one to three meters be studied, he foresaw little enthusiasm among astronomers for the lower end of the spread (which was only a small increase over that used in the Orbiting Astronomical Observatories). Spitzer suggested that they consider an alternative telescope with a primary mirror between 2 and 2.5 meters. Marshall, he cautioned, would also have to take an active and imaginative approach to cost savings, as "the contractors have a certain economic interest in keeping the cost of the LST as high as is politically feasible."[5] The direction given by the Congress to NASA on August 21, 1974, to "pursue a less expensive project" vindicated Spitzer's judgment and forced the Marshall task team, the contractors, and the astronomers to rethink radically the design of the telescope and how the idea of the telescope might be repackaged so as to help in its selling.

On September 18, Marshall presented to NASA Headquarters its initial views on ways to cut costs, some of them drastic. The central changes Marshall suggested were as follows: (1) Drop the size of the primary mirror from three meters into the two-meter class (which would also allow the pointing and stability requirements to be relaxed and thereby permit a less costly pointing system to be used). (2) Chop the number of U.S.-provided detectors for the scientific instruments from two to one, and cut the number of U.S. instruments from three to two. (3) Eliminate or scale down the means of altering the shape of the telescope's primary mirror in orbit. (4) Replace the extendable sunshade by a fixed, shorter one. (5) Reduce the diameter of the Support Systems Module. (6) Simplify the design by canceling the capability for removal and replacement of complete scientific instruments in orbit.

Marshall contended that by taking these measures and a few others, the program's estimated cost could be lowered significantly. Assuming a fiscal 1977 new start, Marshall reckoned the total cost of the restructured program to be $225 million in FY 1973 dollars, as opposed to $320 million for the planned three-meter LST.[6] Noel Hinners, the newly appointed head of the Office of Space Science, told Marshall that the cost range under discussion for the telescope in NASA Headquarters was $300–350 million (current-year dollars), but those were not hard and fast numbers.[7] Indeed, the key point here is that the cost targets were not based on a studious assessment of the likely cost of each section of the telescope. Rather, they were the best guesses of NASA Headquarters managers regarding what size LST program could be sold to Congress. Purely financial considerations were setting the limits within which NASA engineers, contractors, and astronomers had to work. NASA was adopting a specific policy of "design to cost."[8]

It was not, however, purely a matter of reducing costs. In fact, the NASA managers faced a dilemma. As we have noted earlier, if they cut the program and the telescope's capability too much, they would lose the interest and support of the astronomers. In that case, the program would be dead. Yet if Marshall did not reduce the costs substantially, the telescope would most likely die at the hands of Congress.

The only way for NASA to walk that fine line was somehow to devise a new program that, although less attractive to the astronomers than the three-meter telescope, would be one they could accept. That Lyman Spitzer, widely seen as the father of the project, had proposed a telescope more modest than a 3-meter LST made the agency's task easier. At a Marshall staff meeting on August 28, there had been general agreement to switch to what was referred to as a two-meter-class telescope. The following day, O'Dell polled the Science Working Group by telephone and secured their approval for the change.[9] A staffer in NASA Headquarters contacted various optical astronomers to get their opinions on the minimum performance they considered acceptable for a redesigned Large Space Telescope.[10] Some astronomers wanted the telescope to be able to reach the Cepheid stars in the Virgo cluster of galaxies. The Cepheids were important because they might help to fix the distance scale of the universe (see the Introduction). For other astronomers, the minimum was considerably better performance, perhaps ten times better, than could be achieved with ground-based telescopes.

By early October, at a meeting of the Science Working Group, O'Dell and Roman, acting together as project scientist and program scientist, had produced a set of specifications for what they termed the "minimum" LST. The primary elements were as follows: (1) The telescope should be a versatile, long-lifetime observatory; that is, it should have the capability to accommodate a variety of scientific instruments and to vary the set of instruments with time. (2) The telescope should be capable of a resolution of at least 0.1 second of arc. (3) The telescope must work efficiently down the wavelengths around 1,150 angstroms, that is, far into the ultraviolet region of the spectrum, and it must also allow for efficient observations at infrared wavelengths longer than those reached from the ground. (4) The telescope should be capable of carrying four scientific instruments. (5) The telescope must be capable of measuring objects appreciably fainter than those observed from telescopes on the ground.[11]

The minimum LST, though extremely impressive by the standards of ground-based telescopes, clearly represented a less powerful scientific tool than the planned three-meter telescope. At that same October meeting, Hinners warned that although he gave the telescope a high priority, there was a feeling at NASA Headquarters that any program above $300 million would not sell. In fact, that figure was sometimes referred to as a choke limit: Go above that limit, and Congress would choke on it. The gloomy tone at NASA can be gauged from the fact that conversations began turning to what other programs might be pursued if even the minimum-cost telescope proved to be too expensive.

The two Phase B Optical Telescope Assembly contractors, Itek and Perkin-Elmer, had also studied the requirements of the minimum Large Space Telescope and had used them as the basis for their own recommendations on ways of saving money. For both Itek and Perkin-Elmer the central proposals were that the diameter of the primary mirror be reduced to 2.4 meters and that the sunshade be fixed, not retractable. Both contractors argued that with the 2.4-meter mirror, they could make substantial savings in the facilities they would need for building and testing flight hardware.

Why was there such a quick convergence toward the 2.4-meter Large Space Telescope? The central point, made at the Science Working Group meeting, was repeated by NASA in response to congressional questions a few months later: With the 2.4-meter primary mirror, use could be made of existing industrial facilities, with some modifications, whereas the 3-meter mirror would require a substantially larger investment in new facilities. [12]

It was well known that 1.8-meter mirrors had been built for the Manned Orbital Laboratory, a prospective U.S. Air Force program that was canceled in the late 1960s. Moreover, the options that NASA, its contractors, and the astronomers eventually studied for the size of the telescope's primary mirror were 3, 2.4, and 1.8 meters. [13] It is therefore tempting to conclude that 2.4 meters was chosen by NASA simply because it sat exactly halfway between the proposed 3-meter and 1.8-meter sizes, the latter figure being selected because the technology to build mirrors of that size for space use already existed. In other words, the decision to go to 2.4 meters was perhaps a good bureaucratic compromise. Even if that is not correct, it is certain that the optical houses had already built mirrors for space use of a size much closer to 2.4 meters than to 3 meters.

Since the early 1960s, the United States, like the Soviet Union, has flown a variety of reconnaissance satellites, some of which have been kinds of space telescopes. Unlike the Large Space Telescope, these reconnaissance satellites have been designed to observe sites on the earth or monitor radio traffic, not explore the heavens. These spacecraft have become so accepted by both superpowers that their continued use has been written into treaties on nuclear arms limitation.

Although little unclassified information is available about such reconnaissance satellites, there have been reports that during the late 1970s the United States was employing three types of photographic intelligence ("PHOTINT") satellites. The KH-8 satellite was introduced in 1966 and is reputed to be cylindrically shaped and 24 feet long, with a diameter of 4.5 feet and a weight of 6,600 pounds. In 1971 the United States launched the first of its KH-9 ("Big Bird") series. The Big Bird satellite is reputed to be based on the Lockheed Agena spacecraft, about 50 feet long, 10 feet in diameter, and weighing about 30,000 pounds. A later development was TRW's KH-11 satellite, first launched in 1976, and claimed in one report to be 64 feet long and to weigh about 30,000 pounds. [14]

None of those reconnaissance satellites, at least in gross features, shows much resemblance to the Space Telescope (assuming these figures are correct). However, a more recently reported reconnaissance satellite, the KH-12, may bear closer similarities, particularly as its two main contractors have been reputed to be the same as for the Space Telescope. [15] But little is known publicly about the KH-12. In any case, the first such satellite, if the reports are reliable, had yet to be flown in 1988 and so had been a long way from completion, and perhaps even inception, in the middle and late 1970s,

147

the time when the critical decisions on the telescope had to be made.

Another crucial point is that earthward-looking telescopes and astronomical space telescopes pose distinctly different problems for their designers. The central design concern for the Large Space Telescope has been how to direct it to, and keep it locked on, extremely faint objects for many hours.[16] Reconnaissance satellites, in contrast, observe sources on the earth for short times. Also, one of the chief reasons for orbiting the Space Telescope is to observe the ultraviolet emissions from astronomical objects, whereas for reconnaissance satellites, ultraviolet performance is irrelevant — the ozone in the earth's atmosphere blocks an earthward-looking telescope from detecting anything in that region of the spectrum. There must therefore be significant differences between the Space Telescope and reconnaissance satellites.

Nevertheless, as presidential science advisor George Keyworth was to remark in 1985, there "is, however, no question that it would have been a very much more difficult task [to build the Space Telescope] if we had not already acquired considerable expertise in both talent and industrial manufacturing. The [Space Telescope] is new, but it draws upon technologies used in military systems."[17] That point had also been made when the House Authorizations Subcommittee visited the Lockheed Missiles and Space Company in Sunnyvale, California, in early 1974. A Lockheed executive explained to the subcommittee why his company was suited to build the Large Space Telescope:

That support module you see there looks very similar to the low earth orbit satellite we developed for the Air Force and are now flying out of our Western test range. It has the capability in the lower orbit of doing very similar things that you do when you control this telescope. Ten feet in diameter sizing is virtually identical and that is why we are in this business [of defining the LST and] why we think we can make it possible for NASA to keep the cost down on this system. Something below that which you can get by developing a totally new system.[18]

The telescope therefore was derived to a considerable degree from satellites and technology developed for purposes other than space astronomy. For a time, the Department of Defense was concerned that the "high" technology exploited for military satellites might, via the telescope, be transferred to foreign companies, a concern that was no doubt heightened by the proposed substantial European participation in the building of the telescope.[19] That issue, however, seems to have involved not so much specific items of hardware as methods and procedures for developing, fabricating, and integrating the various subsystems and modules to be employed in the telescope.[20]

The crux of the matter for NASA was that with a 2.4-meter primary mirror and the consequently scaled-down telescope, the agency could save money over the 3-meter design and also have more confidence in the accuracy of its contractors' cost estimates.[21] The 3-meter telescope, the argument went, would have gone considerably beyond the state of the art, increasing the uncertainties in the cost estimates and making its construction significantly more difficult than that of a 2.4-meter telescope.

The Science Working Group was facing a quandary. A 2.4-meter Large Space Telescope meant a disappointing loss of scientific capability. First, the light-collecting area of a 2.4-meter telescope would be only two-thirds that of a 3-meter. In consequence, for a given astronomical object, longer exposures would be required with the 2.4-meter telescope. Some objects that could be detected with a 3-meter telescope would not be within the light-grasp of a 2.4-meter Large Space Telescope. Second, the telescope's resolving power, the main reason for orbiting it, would be reduced from 0.06 to 0.1 second of arc. The 2.4-meter telescope would not be able to resolve such fine detail as would a 3-meter, or detect such faint sources. Nor could the 2.4-meter telescope carry as many scientific instruments. This change obviously meant a loss of scientific flexibility, although the level of funding available to NASA even before the congressional directive to save money seemed likely to agency managers to limit the telescope to four scientific instruments.[22]

Astronomers thus had to weigh the reduced scientific performance against the lower estimated cost, which not only helped their chances for congressional approval but also promised a relatively easier engineering task. However, the struggles on Capitol Hill and the direction from Congress to cut costs had boxed in the astronomers and given them little room to maneuver in the negotiations with NASA. They therefore agreed at a Science Working Group meeting in October 1974 that the contractors should study the 2.4-meter LST with the equivalent of 2-meter diffraction-limited performance.

At the same meeting, Hinners asked that between October and December, Marshall study the 2.4-meter telescope. Revised plans could then be presented to Congress in time for a new start in fiscal 1977. Shortly after the meeting, Hinners further decided that the first part of the study contracts with the Support Systems Module contractors (to be let in November 1974) should include analyses of a smaller telescope with a mirror about 1.8 meters in diameter.[23] This telescope could be a possible fallback if Congress declined to fund even the 2.4-meter.[24]

On November 15, NASA announced the selection of Boeing, Lockheed, and Martin-Marietta for parallel contracts of about $700,000 each for preliminary design and program definition studies for the Support Systems Module. The contractors were initially instructed by NASA to pay particular attention to telescopes of dif-

149

ferent sizes: Boeing would examine the 1.8-meter, Lockheed the 2.4-meter, and Martin-Marietta the 3-meter. The aim was to investigate a variety of trade-offs between cost and performance. The three options would then be reviewed by NASA in the light of these studies, and from that point all three contractors would work on the size that had been selected.

By December 13, Marshall was ready to present its revised plans to the Science Working Group. In project folklore, that date won the title "Bloody Friday," for it was there that the cuts aired in September were to be accepted by the astronomers, even if very reluctantly.

When the main features of the 1.8-meter telescope were described to the working group, the members balked.[25] Although they were prepared to compromise, dropping to a 1.8-meter telescope with only two scientific instruments might be a knockout blow, for, the working group protested, it would be a gross political blunder. If the 1.8-meter telescope were built, Congress and even NASA might come to consider it as a valid Large Space Telescope. The construction of a "real" LST might then be delayed by many years.[26]

With the Science Working Group's plea to the Marshall task team to hold the line at 2.4 meters, the revised plan was presented to Hinners at NASA Headquarters on December 18. Marshall project manager Downey announced that although the contractors would not complete all of their cost/performance trade-off estimates for some months, the latest cost estimate for the minimum LST was $335 million (in current-year dollars, assuming 5% inflation), a sizable reduction from the $483 million for the 3-meter telescope.

There was, in addition, a fundamental change in the telescope's design from what had been presented to Hinners as recently as September. The Support Systems Module had been pushed forward and so was wrapped *around* the primary mirror, instead of *behind,* as it had been throughout Phase A and for the first year and a half of Phase B. The main section of the Support Systems Module had come to resemble a doughnut or, more properly, a toroid.

With this design change came several advantages. A crucial improvement was that, compared with the 3-meter, the 2.4-meter telescope could be turned in space more easily, and the demands placed on the pointing and control system were not so great. The chief bonus was that reaction wheels alone could be used for directing the spacecraft's movements around the sky, simplifying the engineering task and reducing costs. Further, reaction wheels work at much lower speeds than do their competitors (control moment gyroscopes) and so do not pose the same problems of vibration, and a key point for the design engineers was to reduce vibration in the spacecraft as much as possible.

The 2.4-meter telescope also held out the promise of what appeared to the Marshall engineers to be a "neater" engineering package. The point was that it would not push the limits of the Space Shuttle's weight and volume-carrying capability nearly so much as would a 3-meter telescope with the Support Systems Module behind the primary mirror. Marshall estimated a 3-meter Large Space Telescope to weigh around 25,000 pounds, but the 2.4-meter only 17,000 pounds. The conventional wisdom is that as a program matures, more weight is almost invariably added to the spacecraft; the smaller telescope offered Marshall engineers the promise of a significant margin of safety. So in addition to the economic benefits to be gained by shifting to a 2.4-meter telescope, there were for Marshall persuasive engineering reasons for the change.

THE BOTTOM LINE

In addition to these potential design changes, there was another, more fundamental issue in play, as revealed in a published interview with NASA administrator James Fletcher.[27] Fletcher argued that expensive space science missions (he cited the Viking probe to Mars to be launched in 1975 as an example) would not be attempted again until the nation's economy improved. Other missions, such as that of the Large Space Telescope, "may have to be trimmed down to a less ambitious effort." The economic state of the country meant that "the number of new starts is going to be held to a minimum." More optimistically, he thought that the situation was only temporary, and not unique to space, as the pinch was on all major new science projects. For Fletcher, a low profile for space science was a tactical necessity for the time being.[28]

Certainly NASA Headquarters was interested in a telescope smaller, simpler, and thus cheaper than a 2.4-meter. As part of the continuing debate on what would constitute an acceptable Large Space Telescope, the members of the Science Working Group were polled in early 1975 by O'Dell and Roman. These two joined forces again to make their own opinion clear:

Both of us feel that the program has already been so drastically reduced that we are at the point that the line must be drawn and that the agency must either allow us to move ahead with at least this minimum LST, or that we are confronted by an agency money problem and put LST into hold until the future. . . .

A further complication is the proposal by Al Schardt [director of the astrophysics division at NASA Headquarters] that we try to reduce program costs by constructing an evolving complexity LST, i.e., one that at initial launch would not meet the minimum performance specifications; but, one that would possess the potential for upgrading later upon successive ground returns to a full LST capability. Although detailed calculations have not been made, it appears that such a plan is plausible and at some reduction in cost to initial launch. The overall cost to eventually reach LST performance with this plan would certainly be higher than the direct approach. Although Dr. Schardt considers this plan very seriously, he will not support a plan that is not supported by you, the ultimate users and justifiers of LST.[29]

The clear message from the working group was that NASA should hold to the agreed minimum Large Space Telescope.[30] Similarly, when in April Hinners asked the working group if the 1.8-meter telescope could not be considered a Large Space Telescope, the response was that it certainly could not. The 2.4-meter telescope was the minimum the astronomers would accept. As one member of the working group recalls, "you felt you were going to be whittled to death, and that's why . . . the language adopted by the Science Working Group was fairly strong, as these things go. It was essentially, 'Well, we can drop down to 2.4, but if that's too big, then forget the whole thing, we're just not interested.' "[31]

There was another force acting to drive down costs, and it derived from the highly influential Space Science Board of the National Academy of Sciences. In 1974, the board's support for the telescope had helped to undermine Congress's claim that astronomers did not accord the telescope a high priority. In January 1975, John Naugle attended one of the board's meetings. It was clear to him that the members were wary and would want to reassess their strong endorsement of the Large Space Telescope if its cost "is so large that it is likely to eliminate several other programs; or alternatively, if the LST NASA proposes is such a small advance over OAO and [the planned International Ultraviolet Explorer] that the scientific return will not merit the projected costs."[32] If the telescope was to be sold, the coalition favoring it would have to be kept as broad as possible and even extended. But, as Naugle recognized, that was a difficult task: Drop below 2.4 meters, and the working group and other astronomers probably would lose interest; make the telescope too expensive, and the support from other space scientists might collapse, which would surely provide its opponents in the Congress the ammunition they needed to kill it.

NEW APPROACHES

The reevaluation of the Large Space Telescope program by NASA, its contractors, and the astronomers had produced another notable change: NASA's approach to building the telescope began to place even more emphasis on the use of the Space Shuttle for in-orbit repairs. Except for the optics, NASA now planned that all of the major components on the telescope would be designed so that they could be replaced in orbit. Much of the cost of a space science program is associated with the exceptional degree of quality assurance

needed to produce hardware for use in space. With the aid of the
shuttle, NASA's claim went, it would be unnecessary to have such
a high level of quality assurance, because the orbiting Space Shuttle
would be available to rendezvous with a satellite and enable astronauts to fix problems in orbit. Thus, less money would be needed
for developing long-lived components and for testing on the ground.
In this way, the shuttle would permit spacecraft designers to accept
greater risks than they might otherwise have done. Adopting this
approach, the telescope could then be constructed at a cost lower
than that of using conventional methods.[33] Such a plan was not
without its difficulties and contradictions, particularly as the agency
had earlier rejected that plan because of the expense of designing
instruments and subsystems that could be replaced in orbit.

NASA's drive to reduce costs also underlined an approach that
might be termed "conservative innovation." In the words of Marshall project manager Downey, this meant that the engineering style
of the program would be

to procure the lowest cost system that will provide acceptable performance.
Furthermore, we will be willing to trade performance for cost, particularly
in areas where significant cost savings can be realized. We are wary of
proposed sophisticated solutions and complicated designs. If one is considering a trade between sophistication and performance versus simplicity and
somewhat reduced performance, it is usually prudent to adopt the simpler
approach. Also, flight proven components and subsystems and standard
components will be used whenever practical. Use of space-proven equipment is doubly beneficial by reducing both development and performance
risk.[34]

Again, as we shall see later, these aspirations were to prove difficult
to fulfill.

By April 1975, Downey was ready to brief Hinners on the latest
studies of the Large Space Telescope. The new cost estimates were
(in FY 1975 dollars) as follows: 3-meter LST, $334 million; 2.4-
meter LST, $273 million; 1.8-meter LST, $259 million. These estimates indicated that the money to be saved by dropping below 3
meters was not nearly as much as NASA had expected. The reason
was that whatever mirror size was chosen, the spacecraft still would
have to be precisely pointed, and the supporting equipment (e.g.,
batteries, solar arrays, associated power system) still would be needed.
Also, the figures showed that there was relatively little to be saved
by switching from 2.4 to 1.8 meters, and there was a high price to
be paid in scientific capability. The strident opposition of the astronomers to this latter size helped NASA managers to rule it out
relatively easily. In addition, Phase B contractors for both the Optical Telescope Assembly and the Support Systems Module had backed
the 2.4-meter telescope. The 2.4-meter size had also been endorsed
by the Science Working Group, as well as by the National Academy

of Sciences' Space Science Board. Given all of these factors, Marshall strongly recommended that the contractors proceed with the preliminary design of a 2.4-meter Large Space Telescope. When, in turn, members of the Office of Space Science at NASA Headquarters briefed the NASA administrator and other top managers on May 15, they, too, identified the 2.4-meter telescope as the best option.[35] Administrator Fletcher agreed.

The 3-meter Large Space Telescope was dead. After a decade as the focus of planning for a large telescope in space, it had become the lamented victim of the struggle to win congressional support, but had been reborn as the revised and repackaged 2.4-meter LST. The program was, however, back on course with its coalition of supporters still intact, with the expectation, Congress willing, of a new start in fiscal year 1977, and launch of the telescope in 1982.[36]

Artist's conception of the 2.4-meter space telescope, as envisaged in 1976. The 2.4-meter telescope had many differences from the originally planned 3-meter telescope. (Courtesy of NASA.)

A BUY-IN?

The switch to the 2.4-meter telescope did not halt concerted efforts by senior NASA managers to produce a still cheaper program. In

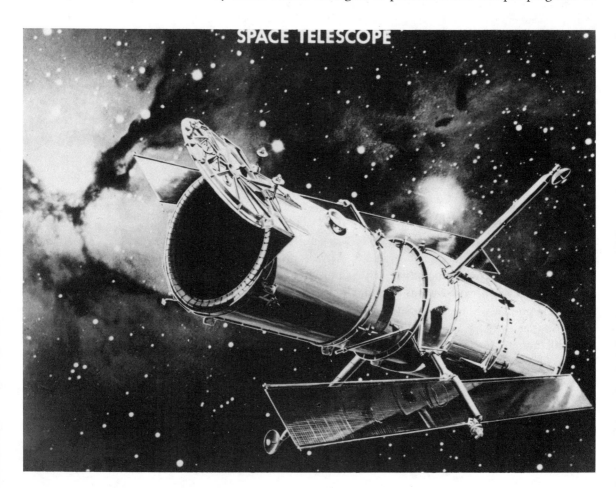

June 1975, NASA deputy administrator George Low called for dramatic savings. Low thought it essential for NASA to have a "balanced" collection of programs and not to focus solely on the shuttle. In his opinion, it "would be a great loss to the space program" if the telescope did not get approved, and he searched for ways to move it ahead. He had spoken with highly placed officials in Boeing, Lockheed, and Martin-Marietta. The message he had conveyed was that "Fletcher and I were concerned that a large space telescope at the costs now being projected would be impossible to include in any NASA budget in the foreseeable future and the project therefore might well be cancelled." Low had pressed the three contractors to see if, by reducing the telescope's requirements, the cost could be cut by about half. "I reminded them of the situation with the Space Shuttle in the 1971 time period and that it was industry and not NASA which came up with the clever designs which essentially cut the development costs of the Shuttle by a factor of two. I urged the same kind of innovative thinking by industry on the LST." At a minimum, Low wanted to ensure that NASA became far more cost-conscious. "I let it be known," Low noted, "that unless we had a very much less expensive telescope, we won't have any." Fletcher and Low wanted a $200 million telescope program.[37]

Naugle also talked directly with industry representatives about cutting costs. Two highly placed Boeing officials told him that they were unsure that it was feasible to save money even by further technical innovations. Instead, they suggested a different management approach: Use a primary contractor, not associate contractors, to build the spacecraft. If that were done, as much as $50 million might be saved.[38]

Fletcher and Low were convinced that despite the design changes, the reductions in funding, and the alteration in approach underpinning the LST program, there was no guarantee that Congress would fund the building of the telescope in fiscal 1977. The congressional hearings during early 1975 had demonstrated that the project still required some hard selling. The House Appropriations Subcommittee, in particular, had taken a tough line. As mentioned earlier, Congressman George Shipley had been the most aggressive questioner. The NASA manager to bear the brunt of the questioning was Noel Hinners, head of the Office of Space Science since the summer of 1974.

Long-limbed, and with a deceptively youthful appearance, Hinners had solid scientific and management credentials. He had been awarded a Ph.D. in geochemistry and geology from Princeton in 1963. Head of the Lunar Exploration Department at Bellcomm for several years, Hinners had joined NASA in 1972, although by then he knew the agency extremely well because he had been intimately involved in conducting studies for the program of moon landings.

By 1975 he was already proving an effective advocate for NASA's science program, but the hearings before the skeptical House Appropriations Subcommittee posed distinct problems.

The focus of the hearings was the telescope's cost, but on one point Hinners held firm. Shipley had chiseled away trying to get a total-cost estimate. "I am very hesitant," Hinners repeated, "to give an estimate, since it would not be meaningful until the study efforts are completed. . . . I have insisted in the LST studies that there be no possibility of a 'buy-in.' . . . I am new enough on the job that I want no problems like that." It was too much for Shipley: "I am old enough on the job to know . . . that estimates invariably are low and you run into cost overruns. Will you know when you complete Phase B what this project will cost?" Shipley wanted to avoid any buy-in by which Congress would approve the Large Space Telescope at some stated cost figure, only to be told a few years later that the actual cost would be much higher; yet because so much money would already have been invested, there would be great reluctance to cancel the project.

Noel Hinners, who became associate administrator in the Office of Space Science at NASA Headquarters in 1974 and soon won a reputation as a champion of the Space Telescope. (Courtesy of NASA.)

Other space science projects had suffered overruns. The most recent example had been the Viking mission to Mars, and the LST testimony had to be put in the context of the other programs that had appeared before the Appropriations Subcommittee. As the subcommittee chairman Edward Boland had pointed out earlier in the hearings, Viking had been projected to cost $364 million, but would reach $1 billion.[39] Hinners's assurance of no buy-in was his promise that it would not happen again. The NASA comptroller also contended that the Viking cost overruns were smaller than they appeared, because the earlier estimate had been only preliminary. Hinners emphasized that it should not have been released to Congress, because it was unreliable. He did not want NASA to make the same mistake again.

These arguments on the telescope and criticisms of Viking, however, revealed only the whitecap of a larger wave of congressional discontent. During the fiscal 1975 and fiscal 1976 NASA hearings, the views of the House subcommittee and NASA were not meshing at all well. At the root of this discord was a widespread opinion in Congress that the agency should do more to apply NASA-developed technology to urgent problems on earth, such as energy and earth resources. In fiscal 1975, for example, the House subcommittee had urged NASA to build a third earth resources satellite (ERTS), later to be renamed LANDSAT.[40] To do so, the subcommittee had pressed the agency to switch funds from other approved programs. The subcommittee's questioning therefore often turned to how the telescope could be used to help solve practical problems on the earth. The responses of the NASA witnesses often were to invoke the lessons of history. At one point, for example, NASA administrator Fletcher contended that

the benefits of astronomy are generally longer term than some of the others we talked about. On the other hand, they are far-reaching in their impact. The benefit from Galileo's experiments was literally the Industrial Revolution. How can you put a value on that? We are on the edge of extremely important discoveries in astronomy.

It [astronomy] is starting to blossom much in the same way as in the days of Galileo and Copernicus. If you ask the two areas the astronomers are most interested in, one would be the quasars, enormous energy sources coming out of those objects, second, the nature of our universe. . . . Is it expanding without end? How did it begin? Is it going to contract again? Understanding those processes will contribute to our fundamental knowledge of science. There could be brand new energy sources downstream, just as nuclear energy came out of Einstein's investigations. By the way, that was astronomy, too. The whole idea of relativity came out of astronomy.

The "historical" arguments presented by Fletcher, though often repeated by others in various forms, were unfounded. It is absurd, for instance, to claim that relativity was somehow derived from astronomy. Congressman Max Baucus of Montana, moreover, was not interested in vague promises. He wanted Fletcher to give solid estimates of the practical benefits to be won by the telescope. Fletcher responded: "Even though you try to put probabilities on realizing benefits, even if it is only 5 to 10 per cent, we are not talking about billions or trillions of dollars, we are talking about the salvation of the world. It is worth a 10 per cent chance to seize the opportunity."[41]

Although that reply was surely hyperbole, there and in other parts of his testimony Fletcher was nevertheless attempting to make a serious claim, a claim often repeated: Scientific developments *can* lead in unpredictable ways to technological breakthroughs. Indeed, that has been a central justification for federal support for science since World War II.[42] One image it conjures up is of a pool of scientific knowledge from which those who thirst after technological innovations can drink. Adding new knowledge to the pool increases the chances of additional technological advances.

Such arguments have regularly been advanced since the nineteenth century. In 1872 and 1873, for example, John Tyndall, professor of natural philosophy at the Royal Institution in London, gave a series of lectures in various cities in the eastern United States. In addition to explaining the wonders of light, Tyndall seized the opportunity to press for financial backing for researchers in pure science. He complained that such support was sadly lacking in the United States, though the pursuit of knowledge for its own sake was the bedrock on which practical applications were built. To drive his point home, Tyndall quoted de Tocqueville, who had chastised Americans for collecting the treasures of the intellect without taking the trouble to create them. In other words, Americans had exploited the researches of scientists in other countries, but had not chosen to add to the common pool so that others might drink.

Although Tyndall's arguments often have been cited in making the case for federal funding of basic science, they raise important problems. In particular, they are similar to arguments made in classical economics. For classical economists, there was a hidden hand at work that translated the pursuit of individual self-interest into the pursuit of the general good. The hidden hand was identified with the mechanism of the marketplace. The arguments promoted by scientists such as Tyndall concerning the benefits of basic science, as well as many later arguments, follow the same sort of structure. If scientists "remained free to pursue their calling as they saw fit, to satisfy their scientific curiosity about Nature, their efforts would inevitably — and without need for conscious intent on their part — contribute to the general good." Whereas classical economists had pointed to the marketplace as the hidden hand, the scientists could point to no such mechanism; "they argued their position as a mere article of faith."[43] As historian of science Derek Price has contended,

historically, we have almost no examples of an increase of understanding being applied to make new advances in technical competence, but we have many cases of advances in technology being puzzled out by theoreticians and resulting in the advancement of knowledge. . . . Again and again, we find new techniques and technologies when one starts by knowing and controlling rather well the know-how without understanding the know-why. We often (but not always) eventually understand how the technique works, and this leads to modifications and improvements, giving the impression that science and technology run hand in hand. But historically the arrow of causality is largely from the technology to the science.[44]

But as Tyndall had done a hundred years earlier in different circumstances, the astronomer-lobbyists often invoked the practical benefits that might accrue from the basic researches to be conducted with the Large Space Telescope. A favorite example they and the contractors frequently cited was how nuclear fusion was discovered in the 1930s to be the power source that keeps the stars shining. Fusion produces prodigious amounts of energy, and since the early 1950s it has been widely viewed as a long-term, but likely, source of heat and light for use on earth. That was a vivid illustration of how research in astronomy offered chances of solving practical problems.[45] As a Lockheed handout on the telescope put it, "today's science is tomorrow's technology."

Fusion was a particularly relevant example in the mid-1970s, when America apparently awoke one morning to an energy crisis. "After

years of seemingly unlimited supplies of energy products at low
prices," according to one account, "it suddenly appeared that the
oil wells had run dry. Gas stations were closed on Sundays and often
early on other days, absurdly long lines formed at their pumps.
heating oil was rationed, plane flights were cancelled for lack of fuel,
farmers had no fuel to plow their fields or to dry their crops, and
truckers, angry at high fuel prices and delays, went on strike."[46]

A decade earlier, in a very different time when funds had flowed
more readily, physical scientists usually had not been driven to make
direct claims to Congress about the technological spin-offs of Big
Science research. Arguing for increased funding for the multi-hundred-
million-dollar Stanford Linear Accelerator in 1964, its director had
stated that

I am not of the school who tries to defend this kind of work through its
byproducts. I believe if you want the byproduct, you should develop the
byproduct. I think you would do it more economically and do it more
effectively. If you want to push high powered radio tubes, then the best
way to do so is to push the development of high powered radio tubes and
not to build accelerators which require high powered radio tubes.[47]

Big Science for the sake of Big Science was not enough to sell the
telescope in the mid-1970s. That was reflected in the congressional
hearings as well as in private conversations between scientists and
lawmakers. "There was," John Bahcall recalls, "excitement and interest among Congressmen and Senators about the science itself . . .
but that did not imply this was something that ought necessarily be
done." Although there were a few exceptions, "it was more often
the connection with industrial development in general or in their
particular district or state or defense related activities. It was under
these sort of ideas that it was possible to get an immediate sympathetic hearing, so that got you over the question of why, and then
only discussed the question of what to do." Only rarely was there
the response that "science is good [and] we want to help you with
that."[48] Other kinds of what can be described as nonscientific justifications were also key to the astronomer-lobbyists' defense of the
Space Telescope. One bonus for them, according to Bahcall, was
that on occasion "we benefited from the confusion between astronomical space satellites and satellites designed to monitor military
installations." Bahcall and the other astronomers, of course, were
well aware that although there were definitely links between astronomical and military satellites, astronomical satellites could not
themselves be used for observing the earth.[49]

But the politicians always returned to the issue that was central
for them: "Why is Space Telescope good for my constituents? Why
should I support this?"[50] Bahcall had learned from Spitzer to look
to the interests of the person they were lobbying:

159

You might call it cynical, but it was usually accurate. You could predict people on the basis of what their interest was very often. Sometimes, of course, their clear interest was to get the most science for the available dollars. And we often made use of that, because Space Telescope was good for almost everybody, and when it was appropriate to argue that it was good for industry in this state we found out what industry it was and we argued that.[51]

The scientific results promised by the Space Telescope simply were not sufficient to win federal funding. While NASA was ostensibly selling in its public testimonies a scientific instrument of prodigious power, Congress and the executive branch were buying much more, and the telescope advocates (among NASA, astronomers, and industry) had shaped the argument Congress was buying.

NASA's goal of making the telescope more salable also prompted a name change. After a small committee had considered about thirty possible names for the telescope, in October 1975, quietly and without a stated reason, NASA deputy administrator George Low directed the switch to "Space Telescope" via an internal NASA memorandum.[52]

Some of the astronomers were alarmed. They saw the renaming as a portent of yet more cuts to come. Science Working Group member Margaret Burbidge, who was soon to be president of the American Astronomical Society, sent a vigorous protest to Low. She emphasized that many "of us have taken some delight in the fact that the initials of [the LST] are the same as 'Lyman Spitzer Telescope.' "[53] Low assured Burbidge that the change of name was not driven by any possible further reduction in aperture size from 2.4 meters. Rather, he felt that "in these days of stringent science funding, we should not use titles such as 'Large' or 'Very Large' which suggest an opulence which neither NASA, science nor astronomy are enjoying these days and which tend to invite reductions in the budgets."[54]

The central task facing the advocates of the telescope was to win approval from a skeptical, sometimes hostile, Congress, and the name change was one move in the revision and repackaging of the telescope program to reach that goal. But before going to Congress again, the program would have to be approved by the executive branch of the government for inclusion in the president's fiscal 1977 budget.

LEFT AT THE POST

NASA had begun its detailed discussions with the Office of Management and Budget on the upcoming FY 1977 budget in the spring of 1975. Out of these exchanges had emerged the planning target for the agency's overall budget. The planning target had been sent

from OMB to NASA in July 1975 and had been used by the agency
to prepare its detailed submission to OMB in late September.[55]

In his letter accompanying the draft budget, NASA administrator
Fletcher told James T. Lynn, the director of OMB, that for a balanced program in aeronautics and space, a number of new programs
should be started each year. Fletcher stressed that for NASA, new
program starts "are simply *replacements* for other projects as they are
completed. . . . The need for NASA to initiate a number of new
programs in FY 1977 is especially acute because of the general moratorium on new program starts in the FY 1976 budget."

Number one on NASA's list of new starts was the Space Telescope. Fletcher explained that it was projected to cost about $400
million (FY 1977 dollars), a cut of $100 million compared with the
earlier estimates for the 3-meter version.[56] The other flight project being proposed as a new start was the "Solar Maximum Mission," a satellite designed to make detailed observations of the sun
during the 1979–80 period of peak solar-flare activity. Its cost was
estimated at $75–90 million and had itself already been deferred in
fiscal year 1975.

In its planning target sent to NASA in July, OMB had made a
specific provision of $20 million for new flight projects to be started
in fiscal 1977. Moreover, OMB recommended that the Space Telescope be funded as a new start. Although the cost was high, OMB
argued that it was roughly on a par with a major space flight mission
to the planets, and after its own checks among astronomers, OMB
had been assured that the scientific potential was large.

However, in a speech at the White House on October 6, 1975,
President Ford, disturbed by a rising deficit and aiming to balance
the books within three years, announced cuts of another $28 billion
from the forthcoming federal budget.[57] NASA had to shoulder its
portion of the burden, and so the new and decidedly unpleasant
problem facing Fletcher and his staff was how to meet OMB's direction to delete $305 million from the budget NASA had submitted
only a month earlier.

OMB wanted to minimize the impact of the cuts on the development of the Space Shuttle. But, as OMB contended initially, to
do so would require extensive "belt tightening" across the whole of
NASA, cancellation of at least one ongoing flight project (probably
"Pioneer-Venus"), and relatively large reductions in the number of
the agency's civil service employees. If that plan were followed, it
still would mean a six-month delay in the Space Shuttle schedule.

The mix of these proposals made too thin a gruel for NASA administrator Fletcher's taste. He protested vigorously to OMB. His
first point was that the cut would be bound to force a major delay
in the shuttle's schedule. That delay would lead, in turn, to "in-

creases in total cost, large reductions in employment in 1976 as well as 1977, possible loss of foreign contributions, and a serious disruption for many months in a program that is the central focus of the entire space program on which all our plans for the future are based." For another two and a half pages Fletcher marshaled arguments why the NASA budget should not be cut. He then turned to what actions the agency would take if OMB did not alter its decision. He emphasized that NASA would not cut or stretch out any major ongoing flight project. If Pioneer-Venus were to be cut, for example, it would mean a total loss of the $65 million already spent. Fletcher did concede, however, that "we would feel obliged to cut in about half the outlays included in our FY 1977 budget for replacement project starts in research and development."[58]

The anticipated new start for the Space Telescope was thus under serious threat. Moreover, if OMB approved NASA's method of reaching the $305 million cut, the Space Telescope would be in direct competition with the much less expensive Solar Maximum Mission for inclusion in the president's budget. The Solar Maximum Satellite was also time-limited; that is, to launch the satellite at a time when solar activity was not at a maximum would defeat its purpose. A delay in the Solar Maximum Mission would be critical; in contrast, the telescope had no particular launch date to meet.

Fletcher, as he described to one of the astronomer-lobbyists a few months later, had fought the removal of funds from the NASA budget for some time and had managed to lower the amount to be cut below $305 million. He had even, he claimed, twice taken the Space Telescope matter to President Ford.[59] But, according to Fletcher, Ford had been concerned about an "indefensible" new start by NASA while he was cutting social programs. Indeed, it was widely expected that the Democratic Congress would find the Republican Ford's reductions in social programs unacceptable and thus vote to increase expenditures in the FY 1977 budget.[60] Moreover, that budget would go before Congress in an election year, 1976. However, if the telescope were delayed until fiscal year 1977, it would not appear in the president's budget until after the November 1976 election.

Fletcher claimed he had also begun to see what, in his opinion, were weak spots in the telescope program. Further possible cost savings were, he thought, coming to light, but had not yet been evaluated. Fletcher's advisors had told him that in view of these problems and the known opposition in Congress – in the person of George Shipley – it would be a mistake to press for the telescope in fiscal 1977. A new-start proposal in fiscal 1978 would be stronger and would stand a better chance of succeeding.[61] This, at least, was NASA's public story.

There was certainly another point against the telescope. The executive branch had accorded an ERDA (Energy Research and Development Agency, soon to be renamed the Department of Energy) proposal to build a $78 million colliding-beam accelerator relatively higher priority than the telescope for scarce new-start funding.[62]

Although OMB regarded NASA's space science program as "well run, well justified"[63] and strongly favored the telescope, the odds were not in the telescope's favor. OMB was also concerned to ensure that construction of the Space Telescope be phased with the Space Shuttle program such that telescope monies would be needed *after* shuttle funding had reached its peak. The Space Telescope would then reap the harvest of what was known in OMB as the "shuttle dividend," the availability of a growing amount of funds that would be freed by completion of the development of the Space Shuttle.[64] Yet however the numbers were juggled, the cut in the NASA budget would compel a delay in the shuttle schedule, undermining, in OMB's opinion, the case for starting the telescope in fiscal 1977.[65]

Another factor in play was that, as deputy administrator George Low noted at the time, if the telescope program went ahead immediately, NASA and its contractors could not "make use of some possibly existing classified systems." The Department of Defense, it seems, had objected to the possible transfer to the telescope of technologies used for national security purposes. From Fletcher's viewpoint, the odds had become overwhelming. NASA decided that too much political capital would be lost by continuing to fight against powerful, well-arrayed forces, particularly as it might not seem wise to start a $400 million program when the rest of the NASA budget was so constrained. The position NASA and OMB reached in their negotiations was that the Solar Maximum Mission would be included in the president's budget. The Space Telescope would not, despite the fact that its omission would save only $4 million.[66]

THE ASTRONOMERS REACT

The president's budget was officially embargoed until it should reach Congress in January 1976, but on December 22, 1975, news of the Space Telescope's omission was leaked to John Bahcall. His reaction, as well as that of many other astronomers when they heard the news, was a mixture of surprise, dismay, and anger. The telescope's prospects looked particularly dismal because there were not even funds in the budget for further Phase B studies.

The Space Telescope had already suffered a year's delay because Congress had directed NASA to reevaluate the program's scope and to seek international participation, and there were still strong critics in Congress. The contractors, too, would be hesitant to risk more of their own resources on a program that appeared to have such flimsy political backing. Some observers feared that the Space Telescope

163

might well suffer the fate of other programs that had repeatedly failed to be approved over a period of several years, that is, eventually to be dropped entirely. Some of the astronomers turned their anger on the Space Shuttle.

Space scientists, on the whole, had never been enthusiastic about a vehicle like the Space Shuttle. Some had attacked it publicly, arguing that the United States had no compelling need for such a spacecraft, a spacecraft that would drain money away from what they believed were worthier projects.[67] Nor had scientists had any real say in the shuttle's design characteristics, and they, after all, would be among its major users. As one prominent space scientist would recall, "scientists were told, 'There's the Shuttle. Figure out how to use it.' "[68]

Bahcall, along with other astronomers, was outraged that whereas NASA had used the scientific payloads it would be able to fly as one of the central justifications for the Space Shuttle, once the shuttle was sold, science apparently was pushed aside. He made exactly that point in a letter to NASA administrator Fletcher, protesting that it was "shocking that out of a total NASA budget of 3.7 billion dollars priorities would not be arranged" so as to fund the telescope at some low level. "I am afraid," Bahcall wrote,

that the scientific community will conclude that NASA is intent on obtaining funds to support its own institutional needs in preference over wider goals. I fear that scientific support for the Shuttle program (over 1 billion dollars in the present NASA budget) will be decreased. Many will conclude that [NASA] cares most about its space vehicle (the Shuttle) and much less about what science it is doing with the Shuttle.[69]

A similar point was made by O'Dell, who warned that the coalition assembled to support the telescope might start to crumble. "Space Telescope has been a strongly supported scientific payload for the Shuttle era," O'Dell reported to NASA Headquarters manager John Naugle.

This is very important at a time when the discontent of scientists being forced to accept the Shuttle development has not really died away. This strong support [for the telescope] was not always present, but is the result of a carefully orchestrated activity over the last few years to educate the ground based astronomers about the potential of ST and to convince them to support this rather than the continued construction of many large ground based telescopes. We are now riding on a crest of their support, but if we fail to capitalize on it, we may lose it.[70]

NASA's top managers were certainly aware that many scientists were wary of, and others hostile toward, the shuttle. For example, when in the autumn of 1974 the agency-funded Pioneer-Venus mission had been threatened with cancellation, Fletcher had warned a White House staffer that any cutback "could be viewed as caused by

Shuttle funding requirements, and thus have an adverse effect on the support for the Shuttle."[71] And the defense of the shuttle was far and away Fletcher's chief concern, as he believed it was the program on which the agency's future hinged.

What, however, were the astronomers to do about the telescope besides putting pressure on NASA by criticizing the shuttle? Bahcall sought a means of restoring the Space Telescope to the budget. He looked to Sidney Drell for guidance. Drell, a veteran of science politics, was an associate director of the Stanford Linear Accelerator, a major scientific facility. During the previous budget cycle, Stanford had expected new-start funding for an addition to the accelerator, with money provided by ERDA.[72] OMB had deleted that funding, but the director of the accelerator and his staff, by appealing directly to Congress, had succeeded in having it reinstated at a low level for fiscal 1976. During their campaign, the physicists had kept their sponsoring agency informed unofficially and informally of everything done. That ensured that no toes in ERDA would be stepped on.

During a telephone conversation with Drell in which they discussed Stanford's lobbying activities, Bahcall fastened on the idea of adopting the maneuvers exploited by the physicists.[73] The first step was to secure approval from friendly managers in NASA to start lobbying. By the end of the day on which he talked to Drell, Bahcall had spoken twice with Noel Hinners at NASA Headquarters and assured him that the astronomers would consider an effort to get congressional restoration of funds only if they had his implied blessing. Hinners agreed. He had already supplied lists of people to contact, and he was in touch with representatives of the contractors to get their reactions to the telescope's omission from the budget.[74]

The contractors, too, had been dismayed that the Space Telescope had not been listed as a new start. They needed firm decisions soon so as to make the best use of the teams they had working on the Space Telescope. The business of the contractors was to make money, and if the telescope was not to go ahead, they would reassign their teams immediately. The contractors' message to Hinners was that a new start should be pursued in fiscal 1977.[75]

Bahcall polled ten leading optical astronomers on whether or not to push for reinstatement of funds at a low level for fiscal 1977. Seeking reinstatement would carry a decided element of risk. If it should fail and the lawmakers become antagonized, that might delay the Space Telescope for a number of years.[76]

But with Hinners's concurrence for a lobbying campaign, and the encouragement of a number of astronomers, Bahcall and the other leading astronomer-lobbyists decided to accept the risk and drive forward. They would appeal directly to Congress to restore funding for fiscal 1977. In so doing they would launch a campaign whose audacious goal, as they saw it, was nothing less than to outflank OMB and the top NASA management.

A COALITION AND AD-HOCRACY

There began a massive lobbying campaign, with articles planted in influential journals,[77] visits, telephone calls and letters to NASA managers, congressional staffers, congressmen, journalists, and anyone else who might help. Such movements derive their power not only from the quality of the arguments presented but also from the number of people involved and their determination. The participation of astronomers and physicists with no direct interest in the Space Telescope was seen as especially helpful by the astronomer-lobbyists because it demonstrated a broad base of support. Hence the invocation by the campaign's leaders that the volume of mail, telephone calls, and personal visits would be extremely important.

At no point, however, did the leading astronomer-lobbyists – Bahcall, George Field, Spitzer, and George Wallerstein of the University of Washington – act to form a committee to run the campaign. Although they circulated memoranda to each other, most of their links were by telephone. One of their most important tasks was to contact other astronomers and prompt them to join in the campaign. Once that had been done, and with a little coaching, the astronomers generally took the initiative themselves.[78]

The 1976 campaign was more of a grass-roots enterprise than the hastily arranged effort of 1974 had been. Although not formally constituted as an interest group, the astronomers were in many ways acting as one. "Of course," one writer points out, "Americans' love affair with interest groups is hardly a new phenomenon. From abolitionists to abortionists there has never been a lack of issue conscious organizations."[79] Nevertheless, by acting as they did, the astronomers' behavior mirrored changes within the wider political culture of the United States in the 1960s and 1970s, for it was during that period that there was an explosion in the number of groups lobbying in Washington, and a greater involvement in politics by the American middle class.[80] The astronomers had stumbled blindly into the political process, but with the aid of NASA and its contractors, their eyes had been opened quickly.

The interests of the industrial contractors and those of the astronomers had become merged, for both groups had the common short-term goal of securing funds for the Space Telescope. The contractors conducted their own lobbying, but also offered the astronomers background information on the important people to be contacted, as well as advice on their political preferences and the best ways to gain entry to their offices. Speaking in 1983, Bahcall recounted that

Lyman [Spitzer] and I met on several occasions in this office [at Princeton] with contractual people . . . and they were just enormously well informed, and they were using us in the same way we were using them. . . . The crucial thing for us was information, how do we get to talk to people, who are the people that are in the offices, what are their interests, how do we get into the offices, how do we approach them, what are the things that are going to sell to this Congressman or this Senator? The best information we had on that was from professional lobbyists for these companies. They were really good.[81]

Names of firms likely to receive Space Telescope contracts, arranged by state and size, were provided to Bahcall, for example. Those lists were then exploited to persuade members of Congress that their constituents would be well served by the construction of the Space Telescope. The contractors' lobbying efforts were also loosely coordinated at NASA Headquarters by T. Bland Norris, recently appointed head of the Astrophysics Division in the Office of Space Science at NASA Headquarters.

The funding problems had energized and brought into closer relationship the various elements of the coalition that favored the telescope. At its heart were astronomers pursuing professional interests and contractors pursuing economic interests. There was nevertheless significant NASA input and some measure of oversight, as well as firm support, indeed, at times, highly enthusiastic support, from a number of people in Congress, the White House, and the media.

What was the nature of the process by which this coalition sought to win approval for the telescope? One model of how policy is made is the leadership model. In that model, a leader in a powerful position in the political structure sets out to build a coalition for a project in which he or she believes. An example of the leadership model is provided by President Kennedy's decision to commit the United States to go to the moon. That model clearly is not applicable to the case of the Space Telescope.

Some political scientists have argued that a major feature of the American political process is the formation and operation of so-called iron triangles. Iron triangles are informal but enduring links among executive bureaus or sections of the federal bureaucracy (NASA, for example), congressional committees, and interest groups with a stake in a particular program. An iron triangle can exert powerful control over the link between politics and administration, between policy goals and their implementation. Other political scientists have referred to related concepts: subgovernments, policy subsystems, and policy "whirlpools." All these terms have been used to answer how decision making is segmented into different arenas and why policy supposedly becomes made by limited, predictable, and specialized sets of participants.[82]

These concepts, however, do not really fit the case of the Space Telescope once the policy-making had become controversial. For the telescope, the organization was much more fluid than that of an iron triangle or similar ideas. In the case of the telescope, people moved in and out of the selling process, and a coalition had to be carefully assembled; that required a structure perhaps better described by what political scientist Hugo Heclo has termed an "issue network," a loose grouping whose members share knowledge about some aspect of public policy. For Heclo, there is a shifting pattern of participants for different issues and for the same issue over time.[83] In the case of the Space Telescope, one might argue that those in the issue network, that is, those interested in the pursuit of space astronomy, had deliberately sought to draw in outsiders. As one aspect of that process, there had coalesced a group whose members shared the belief that the Space Telescope should be built, and they had started to lobby for it.

But a development of Heclo's notion, what has been termed an "ad hoc" model, best captures the policy-making on the Space Telescope. By ad hoc policy-making is meant a process in which a question such as whether or not to approve the Space Telescope is constantly being reframed, and the issues surrounding the decision reshuffled and repackaged. Certainly an appeal to ad-hocracy reflects better than does an issue network that the telescope designs and scientific objectives and the planned program to build the telescope were continually being revised, in part because of the demands of coalition-building and the need to extend and strengthen the coalition that favored the telescope.[84]

In keeping with ad-hocracy and the need to create and seize opportunities to further the telescope, the lobbying campaigns launched by the coalition of astronomers, contractors, and their supporters to get the telescope reinstated in the budget were just what Hinners had hoped for. He was eagerly searching for ways to strengthen that coalition. As we shall see in Chapter 6, Hinners's support for a "Space Telescope Institute" to run the telescope's operations in space derived in part from this motive. Hinners knew the institute was the scheme backed by ground-based astronomers, and he was eager to integrate them more firmly in the coalition of supporters. Also, when the telescope had been removed as a new start, Hinners, unknown to the astronomers, had gone even further and helped to delete all Space Telescope funding. As he recalls, "the money wouldn't have done much for us. So my strategy was to put nothing in, and get the outside world really antagonized. And it worked. Panic!"[85]

The use of an apparent cancellation of a project to strengthen its political position is not an uncommon maneuver by federal agencies in dealing with Congress and the executive branch, nor by Congress in its interactions with the executive branch. The maneuver even has a name: the Washington Monument game.[86] That name, so the story goes, derives from a time when the National Park Service's budget was to be cut. The Park Service's response was to close the elevator inside the Washington Monument on the mall in Washington, D.C., a highly visible demonstration that the Park Service believed it needed more money, and a tactic likely to win public sympathy. Hinners, playing a version of the Washington Monument game, proposed to remove all Space Telescope funds as a clear sign to astronomers and contractors that the program was in desperate trouble. They would then, Hinners hoped, rise up and vent their wrath. In this apparently contradictory manner, Hinners sought to rally support for the telescope, strengthen the coalition that supported it, and bolster the lobbying campaigns. Hinners's tactic had fanned what a note in the OMB files referred to as a "firestorm" among the astronomers, and that fire had been sparked by the omission of the Space Telescope as a new start.

PERSUADING FLETCHER

Some of the lobbying was aimed directly at Fletcher, whose motives were coming to be widely distrusted by the astronomers. As the astronomers were well aware, Fletcher's attitude toward the tele-

NASA administrator James Fletcher explains a model of the Space Shuttle to President Ford in the Oval Office of the White House in 1976. (Courtesy of Gerald R. Ford Library, photograph #B-1424-25.)

scope was crucial. Born in New Jersey in 1919, Fletcher had grad-uated from Columbia University in 1940 with a B.A. in physics. After wartime service, he had received a Ph.D. from Caltech in 1948 and then joined the Hughes Aircraft Company as director of the Theory and Analysis Laboratory in the company's Electronics Laboratory. After a highly distinguished career in the aerospace in-dustry, during which he had received patents in areas such as sonar devices and missile guidance systems and had been involved in es-tablishing and developing companies, he became president of the University of Utah in 1964. During his career he also served on numerous national technical and scientific committees, chairing sev-eral of them.

One member of the Science Working Group who knew Fletcher well from his work as chairman of NASA's influential Physical Sci-ences Committee was George Field, director of the Center for Astro-physics of the Harvard College Observatory and the Smithsonian Astrophysical Observatory. A brilliant and politically astute astron-omer, Field had close ties to both Bahcall and Spitzer. He had writ-ten his Ph.D. dissertation at Princeton under Spitzer, whom he had come to know extremely well. Field had even taken Spitzer as a model for his scientific career.[87] As a member of the Science Work-ing Group, it had been natural for Field to be at the heart of the lobbying activity.

Field's senior position as an astronomer and member of NASA's advisory apparatus meant that he could be fairly direct with Fletcher. Indeed, Field posed the central question when he wrote to the NASA administrator: Why have the astronomers reacted so strongly? "As-tronomers who are normally quite restrained have become passionate because they feel so frustrated." Field pointed tactfully to what he saw as failures in NASA's handling of recent events. "As things are now," he reported, "there are serious doubts among the senior as-tronomers as to NASA's true intentions." Field knew that Fletcher had offered a strong endorsement of the Space Telescope at an "In-ternational Discussion of Space Observatories" in January 1976, but that message had not been conveyed to the astronomers.[88] Field con-fided, "most of all, Jim, the astronomers need to meet with you face to face so that you can allay the fears which cripple effective action and so that your own interest and excitement about ST can be com-municated to all."[89]

Within a few more weeks, plans were made for a meeting on May 19, to be attended by Fletcher, several other NASA officials, and some twenty-nine leading astronomers, carefully chosen by Field and others to present a good mix of young and old, male and female, East Coast and West Coast, and so on. There were to be formal presentations on the benefits of astronomy to society and on the science and technology of the Space Telescope.[90] At that meeting,

Fletcher announced to the astronomers that he was determined to press ahead with a new start for the telescope in fiscal 1978. His audience was far from mollified. From their perspective, Fletcher appeared to show little grasp of the telescope program. Bahcall, for one, thought that he had been "astoundingly ignorant."[91] Even more alarming, in the opinion of the astronomers, was that Fletcher, during the questioning period, had revealed himself possibly to be willing to drop the Space Telescope program in favor of two more Space Shuttles, to be used by the military, but paid for by NASA.[92]

ON THE HILL

The meeting with Fletcher was held in May, but by that time a great deal had already happened on Capitol Hill. The astronomer-lobbyists had first sought a small amount of money to begin the Space Telescope program in fiscal 1977. They had met with success in the House Authorization Subcommittee, chaired by aerospace enthusiast Don Fuqua, a subcommittee that had always backed the telescope. In late February the subcommittee recommended a new start, with funding of $3 million reprogrammed from within the NASA budget.

The Senate Authorization Subcommittee disagreed. It committed itself only to stating that the Space Telescope should receive "highest priority" for a fiscal 1978 new start. Faced with a discrepancy between the Senate and House authorization bills, the astronomers' next step was to attempt to gain full authorization when the House and Senate resolved their differences. To do that, the astronomers had to find an advocate in the Senate. Bahcall and Spitzer turned, as they had in 1974, to Republican Charles Mathias of Maryland.

On March 31, Bahcall and Spitzer visited several staffers and two senators, one of whom was Mathias. They were accompanied by astrophysicist Richard C. Henry of The Johns Hopkins University in Baltimore, Maryland, to add "local color," a typically careful touch.[93] Bahcall had ready a draft of remarks for Mathias to read in the full Senate session of April 1. There Mathias requested that the Senate follow the House in providing the $3 million authorization from reprogrammed funds.[94]

Senator Charles Mathias.

A few days after the meeting with Mathias, Fletcher wrote to a highly placed OMB official:

It appears we had underestimated the commitment to the Space Telescope that had grown over the years among scientists, industrial organizations, and members of Congress. We have been inundated with severe criticism from virtually every academic institution associated with astronomy. We have been urged forcefully by the aerospace and optical contractors to do something to alleviate the high costs of their holding together effective engineering teams in order to be able to bid on the telescope project if and when it were authorized and funded.[95]

Fletcher, it will be remembered, had been sworn in as NASA administrator in April 1971, a difficult time for the space agency, as it sought to adjust to a loss of some of its political support and a budget much diminished from the heady days of the early 1960s and the race to the moon. Here it is perhaps worthwhile to refer to a memorandum sent to Fletcher in late 1971. Homer Newell, a highly placed and experienced NASA science manager, had described to Fletcher some recent problems the agency had had with the scientific community. In Newell's opinion, those problems had arisen in part because the agency had pushed projects before the scientific community had a full appreciation of what they involved. Newell had even suggested that when the time came to consider building the Space Telescope, the agency should wait until the pressure for it "is so great that we can hardly fail to accede to it." In his marginal comments, Fletcher had written "good point."[96]

Had Fletcher decided in mid-1976 that Newell's criterion had at last been met, that the support for the telescope had indeed become overwhelming? Such a view is supported by comments made by John Naugle at a Science Working Group meeting in February 1976. When the astronomers asked if NASA's top management was willing to fight for the telescope, he replied that in August 1975 Fletcher had had doubts about getting the telescope through Congress. However, more recently Fletcher had become "confident" of the telescope's chances of success because of the level of support in Congress after its omission from the fiscal 1977 budget.[97]

In any case, what Fletcher was proposing, and what the agency took as its line on Capitol Hill, was that the request for proposals for the Optical Telescope Assembly and the Support Systems Module be issued later in 1976, and that the competition for the positions of principal investigators to build the telescope's scientific instruments also be initiated in 1976. However, the decision on which contractors were to build the Space Telescope would not be made until after the budget cycle for fiscal 1978 had been completed. That action, which would not commit the Ford administration in advance of the budget process, had the advantage for Fletcher that it would help to dissolve some of the pressure on the agency. The momentum of the Space Telescope program would also be preserved and "will place us in a sound position to proceed with the project if it is approved during the coming [budget] cycle. Conversely, we feel that to take no steps until mid-1977 would result in serious program discontinuities and even some inequities."[98]

The House Appropriations Subcommittee was still the major stumbling block for the Space Telescope's supporters. By 1976, Ed-

ward P. Boland, the chairman, had swung firmly against the telescope because of its cost. A Democrat from Springfield, Massachusetts, Boland was a twenty-three-year veteran of the House and had
solid credentials as a supporter of federally funded research and development. He also had vast experience in the politics of science and
technology. Though his style was quiet, he was very powerful. At
times devastating and often witty in his probing questioning of witnesses in committee hearings, Boland's views carried weight in his
subcommittee, and he was to prove a formidable critic.

Congressman Edward P. Boland.

Although the chairmen of appropriations subcommittees rarely
make headlines, in the mid-1970s they were, despite changes in
Congress in the early 1970s, to a large extent the people who determined how much money would be spent on national defense, on
public works, on agriculture, and on countless other federal programs. They were, as one experienced committee member has described it, "lords with their fiefs and their duchys – each with power
over his own area of appropriations."[99] A chairman has particular
power because most members have to sit on a number of subcommittees. "As a result, it is very difficult for them to attend every
meeting of each or to keep up on all complex legislation before each
one. Thus they rely upon the chairman to concentrate on every session of his subcommittee and all aspects of the bills before it."[100]

Boland's reservations about the telescope had been highlighted
during NASA hearings in February 1976. Those hearings featured
the most trenchant exchanges on public record between opponents
and supporters of the Space Telescope. At one point, Boland asked
Hinners, who was known to be an unabashed advocate, "If we don't
have the Space Telescope, the world is not going to come to an end
anyway, is it?" When Hinners replied, "No sir," Boland continued:
"I think the problem that many members of this subcommittee have
with the Space Telescope is basically one of priorities. . . . The
Large Space Telescope is unquestionably a fascinating and interesting project, but it might get up to $500 million. You know we
could fund many different applications satellites that would have a
more immediate payout." Later in the hearing, Boland added, "You
know, someone said, what good is it going to do when you get it
out there? What does it do? What does it find? And somebody says
'more of the same.' Is that correct?" Hinners defended vigorously:
"No sir. Every time we've looked we find more of the unknown and
the new, and we are not going to find more of the same. If it were,
I would not waste my time with it. It would not be worth it. . . .
I fail to understand why the acquisition of basic knowledge and the
intellectual stimulation are not regarded as applications to the human endeavor." But, Boland complained, "[haven't] we spent
enough?"[101]

Some elaborate explanations for Boland's opposition were considered by the lobbyists, but it does seem that his attitude toward the Space Telescope was shaped by the points he made publicly, not by any congressional logrolling. Bahcall and the other astronomers came to regard Boland as "a very honest and able adversary, and no one questioned his sincerity, his motivation or his honesty . . . it came through very clearly because what he did was so much a matter of principle and not a matter of convenience for him . . . we deluged him as a focal point of our activities. But he stood firm. . . . He did what he thought was right despite external pressure."[102]

Boland's central point was that the telescope would be expensive, and large sums had already been spent on astronomy and space science. So, in a time when federal expenditures had to be reduced, why not delay its start? Indeed, when he dedicated a new radio astronomy observatory at Amherst in his home state of Massachusetts, he went out of his way to emphasize that he thought astronomy had not been shortchanged by Congress. Boland maintained that funding for the Space Telescope had to be weighed against the development of applications satellites. These, he thought, would have a swifter and broader impact on the quality of life on earth. According to one secondhand report well known to the astronomer-lobbyists, he went even further and asserted that the LST, if approved, would be approved over his opposition as long as he was in Congress. And those comments, remember, were made to a group of astronomers.[103]

Those sentiments were echoed by Richard N. Malow, an influential staff member for Boland's subcommittee on whom Boland relied heavily for advice. Malow was strongly opposed to the Space Telescope, and as George Low had discovered in April 1976, "he doesn't see such a telescope at all for quite a few years. He states that the Shuttle will fly in 1979 or 1980, and at that time we should be in a position to fly a large number of payloads which appeal to the Congress and to the general public, and that the telescope is not such a payload." Malow wanted more funding for applications satellites and other areas, "in fact," Low lamented, "in any area other than the space telescope."[104]

From the scientists' point of view, Boland appeared completely immovable. His subcommittee would not approve a new-start appropriation. When the House and Senate came together to iron out differences in their respective NASA appropriations, he remained insistent that NASA not be allowed to release any request for proposals to the contractors for building the telescope *before* a new start was authorized. The Senate Appropriations Subcommittee, too, opposed selecting contractors in that unorthodox manner. But again the lobbyists won a partial victory when a compromise was struck to the effect that the request for proposals could be released as soon as the fiscal 1978 budget was submitted by the president to Congress in January 1977. That did not mean that new-start approval would be granted automatically. A tough fight still lay ahead in fiscal 1978.

ANOTHER TRY

NASA's upper management regarded the fiscal 1978 budget as crucial for the agency. Writing in 1976, one well-placed commentator noted that in

January 1972, after several years of study, the development of the space shuttle was approved as the major technological focus of the space program for the 1970s. Recognition of the need to maintain a "balanced" program of space science and applications during the Shuttle development period led to an agreement in the Executive Branch to a "constant level" budget for NASA; that is, NASA would both plan and manage both shuttle development and other elements of the NASA program at the then current total NASA budget level measured in constant dollars, $3.4 billion in 1971 dollars. Since that time, however, "ratchets" have been applied repeatedly in the Executive Branch budget actions, requiring some slippage and disruption in shuttle development and constraining the amounts available for the balanced program. NASA estimates the FY 1977 budget is about $1 billion below the "constant level" of $3.4 billion in FY 1971 dollars.

The hunt for money, support, and a well-defined role in the post-Apollo era had led senior NASA managers in unexpected directions. In 1971, for example, there was some talk about securing for NASA government-wide responsibility for the application of technology to national needs. In 1974, administrator Fletcher and deputy administrator Low had even tested opinion on NASA reporting to the National Security Council in the White House, not the Domestic Council. As Low told Brent Scowcroft, Henry Kissinger's deputy on the National Security Council, he thought the country might be better off and have a better space program if NASA "were more closely associated with Defense and other security agencies as opposed to being associated with health, welfare, etc." Although that initiative did not take hold, NASA's most senior managers became increasingly more concerned that unless the agency's funds were increased substantially, it could not run a balanced space program.[105]

In June 1976, Fletcher wrote to President Ford that

as a matter of conscience and duty, I must inform you of the steady erosion of the United States space capabilities and of the dangers this poses. Over the past five years, we have not been permitted to maintain the program breadth of momentum necessary for continued contributions to national security, international policy, and technological progress. . . . In my view, we have reached a breaking point. . . . In blunt terms, if we cannot extend the scope of NASA's activity, the civil space program will be irreparably damaged.

Fletcher also explained NASA's five-year plan to Ford, one goal of which was an integrated approach to the scientific exploration of the universe. And the "most critical and immediate need in remote exploration . . . is the 2.4 meter Space Telescope, a permanent man-tended orbital facility that can quadruple the reach of man into the universe, can find the planets around nearby stars, can look back into time some 15 billion years, and can help decipher the now unexplained energy generating mechanisms of stellar systems and objects."[106]

On September 8, Ford met with Fletcher. The first item of business was the naming of the first Space Shuttle orbiter to be the *Enterprise* (in response to thousands of calls from "Star Trek" fans),[107] but Fletcher then had the chance to press his concerns about the future of the civilian space program. To a degree, Fletcher was preaching to the choir. OMB judged that "Dr. Fletcher's letter and paper include a fair amount of 'eyewash' and a number of dubious assumptions." Nevertheless, OMB agreed that certainly there was something to Fletcher's claims, and OMB was in sympathy with the idea of increasing NASA's overall funding above the level of inflation.[108] There was at that time a strong feeling in the executive branch that the NASA budget should indeed be raised and that several new programs should be started.

It is worth emphasizing that at the time that Fletcher met with

The B.C. cartoon much used by advocates of the Space Telescope in the mid-1970s. (Courtesy of Johnny Hart and Creators Syndicate, Inc.)

Ford, major development problems on the Space Shuttle had not become apparent; thus, it was a time when NASA was publicly very optimistic about what the shuttle and its claimed regular and cheap access to low earth orbit would mean for space activities. OMB, usually seen as a nay-sayer, was examining ways to cash in on the investment that the United States had made in the shuttle. One way to do so would be to design spacecraft to exploit the shuttle's capabilities. The Space Telescope was an ideal candidate.

In contrast to the previous year, in 1976 the Space Telescope sailed through its OMB reviews. There are even stories of meetings – unusual, certainly, by OMB standards – that were brought to a halt as staff members pondered what it meant to peer back many millions of years in time with the Space Telescope. At the OMB review of the NASA budget on October 23, 1976, the first item that OMB director James T. Lynn saw was a cartoon, placed there by Rondel Konkel, an OMB staffer and telescope advocate. It shows two cavemen sitting on a rock admiring the heavens. One caveman tells his companion that the secrets of creation could be mankind's, except that "the key is not ours to have." When asked "How come?" he replies, "It ain't in the budget." The joke was not lost on Lynn.

Copyright 1981 by Herblock in the *Washington Post*.

The telescope had even become something of a favorite for both Lynn and President Ford. Ford had raised no objections when it had been deleted from the budget the previous year, but after a meeting among President Ford, OMB director James T. Lynn, and presidential science advisor H. Guyford Stever, at which the discussion had turned to the Space Telescope, Stever jokingly complained that because Lynn and Ford had enthused so over the telescope, he could not get a word in edgeways![109]

Another bonus for the Space Telescope was that it meshed extremely well with an initiative launched during Ford's presidency to increase spending on research and development in science and technology. Although Ford was perhaps the most conservative president since Calvin Coolidge, government funding for science met with more sympathy from Gerald Ford than from Richard Nixon.[110] Ford had served on the House Select Committee on Astronautics and Space Exploration, and during his presidency he took an active interest in many science policy issues. In 1976 he had reestablished the Office of Science and Technology Policy in the Executive Office of the President, an office that Nixon had abolished. James Lynn was not only Ford's OMB director but also a close advisor, and Lynn was a strong supporter of federal funding for science and technology. As Ford explained in his fiscal 1978 budget message, "in spite of the financial pressure on the Federal Budget . . . I am again proposing real growth for basic research and development programs this year because I am convinced that we must maintain our world leadership in science and technology in order to increase our national produc-

OMB director James T. Lynn seated next to President Ford before a budget meeting in July 1976. Under Lynn, one of Ford's closest advisors, OMB moved to increase federal support for basic science. (Courtesy of Gerald R. Ford Library, photograph #B-0800-11.)

tivity and attain the better life we want for our people and the rest of the world."[111]

Indeed, it was because of a policy initiative originating in the Ford administration's OMB that a slight decline in federal research and development spending that had begun in the late 1960s was reversed, and the curve of funding resumed the upward climb that has characterized it since World War II.[112] Although the funding for basic research had also been increased in fiscal 1977, the major thrust had been to benefit the National Science Foundation, whose budget had risen by some 19.5 percent (or around 12 percent in real terms). The NASA space science budget had actually fallen that year, and that decrease, as we saw earlier, had included the Space Telescope being omitted from the president's fiscal 1977 budget.

A number of important developments during the Ford presidency

had combined to create a favorable climate for initiation of the telescope program: the boost in federal funding for science and technology, the high-level support for the telescope in the executive branch,
and the opportunity the executive branch saw in the telescope to
exploit the Space Shuttle. If the telescope had reached a comparable
stage in its selling campaign two years earlier or two years later,
these factors would not have been in play to the same degree, and
the telescope's chances of success would surely have been diminished. In fiscal 1978, with the firm backing of the White House,
there were no late hitches.

Despite its support for the telescope, OMB had nevertheless sought
one important change in NASA's preferred program. OMB had
stressed that the agency should seek to maximize international participation in, and financial contributions to, the Space Telescope. In
the light of the earlier congressional criticisms and demands to seek
foreign partners, this was a central point.

By late 1976, NASA and the European Space Agency (ESA) had
gone far toward reaching an agreement on a cooperative project. The
deal was that ESA would pay about 15 percent of the costs. ESA
would provide one scientific instrument (the Faint Object Camera),
provide the solar arrays to power the Space Telescope, and recruit
some staff members for the Space Telescope Science Institute (as
detailed in the next chapter). Even so, NASA's budget submission
did not meet the OMB's hopes. OMB instead proposed that unless
NASA objected strongly on technical grounds, the appropriate national policy should be to encourage maximum international cooperation on such large-scale scientific endeavors as the Space Telescope. NASA objected. From the agency's viewpoint, a larger
European share of the program would result in a more complex management task because of the larger number of interfaces, in addition
to increasing the problems inherent in the transfer of U.S.-devel-
oped technology to the Europeans. Hence, as it had done with Congress, NASA argued against going above the 15 percent level.
Moreover, it would be difficult for ESA to provide more than that
amount, because to do so would require special permission from the
various member states, and one such specially organized program
was already under way: Spacelab.[113]

With the Space Telescope set as a joint NASA/ESA project, and
the European contribution fixed at around 15 percent, the battle-
ground shifted. Once more the focus became the House Appropriations Subcommittee.

BACK TO CONGRESS

Given the adamant opposition of Congressman Boland and Malow,
one of the chief goals of the Space Telescope advocates was to win

support from the other members of the House subcommittee. The task was not as daunting as it might have been, for George Shipley, a robust critic of the Space Telescope, was no longer a member. Also, what for the astronomers, at least, had begun in 1974 as a rough-and-ready lobbying operation had become far better coordinated and more professional, for the astronomer-lobbyists had rounded up some energetic, skilled, and experienced public leaders to side with the contractors and NASA. Bahcall, for example, was now so well informed that he could issue progress reports, with suggestions for action, together with an assessment of which way any particular congressman was likely to vote.[114] Other members of the astronomy community were also far better equipped to participate in the political struggle after their experiences of 1974 and 1976. And participate they did.

At one House Appropriations Subcommittee hearing, Congressman Bob Traxler of Michigan was moved to comment that many "of us have had our rugs and carpets worn out by astronomers interested in the project."[115] When George Field presented testimony to the subcommittee two days later, Traxler admitted that he "didn't know there were so many astronomers and astrophysicists in the United States, frankly, until the last six months. You will be pleased to know they are verbal. They all have typewriters, also. And they can find their way to the Post Office."[116]

In addition to their regular, ordered lobbying activities, the astronomer-lobbyists seized opportunities to push the telescope issue whenever they could, particularly with Boland's subcommittee. The great prize, of course, would be to influence Boland himself, and so he was subjected to all sorts of tactics. Marshall arranged for Democratic Congressman Ronnie G. Flippo – whose district included the Marshall Space Flight Center – to appeal to Boland. In a letter drafted for him at Marshall, Flippo emphasized the benefits of basic science for economic developments on earth. He also played the national security card: "Clearly the Space Telescope and the NASA mission do not directly affect national security; however, the development here of a technology for precise imaging telescopes in space has obvious applications for national security."[117]

There was also talk of a visit to Boland from a union official to spotlight the jobs to be secured in building the telescope. Another supporter of the telescope who had influence was William L. Putnam. Putnam was chairman of the board of Springfield Television Broadcasting Corporation, owner of WWLP, channel 22, in Springfield, Massachusetts, Boland's home district. As Putnam wrote to Boland, his longtime friend, "I have some friends in the astronomical business, a relationship that goes back many years through my father and his uncle Percy." "Uncle Percy," in fact, was Percival Lowell, famous proponent of the theory of canals on Mars and founder of Lowell Observatory in Arizona, the longtime trustee of which was Putnam's brother. William Putnam had been approached the previous year by Lowell astronomer William A. Baum to intercede with Boland, and after he had been persuaded by Baum of the telescope's worth, he had done so. So in 1977 he again pressed Boland to receive a delegation of astronomers and hear their arguments.[118]

The attempts to sway Boland sometimes took unexpected turns. George Field told Bahcall of how, in the Delta Airlines waiting area of Washington's National Airport, he had run into the president of Massachusetts Institute of Technology and a former presidential science advisor, Jerome Wiesner. Wiesner was speaking with Thomas P. ("Tip") O'Neill, speaker of the House of Representatives, and Field was introduced to him. O'Neill had shared an apartment in Washington with Boland for some twenty-four years and was exceptionally close to him both personally and politically. The conversation turned to the Space Telescope. When Field explained that Massachusetts might benefit by as much as $50 million in contracts, that seemed to please O'Neill, whose own district was, of course, in Massachusetts. O'Neill ended the talk with words to the effect that he would "discuss it with Eddie," who was shortly to join O'Neill and his wife for a vacation to Cape Cod. The sheer good luck of that chance encounter spurred Field to close his letter by urging Bahcall to "fly Delta!"[119]

George Field, one of the leading figures in the lobbying campaign of the astronomers. (Courtesy of George Field.)

NASA, too, had stepped up its efforts to inform the lawmakers about the Space Telescope. A special booklet, *Why Space Telescope? The Eye of Space Ship Earth,*[120] had been prepared. Hinners worked actively with the astronomers, and the director of the astrophysics division at NASA Headquarters, T. Bland Norris, played a similar role with the contractors.[121]

The Office of Science and Technology Policy, next headed by President Jimmy Carter's science advisor, Frank Press, also strongly backed the Space Telescope. Press had known Noel Hinners at Caltech and was quite familiar with NASA, as he had served on NASA's Lunar and Planetary Missions Board. During congressional hearings to confirm Press as presidential science advisor, the Space Telescope came up in the questioning. In fact, Bahcall had primed the subcommittee chairman to ask about the telescope. Bahcall, acting on a suggestion from Caltech astronomer Jesse Greenstein, had sent Press a one-page fact sheet on the Space Telescope. Thus, a triumph for Bahcall's lobbying: He had in effect helped orchestrate both the congressional questions *and* the answers on the Space Telescope.[122]

Press saw the Space Telescope as an integral part of President Carter's science policy. He thus used his office to pressure Democratic members of Boland's subcommittee and request Tip O'Neill to intervene with his old roommate, Edward Boland.[123]

A memorandum by Press in April 1977 gives us one view of what the executive branch thought it was buying by funding the development of the Space Telescope. The points made there mirrored closely some of those being expounded publicly by the astronomer-lobbyists and sympathetic journalists: (1) The telescope would represent a major scientific advance. (2) The telescope would help maintain U.S. preeminence in astronomy: "The U.S. is in a position to capitalize on the technology we gained in the space program to take this step into space-based astronomy, and it would ensure our leadership for years to come." (3) The telescope would demonstrate the utilization of the Space Shuttle. It made no sense, Press repeated, to develop the shuttle and then not support the projects that were planned to capitalize on it.[124]

Although, as noted earlier, all sorts of maneuvers were employed to try to win Boland's support, it is not clear that he was ever persuaded. But even if Boland did not favor the telescope, it appeared that the lobbying probably had garnered enough support to win out, particularly as Bahcall had secured the articulate and spirited approval of another Democrat on Boland's subcommittee, Lindy Boggs from Louisiana.

It was natural for Bahcall to approach Boggs, because Bahcall had been born and raised in Louisiana. He found that Boggs and her congressional staff had "a real respect for science, and were very proud of Louisiana scientists."[125] A daughter of Congresswoman Boggs, it so happened, lived in Princeton – and, indeed, was later to become its mayor – and had been able to reinforce the initial contact between her mother and Bahcall, who, of course, was based at Princeton's Institute for Advanced Study. Boggs was new to the subcommittee, but she put special effort into convincing her colleagues of the telescope's worth.[126]

Congresswoman Lindy Boggs.

SPACE TELESCOPE AND *GALILEO*

Despite all the points they had raised in favor of funding the Space Telescope, there was a nagging worry for its proponents. This anxiety arose from the fact that the Space Telescope was appearing for a new start in the fiscal 1978 budget, along with the Jupiter Orbiter Probe (JOP), an expensive major mission to Jupiter. It was planned that the Jupiter Orbiter Probe would be managed by the Jet Propulsion Laboratory (JPL), the leading institution in the United States for planetary science. JPL is a NASA center, but unlike the other centers, it is not a civil service establishment. The facilities are government-owned, but the staff members are all employees of the California Institute of Technology. Whereas the directors of other NASA centers felt that it was not their privilege to walk into the OMB or onto Capitol Hill and lobby directly and openly for their programs, the director and staff of JPL had much more room to maneuver.

Because the Jupiter Orbiter Probe was a JPL project, there was much unease at NASA Headquarters that JPL and the planetary scientists might fight so hard for the Jupiter Orbiter Probe that they would knock out the Space Telescope in the process, or the telescope's advocates might wreck the Jupiter Orbiter Probe's chances of funding.

In 1975 there had been an example of the kind of damage that could be done if one group started promoting its own project at the expense of another. In 1975, some Space Telescope advocates had criticized the NASA space science budget as unbalanced. They protested that a disproportionate number of NASA dollars had been spent on planetary science missions, whereas relatively little had been allocated for study of the stars and galaxies, the principal mission of the Space Telescope. Those remarks had been used by Boland's subcommittee to justify cutting $48 million from a planetary mission, the Pioneer-Venus project, pending a trade-off involving the Space Telescope and Pioneer-Venus the following year. [127] As a report at the time in *Science* magazine put it, "public internecine warfare over budgets among scientists of different disciplines is not unheard of, but it has been relatively rare in recent years, especially among basic researchers." [128] Another such outbreak might kill off the Space Telescope or the Jupiter Orbiter Probe.

A 1975 NASA report had pointed out that the Space Telescope could perform planetary science from earth orbit, but was "in no way" a substitute for a program of planetary flybys, the method planetary scientists strongly preferred for pursuing their research. [129] NASA Headquarters had nevertheless pressed to make the Space Telescope more attractive to planetary scientists, often in the face of opposition from astronomers, who saw it very much as a tool for the study of stars and galaxies. The most public expression of this attitude came in an after-dinner speech George Field delivered to a meeting of the Division for Planetary Sciences. As recalled by William Baum, chairman of the Division for Planetary Sciences and an astronomer with a foot in both the planetary science and stellar and galactic astronomy camps, Field did not grasp the sensitivity among planetary scientists to their lack of new space flight missions and "came across, in effect, saying 'You've had yours, now it's our turn.' " [130] That did not go over at all well with the planetary scientists and required a great deal of fence-mending, particularly because in 1976 there had been, as Baum told Nancy Roman in June of that year, "agitation over NASA's reduction of funds for planetary science." [131] Thus, in mid-1976, Roman, who had long considered it important to make the telescope useful to planetary scientists in order to make it more feasible politically, had established a small group to examine how appropriate the proposed scientific instruments for the Space Telescope were for planetary researches. [132] Roman and others in NASA were well aware that if planetary scientists should support the telescope, that would broaden and strengthen the coalition in its favor. The move being discussed, then, was to shift the scientific objectives of the telescope to make it more attractive to planetary scientists, in part in order to aid its chances of being funded. Nor were these activities designed simply to revise the rhetoric about the telescope's scientific potential. NASA, as we shall see in Chapter 7, was also shifting its ground on some hardware choices so as to make the telescope's main camera more appealing to planetary scientists.

NASA Headquarters also worked with the supporters of the telescope and the Jupiter Orbiter Probe to prevent an outbreak of verbal warfare. Noel Hinners told JPL

not to get any activity going anywhere that was aimed at shooting down the Space Telescope, and then working with the planetary scientists themselves, the leaders, the lobbyists, to really get them to support actively their program, but don't attack the other one, or you'll start fratricide. And on the other side, working with the astronomy group, John Bahcall, George Field, Lyman Spitzer, and the industry, telling them the same thing, to not take aim at the planetary program. And pretty well worked that way.[133]

As one of the leaders of the lobbying effort for the Jupiter Orbiter Probe — which was to be renamed *Galileo* — recalls,

there was an awareness all along in this effort that one should be very careful about questions, which certainly did materialize from both the press and from representatives of Congress, about trying to prioritize. You know, do you want Space Telescope or Galileo? That was the danger that we faced, and questions that seemed to raise that spectre would occasionally come down, and so it was a very clear point that in no way was that to be regarded as a legitimate question. We needed both of them. They addressed different questions and were different kinds of instruments . . . we were fully cognizant of the ongoing ST effort, not merely neutral, but supportive. This is a project that's going to be useful to planetary [science] as well, but [it's] addressing astrophysical problems that are very important.[134]

In fact, Boland's subcommittee supported the Space Telescope and directed its fire at *Galileo*. The leading Space Telescope advocates rallied around the *Galileo* mission. George Field, chairman of NASA's Physical Sciences Committee and long an advocate of the Space Telescope, organized a letter of support for *Galileo* from all the members of his committee. A lobbying campaign was mounted to save *Galileo,* and after a hard floor fight in the House in which Boland's subcommittee's recommendations were decisively and dramatically rejected, the *Galileo* mission was reinstated in the NASA budget.[135]

The last troubling moment for the telescope's promoters came

when William Proxmire, chairman of the Senate Appropriations
Subcommittee, tried to remove all the telescope's funding. If that
vote had been carried, then the fate of the Space Telescope would
have rested on the outcome of the resolution of the House–Senate
differences, probably in a conference between the two bodies. But
Proxmire had little influence in his subcommittee, and he was not
seen by NASA or the lobbyists as a real threat. That proved to be
accurate when he was voted down 4 to 2 by his subcommittee.

Following the congressional approval, NASA and ESA signed a
"memorandum of understanding" detailing their collaboration and
establishing the Space Telescope as a joint NASA/ESA program.[136]

AFTERTHOUGHTS

For most of the preceding three years it had appeared to many of the
project's participants that the Space Telescope was clinging to life
by a very thin thread. For some of the participants, in fact, the fiscal
1978 campaign was a bit anticlimactic after the earlier life-and-
death struggles. That the telescope had survived was largely because
of the solid backing of OMB and a number of legislators, as well as
the political skills and commitment of the astronomers, the contractors, and some NASA managers in crucial positions. After 1974,
when the agency had grossly underestimated the amount of work
necessary to sell the telescope, even though it was included in the
president's budget only for Phase B funds, the Office of Space Science in NASA had adroitly aided the astronomers and contractors in
nursing the Space Telescope to a new start. That goal was reached
despite sometimes vigorous opposition in Congress, as well as doubts
and sometimes a lack of resolute support at high levels within the
agency itself. Indeed, some of those involved in the funding battles
would look back on the roles they played in those battles as among
their major achievements on the Space Telescope program.[137] Reflecting in 1983 on the struggles to initiate the project, John Bahcall noted that before the lobbying for the Space Telescope had begun, optical astronomers did not act as a group, but "if we had
criticized the actions in our laboratories but not taken external leadership, we also wouldn't have had a Space Telescope."[138] The coalition that backed the telescope, in which the astronomers played a
central part, had indeed been crucial in winning its approval.

In the course of the maneuvers and negotiations to win political
approval, the Space Telescope, and the program to build it, had
been much changed. Most important, the Large Space Telescope had
become the Space Telescope, the Space Shuttle and the telescope had
been packaged together even more closely by NASA, the aperture
had been reduced from 3 meters to 2.4 meters, the number of scientific instruments had been reduced, an international partner (ESA)

had been added, the drive to lower costs had led NASA to alter the engineering approach for the design and development stage, to help attract ground-based astronomers the agency had looked favorably on a Space Telescope Science Institute, and the desire of NASA managers to extend the telescope's base of support among other reasons led to the attempt to accommodate the interests of planetary scientists in the telescope's scientific objectives and the design of the instruments.

The arguments and debates with Congress and the executive branch had been hugely frustrating and often downright confusing to the project members. The construction of the 2.4-meter Space Telescope should, everyone agreed, be a considerably easier task than construction of a 3-meter Large Space Telescope, and it was only because of the interactions with Congress that really serious thought had been given to the cost/performance trade-offs for different aperture sizes. In that process, a telescope program had been shaped that was politically feasible, but a price had been paid. The price was that the telescope program had been both oversold and underfunded. As we shall see in later chapters, the course of the telescope's design and development has to be interpreted to a large degree in terms of a program trapped by its own history.

6

Making an institute

So to me, an institute could solve two problems; one, pacify, if you will, the ground based astronomy community, so that they'd be all the more supportive of the Space Telescope, and two, really provide an external advocate for a good operations program.

Noel Hinners

Two of the central themes in the history of post–World War II science in the United States are, first, the remarkable growth in government support for the scientific enterprise and, second, the desire of scientists to retain as much authority as possible over the direction and management of research. Certainly, underlying many scientists' activities is the tenet that scientists are the people best qualified to manage science. Not surprisingly, the politicians and bureaucrats who have dispensed to scientists such large amounts of taxpayers' money have also often wanted to ensure strong measures for oversight of their investments. There have thus been tensions, sometimes barely noticeable and, as we saw in the last two chapters, on occasion highly charged indeed, between the scientists and their patrons.

Space science is one kind of research that has flourished because of the availability of government support, most important being support from NASA and the military. Although the union between NASA and scientists undoubtedly has been immensely fruitful, it has also been characterized at times by the second theme, the struggle for control over the direction of scientific research and the management of that research. To Homer Newell, a NASA science manager with fifteen years of experience in senior positions in the agency, this is "best described as a love–hate relationship."[1] The new source of support handed space scientists new opportunities for research, but it also constrained them in various ways, ways they often disliked. However, the key point about this relationship between NASA and space scientists, the battles and skirmishes notwithstanding, is that a partnership was forged.

In many respects, NASA's interactions with astronomers over the scientific operations of the Space Telescope match the agency's broader ties to space scientists. Here we should remember that the Space Telescope's task is to enable astronomers to conduct research. The telescope of itself does not do astronomy; rather, it is the Space Telescope/people *system* that does, for it is people who decide which scientific observations should be made, in what manner, and by whom. There is therefore a crucial link, in terms of scientific operations, between the telescope and its ultimate users, the astronomers. How, then, to fashion and manage that link?

For some of the influential astronomers concerned with the Space Telescope it became important that they, not NASA, control the telescope's scientific operations. Such a concept was not readily accepted by the space agency, and it was embedded in the set of NASA's bureaucratic aims only with great difficulty, as we shall see in this chapter. But in order to better understand the agency's responses, we shall first backtrack to NASA's early years and examine the changing pattern of the agency's relationships with its scientist clients, particularly space astronomers. These relationships shaped fundamentally the solution both the agency and the astronomers eventually accepted for their link to the telescope, a solution that also has to be interpreted in terms of coalition-building and the need to strengthen the coalition that backed the telescope. In earlier chapters we have seen astronomers acting as advocates for a new scientific tool, the Space Telescope. Here we shall see them acting as advocates for what was, for NASA, a new kind of institution: the "Space Telescope Science Institute."

EARLY PLANS AND CLASHES

The National Academy of Sciences was founded by an act of Congress in 1863 to provide scientific advice to government agencies. Ineffectual for much of its life, its prestige rose enormously after World War II because of the way scientists had demonstrated their central relevance to national defense.[2]

Shortly after the launch of *Sputnik* in 1957, the academy decided to make its voice heard in the shaping of space policy by creating the "Space Science Board," a panel of leading scientists. The board would provide recommendations and advice on the nation's program of space science. To begin with, some of the board's members even hoped to position themselves as a kind of board of trustees to oversee NASA and to set its goals. Matters would prove to be much more complicated.

On its inception in 1958, the Space Science Board invited proposals and suggestions from U.S. scientists for projects to succeed the program of satellites for the International Geophysical Year (see Chapter 1). After reviewing about two hundred replies, the board sent its recommendations to NASA (at that time, just about to take form as an agency). Although these were incorporated into program planning,[3] NASA managers balked at the implication that the scientist members of the Space Science Board would dictate to NASA what missions and experiments to fly.

As Homer Newell has emphasized, "NASA's position was that operational responsibilities placed by law upon the agency could not be turned over to some other agency. Moreover, decisions concerning the space science program could not be made on purely scientific grounds. There were other factors to consider, such as funding, manpower, facilities, spacecraft, launch vehicles, and even the salability of projects in the existing climate at the White House and on Capitol Hill – factors that only NASA could properly assess." In contrast, the scientists pressed for NASA to run its scientific programs in the same way that programs were run by the National Science Foundation and those government agencies responsible for high-energy physics accelerators. In the development and operation of accelerators, as well as large ground-based telescopes, the scientists had been given significant measures of management responsibility. NASA, however, would be loath to follow the same path.

The evolution of government funding for science in the United States has in fact produced two distinct institutional approaches to federal support. The first kind might be characterized as the "direct" approach, the approach adopted, for example, by the National Science Foundation. The second might be termed "indirect," as it entails funding derived largely from the military or through mission agencies. Each approach involves a distinct set of administrative and political realities, and however much space scientists pressed and cajoled NASA to follow the direct approach, the agency would not do so.

Certainly, Newell and other NASA science managers refused to take their marching orders from the Space Science Board; they adopted the position that the space agency would be intimately involved in the science activities it would fund. NASA also maintained that if it was to deal properly with non-NASA scientists, the agency would have to have its own body of scientists.

The fight between NASA and the Space Science Board was sharp, but short-lived. NASA emerged from the tussle in firm control of space science.[4] The clout of the National Academy of Sciences could nevertheless not be dismissed, and the Space Science Board would come to assume a powerful position, with its approval virtually essential if any proposed space science mission was to win funding. To confuse the issue, Congress had not made it clear exactly what

NASA's responsibilities were for space science, and NASA's view prevailed only after a spirited contest with the Space Science Board, as well as skirmishes with university scientists, other government agencies, the military, and industry.[5]

Despite the element of competition with the scientific community, NASA tried to involve senior scientists in its programs. The agency believed that policy not only would ensure the highest-quality science possible but also would help to defuse opposition to NASA's plans.

Many scientists resisted. Some, as mentioned in Chapter 1, had been shocked at the cost of the Apollo program to land men on the moon and so were not enthusiastic about NASA. They believed that the money would be much better used to support research nearer to home. For some ground-based astronomers, the thought of the millions of dollars spent on one space shot was galling, given, for example, that the Palomar Mountain 200-inch telescope had cost a fraction of the price of a single manned space flight and had helped to produce high-quality science for well over a decade.

Even those astronomers who were eager to work with NASA and did become active in the space program found that they had to adjust their usual methods of doing business. Optical astronomers typically have placed a premium on working as individuals or with a few colleagues. In conducting astronomy from satellites, that small scale of enterprise would no longer be the case. Instead, the emphasis in satellite astronomy would be on Big Science methods and interdisciplinary teams of scientists and their associated armies of engineers and technicians. Astronomers also had been accustomed to managing their own endeavors. But given NASA's manner of working, the final authority for decision making would rest in the hands of NASA managers, often to the disappointment of the scientists involved. The pursuit of astronomy from satellites was thus to place new demands on astronomers, not only in terms of the reliability of their instruments and their methods of work but also how that work was to be directed and controlled.

The principal investigators on the Orbiting Astronomical Observatories program, for example, were at first surprised at how the program was to be managed. Even Lyman Spitzer, who had participated in a variety of government-funded programs, was taken aback. In March 1960, he wrote to Nancy Roman at NASA Headquarters to complain.

Spitzer believed that a complex scientific system would be most effectively designed if overall responsibility were given to one organization, and preferably one individual. For the earlier and simpler scientific payloads it had been feasible for the scientists to prepare "little black boxes," the engineers to design the satellite, and the NASA project manager to then coordinate the boxes with the satellite. For sophisticated satellites, such as the Orbiting Astronomical Observatories, Spitzer thought it was no longer efficient to separate the scientists from the engineers. Instead, it would be better

> to have the scientific experiment and the engineering design integrated under one individual (or institution). Then the scientific requirements can be altered to meet the detailed engineering difficulties as they arise, and the engineering plans can be guided to yield the maximum scientific usefulness. . . . It is obvious, I believe, that the overall responsibility to which I refer must be given to someone who is intimately familiar with the objectives of the system, and with the possible uses of the scientific equipment. Hence this responsibility should be given not to an engineer but to an astronomer, preferably a senior astronomer who has had substantial experience in the coordination of large engineering programs.[6]

In June 1960, NASA completed a detailed plan for the organization of the Orbiting Astronomical Observatory program. The agency had responded to some of Spitzer's points by offering consultations on the major technical problems of the spacecraft. But NASA still refused the experimenters official responsibility for the overall aspects of the observatories. As far as NASA was concerned, to have done otherwise would have meant a serious loss of control over the Orbiting Astronomical Observatory program, and it was the agency that was responsible to Congress, not the astronomers. But the issue of control would continue to be central for space astronomers in their dealings with NASA for the reason that it was so basic. What, after all, did it mean to be a space astronomer, and what did it mean to pursue space astronomy? Was an individual space astronomer to be able to determine the direction of his or her research? Where were the boundaries of authority to be drawn between NASA and a space astronomer? These questions, we shall see, were to be fundamental for those astronomers advocating a Space Telescope.[7]

SEEKING ADVICE

Although NASA wanted to maintain firm control over individual projects, it also sought to draw in advice on future programs from the scientific community in a more intimate and frequent fashion than could be achieved with the Space Science Board. One result was the creation in 1959 of a committee and associated subcommittees composed of scientists drawn largely from outside the agency, but reporting directly to NASA.[8]

Within a few years, however, the scientific advisors had become unhappy at their lack of influence. In contrast, many not on the advisory committees believed that those groups exerted too much influence.[9] There were thus growing tensions between NASA and the scientists concerning several factors: the place and scope of the space science program (particularly given the vast amount of money going to Apollo, ultimately to cost $26 billion), dissatisfaction among the NASA advisors with their role and lack of clout, and the increasing NASA involvement with universities, inspired by NASA administrator James Webb's vision of a university-industry-government complex.

Following a major meeting at Woods Hole in 1965 on the future of space science (see Chapter 1), Webb was prompted to ask a veteran of science politics, Professor Norman F. Ramsey of Harvard University, to assemble and chair a committee. Webb wanted the committee to advise NASA on the conduct of certain major new projects NASA had under way or under serious consideration. In the past, Webb wrote, scientists had been engaged in the space science program by conceiving, designing, and building their own experiments. The experiments were then integrated into a spacecraft and flown. But the new projects would be of much larger scale and would require new methods. "We in NASA," Webb told Ramsey, "think it is essential that competent scientists at academic institutions participate fully in the next generation of space projects, and we believe that we will need new policies and procedures in order to enable them to participate." In other words, how was NASA to enlist more of the scientific community in the execution of the large projects, such as the 120-inch Large Space Telescope, recommended by the 1965 Woods Hole study? "We have," Webb noted, "a nucleus of competent astronomers and engineers at the Goddard Space Flight Center; however, we expect that nucleus will have to be strengthened."[10] How, then, to attract astronomers to NASA and its programs?

Ramsey's group included prominent biologists, physicists, and geologists, as well as astronomers. They met for a total of eleven days during 1966 and heard more than twenty-five guest consultants and speakers. Their final report was sent to the NASA administrator in August of that year.

Although that committee indeed discussed a large orbiting telescope, they chose to place it in the context of several other telescopes for observing different wavelengths.[11] The Ramsey committee's central point was that they wanted strong scientific oversight of the orbiting telescopes project. To achieve that end, the committee proposed that an organization be formed of universities active or interested in the fields of optical, radio, x-ray, and gamma-ray astronomy. Named STAR, Space Telescopes for Astronomical Research Inc., it would develop preliminary specifications for the orbiting telescopes. These would be sent to NASA Headquarters, which would in turn direct the appropriate NASA field centers to prepare alternative and competing designs and engineering approaches to meet the specifications. STAR would then help NASA and the industrial contractors to evaluate the designs and conduct trade-off studies (Phase

A and B studies in the terminology used earlier). The scientific program and the scientific operations of the telescope would also be under the direction of STAR, whereas operational control of the space telescopes would be the charge of the NASA centers. Scientists would run the scientific operations.[12]

That the Ramsey committee should argue that the scientific community should, via STAR, have scientific direction and control of the orbiting telescopes project was in line with one of the key developments in postwar American science: the emergence of groups of universities as advocates for and managers of new and expensive research facilities. By the mid-1960s, the use of university consortia to manage large-scale civilian scientific research had become the accepted way of transforming the concept of "national" laboratories or facilities (national in the sense of being open to all qualified scientists) into working institutions.

Associated Universities Incorporated (AUI) was the first such university consortium. At first a grouping of nine universities, AUI had been brought into being in 1946, at the initiative and largely under the direction of nuclear physicists, to manage what became the Brookhaven National Laboratory for high-energy experimental physics.[13]

The National Science Foundation had then contracted in 1956 with AUI to build and operate the National Radio Astronomy Observatory, a process that had alerted astronomers to the value of creating a university consortium.[14] The Kitt Peak National Observatory, an observatory for optical astronomy composed of telescopes, workshops, and laboratories, was founded shortly afterward. It was the responsibility of the Association of Universities for Research in Astronomy (AURA), a newly created consortium.

By the mid-1960s, astronomers were therefore no strangers to the concept and use of national facilities. Optical astronomers were thoroughly used to working with Kitt Peak and often looked to it as a model of how they should conduct their affairs. Kitt Peak, moreover, was funded by the National Science Foundation. In the words of the foundation's most recent historian, its establishment "was in the main a successful example" of the National Science Foundation behaving as the foundation's director said it should, as leading astronomers "largely controlled events with [the foundation's] blessing."[15] Hence, the optical astronomers wanted to fashion an organization that they controlled to take charge of the scientific operations of the Space Telescope.

NASA, as we have noted, operates in a fashion much different from that of the National Science Foundation. Scientists in general propose projects to the National Science Foundation; NASA often proposes projects to scientists. NASA, as fundamentally a technical management agency, also very much assumes that it should be in tight control of the activities it funds. The idea of STAR, and the creation of a rival for some of NASA's activities, therefore met with a frosty reception at NASA.

In an "interim response" drafted by Newell and another NASA staff member, there were a number of critical points. They foresaw that "a strong STAR with a permanent staff of highly competent astronomers could become a strong competitor with universities, observatories, and NASA Centers for scarce scientific manpower and observing time on the observatory when it is established. Such a competition is not desirable from NASA's or a national viewpoint and is not likely to prove desirable from an academic viewpoint." One problem for NASA, for example, was this: How could STAR be a consortium of universities that would both advise NASA as to the content of the astronomy program and contract with NASA, thereby accepting public money to perform the work to design and develop hardware to be used in the program?[16]

Following the report of the Ramsey committee, two of its members, Leo Goldberg, director of the Harvard College Observatory, and Princeton professor Martin Schwarzschild, made a number of visits to NASA Headquarters over a period of several months to discuss implementation of the committee's recommendations. Those talks involved NASA staff, including, on occasion, NASA administrator Webb. Webb, and some others in NASA, as Goldberg recalls, were quite favorably inclined to the idea of space science institutes. For Webb, such institutes might be "homes away from home" for scientists working with data from space missions. Goldberg also discussed an astronomy institute with Webb. It would be built in proximity to Goddard, but outside of Goddard's gates and therefore free of the usual restrictions on government laboratories.[17] But rather than establish a body similar to the Ramsey committee's STAR, Newell proposed an alternative and quite different scheme: that the NASA administrator appoint an "Astronomy Missions Board" composed of ten to fifteen astronomers drawn primarily from universities. These astronomers would give advice on the objectives and strategy for the NASA astronomy program. A parallel body, the Lunar and Planetary Missions Board, had already been established for planetary and lunar science and could serve as a model.

The Astronomy Missions Board was duly formed and, chaired by Goldberg, met regularly for over two years. By the end of that period, however, Goldberg complained to the NASA administrator that the spirit of the board members had been sapped by the agency's inability to act on much of their advice. The chief problem was a lack of funds. If there was no money available to conduct the recommended programs, why bother to make the recommendations?[18]

By 1970, NASA's relationship with the scientific community was plumbing new depths. A crisis had arisen because of scientists' massive discontent with the direction of the agency's long-range planning.

In February 1969, newly inaugurated President Richard Nixon had established a Space Task Group. Chaired by Vice-President Spiro Agnew, the group's charter was to make proposals on the course the space program should take once the Apollo program was over. Its report, issued in September 1969, presented a variety of options, but in all of them the ultimate goal was to land astronauts on Mars. The only real question was how soon this should be done.

When the Space Task Group's findings became known, the members of NASA's Lunar and Planetary Missions Board were at first inclined to resign en masse. For the board's members, the task group's report showed how little effect they had on NASA planning. For a decade NASA had been preoccupied with the Apollo program, which had meant a preoccupation with engineering and management, not science. But from the continuing emphasis on manned space flight and large-scale engineering enterprises, the Lunar and Planetary Missions Board judged that NASA still accorded science second-class status. The Astronomy Missions Board, too, was disturbed at the poor relative standing of science within NASA.

The discontent, moreover, went much further than the members of a couple of advisory boards. As Alex Dessler, a space scientist sympathetic to NASA, reported to Newell, the criticism was serious. Some of the most outspoken critics of the agency's program of space exploration were those who initially had been strong supporters of NASA in general and manned space flight in particular. Also, those who had been quiet before were becoming emboldened to speak out. Dessler contended that although funding for space science was not being cut as severely as for nuclear and solid-state physics, the unhappiness among space scientists was nevertheless greater because they felt they had little power in the decision-making processes in NASA.[19] By May 1970, things had reached such a state that Newell was preparing a "Review of Present Ferment and Foment in the Scientific Community."[20]

Writing to the NASA administrator from the vantage point of December 1971, Newell recalled that

relations with the Space Science Board, and also with our own Boards and Committees, began to come apart about the time the Space Task Group Report was published. Strains developed because the Boards and Committees felt they were not being effective or listened to by NASA. The budgets in the Space Task Group Report were regarded as appallingly high. The emphasis given to very large scale programs, – space shuttle, space stations and space bases, lunar bases, nuclear shuttles, Grand Tours, and manned missions to Mars, – had a very negative effect. [It was in] this period of turmoil and reaction . . . that disenchantment with [the] Viking [program to Mars] and active concern over large scale planetary projects like Grand Tour began to develop.

By 1971, NASA had revised its advisory committee structure, creating a Space Program Advisory Council to advise the agency on the entire space program.[21] Newell also thought it necessary to establish a better relationship with the Space Science Board, and for NASA and the board to avoid seeing each other as adversaries. Newell nevertheless cautioned that the "form of our relationship with the Space Science Board and the scientific community will be of no real significance unless it also produces a program that the Board and the scientific community can believe in and support."[22]

It is against this backdrop of NASA's sometimes stormy efforts to establish good working relationships with the scientific community that we must see the negotiations and arguments on how to manage the scientific operations of the Space Telescope. For some time, NASA and the astronomers groped for an appropriate form of organization to take responsibility for scientific operations. Certainly the eventual solution, the Space Telescope Science Institute, did not appear suddenly and in a well-defined shape. Instead, it was the complex product of contrasting and conflicting interests and power struggles. It would, in fact, emerge as the concrete answer to an array of problems and concerns, the central one for the astronomers being how they could exercise control over the telescope's scientific operations.

TERMS OF THE DEBATE

Soon after Bob O'Dell arrived in September 1972 at the Marshall Space Flight Center as project scientist, he composed a "Science Management Program" for the Space Telescope. As noted in Chapter 3, this was in large part focused on Phase B activities. O'Dell had, nevertheless, also cast his eyes toward the operations of the observatory in orbit.[23]

Relatively little public attention was paid to this issue until a Science Working Group meeting in November of 1973. There a Goddard staffer presented the results of some Goddard studies in which it was assumed, as had been inherent in NASA's planning to that point, that science operations would be directed by Goddard. George Field disagreed. He called for a close tie between the operations and management of the Space Telescope and the universities, and, by implication, separation of these elements to some degree from NASA.[24] Although probably not seen as such at the time, that was the first public shot in a lengthy and at times bitter and hard-fought debate between Goddard and non-NASA astronomers on the issue of how the scientific operations of the Space Telescope should be managed.

What were the terms of that debate? Everyone agreed that there should be a control center to tend to the engineering needs of the telescope in orbit, to check that components were not overheating, to ensure that it was pointed in the right direction, and so on. The key disagreement concerned who was to be responsible for the scientific operations. Were these to be run by NASA or by the astronomers? Also, were the science operations to be conducted from the same site as orbital operations? That was a major issue, because orbital operations would surely be directed by a NASA center.

The astronomers on the Science Working Group took the initiative by constructing their own detailed plans. At a meeting of the working group in February 1974, the talk turned to ways of contracting for the scientific instruments to be flown aboard the Space Telescope. In order to investigate the various possibilities, as well as explore ways of operating the telescope, O'Dell appointed a small group led by Arthur D. Code, one of the leading and most experienced exponents of optical space astronomy.[25] Known as the "uncommittee" because of Code's aversion to committee work, a report of its deliberations was presented to the working group in April 1974.

The Code uncommittee concluded that the best way to procure flight instruments was to use principal investigators, but it was nevertheless concerned that many people who might qualify as principal investigators for the Telescope were not willing to become involved. How to attract these people? The solution suggested by Code's group was that the telescope be operated by an observatory staff. To reward the principal investigators and their teams for building the scientific instruments, they would then constitute that staff and be allotted some substantial fraction, say 40 percent, of the observing time of the telescope when in orbit.[26]

ENTER THE INSTITUTE

The following month, Code's uncommittee began to address explicitly the issue of what sort of institution was appropriate for the observatory staff. In this way, they were led to consider a "Space Telescope Institute," an organization to be run by non-NASA astronomers and in control of devising science policies, as well as planning the observations to be made by the telescope. The uncommittee began to seek answers to these questions: Should the institute be at a leading university? What facilities should it contain? What size of permanent staff would it need? Should it be operated by a university consortium or other governing body?[27]

The summer of 1974, as we saw in the last two chapters, was a period of confusion and doubt for all associated with the Space Telescope as its supporters battled to ensure its survival. By the fall, however, the program was back on course, even if it was rather different from that charted earlier in the year. At the October 1974 Science Working Group meeting, the astronomers could therefore look beyond the immediate political struggles and spend some of their time debating the merits of the concept of a Space Telescope Institute. It was to be a pivotal meeting.

The Code uncommittee had viewed the Brookhaven National Laboratory and the Kitt Peak National Observatory as the models for the operations of the Space Telescope. As was the case with those two national facilities, it strongly favored placing the institute under the charge of a consortium of universities, such as AUI or AURA. That proposal fell on receptive ears, as many on the Science Work-

Telescope domes at the Kitt Peak National Observatory in Arizona in the mid-1970s. Ground-based astronomers were familiar with the concept of national observatories and sought the same kind of arrangement for the scientific operations of the Space Telescope. (Courtesy of National Optical Astronomy Observatories.)

ing Group had experience with AURA: Code had sat on its board (1958 to 1969, and 1971 to date), as had O'Dell (1971 to 1972), Spitzer (1959 to 1969), Bob Danielson (1969 to 1973), and William F. van Altena (1973 to 1974); Margaret Burbidge had just joined the board. George Field had also sat on AUI's board.

The Data Management and Operations Team, under team leader Robert Bless, had also examined how the telescope should be operated. It, too, advocated an institute. Bless's team nevertheless deemed it best to place the institute and the control center at the same site; such closeness would help ensure that the staff of the operations center and institute would interact well, understand each other's problems, and thus produce an efficiently functioning spacecraft.[28]

While the Science Working Group had been developing its plans, Goddard, too, had been active. Goddard already had extensive experience in operating scientific satellites. *OAO-II,* for example, had been run wholly at Goddard (although the scientific operations had been planned and implemented by astronomers from the University of Wisconsin and the Smithsonian Astrophysical Observatory), and the scientific operations for *Copernicus (OAO-III)*, launched in 1972, were split between Goddard and Princeton. Operating those astronomical satellites had not involved separate institutes.

For the Space Telescope, Goddard focused on three options: (1) the institute, (2) a "central facility" composed of a Mission Operations Center and a Science Operations Center, both sited at Goddard, and (3) a system based on a smaller central unit at Goddard in association with "regional science centers" at various points around the United States. But Goddard's conception of the institute differed fundamentally from that held by the Science Working Group, and in all of Goddard's plans there were strong measures of government oversight and control.[29]

With two such different visions of what the institute might be – that proposed by the Science Working Group and that by Goddard – it is not surprising that there was intense discussion. The astronomers on the working group were basically for an independent, non-NASA institute headed by an astronomer. In contrast, George Pieper, Goddard's director of earth and planetary sciences, proposed the use of civil service staff and a civil service director.[30] John Bahcall, one of the most enthusiastic supporters of an independent institute, retorted that this ran in the face of history and would not win the favor of astronomers. When O'Dell pressed Pieper whether or not the independent institute was a feasible option, Pieper replied that Goddard would do anything NASA Headquarters told them to. He nevertheless much preferred a Science Operations Center within a government installation – which everyone present must have known would almost certainly be Goddard because of its experience in operating astronomy satellites – with several science centers spread around the country.[31] In making this argument and in advocating regional science centers against an institute, Pieper touched on one of the central science policy issues raised a few years earlier during the presidency of Lyndon Johnson: pluralism versus elitism. Johnson had attempted to break the power of approximately twenty elite universities, mostly on the East Coast and West Coast, to abstract about half of the federal monies for research. He sought to spread funds around the country more evenly, to strengthen science in a number of universities rather than give most of the funds to the best science universities. Most of the non-NASA members of the Science Working Group, however, were associated with the so-called elite universities, and their pressure was for an institute that would make no concessions to pluralism.

Several members of the Science Working Group also contended that only an independent institute would be capable of attracting the quality of personnel required for the scientific operations of the telescope. The claim of non-NASA astronomers that the best astronomers should run the science operations of the Space Telescope and that such could not be attracted under civil service conditions was to underpin much of the debate on the institute. Those feelings were compounded by the view some non-NASA astronomers held of Goddard astronomers as major competitors for scarce resources in space astronomy. Giving control of the telescope's operation to Goddard, the assertion went, might lead to the non-NASA astronomers getting cut out.[32]

When Science Working Group member Robert Bless went on a speaking tour for space astronomy in 1969, he was struck by "this strong anti-Goddard feeling, expressed vehemently by people who had never been within five miles of Goddard's gate." In Bless's opinion, that was one of the factors "which led to the feeling by a lot of people, a lot of committees, that when [the Space Telescope] comes along, there should be an independent institute to run it. The standard reason given was, 'We don't want any appearance that Goddard or NASA is running the Space Telescope. NASA's so big and powerful and wealthy that it will just willy-nilly dominate it if the institute is sitting in its back yard.' " Another reason had to do with the way some astronomers judged NASA. For them, as Bless recalls, the concern was that NASA would "just foul everything up if they're allowed to have much of a say about it."[33]

When O'Dell polled the members of the Science Working Group on their preferences, of the eight non-NASA astronomers present, six voted for an independent institute. One[34] thought the discussion irrelevant (cost was the big problem), and although Spitzer, who had worked closely with Goddard on the *Copernicus* satellite, abstained, he noted that the astronomical community was skeptical of NASA's responsiveness to scientific needs. In contrast, the three

Goddard astronomers thought that the institute could be managed successfully within the government. O'Dell, who to a large degree saw himself as the representative of non-NASA astronomers, despite being the NASA project scientist, preferred a separate institute, but he judged that it should be close to the Mission Operations Center. There had thus been a fairly clear-cut division: Except for O'Dell, the NASA astronomers favored government control of science operations; the non-NASA astronomers desired an autonomous institute run by a consortium of universities.[35]

Although the institute idea had won favor with many astronomers outside of NASA, in late 1974 O'Dell was perhaps the only fully committed supporter of the concept within the agency. As, in effect, the agent of the Science Working Group within NASA, his opinions carried considerable weight, and he had been energetically promoting the science institute idea behind the scenes. O'Dell strove to steer it to concrete reality through what he and most of the other astronomers saw as a miasma of bureaucratic inertia and political indifference and hostility. But he and the working group were making headway. They had persuaded Marshall's project manager, James Downey, to tell Goddard that "an autonomous LST Science Institute is the plan of LST science operations that is preferred by the body of potential users," and so Goddard should use that approach in all studies of Space Telescope operations. Focusing on the institute, Downey noted, was important because "it will help to establish within the scientific community confidence that we are responsive to their wishes."[36] In March, Downey further told Goddard that the center should drop studies of other options and concentrate instead on a science institute and a Mission Operations Center in close coordination, but as separate, distinct elements. By that time, Marshall also assumed that the science institute would be staffed and operated by nongovernment personnel.[37]

After more private lobbying, O'Dell used his position as project scientist to crank up the pressure from the astronomers for an institute. At the April 1975 Science Working Group meeting, he again forced the issue by polling the members. O'Dell posed the following question: Should the operations of the Space Telescope be the responsibility of an independent nongovernment organization, under a director who is a scientist, operated by a consortium of universities, or should it be a government-operated laboratory with government scientists? All ten non-NASA astronomers in attendance voted for the science institute, and the absent member had already indicated the same preference by letter. Alois Schardt, director of the Astrophysics Division at NASA Headquarters, also pointed out that he expected Marshall to recommend a scheme for science operations that had the backing of the Science Working Group. He supported the institute idea, but expected NASA's upper management to resist it. Roman, who had initially been opposed to the institute concept, but had switched her position,[38] argued that it would be more appropriate to accept the institute after the telescope had become an approved program (remember that it was 1975, and approval from Congress was still two years away). The working group nevertheless recommended by a unanimous 13-to-0 vote that "the science programs of the LST be under the direction of an LST Institute, a nongovernmental organization, organized by a consortium of universities."[39]

From correspondence between two of the leading advocates of the institute we can also gain insight into the depth of feeling on the issue. Shortly after the April 1975 Science Working Group meeting, Bahcall protested to O'Dell that

NASA seems to be controlling the [Space Telescope] program to an extent that would be unthinkable in their successful high-energy programs (or even their [Orbiting Astronomical Observatory] programs). For example, there are no full-time non-NASA employees on the program (contrast this with the stable of fine young scientists that work for Giacconi [on a high-energy satellite] and Spitzer [on *Copernicus*]). NASA officials (Roman and Schardt) participate in our working sessions (inconceivable in the high-energy programs).[40]

O'Dell, however, thought that on the whole it was better to have Schardt and Roman at the meetings:

First, I don't think either of them really influence our conclusions. On the other hand they are the enemy and the better we know the enemy, the easier it is to beat him. Enemy is a strong word, but appropriate for unless they do the whole program right, it shouldn't be done, for all of us are spending a quantity we can never replace – our best years. . . . These bureaucrats can immolate [the Space Telescope] if we let them, and letting them will occur if we don't make them face us in person. Otherwise, they'll stay away and do things their way.[41]

For Bahcall and O'Dell, as well as for others, establishing an institute was a central aspect of doing the program right.

As much as some astronomers wanted an institute, the idea was greatly disliked and trenchantly resisted at Goddard.[42] Goddard had extensive experience of operating scientific satellites, and the pressure for an institute ran counter to a strong Goddard tradition and the interests of a substantial group of Goddard's staff. The result was a running fight on the issue: the astronomers and, to a lesser degree, Marshall aligned against Goddard. There had, for example, been a skirmish on the "guidelines" on the planning for science operations sent by Marshall to Goddard in March 1975. When Goddard's study manager, George Levin, replied that the cost of an institute remote from the Mission Operations Center would be too high, Marshall altered the guidelines to delete that option. But when

Levin told Marshall that he still could not accept the new guidelines and that Goddard was revising them again, the astronomers interpreted that as an effort to downgrade the science institute to an organization resident at Goddard, giving only general directions, while Goddard really ran the science operations for the Space Telescope.[43] O'Dell judged the matter sufficiently desperate to require an end run to a high level in the agency.

As he had in similar situations, O'Dell turned to John Naugle, who had become associate administrator in NASA Headquarters. O'Dell protested that from discussions within the Science Working Group and with other astronomers, "it is clear that an autonomous LST Science Institute is the only acceptable method of operations for the LST." In his opinion, support for the science institute was founded on three basic points: (1) The telescope must be freely accessible to all U.S. astronomers. (2) "NASA has not and probably cannot hire the caliber of astronomer that can be entrusted with the responsibility for the scientific success of LST." (3) Over the long run, people who work within a certain system end up conforming to that system and so lose their ability to judge a situation on an "absolute" scale of values. Moreover, the operations plans were "being so corrupted as to undermine the original goals of the program," and because Goddard's resistance had shaped the attitudes of managers and scientists at NASA Headquarters, almost every element within NASA had opposed the concept of the science institute.[44] As a coda, O'Dell issued a threat:

I have observed the program being eroded over the last two and a half years due to money constraints, which my colleagues and I can accept as long as it is still an LST and does not slip beneath the breaking point; however, we see the program being eroded through Center parochialism and unwillingness to address financial problems directly and this we cannot accept. Inevitably this situation will cost loss of support for the program from the potential users and . . . myself, with the departure of both.[45]

O'Dell, however, was not as alone in NASA in favoring an institute as he thought. In fact, the astronomers' arguments had started to bear fruit where it mattered most: with Noel Hinners, associate administrator for space science, whose discussions with John Bahcall were to be particularly influential in convincing him of the institute's worth.[46]

As late as March 1975, Hinners had been relatively ignorant of the science institute idea. Indeed, when it was first brought to his attention, he had suspected that the support for it had much to do with three factors: (1) those astronomers who were not used to working with NASA had several negative perceptions about NASA scientists and so were "paranoid" about NASA running the science operations for the Space Telescope; (2) a desire to work in a manner the ground-based astronomers were used to, that is, with a facility run by a consortium of universities providing the link between the telescope and the astronomers (as was the case with the telescopes at the Kitt Peak National Observatory); (3) "a belief that NASA sees itself as a builder of projects to the detriment of long-term operations."[47]

In Hinners's opinion, the first factor was playing a central role, and such views were, he thought, taken by some non-NASA astronomers to represent basic truths. Those perceptions boiled down to two major perceptions: First, NASA scientists were not as competent as their academic counterparts, and, second, they received privileged treatment. (By supposedly having all their time available for research, they could, for example, exert undue influence on NASA Headquarters in terms of scientific priorities and objectives, and thereby secure the money for projects that might otherwise have gone to scientists outside of NASA.)

Hinners had at first been inclined to chalk up these views to "sour grapes and a lack of visibility into the workings of NASA." But he and others in NASA Headquarters, though regarding the more extreme opinions of the outside scientists as gross exaggerations, had come to harbor suspicions that some of the criticisms were valid. In addition, following his own involvement in planning the landing sites for NASA's missions to the moon, Hinners held doubts about the strength of NASA's commitment to fund the operations of the telescope for the planned fifteen to twenty years of its life. Hinners knew that there had been some feeling in NASA to cancel the series of lunar visits after *Apollo 11,* the first moon landing. The proponents of that idea had wanted the agency to move on to the next major engineering enterprise, to make something new and get it to fly. Thus, one of Hinners's goals was to ensure that the agency would not be tempted to downplay science operations once the telescope was in orbit. A Space Telescope Science Institute could help prevent that by being a powerful advocate outside of NASA for the science to be performed with the telescope. By 1975, Hinners had also concluded that the lobby pressing for an institute "wouldn't go away, that there was a part of the outside world agitating for an Institute."[48] Mid-1975, remember, was also a time when the telescope's backing among astronomers was still relatively flimsy. The telescope's supporters, such as Hinners, were trying to strengthen the coalition that favored it, in particular by drawing in ground-based optical astronomers who were still wary of space astronomy. One way to do that would be to devise a way of running the telescope's scientific operations with which the ground-based astronomers would be familiar and happy. An institute run by a university consortium was the obvious way to achieve that end. To Hinners, it appeared that the institute could help solve two problems, one long-term and the other short-term: one, "pacify, if you will, the ground-based astronomy community, so that they'd be all the more supportive of the Space Telescope, and two, really provide an external advocate for a good operations program."[49]

Hinners did not want to show his hand too early, however, and so the institute's promoters had not understood his real position as he slowly nudged the agency toward acceptance of the concept. Nevertheless, by November 1975, Hinners had declared publicly that the Space Telescope's science operations would be managed and conducted by a "Space Telescope Science Institute," although he had yet to decide who would control the institute, the nub of the issue for the astronomers.[50]

BROADER INSTITUTES

It was not only the supporters of the Space Telescope who wanted an institute. In the mid-1970s, a number of leaders in the high-energy astrophysics community, particularly Harvard-Smithsonian-based Riccardo Giacconi, were lobbying NASA hard for a "National Institute for X-Ray Observatories."[51]

In 1975, a group of those working in high-energy astrophysics met to define a management approach for the design and operation of what they hoped would be national x-ray astronomy observatories. Following the talks, also attended by the presidents of two university consortia, the group wrote to NASA administrator James Fletcher to urge that an institute be created for x-ray observatories.[52] As the Ramsey committee had argued in 1966, such an institute not only would concern itself with operating the spacecraft in orbit but also would play a leading role in designing and developing the spacecraft. The suggested management approach would also assure "direct control by an independent scientific group, under the control of the research community [and it] requires diverting only a small fraction of the total mission cost to the Institute."[53]

Yet the issue was even wider than one of an institute for the Space Telescope or x-ray astronomy. In August 1975, O'Dell and George Pieper of Goddard made presentations on a Space Telescope institute at a meeting of the Council of the American Astronomical Society. The discussion was somewhat anticlimactic, as Goddard, under pressure from NASA Headquarters, had shifted its position on the institute much closer to that adopted by Marshall.[54] But the discussion had also moved on to a "Space Astronomy Institute." Some speakers suggested that such an institute could perform the same role for space astronomy as the Kitt Peak National Observatory played for ground-based astronomy. After what a participant called a "lengthy, frank, and freewheeling discussion," the council passed a resolution that NASA should establish an independent "Space As-

During the mid-1970s, Riccardo Giacconi, then based at the Harvard-Smithsonian Astrophysical Observatory, was a forceful advocate of the use of institutes for the pursuit of space astronomy. Giacconi (here seen in 1985) also worked with other x-ray astronomers to press for an x-ray institute. (Courtesy of Space Telescope Science Institute.)

tronomy Institute" to run the scientific operations of the Space Tele-scope.[55] However, as George Field noted in a letter to Spitzer, "I have not made up my mind on these issues, nor, I believe, have others in the community."[56]

If most astronomers were unsure about a broad institute, the pro-posals for institutes other than a Space Telescope institute met with a frigid response from NASA. The issue of an x-ray institute never-theless reached as far as a meeting between a potential sponsoring university consortium (Universities Research Association) and NASA administrator Fletcher. Hinners had even attended a later meeting at Fermilab, the large physics laboratory run by Universities Re-search Association.[57] But Hinners was only going through the mo-tions. His blunt reaction was, he recalls, "Buzz off. I didn't want any part of an X-ray institute. I discouraged them as much as I could."[58] He saw no pressing need for an x-ray science institute. Moreover, such issues raised unwanted complications for those who, like Hinners, were seeking to create an institute for the Space Tele-scope. Selling one institute within NASA would be, at best, an uphill slog; two might well be impossible. Hinners and others at NASA Headquarters judged it overwhelmingly preferable to get the Space Telescope under way and its science institute in place, and then, and only then, think about other institutes for other kinds of orbiting telescopes.

NASA had already brought two institutes into being, but only one had much resemblance to the proposed Space Telescope insti-tute. The Lunar Science Institute in Houston had been designed to make the best use of the moon rocks and other kinds of lunar data collected during the Apollo landings. Sited next to the Johnson Space Center, the Lunar Science Institute was administered by a university consortium, University Space Research Association (USRA), that had been founded especially to oversee its running. NASA's declin-ing budget in the late 1960s and early 1970s, however, meant that it had not had a smooth existence. In fact, NASA's space science managers had several times considered withdrawing financial sup-port.[59] So, according to Homer Newell,

while the Lunar Science Institute could not be called a failure, its success in the severe [budgetary] climate in which it was launched, was an uneasy one. There could be little question that when time came to consider estab-lishing an astronomy institute in support of an orbiting astronomical facil-ity, or a planetary institute in support of more intensive exploration of the solar system, such propositions would receive long and searching scrutiny before being implemented.[60]

Hinners agreed, and he recognized that the novel concept of a Space Telescope institute would take a lot of explaining within NASA, not to mention OMB and Congress. As part of the process of bringing the institute slowly into being, he therefore sought the advice and approval of the National Academy of Sciences.[61] The academy agreed to organize a study for a "Space Telescope Science Institute," and Donald F. Hornig, who had been President Johnson's science advisor, was selected as the chairman in 1976.

HORNIG COMMITTEE

The Hornig committee's charge was to consider the general principles that should govern the links between large space observatories and their scientific users, both inside and outside of NASA. The study was to be directed primarily toward the Space Telescope, but the committee was to examine questions sufficiently general that its findings could be applied to other space observatories.[62]

Most of the seventeen committee members were astronomers from universities, but there was a leavening of space scientists, together with representatives from industry, to provide expertise in data operations. The committee heard three days of presentations before retreating to bucolic Woods Hole in Massachusetts for an intensive two weeks of discussion. Out of those interchanges emerged a set of twenty-seven "Conclusions and Recommendations." Among them were the following: (1) The best scientific use of the Space Telescope would require the participation of the astronomical community. (2) The best way of providing long-term guidance and support for Space Telescope science would be a "Space Telescope Science Institute." (3) The institute should be run by a broad-based consortium of universities and nonprofit institutions under contract to NASA. (4) The institute should have a staff of the highest professional stature. (5) The institute should be of sufficient size, in facilities and staff, to carry out its functions, but should not become so large as to absorb an inordinate fraction of the resources devoted to astronomical research. (6) There were no compelling data-handling, managerial, or cost reasons for locating the institute at an existing NASA center. Another important point the committee made was that whereas "the cost effectiveness of the Institute will depend on many different factors, it is our judgment that the Institute approach is the one most likely to provide an optimal scientific return for a given dollar investment in the [Space Telescope] program."[63]

The Hornig committee members had then, to a large degree, fleshed out the basic concept of an institute advocated by the Science Working Group and had underlined the working group's reasons for bringing the institute into existence. It, too, in thinly veiled language, strongly urged that the institute not be located at Goddard. Placing the institute in a very large organization, the Hornig committee argued, "whose goals are not primarily astronomical research might well result in distractions from carrying out its tasks with maximum effectiveness."[64] The Hornig committee also contended that it would be easier to recruit leading astronomers if the institute were not based at Goddard.

By mid-1976 there was certainly a great deal of support among astronomers for a "Space Telescope Science Institute,"[65] and the Hornig committee's report, *Institutional Arrangements for the Space Telescope,* published in December 1976, reflected that position. That report was well received by many astronomers. In part, that was because, as John Bahcall noted in 1984, it "expressed all of our prejudices and very eloquently,"[66] although the Hornig committee had also pointed to many positive reasons for creating an institute. But perhaps most important, publication of the Hornig report meant that the authority of the National Academy of Sciences was behind the concept of an independent institute, a significant development and one that would make it easier to sell the institute in NASA and outside.

That the Hornig committee would argue strongly for an independent institute came as no surprise to Hinners. "When we asked the Space Science Board to conduct that study," he recalls, "it was pretty clear what the answer would be. I don't think there was any doubt. If we didn't want an Institute, or if I didn't want one, I would never have supported the Space Science Board doing that study."[67] With that influential stamp of approval in hand, Hinners could begin to move the institute a step further within NASA. He established a special working group to review, as well as seek ways and means to effect, the Hornig committee's recommendations. It was chaired by Warren Keller.[68]

Born in Knoxville, Tennessee, in 1931, Keller had earned a B.S. in engineering physics from the University of Tennessee in 1953, and after serving in the U.S. Army and working for Convair, he joined NASA. By 1977 he was a seventeen-year NASA veteran, fifteen of them spent at NASA Headquarters. As manager for various large planetary missions, Keller well knew what it took to get a program off the ground, and had earned a reputation as a leading program manager in NASA Headquarters. In 1976 he had become the headquarters manager for astronomy payloads, although from the start Keller was, in effect, the telescope's program manager. In that position he had quickly established good relationships throughout the program.[69]

Although the Keller group's charter did not stipulate that they had to assume a science institute in their planning, the working group had a clear charge to define what such an institute should be like, if NASA decided to create one.[70] Indeed, Keller's group took the Hornig report as its starting point.

The Keller group's own conclusions were presented to Hinners on July 21, 1977, and to a large degree followed those of the Hornig committee.[71] There were, nevertheless, some significant changes. Unlike the Hornig committee, the Keller group proposed that NASA retain responsibility for developing and procuring the telescope's scientific instruments. So whereas the institute would retain a considerable degree of independence and be the organization to which NASA would look for dealing with the scientific community, it was some distance from being the sort of institute that the Space Science Board had recommended, particularly in regard to the role the institute would play in hardware and software development.

Donald Hornig and Alastair Cameron of Harvard University, chairman of the Space Science Board, had been invited by Hinners to the July 21 presentation of the report of the Keller committee. Cameron had a few objections, but on the whole he was enthusiastic. "I was delighted," he told Hinners, "and I am sure that members of the Space Science Board will be also, at the degree of convergence of views on the specifications for [the] Space Telescope Science Institute between the NASA Working Group and the [Hornig] study."[72] The Keller group had, in fact, kept the Space Science Board fully informed of what NASA was doing and why. Prior to the presentation to Hinners, the Keller committee had even reviewed with the Space Science Board the differences between its findings and those of the Hornig committee.

The July 21 briefing for Hinners was also attended by Tom Owen, a representative acting on behalf of the Office of Science and Technology Policy. In his view, NASA was proposing "the establishment of a Space Telescope Science *Service* Institute as contrasted with the proposal made by the Space Science Board. Strangely enough, Dr. Hornig and Dr. Cameron agreed with my [Owen's] interpretation and indicated that this was the proper role for any such Institute to play." Owen's conclusion, nevertheless, was that "under the circumstances, NASA is recommending a very sensible organizational arrangement."[73]

Even before the Keller group had started work, the institute was looked on favorably by OMB. As OMB staffer Memphis Norman noted, "an Institute is a good idea, particularly the involvement of ground-based astronomers. We have often talked about the need for coordination and a comprehensive strategy for astronomy. The Institute may be a beginning." "It appears that NASA," Norman continued, "is correct in sensing that astronomers (particularly ground-based) are afraid of NASA. We have heard numerous accounts before from [the president's science advisory committee] members and [the National Science Foundation] – perhaps there are good reasons for fear, particularly about the Marshall center which will manage ST development."[74]

After some minor revisions, the report of the Keller group was completed in November 1977.[75] Goddard's director, Robert Cooper, was not prepared to accept its conclusions. In December 1977, Cooper told his deputy director that Goddard should lobby Hinners to support another scheme. In that proposed alternative, science policy for the telescope would be decided by a non-NASA group, science operations would be controlled by Goddard, and there would be a variety of regional centers to allow astronomers to use data from the Space Telescope.[76]

In fact, NASA took several more months to make its final decision on the institute. During congressional testimony in February 1978, for example, Hinners noted that prior to coming to a final decision, "we must assure ourselves and the Congress, that if we go the Institute route, we are creating an institution, which we will be able to control, to see that it does the right job and that we are not creating something that controls us. It must serve the scientific community and not become something unto itself."[77]

In April 1978 there were two crucial meetings at NASA Headquarters. At the first, Marshall and Goddard aired their objections to the NASA Headquarters planning on the institute. Cooper opposed outright the creation of an institute, arguing that it would become a millstone around the agency's neck.[78] Hinners and Roman stoutly defended the institute concept, and Hinners overruled the Goddard objections. He wanted an independent institute, but one that would be concerned primarily with science operations; the spacecraft operations center would be at Goddard. Goddard would also run the contract for the institute.

Later the same day, Warren Keller made a presentation on the institute to NASA administrator Robert A. Frosch. A scientist by training and background, Frosch was no stranger to the concept of the institute and had discussed it at length with, among others, Bahcall, Field, and Spitzer.[79] According to the notes of one eyewitness, Frosch joked that "no one is more cynical about scientists than me, I've worked in big science all my life."[80] Nevertheless, Keller's well-honed case won out. Although no final decisions were taken, Frosch agreed that the agency should begin detailed planning for an independent science institute and requested that Hinners present NASA's plans to the president's science advisor, Frank Press, as well as to the National Science Foundation. Hence, pending approval by the White House and Congress, there would indeed be a science institute.[81]

Before contacts were made with Press and the National Science Foundation, a meeting was held to complete NASA's plans. In a last-ditch stand, Goddard's director again pressed for his center to run the telescope's science operations, but the proposal was scotched by Frosch's deputy.[82]

Eight days later, Hinners announced on Capitol Hill NASA's scheme for an institute. The Space Telescope was the first long-life scientific satellite NASA had planned, and so previous procedures did not necessarily apply, Hinners cautioned. In particular, whatever plan NASA adopted would have to allow for the fact that the Space Telescope was to be in orbit for some fifteen to twenty years. National laboratories had already shown themselves to be responsible to their user communities and had worked well with their funding agencies. That was the model NASA would follow. Hinners nevertheless stressed that NASA would still retain control of operating the telescope in orbit.[83]

ESTABLISHING THE INSTITUTE

Once NASA had opted for an institute, the agency had to decide exactly how and where it should be established. Should NASA itself pick a site and then select the organization to run the institute, or should NASA select the organization and have it choose the site? That was a crucial point, because the locations of the scientific facilities often have been decided as a result of political pressure. With such facilities come jobs and other benefits to a congressman's district or state. The government agency choosing the site is thus open to political arm-twisting. One example is provided by the LANDSAT program to take pictures of the earth for civilian use. There the issue was where to site the data center for the Department of the Interior's use of LANDSAT. Technical reasons dictated that it should be somewhere in the Midwest, but exactly where? "The process of choosing a site," historian Pamela Mack argues, "shows the most elementary level of the politics of technological projects. Congressman Ben Reifel and Senator Karl E. Mundt of South Dakota heard of the search for a midwestern site and urged community leaders in various cities to suggest South Dakota as the ideal location." And because of local efforts in Sioux Falls and "Senator Mundt's bargaining power with the Nixon administration, the other cities under consideration stood little chance."[84]

Despite the likelihood of political interference, most of the university consortia that were interested in operating the Space Telescope Science Institute wanted NASA to choose the site for them.[85] Indeed, after one meeting in 1977 with the leaders of several university consortia, Hinners had been inclined to agree with them. However, Hinners's staff, in particular Warren Keller, had argued vigorously against that line, urging that the consortia themselves should negotiate the hurdle of site selection. If NASA were to choose the institute's site, it would surely become a politically charged issue. However, if the organizations that aspired to operate the institute had to make the choice, that would be less likely. The selection process would, moreover, be a robust test of an organization's management mettle. If a consortium could not make a good job of picking the institute's site, that would hardly augur well for its future supervision of the institute. In consequence, NASA decided to let the various consortia suggest their own sites.

By 1977, there were already several consortia interested in running the institute, and the two with the closest links to astronomy were AURA (Association of Universities for Research in Astronomy) and AUI (Associated Universities Incorporated). AURA, as noted earlier, operated, among other facilities, the Kitt Peak National Observatory in Arizona, and AUI operated, for example, the National Radio Astronomy Observatory in West Virginia.[86]

AURA had long been interested in space telescopes. In 1959, AURA representatives had visited the U.S. Army Ballistic Missile Agency in Huntsville, Alabama, to discuss AURA's plans for a fifty-inch orbiting telescope (see Chapter 1),[87] and for over a decade AURA had operated a Space Division at Kitt Peak. Before the Hornig committee met in the summer of 1976, a Kitt Peak Observatory group had produced its own paper on the possibilities of a Space Telescope Science Institute. A considerable portion of that paper subsequently found its way into the final report of the Hornig committee.[88]

However, when in 1977 John Teem became president of AURA, it was far from certain that AURA would bid for the institute. AURA's policies at Kitt Peak — the consortium's most visible institution — had fluctuated between emphasizing service and emphasizing research. Those swings had led to some disruption at the observatory, as well as criticism of its activities. Teem therefore found a certain sentiment on AURA's governing board that AURA should focus its attention on ground-based astronomy, get its house in order, and not become involved in space projects. In contrast, AUI was widely seen by astronomers as a generally well-run consortium whose board contained few astronomers, and therefore, in contrast with AURA, its board let the director of the National Radio Astronomy Observatory get on with matters with little interference.

AURA was unsure whether or not to propose for the institute, and so in late 1978 the consortium issued a special newsletter posing the question: "Should AURA Aspire to the Management of NASA's [Space Telescope Science Institute]?"[89] The majority of astronomers who replied thought that AURA should at least bid for the institute. A significant minority nevertheless made negative comments, such as, for example, that if AURA won the competition, too much authority over U.S. astronomy would be placed in one set of hands.[90]

But taking the responses overall, AURA president John Teem judged, and the AURA board agreed, that AURA had an obligation to "our constituents to put the best proposal together we could."[91] Teem was nevertheless ambivalent. His involvement with contract solicitations from NASA and the Department of Defense during his period as an employee of the Xerox Corporation had alerted him to what an enormously laborious and complex task it would be to re-spond to NASA's request for proposals for the institute.[92]

As the date approached for issuing the request for proposals, the Hornig committee was reconvened in late 1978 to review NASA's plans for the institute, and its views were sent to NASA in April 1979. There was general agreement on the agency's approach, al-though the Hornig committee did appeal some of the points NASA had earlier rejected, in particular the need to provide modest labo-ratory facilities at the institute, and computer terminals to be dis-tributed around the United States, but connected to the institute. The Hornig committee also stressed that NASA should issue the request for proposals quickly so that the institute could be involved in procuring the telescope's ground system.[93] In the Hornig com-mittee's opinion, too, the overriding concern for the institute should be scientific excellence.[94]

By April 1979, the time of the Hornig committee's response to NASA, the university consortia that intended to propose for the institute were gearing up their activities. Perhaps the most extensive plans were being laid by AURA. In 1978, AURA had established an ad hoc committee to prepare and coordinate the consortium's proposal for the institute. The first and most important item of business for the committee was where the institute should be sited if AURA won the competition. AURA's scheme was to examine several possible sites. Then, in a version of the decision-making pro-cess that NASA itself would follow, each institution would develop its own section of a draft proposal, and that would be added to a common AURA-produced portion of the draft proposal. The various complete draft proposals would then be scrutinized and graded by an independent group of astronomers. To begin with, AURA had decided to approach several universities that might be potential sites for the institute: the University of Colorado, the University of Chi-cago, the University of California–San Diego, the University of Maryland, and Princeton University.[95] Princeton was widely as-sumed by astronomers to be the front-runner in the institute stakes. Certainly Princeton had had a long connection with the Large Space Telescope and then the Space Telescope, and John Bahcall, of Princeton's Institute for Advanced Study, and Lyman Spitzer, of Princeton University, had, through their efforts to sell the telescope to Congress (see Chapters 4 and 5), become closely linked in the minds of many astronomers with the telescope. There had even been an effort, fronted by George Field, who was based at Harvard but who had very close ties to Bahcall and Spitzer, to preempt the com-petition and ensure that Princeton would be the site for the insti-tute. That move had taken the form of a campaign to enlist many influential astronomers in Princeton's cause. In the face of NASA's refusal to select a site, the effort had crumbled.

It was nevertheless reborn in another form as Princeton sought to become the site selected by each of the consortia that were to bid for the institute. Princeton was successful in three of five cases: AUI, USRA, and Battelle Memorial Institute each picked Princeton as its prospective site. URA (Universities Research Association) almost followed suit.

URA, a consortium of fifty-three universities throughout the United States and Canada, ran Fermilab. It also had a long-standing interest in astronomy institutes because of its involvement in plans for a possible x-ray institute, as mentioned earlier, and its board was well acquainted with the idea of the Space Telescope Science Institute. The board also viewed it as appropriate for URA to embark on new initiatives, and so in March 1979 it decided to "authorize the President of the Corporation to take whatever steps necessary to prepare and to submit to NASA a proposal in response to the expected NASA [request for proposals] for the management of the [Space Telescope Science Institute]."[96] URA then established a group of astronomers to help it select a site. After reviewing six potential sites, the astronomers picked Princeton first, only slightly ahead of The Johns Hopkins University and San Diego. That decision, however, was overturned by the board in favor of a site at Fermilab, which had placed fourth in the URA competition.[97] The change was controversial, particularly with the astronomers: Why, it was asked, form a site selection committee if its recommendations are going to be rejected?

The Johns Hopkins University had placed a close second in the original URA selection. Although Hopkins entered the competition late, it was to do even better with AURA.

AURA had wanted its list of possible sites to include one near the Goddard center in Greenbelt, Maryland. Such a site, it seemed to the AURA committee studying the issue, would offer the potential advantage of facilitating a close working relationship with the operations staff at Goddard. Yet it could be sufficiently distant from Goddard to assure astronomers that it would not be dominated by Goddard, a central reason for creation of the institute in the first place. The University of Maryland seemed to fit the bill: It was only a few miles from Goddard and had a large astronomy department. But when AURA made an approach, it found that Maryland was already tentatively teamed with another consortium, USRA. AURA then chose to look elsewhere.

In February 1979, a meeting was held with astronomers and representatives of the university consortia to debate the sort of institute that NASA should establish, as well as to apply pressure on NASA to ensure it did not backtrack from the recommendations of the Keller report.[98] One of those to attend was astrophysicist Arthur Davidsen of The Johns Hopkins University. Davidsen had been impressed by the Hornig report and the case for an institute. "It seemed like an extremely good idea, well thought out and nicely presented; and I just embraced the concept immediately that, indeed, such an Institute should exist for the Space Telescope."[99]

The members of the astrophysics group in the Physics Department at Johns Hopkins were well versed in space astronomy and had earned a good reputation. Two of them were also closely associated with the Space Telescope: Davidsen was a member of the science team for one of the telescope's scientific instruments, the Faint Object Spectrograph, and William ("Bill") Fastie, in his position as telescope scientist, was a member of the Science Working Group. In addition, Davidsen and Fastie had been among the three members of the Hopkins astrophysics group that in 1977 had flown a rocket-borne telescope to make the first observations of the ultraviolet light of a quasar. That experiment had been described as "remarkable" in the pages of the prestigious science journal *Nature*.[100] However, when the notion of Hopkins hosting the institute had first been raised at one of the group's regular lunchtime discussions, it had fallen flat, the Hopkins people assuming that their group was far too small to stand a chance.[101]

AURA chairman Arthur Code had also discussed Hopkins as a potential site with Fastie during a meeting at Goddard. But when Code asked Fastie what he thought of the idea, Fastie's response had been unenthusiastic: "Well, we don't have an astronomy department. We have an astrophysics group in the physics department, but it's rather small. And I doubt very much that we would be a viable candidate for that."[102]

Teem, however, was still anxious to include a site reasonably close to Goddard among AURA's list of candidate sites. So after the February meeting on the institute, he asked Davidsen, over dinner at the Cosmos Club in Washington, D.C., if Johns Hopkins might be interested in being the site.[103] Following that approach, the Hopkins astrophysicists got down to some hard thinking. As one of them recalled, "suddenly we had to take ourselves seriously, then for the first time our brains got in gear and started thinking about it. Then as soon as we thought about it, we realized that it wasn't ridiculous, it was perfectly conceivable."[104] After winning the approval of the Physics Department and gaining the enthusiastic support of the Hopkins administration, the Hopkins astrophysicists decided to make themselves available to whichever consortium wanted to propose Johns Hopkins. Davidsen became extremely busy, as he joined the AURA board and also led the Hopkins effort to play catch-up and gain ground on those universities that had the advantage of having started their planning earlier. That entailed a great deal of work, including selecting a location for the institute on the Johns Hopkins campus, commissioning an architect's studies for the building, developing arguments in favor of Hopkins as a site, and hosting site visits from consortia. Hence, the Hopkins astronomers, like astronomers at other potential sites for the institute, had become advocates for an institution and a university, another of the numerous roles sometimes demanded of scientists by Big Science.

In the summer of 1979, the panel of astronomers AURA had invited to help pick the institute's location started grading the draft proposals from the various candidate sites. The efforts of Davidsen and the other Hopkins people were rewarded as Johns Hopkins placed first in AURA's selection process, slightly ahead of Princeton. Although four of the eight reviewers opted for Johns Hopkins and four for Princeton, both AURA president John Teem and AURA chairman Arthur Code preferred Johns Hopkins. As Code suggested to the AURA committee on the institute, "it would probably be in the best interests of the astronomical community, since the Princeton site was clearly going to be proposed by two other bidders, for AURA to choose the Johns Hopkins site and provide NASA [a choice]." In Teem's opinion, AURA probably could produce a "better proposal with Johns Hopkins than with Princeton; one that would be unique." The committee agreed. If AURA should win the competition, Johns Hopkins was also able to offer inducements in the form of $2 million to go toward the cost of the building for the institute, as well as a professorship for the institute's director in the Hopkins Physics Department.[105] So, following ratification of that decision by the full AURA board, AURA would propose Hopkins.

For the various consortia, the summer and fall of 1979 was an active period as they chose their sites and looked ahead to the release of NASA's request for proposals. A draft of that request was issued in September and helped to focus the energies of the consortia and their proposal-writing teams.

Earlier in the year, AURA had decided that compiling the proposals would be such a time-consuming task that AURA could not possibly be competitive in the selection process if it drew only on staff from AURA and the host institution. AURA had therefore entered a teaming agreement with Computer Sciences Corporation (CSC), a corporation with considerable experience in operating scientific spacecraft, for the proposal writing.[106] Some AURA astronomers also became closely involved. For example, astronomer Barry Lasker of the AURA-run Cerro Tololo Inter-American Observatory in Chile joined the AURA proposal team in the fall and proceeded to work full-time on the proposal for the next few months.

On December 14, 1979, NASA issued its "Space Telescope Science Institute Request for Proposals." The 360-page document laid out in great detail the items to be addressed by the consortia in their proposals, completed versions of which were due by March 19.

The AURA team was perhaps the largest, drawn as it was from

AURA staff and those connected with the consortium (including AURA chairman Arthur Code and some who worked on the proposal for only short periods), CSC staff, and members of The Johns Hopkins University (including some from the university's Applied Physics Laboratory), with Arthur Davidsen playing a key role as the chief liaison to the university.[107] Despite the team's manpower, it got off to a stuttering start, and the writing failed to gel. There were some problems as the astronomers and operating staff of CSC sought to get on the same wavelength and to understand what was expected from each group.[108] Some of those involved wanted a stronger management structure. A means had to be found to take hold of the proposal and drive it in an appropriate direction. As Barry Lasker recalls,

Johns Hopkins astrophysicist Arthur Davidsen played a key role in the bid AURA, Johns Hopkins, and CSC developed for the Space Telescope Science Institute.

All of this was going on over Christmas and January and February: frantic editing; pieces of it absolutely not falling into place. And one of the big things that happened in . . . late December I think, Bill Fastie [of Johns Hopkins] took [AURA president] John Teem aside and said: "John, there's only one way we're going to get this thing . . . done. You've got to come and join the team." So John moved to [CSC Headquarters in] Silver Spring in January and stayed there until March; and was in there with the foot soldiers. John has incredible stamina.[109]

The last months of proposal writing were incredibly intense and hectic as AURA and the other consortia hurried to meet the March 19 deadline.

AUI, which had selected Princeton as its site, had the assistance of OAO Corporation in its proposal preparation. But two other consortia had also chosen Princeton, and so the efforts of the Princeton astronomers had to be distributed among three groups. Also, because of Princeton's divided loyalties, AUI chose to develop the final drafts of its proposal itself rather than leave the job to Princeton.

THE DECISION

In the end, five consortia sent proposals to NASA. NASA's Source Evaluation Board then sifted through the proposals in order to prepare its report to the NASA administrator, for it would be the administrator who would make the final decision.

By September 1980, the field had been narrowed down to two: AUI and AURA. Oral defenses of the surviving proposals followed, with NASA staff questioning representatives of the two consortia. Those defenses had their lighter moments. Barry Lasker had been concerned about chapter four of the technical volume of AURA's proposal. As he jokingly recalls, the NASA examiners reported that " 'We have read your chapter four as carefully as we can. We just can't figure out what it is you are going to do.' I remember thinking

217

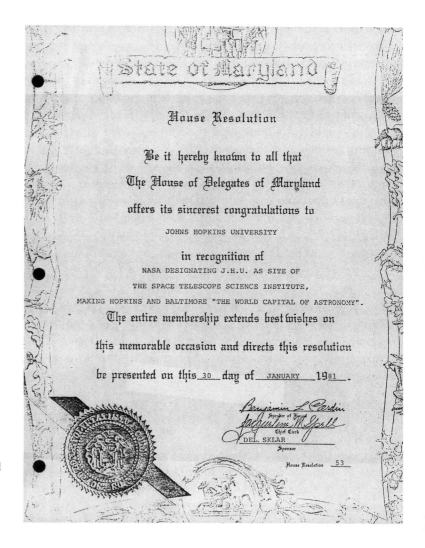

State of Maryland

House Resolution

Be it hereby known to all that

The House of Delegates of Maryland

offers its sincerest congratulations to

JOHNS HOPKINS UNIVERSITY

in recognition of

NASA DESIGNATING J.H.U. AS SITE OF
THE SPACE TELESCOPE SCIENCE INSTITUTE,
MAKING HOPKINS AND BALTIMORE "THE WORLD CAPITAL OF ASTRONOMY".

The entire membership extends best wishes on

this memorable occasion and directs this resolution

be presented on this _30_ day of _JANUARY_ 19_81_.

Speaker of House
Chief Clerk
DEL. SKLAR
Sponsor

House Resolution ___53___

Copy of resolution sent to Johns
Hopkins University by the Maryland
House of Delegates following
NASA's decision on the institute.
(Courtesy of Space Telescope Science
Institute.)

. . . 'Well, we can't, either.' But somebody gave an answer that was
sufficiently long and complex to be incomprehensible, and some
patching was done in the written response."[110]

John Bahcall was deeply interested in NASA's decision. Only a
few days before NASA's announcement of the winning consortium,
AUI had decided that he would be the institute's director if AUI's
bid were selected. But NASA administrator Frosch's choice, announced on January 16, 1981, was that the AURA/Johns Hopkins
proposal was the winner. The Johns Hopkins University would
therefore be the site for the institute. As Bahcall told Teem, "congratulations to you and your whole team. You did a great job and
wrote a really superior proposal (against what I like to think was
really tough competition)."[111] As Bahcall remembers, in a way he
was very relieved, because he knew that if he did become director of
the institute, he would have to devote himself almost full-time to

218

administration. When the decision was made public, "everybody [at Princeton] was surprised. But for me there was no longer this terrible dilemma, science or no science. And suddenly it was an irrelevant question."[112]

Despite the AURA/Johns Hopkins victory, John Teem, who suspected that the task of building an institute from the ground up would be enormous, harbored doubts: "I said many times," Teem notes, "that the one who would win this competition would be the one who came in second, and in fact, when we did win, I called Bob Hughes [of AUI] and I said 'Congratulations!' He said, 'What? I should be congratulating you.' I said, 'No, the winner of this one's the one who came in second. You won it.' "[113]

There were, perhaps inevitably, rumors that political strings had been pulled in the decision making. One story was that Hopkins had won because it was based in Maryland, one of the states to vote for President Jimmy Carter in his losing campaign of 1980. Thus, as a reward for services rendered, so to speak, the institute went to Baltimore. There seems to be no worthwhile evidence for such surmises. As Bahcall remembers, "I think we did everything we could, and everybody I know . . . did everything they could, to see that there were no political forces brought to bear at the time of the decision, because we felt that if there were, it would be disastrous for astronomy. It would go to some place which had absolutely no relation to astronomy."[114] But as Bahcall also points out, "that's not the universally held opinion at Princeton, but it's my very strong opinion. There are some people in Princeton who can't understand how you can choose any place outside of Princeton . . . for any objective reason."[115]

In fact, the contest between AURA and AUI was not really close, for the Source Evaluation Board had placed AURA in front by "a wide margin." And when NASA administrator Frosch and senior officials from NASA Headquarters and Goddard had met with the Source Evaluation Board on January 7 and 16 to discuss the choice, AURA, teamed with Johns Hopkins and CSC, had romped home. The AURA/Johns Hopkins proposal was judged the stronger because of the way it had shown how to optimize "the important program objectives at a reasonable cost." In what was by far the most important set of selection criteria, "mission suitability," the Source Evaluation Board had placed AURA first on nine of the ten counts.[116]

The institute's controlling consortium had been selected, and the institute itself was about to be formed. As a concrete response to an array of problems – for the astronomers, most particularly, how to wrest control over the telescope's scientific operations from NASA – the institute in many ways was a compromise between the requirements of its supporting agency, NASA, and academic traditions and values. How the demands of its sponsor and users would play out in practice, we shall see in Chapter 9. Following its victory, AURA had the responsibility of negotiating a contract with NASA for the institute, as well as the daunting task of establishing a fledgling institution beset by powerful vested interests – all during a period when, as we shall see in the next two chapters, money and time for the Space Telescope Project were in short supply.

7

Up and running

Jim Gunn came into my office, closed the door, stood by the side of the blackboard, looked me dead in the eye, and said "we've got to build a wide field camera for the Space Telescope." I said "you're out of your bloody mind."

James A. Westphal

On January 28, 1977, Lockheed engineer Domenick Tenerelli was at the Marshall Space Flight Center. NASA was to issue its requests for proposals to industry that day, and Lockheed wanted a jump on the company's rivals to build the Space Telescope's Support Systems Module. Tenerelli was to obtain copies of the proposals, immediately telephone the details back to his company, and then catch a plane from Huntsville to return with the proposals to Lockheed in California.[1]

When, two months later, the contractors had completed their responses, NASA would have to evaluate them. In so doing, the agency would select who was to build the Optical Telescope Assembly and the Support Systems Module. NASA would also have to establish a project organization, an immensely important job.

NASA and its contractors had to put into place the Space Telescope's "marching army," the thousands of people divided into many interlocking and multidisciplinary teams who would design and construct the Space Telescope. That task, given the size of the enterprise and the variety and diversity of the institutions and groups involved, together with the diversity of interests to be accommodated, was far from straightforward, particularly as the telescope was planned to be the most sophisticated scientific satellite yet constructed. It also meant that NASA managers had to devise a scheme of systems management for the telescope program, a means of securing information on the program's status. Nor were issues of organization and systems management divorced from the design, development, and choice of the specific technologies that would constitute the telescope and the means of operating it from the ground. In earlier chapters, we have seen how the selection of particular institutions, groups, and individuals, and the power relationships between them, helped to fashion telescope designs, scientific objectives, and the manner in which NASA planned and conducted the program. So it would be for the detailed design and development phase as well, as we shall see in this and the next chapter.

PICKING CONTRACTORS

NASA had written the requests for proposals for the Optical Telescope Assembly and the Support Systems Module very carefully.[2] Composing them had been a lengthy and often subtle job, for the requests were in many respects a distillation and synthesis of the findings of Phase B. For example, a statement in the request for proposals on the telescope's mirrors read that "primary and secondary mirror blanks shall be constructed of ultra-low expansion material, titanium-silicate glass, Corning Code 7971."[3] That single sentence was the visible product of a great amount of work by both the agency and the contractors in the earlier planning stages, which had included the formation by Marshall of a special review group to investigate the choice of mirror material in depth.[4]

The requests for proposals also represented NASA's view of the performance the Space Telescope should achieve. The proposals were not, however, intended to provide the definitive design for the Space Telescope — that was what the detailed design phase would be all about.

The authors of the requests for proposals had also arranged them in such a way that none of the five competing contractors was needlessly cut out of the competition. "Now the secret of it," argues William C. Keathley, the Space Telescope project manager, in 1977, was to write specifications for the two requests for proposals — one for the Optical Telescope Assembly and one for the Support Systems Module — "that will not preclude any of those five contractors or any combination thereof . . . from winning the competition. For instance, you simply couldn't put in a [specification] that only Boeing could meet [for] then you would by definition be eliminating those other two [Support Systems Module] contractors from even fulfilling the requirements."[5] So in writing the proposals, Marshall had built into them performance specifications and interface specifications to cover all possible final combinations between the contractors. Also, in Phase B there had been numerous studies — particularly of what Marshall had regarded as the high-risk areas of the metering truss, the primary mirror, and the fine-guidance sensors — but they had by no means settled all of the questions about the telescope's configuration, another fact that precluded opting for one particular design right at the start of Phase C/D and sticking to it rigidly.[6]

The requests for proposals therefore incorporated what might be described as umbrella specifications. Only after the contractors had been chosen and detailed work had begun on the telescope's design and the requirements were better understood could the general, or umbrella, specifications be converted by the contractors into more specific terms. For example, NASA required the telescope to have a pointing stability of 0.007 second of arc (later changed to 0.012 second of arc). The Support Systems Module contractor would then have to break down that requirement into more detailed specifications for the various elements of the pointing and control system. Those specifications would then drive the design of the different items of hardware this incorporated.

The two requests for proposals were issued in January 1977, and the responses from industry were received in March. NASA's next exercise was to review and grade the responses. The agency usually chooses its contractors through use of a "Source Evaluation Board." Such boards had been introduced into NASA's scheme of working in 1961. Composed of NASA engineers, scientists, and managers, a board's job is to rate a contractor's proposal in terms of technical competence, feasibility, cost, and management capability.[7] Space Telescope Source Evaluation Boards were created to report to the NASA administrator on the contractors for the Optical Telescope Assembly and Support Systems Module.

Three companies had bid to build the Support Systems Module: Boeing, Martin-Marietta, and Lockheed. These were the same three who had been awarded Phase B contracts. In contrast, the competition for the Optical Telescope Assembly received a late entry, and NASA received a welcome surprise. Itek and Perkin-Elmer had been awarded the Phase B study contracts in 1973; indeed, they had been the only two optical companies to propose. However, in July 1976, NASA administrator James Fletcher received a letter from a senior executive at Eastman Kodak announcing that Eastman Kodak intended to bid for the Optical Telescope Assembly.[8] From Fletcher's notes on the letter it is clear that NASA had deliberately sought to stimulate Eastman Kodak's interest, a tactic consistent with the NASA Headquarters strategy to have as strong a pool of candidates as possible for the Phase C/D contracts. In fact, giant Eastman Kodak prepared a joint bid with the relatively small Itek for the Optical Telescope Assembly.

One major participant has recalled that the selection of the Support Systems Module contractor "came down to a question of subtleties, and which one knew more about the job it had to do."[9] The competition for the Optical Telescope Assembly, too, was reported

223

to have been close. One reason for that situation was that the Department of Defense "tries to maintain technical viability in all these companies" in order to ensure that there will be several capable of building precision optics, the basis for many military systems, including reconnaissance satellites.[10]

Their deliberations complete, on July 20, 1977, the Source Evaluation Boards made their reports to James Fletcher's successor as NASA administrator, Robert Frosch. Frosch chose Perkin-Elmer to build the Optical Telescope Assembly and Lockheed to build the Support Systems Module.

NASA later reported that the biggest factor in Perkin-Elmer's success was the company's proposed "Fine Guidance System."[11] Without a means of locking on its targets, the Space Telescope would be useless, and so an effective Fine Guidance System would be essential. The Fine Guidance System had also proved to be a particularly tough challenge in Phase B, and the designs advanced for the system had therefore weighed heavily in NASA's decision making.

The experience of Perkin-Elmer and Lockheed with photoreconnaissance satellites surely played a part too. In 1955 the U.S. Air Force issued "General Operational Requirement Number 80," the aim of which was to secure a photographic reconnaissance system. As a result of its response, Lockheed was awarded a contract in 1956 to develop "Weapons Systems 117L," an advanced reconnaissance system. In so doing, Lockheed would devise three fundamental types of earth observation systems, and each of these would remain in use for the next thirty years. Lockheed, which, as one writer has put it, "already monopolized the existing pool of satellite research personnel and was also extensively involved with the U-2 high altitude reconnaissance aircraft programme," thereby acquired a vast store of expertise with reconnaissance satellites.[12] Perkin-Elmer, according to numerous public reports, had also worked extremely closely with Lockheed on the KH-9 ("Big Bird") photoreconnaissance satellites, the first of which was launched in 1971. Thus, a partnership that had been established for national security purposes was being turned to space astronomy.

The choice of contractors underscores how radical a break the Space Telescope would be from what had gone before in space astronomy in optical and ultraviolet wavelengths. If Grumman had won approval to build a series of spacecraft linking the Orbiting Astronomical Observatories to the Space Telescope, matters would no doubt have been different. But with its choices of associate contractors, NASA had selected one, Lockheed, whose methods, resources, institutional memory, and expertise had been shaped almost entirely by its work on satellites for various national security purposes, and for the other, Perkin-Elmer, a similar, if less strong, claim could be made. It is not surprising, then, that the presidential science advisor

George Keyworth could comment in 1985 that the Space Telescope
is new, but "draws upon technologies used in military systems,"
and in 1974 Lockheed had explained to Congress that its work on
the Support Systems Module was based on a system it had built for
the air force.[13] Although the Orbiting Astronomical Observatories
had preceded the telescope in time, the telescope was by no means
based on them. The Orbiting Astronomical Observatories, in fact,
were largely irrelevant to the technical development of the Space
Telescope.

PROGRAM DESIGN

A week after the two associate contractors were selected, one of the
astronomers was brought up to date on recent events. "Probably the
most happy note about this," O'Dell told him, "is the fact that the
bids ran only about 85 percent of the Project Phase B estimates,
even after our putting in certain adjustments upward."[14]

After the contracts had been negotiated, Perkin-Elmer's contract
amounted to $69,420,000, and Lockheed's to $82,725,000 (both
fiscal 1978 dollars).[15] When allowance was made for the money to
be devoted to the scientific instruments, and the changes that would
result from the detailed design work, those sums still left a substantial reserve in the cost range of $425 to $475 million (in FY 1978
dollars) approved by Congress. So even if the contractors' bids should
prove too low, there would be, it seemed, a sizable reserve available
to meet any shortfall.

For many people, the Space Telescope's Phase C/D began on an
upbeat note. From the perspective of the scientists, there was a distinctly bullish air throughout the project. As one astronomer remembers, for "a long time the word around Marshall was that, oh,
you know we're going to come in at 450 million or thereabouts, we
have a huge reserve, we're just in great shape, nothing will get in
our way."[16] Goddard's instrument scientist, David Leckrone, also
recalls a spirit of aggressive optimism: "I was really bright eyed and
bushy tailed and didn't think there was anything on ST we couldn't
do; I felt very enthusiastic about the technical challenge; I honestly
felt we could be done by 1983 and ready to launch."[17]

There were, nevertheless, some obvious and formidable management and technical challenges for NASA, the contractors, and the
astronomers. One was posed by the very size of the program and the
large numbers of institutions, groups, and individuals it involved
and the wide range of physical resources – engineering and scientific
laboratories, manufacturing plants, and so on – that the program
would engage. Its size placed a high premium on how well the
diverse elements of the program fitted together, that is, how the
program itself was designed. The task for NASA managers was to

integrate an enormous range of people, institutions, and physical resources into a coherent whole, yet at the same time ensure that as the telescope's design and development went through various stages the system was sufficiently flexible to allow changes easily.

The overall program design was, as it turned out, quite elaborate. First, many parts of NASA were engaged. In addition to NASA Headquarters, there were two NASA field centers intimately involved – Marshall and Goddard – with strong input from another two: the Johnson Space Center in Houston, Texas, for shuttle operations, and the Kennedy Space Center in Florida, for the launch of the Space Shuttle that would eventually carry the telescope aloft. Nor was it simply a question of the elements of one space agency meshing together, for this rich mix was further spiced by the European Space Agency, which would provide one of the scientific instruments, the solar arrays, and some of the staff for the Space Telescope Science Institute. Also, there were many scientists at work on the design of the telescope and other aspects of the program – acting as consultants and advisors to the contractors, sitting on the Science Working Group, and serving as members of the teams to build the scientific instruments. Somewhat later, other astronomers would join the program as staff members of the Space Telescope Science Institute. In total, when the use of two associate contractors (of which more later) was taken into account, there were some fifteen major contractual hardware and software interfaces.[18]

The building of the Space Telescope would also require members of disparate "cultures" – institutions, groups, and individuals with different values, beliefs, and attitudes – to work together. As the long-time chief engineer for the telescope, Jean Olivier, commented, "each organization has its own way of doing things, has its own fundamental long-term goals and objectives, and as you try to hammer out one project among all these different organizations, it's not going to be totally coincident with the best interests of all these organizations." Nor did the organizations all do business in identical ways.[19] Even the two major NASA centers engaged on the telescope had their own distinctive approaches to engineering tasks, as well as to management. Illustrations of these differences were seen in Chapters 2 and 3 in the telescope designs Marshall and Goddard had independently generated in the Phase A competition for lead center. Aerospace contractors, too, had their own particular ways of doing business, and it would take a while for NASA to get to know the contractors working on the telescope and for the contractors to get to know NASA.

The astronomers had a somewhat uneasy position in the Space Telescope culture. Often astronomers think about and tackle problems in ways fundamentally different from those used by engineers, and their manner of working can be much less formal than that

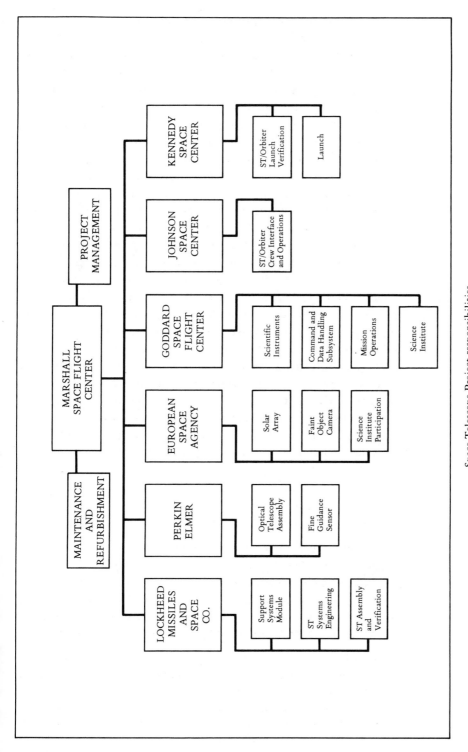

Space Telescope Project responsibilities.

common in most large engineering organizations, and far less geared to working in large teams. The common caricatures of the kinds of spacecraft that scientists and engineers would produce if each group were left to its own devices were as follows: The scientists would design a superb set of scientific instruments in a spacecraft held together by sealing wax and string, whereas the engineers would design a splendidly reliable spacecraft in which extra redundancy and safety factors would be built in at the expense of space and weight needed for the instruments. But those were only caricatures. Space scientists often have assumed, quite effectively, engineering and managerial roles in major space science projects, roles much wider than simply the designing of, and managing the development of, scientific instruments.[20] In high-energy physics, it has been the physicists who have taken the lead in designing and managing their facilities. But as we saw in the last chapter, NASA generally resists any loss of control over its projects. In fact, apart from the project scientist and a few of the members of the Science Working Group, it would be several years before the scientists would have much of a say in the wider decision making of the Space Telescope program. Of course, for a successful telescope, the scientific goals and engineering requirements had to mesh. The division between astronomers and engineers would nevertheless lead them to see problems from somewhat different perspectives, and inevitably there would be tensions concerning the aims and demands of the two groups, a split that was institutionalized by the program design and that loaded the power relationships against scientific interests.

PROGRAM DESIGN: MARSHALL AND GODDARD

The program organization was complicated. At its core were two NASA centers, Marshall and Goddard. But Marshall and Goddard had worked badly together in Phase B. That much was obvious even to the astronomer members of the agency's advisory group in shuttle astronomy. In January 1976, the group sent a unanimous resolution to the head of NASA's Office of Space Science, arguing that

the division of responsibility for the Space Telescope between Marshall and Goddard has led to unnecessary confusion in the past, and could potentially jeopardize the Space Telescope program in the future. The ST project has the good fortune of having a first rank astronomer as project scientist, and a reorganization of the management of the project can increase his effectiveness. We therefore recommend that NASA review the present division of responsibilities with the goal of streamlining the ST management.[21]

The recommendation was not accepted.

The Marshall–Goddard relationship was still poor in 1977, even hostile at times. During the preparations for the start of Phase C/D, the two centers had negotiated at some length a revised "Memorandum of Agreement."[22] That memorandum would determine how the two centers were to divide up their responsibilities and which activities each would manage in Phase C/D. The bargaining was tough, sometimes harsh. Each center pressed hard for the share of the project it wanted and attempted to impose a management structure of its own choice.

From the vantage point of NASA Headquarters, it was not just the competition for resources that underpinned the negotiations. Goddard also preferred to work directly for NASA headquarters, not for another center, in this case Marshall. But for Goddard to have reported directly to headquarters would have undermined Marshall's control over a project for which Marshall was ultimately responsible. From the perspective of headquarters and Marshall, such an arrangement would have been unwieldy and unworkable, for it would have thrust headquarters into arbitrating day-to-day technical, scientific, and managerial disagreements between Marshall and Goddard. That was a refereeing position headquarters devoutly wished to avoid, as it did not possess the number of staff or the breadth of engineering and scientific expertise for such a function.

The negotiations between the centers dragged on to such an extent, and from Marshall's viewpoint were so unsatisfactory, that for a time Marshall angled for the right to manage the scientific instruments itself, a responsibility that had long been expected to be Goddard's. For a time, that notion was coupled with Marshall's suggestion that Caltech's Jet Propulsion Laboratory be responsible for the telescope's science operations. The Jet Propulsion Laboratory was highly experienced in science operations because of its dominant place in the NASA planetary science program. In that proposed scheme, the Jet Propulsion Laboratory would supplant Goddard and be responsible for operating the telescope in orbit, thereby cutting Goddard out of the Space Telescope program entirely. Attractive as that idea might be to Marshall, the final decision rested with headquarters. When headquarters judged the use of the Jet Propulsion Laboratory to be too expensive, the idea of coupling Marshall and the laboratory was dropped.[23]

The deadlock between Marshall and Goddard was finally broken when the director of the Astrophysics Division in NASA Headquarters, T. Bland Norris, visited Goddard. Headquarters had rewritten the memorandum of agreement between Marshall and Goddard. Norris, speaking to the Goddard director, painted a clearer picture of the difficulty the disagreement posed for NASA Headquarters. It was impossible for headquarters to make day-to-day engineering decisions on the project. What was to be done? Perhaps responsibility for managing the scientific instruments might even be given to Marshall. The Goddard director then relented and signed the memorandum of agreement.[24]

The end result was that whereas Marshall would have overall authority for the telescope, Goddard would be responsible for the scientific instruments and the operations of the Space Telescope — responsibilities constituting a sizable chunk of the whole program. The signing of the agreement, however, did not bring the hostilities between the two centers to a halt.

PROGRAM DESIGN: MANPOWER LIMITS

The arrangement with Goddard was not the only part of the program design that sat uneasily with Marshall. Marshall had a "cap" on the manpower it could employ on the Space Telescope, with the number of Marshall personnel being limited to a total of 72 in the first year, although an increase was planned to a total of 116 in fiscal 1981, when activities on the telescope were expected to reach a peak.[25] Those figures were certainly lower than Marshall had anticipated during Phase B, and that influenced the manner in which Marshall was able to manage the telescope's development.

The capping was in part the result of an agreement between NASA and the Department of Defense.[26] It was the department's way of limiting the number of NASA staff who would "penetrate" the laboratories of contractors who were also working on Department of Defense contracts. "Secrecy," one historian has pointed out, "is as old as warfare; compartmentalization is its modern version."[27] As General Leslie Groves, the leader of the vast Manhattan project to build the first atomic bomb, wrote in his memoirs, compartmentalization was the principal means of avoiding inadvertent disclosure of information. "The possibility of betrayal thus became," General Groves noted, "directly proportionate to the number of people employed, and to the amount of knowledge possessed by each." Compartmentalization was, to a degree, being practiced on the Space Telescope program to lessen the risk of the transfer of sensitive knowledge.[28]

There were other justifications for the manpower limit. As a 1983 congressional report put it, NASA Headquarters was inclined to regard Marshall as having a tendency to "overburden" contractors "with requests for reports that were not necessary and were expensive — both in time and money."[29] Also, as a leading Marshall engineer has recalled, the headquarters attitude was that "we've observed the way Marshall Space Flight Center works over the years, and when [Marshall's Science and Engineering Directorate] gets in it they drive the cost of a contract sky high. You've got all these engineers, you just turn these engineers loose on the contract, and

they'll start goldplating everything. We want to avoid that in Space
Telescope. This is a low cost program."[30]

Marshall had to adjust its normal method of working with its
contractors to comply with the manpower limit. Normally, NASA
seeks, in government contracting terms, to "penetrate" its contractors and dig deeply into a project's fine points. Certainly that was
Marshall's usual style, a style intended to produce a partnership with
the contractor. But in the case of the Space Telescope, Marshall
would not, by deliberate design of NASA Headquarters, play nearly
so close a management role, nor would the engineering laboratories
at Marshall become as involved in the program as they otherwise
might have been. Rather, the center would retreat somewhat toward
the wings and leave the contractors firmly in center stage. The design and development of the Space Telescope was to be very much
the responsibility of the contractors, more so than would have been
the case under the usual NASA way of doing business.

The manpower limitations on Marshall's managers also shaped
several other important decisions on how the program was to be
conducted. One of those centered on the crucial questions of the
systems engineering for the Space Telescope. Systems engineering
for a spacecraft begins with the process of converting mission requirements into project specifications, that is, taking aims and goals
and turning them into engineering details. Systems engineering also
involves the development of all the mathematical tools to ensure
that the design truly will meet the mission requirements. Then,
once testing of the spacecraft and its sections has begun, the test
data are fed back into a variety of mathematical models of the spacecraft to see if the design has worked as intended. Also indispensable
to systems engineering are "audits." Audits of weight, for example,
are intended to ensure that the weight of a spacecraft and its various
subsystems stay within the design limits.

Systems engineering also has another aspect. The reference sometimes made by Marshall engineers to the telescope as a "lily in a
pond" captures this point, in that to touch one small part of the lily
causes the rest of the lily to move. Thus, design changes in one part
of the telescope can have ripple effects, mandating changes in other
parts, often in intricate but significant ways. To take one central
example, the signal from the Fine Guidance Sensors, built by
Perkin-Elmer, is fed into the Pointing and Control System, built by
Lockheed. This signal greatly affects the Pointing and Control System's ability to meet the Space Telescope's requirement for pointing
stability, and the pointing stability affects the telescope's image quality
and ultimately the science that can be done.

What can be termed "systems-level" problems also occur in
spacecraft. That is, whereas items in the spacecraft may work fine

when tested individually, problems can arise in one or more of the
many areas in which the individual items interact. It would therefore have to be systems engineers who would "pull" the telescope
together to ensure that the spacecraft's various elements were accurately related, that everything worked in harmony, and that the
telescope was truly an astronomical observatory. The Space Telescope would indeed have to be much more than the sum of its parts.

How was the systems engineering to be organized? Marshall's
project manager, William Keathley, saw three options: one "was to
give it to the spacecraft contractor and let him do systems engineering for the whole project. Another option I had was to do that with
civil service manpower. Well, that was out because I didn't have
enough civil service manpower to do it with. The third option I had
was to get an outside, a separate, contractor, like Aerospace for instance, to do systems engineering for the total project. . . . I'd had
some very bad experiences with that mode so I crossed that one
out."[31] Keathley was left with giving the spacecraft contractor,
Lockheed, responsibility for providing support to Marshall for systems engineering. Lockheed, however, would not be accountable for
the systems engineering of the focal-plane structure of the Optical
Telescope Assembly. That was to be performed by Perkin-Elmer.

The limited number of project personnel at Marshall gave Keathley what he saw as little choice in how he deployed his own staff.
Because of the small number of people available, Keathley decided
not to set up separate offices for each of the major units of the telescope, such as the Optical Telescope Assembly and the Support Systems Module. Instead, Keathley opted for a so-called functional organization, with a Design and Development Office, Assembly and
Verification Office, Office of the Chief Engineer, and Office of the
Project Scientist, all at Marshall, as well as a Mission Operations
Office and Scientific Instruments Office staffed by Goddard personnel based at Goddard.[32] Keathley's aim in dividing responsibilities
in that way was to ensure that a person in, say, the Design and
Development Office would know as much about the Optical Telescope Assembly as about the Support Systems Module. Keathley
judged it much better to have people occasionally work on the same
issue, rather than risk letting an issue slip through the cracks between the offices. In part, that was management by conflict; yet
because of the relatively small number of people, "I had to overlap
gray areas a bit, create those conflicts, so I could wrestle with them
and settle them in a timely manner."[33]

PROGRAM DESIGN: MANAGEMENT TOOLS

Another major aspect of NASA's program design was the creation
of working groups.[34] Composed of contractor and NASA staff, but

chaired by representatives of the contractors, these working groups were intended to be the major conduits for exchange of information between the various designers and builders of the telescope, as well as the places where many technical issues would be resolved.[35] NASA's aim in placing the contractors in a leading position in the working groups was to press its systems engineering contractor, Lockheed, to develop a good understanding of the interfaces between the Support Systems Module and other sections of the spacecraft, as well as what lay on both sides of the various interfaces.[36]

Designing and constructing the Space Telescope would involve thousands of people and many tens of institutions. How was NASA to keep track of all of these complex and highly diverse activities? The basic management tool NASA adopted was a form of "PERT" ("Program Evaluation and Review Technique"), a system developed in the late 1950s.[37] Adopted by NASA in 1961, it had been one of the main means by which the agency had kept track of the gargantuan and incredibly complicated Apollo program.

It is, in essence, a method of arranging and scheduling all of the tasks that have to be accomplished if a project is to be completed successfully. PERT imposes discipline on a project in that it requires that the tasks, or "key events," be arranged in sequence, together with estimates of how long each activity is to take. This results in a "row of parallel paths charted from the beginning of the project, running past critical points and converging towards the end as the finished and tested product," in this case, the Space Telescope. These critical points are known as "milestones" and provide a ready means of gauging the project's progress. If milestones are not reached on the appropriate date, it is clear that something is amiss.[38] By 1977, the PERT system was widely used in the aerospace industry and posed little in the way of a new challenge.

Another management tool, and a way of imposing order on the potentially sprawling Space Telescope program, was the development and use by the contractors and NASA of a variety of documents. Particularly important would be those detailing the links, or interfaces, between the items — both hardware and software — built by different contractors. NASA's aim was to ensure that a contractor in, say, Florida could, by examination of the appropriate documents, know how its own components related to those built by a string of other contractors in the United States, as well as perhaps in Europe. Any changes to the interfaces would have widespread repercussions; so they would have to be carefully controlled, and notices of alterations widely disseminated. In many ways, the documents would become the glue holding the huge Space Telescope program together.

To sum up, then, NASA intended its management approach to have four major features, although only the first was novel: (1)

233

streamlined project office and streamlined technical monitoring, a minimum parallel in-house engineering effort, no support contractors at Marshall, and small NASA offices at Lockheed and Perkin-Elmer; (2) frequent reviews; (3) rigid control of contract changes; (4) continuous evaluation of cost, schedule, and technical performance.[39] Events would eventually compel a drastic change in the first of these features and reveal that NASA had based it on a grossly unrealistic view of what it would take to design and build the Space Telescope. Nor would NASA's attention to the other features prevent large cost, scheduling, and technical problems plaguing the program.

PROGRAM DESIGN: THE CONTRACTORS

In addition to the question of how NASA was to manage the program, there was the issue of how the contractors were to be linked, in managerial terms as well as via the hardware and software they were to build. In the Phase A designs for the three-meter Large

Exploded and cutaway views of the Space Telescope. The telescope's basic design remained quite stable during Phase C/D.

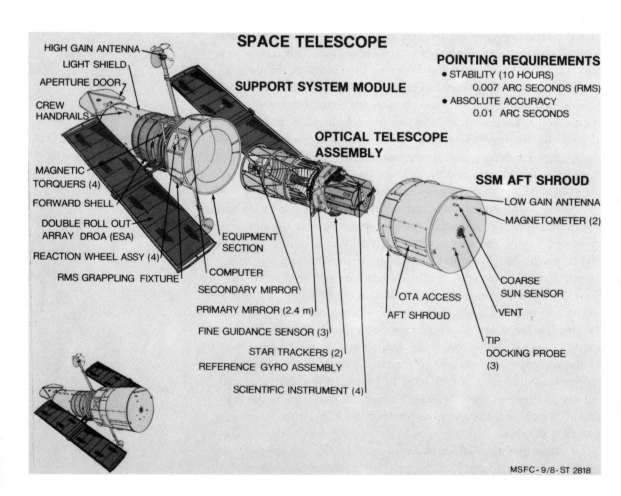

SPACE TELESCOPE

POINTING REQUIREMENTS
● STABILITY (10 HOURS)
 0.007 ARC SECONDS (RMS)
● ABSOLUTE ACCURACY
 0.01 ARC SECONDS

SUPPORT SYSTEM MODULE

OPTICAL TELESCOPE ASSEMBLY

SSM AFT SHROUD

HIGH GAIN ANTENNA
LIGHT SHIELD
APERTURE DOOR
CREW HANDRAILS
MAGNETIC TORQUERS (4)
FORWARD SHELL
DOUBLE ROLL OUT ARRAY DROA (ESA)
REACTION WHEEL ASSY (4)
RMS GRAPPLING FIXTURE
EQUIPMENT SECTION
COMPUTER
SECONDARY MIRROR
PRIMARY MIRROR (2.4 m)
FINE GUIDANCE SENSOR (3)
STAR TRACKERS (2)
REFERENCE GYRO ASSEMBLY
SCIENTIFIC INSTRUMENT (4)
LOW GAIN ANTENNA
MAGNETOMETER (2)
OTA ACCESS
AFT SHROUD
COARSE SUN SENSOR
VENT
TIP DOCKING PROBE (3)

MSFC-9/8-ST 2818

Space Telescope, the Support Systems Module had been located *behind* the Optical Telescope Assembly. By 1977, after the detailed engineering studies and the drive of the mid-1970s to cut costs, the Support Systems Module had been redesigned to *envelop* the telescope's optics, a change made when the size of the telescope's primary mirror had been reduced to 2.4 meters. In fact, the redesigned Support Systems Module consisted of a forward light shield, a forward meteoroid shield, an aft shroud, and an equipment section.

If the Support Systems Module was not to be bolted onto the back of the Optical Telescope Assembly, as had been the essence of the Phase A plan, how were the two to be linked? That was a pivotal and subtle question. If the expansions and contractions of the Support Systems Module – which were bound to occur as the telescope swung every forty-five minutes or so between the daytime and nighttime parts of its orbit – were transmitted to the optical system, it would be difficult, if not impossible, to keep the images in focus. In Phase B, Lockheed engineers had investigated how the Support Systems Module's temperature changes would influence the

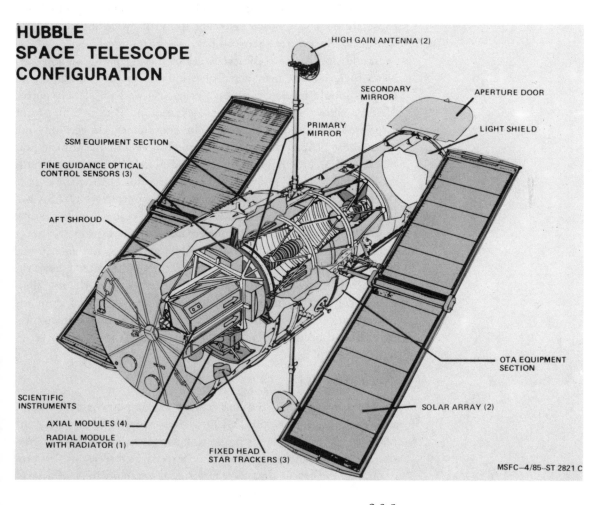

HUBBLE SPACE TELESCOPE CONFIGURATION

HIGH GAIN ANTENNA (2)

SECONDARY MIRROR

APERTURE DOOR

PRIMARY MIRROR

LIGHT SHIELD

SSM EQUIPMENT SECTION

FINE GUIDANCE OPTICAL CONTROL SENSORS (3)

AFT SHROUD

OTA EQUIPMENT SECTION

SCIENTIFIC INSTRUMENTS

SOLAR ARRAY (2)

AXIAL MODULES (4)

RADIAL MODULE WITH RADIATOR (1)

FIXED HEAD STAR TRACKERS (3)

MSFC—4/85—ST 2821 C

Optical Telescope Assembly. The disturbing conclusion they reached was that without a special kind of mount, the deformations of the Support Systems Module would cause the whole telescope to bend. The primary and secondary mirrors would then lose their accurate alignment, an alignment that was essential to in-focus images. Lockheed had calculated the effects to be of the order of 100 arc seconds, a disastrously high number given that the aim was to stabilize the telescope's pointing to a far smaller amount.[40]

The solution adopted by the contractors and NASA for Phase C/D was essentially to isolate the Optical Telescope Assembly from the Support Systems Module. That would be done by making the joints between the Optical Telescope Assembly and Support Systems Module "kinematic," that is, designing the joints so that there would be a minimum of coupling between the two major sections of the spacecraft. Kinematic mounts would be employed for the same reason elsewhere in the spacecraft: to isolate the primary mirror from the rest of the Optical Telescope Assembly, to isolate the scientific instruments from the Optical Telescope Assembly, to isolate the Fine Guidance Sensors from the Optical Telescope Assembly, and to isolate the entire Space Telescope from the Space Shuttle when it was inside the orbiter's payload bay.

In addition to the technical issues of linking hardware and software, there was the managerial question of how the two major contractors should interact with each other. As we have already noted, NASA's chosen solution was for Lockheed and Perkin-Elmer to be associate contractors. Using associate contractors was certainly not unusual for NASA, and such a scheme had eventually won support in Phase B at NASA Headquarters and Marshall as potentially the most economical. We saw in Chapter 3 that for NASA, the crux of the matter was that by opting for associate contractors, NASA had been able to select what it judged to be the best proposal from an Optical Telescope Assembly contractor and the best proposal from a Support Systems Module contractor. In so doing, the agency had itself been able to link the two contractors, rather than have the contractors themselves do the teaming. Perhaps, too, the use of associate contractors meshed with the possible strategy of compartmentalization that we discussed in connection with the manpower cap that had been imposed on Marshall.

EARLY GOALS

On October 19, 1977, the contracts for the Support Systems Module and the Optical Telescope Assembly were signed. Some thirty-one years after Lyman Spitzer's pioneering paper for RAND on the "Astronomical Advantages of an Extra-Terrestrial Observatory," a program was underway that took as its goal the launching of a large optical astronomical telescope into space. If the planned schedule was maintained, the Space Telescope would be aloft by December 1983. But if the program's early milestones were to be met, there had to be a rapid start.

The Optical Telescope Assembly, in many respects the heart of the telescope, was to be constructed by Perkin-Elmer, a corporation that had begun business in the late 1930s by building astronomical telescopes. In addition to extensive work in optics for the Department of Defense, by 1977 Perkin-Elmer had supplied telescopes with large mirrors to over forty major observatories.[41] The company was also experienced in space astronomy and had forged a particularly close link to the Princeton space program. Perkin-Elmer had, for example, built the special lightweight mirrors for the Stratoscope balloon-borne telescopes, as well as the thirty-two-inch mirror for Princeton's experiment package on *OAO-C*, the *Copernicus* satellite.[42]

NASA accepted that a number of demanding technological problems would have to be overcome in order to build the Optical Telescope Assembly and that the problems confronting Perkin-Elmer probably were more difficult than those facing Lockheed on the Support Systems Module. One was the polishing of the Space Telescope's primary mirror to the required accuracy. In Phases A and B, that had been, for some people, the project's "tall pole," the technical problem holding up the rest of the "tent" or project. So critical was the primary mirror deemed to be by NASA that Eastman Kodak was contracted to polish a back up 2.4-meter mirror in case of difficulties arising with the Perkin-Elmer mirror. As an extra measure of insurance, Eastman Kodak planned to polish its mirror by conventional, well-tested techniques. In contrast, Perkin-Elmer would adopt a new approach, with a polisher controlled by a specially devised computer system.

A particular problem was how to grind and polish on earth a very precisely figured mirror that was finally to operate in orbit. Even a rigid mirror will sag a little under the pull of gravity on earth, and when sent into the weightlessness of space, it will relax to a different, undeformed, shape. Perkin-Elmer therefore chose to polish its mirror on a special mount that would simulate zero gravity. As one writer put it, Perkin-Elmer's approach was "to build a support mechanism of 138 rods with each rod exerting a precisely known force on the back surface of the mirror. Each rod simulated zero gravity in a small area of the mirror by off-loading a local region of the mirror by exactly its own weight. The sum of all the upward forces exactly equalled the weight of the mirror, thereby negating the downward pull of gravity."[43]

Figuring mirrors on earth for use in orbit was, of course, not a new problem for a company that had built many mirrors for use in space. What was perhaps novel was the size of the mirror being figured and the degree of precision to which Perkin-Elmer had to work. Also, as important as the methods of shaping and polishing the primary mirror were, those for testing the accuracy of the mirror's surface were just as important.

As Perkin-Elmer's project manager explained to a congressional subcommittee in July 1978, the normal method of polishing such a mirror is to polish, clean, measure, polish, clean, measure, and repeat the cycle many times, slowly edging toward a mirror with the desired precision. To aid that process, NASA and Perkin-Elmer decided to move the fabrication, measuring, and polishing facilities into one area. That shift, they judged, would lessen the risk to the mirror as it was switched from being polished to being tested.

Two 2.4-meter mirror blanks had already been ordered from the Corning glassworks in New York, one for Perkin-Elmer and the other for Eastman Kodak's backup mirror. Usually, no hardware that will eventually be flown is procured by NASA until after a "Preliminary Design Review" (PDR). The Optical Telescope Assembly's PDR was scheduled for mid-1979, but given the extremely

The 2.4-meter Space Telescope mirror, being prepared for tests of the accuracy of its surface, inside a thermal-vacuum chamber at the Perkin-Elmer Corporation in Danbury, Connecticut, in 1980. (Courtesy of Perkin-Elmer Corporation.)

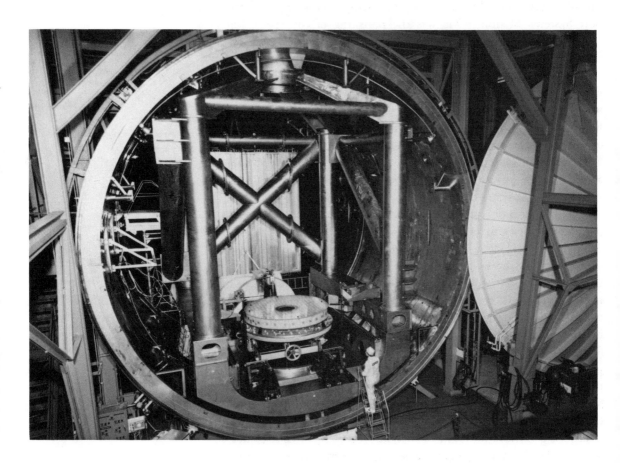

long times involved in fabricating and polishing big telescope mirrors, NASA decided the mirror blanks had to be purchased earlier than usual, in this case even before the PDR.[44]

The blanks from which the mirrors were to be fashioned were made of ultra-low-expansion glass. One property of this material is that thin slabs can be fused together. The Space Telescope's primary mirror initially came in four pieces: the core, the faceplate, the backplate, and a large ring. The faceplate would be fused onto the front of the core, the backplate onto the rear, and the ring fused around the outside. Fusing the plates onto the mirror's honeycomb core would mean a big weight saving over employing a solid slab of glass. It would result in a disk weighing one ton, a quarter of what a solid slab would weigh.

The 2.4-meter mirror blank would not, however, arrive at Perkin-Elmer from Corning for about a year. In the meantime, Perkin-Elmer would continue to check and refine its polishing, grinding, and measuring tools and techniques on a sixty-inch-diameter test mirror.[45]

Designing and constructing the Optical Telescope Assembly, of course, involved much more than simply the primary mirror. Boeing, for example, was contracted to build the assembly's metering truss. The sixteen-foot-long truss was designed by Boeing to maintain the distance and alignment between the primary and secondary mirrors to a tolerance of less than a ten-thousandth of an inch. It would be made of graphite epoxy, a relatively new composite material,[46] but one that when subjected to varying temperatures, holds its dimensions excellently, an essential point if the telescope's images were to remain in focus.

Perhaps the most exacting task for Perkin-Elmer was the design and construction of the Fine Guidance Sensors. For the telescope to work as planned, the sensors and the associated Pointing and Control System often would have to keep the telescope precisely directed at faint astronomical objects for long periods of time, up to twenty-four hours in some cases. Marshall regarded the Fine Guidance Sensors as so crucial that in the early stages of Phase C/D, it purchased some "insurance" for the Perkin-Elmer design. That insurance came in the form of a study by the Jet Propulsion Laboratory (JPL) of an alternative Fine Guidance Sensor. Progress, however, was slower than Marshall would have liked, and by June 1979 Marshall had decided that the alternative sensor was receiving little management attention at JPL and was unlikely to be ready for the planned 1983 launch. The alternative sensor was in consequence dropped.[47] For better or worse, the project was dependent on Perkin-Elmer's design for the Fine Guidance Sensors.

The Fine Guidance Sensors also illustrate a general point about

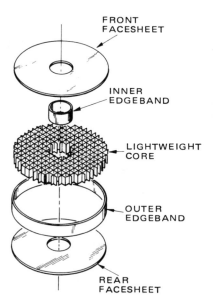

FRONT
FACESHEET

INNER
EDGEBAND

LIGHTWEIGHT
CORE

OUTER
EDGEBAND

REAR
FACESHEET

The Space Telescope's primary mirror, assembled from a lightweight core fused between faceplates of ultra-low-expansion silica glass. (Courtesy of Perkin-Elmer Corporation.)

designing and building the Space Telescope: As was the case for the development of much of the telescope's technology, technical issues associated with the sensors often were entwined with, and in some ways indistinguishable from, other factors, including management, scientific, economic, institutional, and political factors. The Fine Guidance Sensors, for example, were by no means simply engineering devices for aiming the Space Telescope. They were also intended to be used to measure to an unprecedented accuracy the separation of objects in the sky, that is, to perform astrometry. The telescope design featured three Fine Guidance Sensors arranged around the circumference of the focal-plane structure. In order to keep the telescope directed to an astronomical target, two of the sensors would lock on so-called guide stars, stars in the same vicinity in the sky as the target. The separations between the guide stars and the target would be known from measurements of photographs of the area made beforehand. So, provided two Fine Guidance Sensors remained pointed at their respective guide stars, the telescope would remain on the target.[48] Meanwhile, the third Fine Guidance Sensor would be free to scan and measure the angular distances between the various objects in its field of view. A design goal for Perkin-Elmer, then, was for the Fine Guidance Sensors to provide, in effect, a sixth scientific instrument. Hence, in addition to the multidisciplinary team of engineers involved in the sensors' design, there would be an "Astrometry Science Team." It would meet regularly to consider the Fine Guidance Sensors' potential scientific performance and the astrometry the telescope could perform. Any changes in the design of the Fine Guidance Sensors would have implications for the science to be performed by the telescope, going beyond simply the ability of the sensors to stay locked on guide stars. And, as we have just seen, there were also issues concerning how much time and effort NASA should devote to the back-up Fine Guidance Sensor studies.

THE SCIENTIFIC INSTRUMENTS

However wonderful the spacecraft produced by the contractors might be, the telescope would be useless without scientific instruments, for it would be the scientific instruments that would analyze the light collected by the telescope. Thus, when the requests for proposals had been issued and the Source Evaluation Boards for the Support Systems Module and Optical Telescope Assembly had been meeting, the scientists interested in constructing instruments for the telescope had also been extremely busy.

The Phase B Science Working Group had debated at length the scientific instruments and detectors that should be flown. At the start of Phase B in 1973, the astronomers and NASA had hoped to include seven instruments; as we have seen, that plan had been sunk by lack of money and the difficulty of arranging so many instruments around the telescope's focal plane. Further, if NASA and the European Space Agency should reach their expected agreement on jointly building the telescope, one instrument, the Faint Object Camera, would be provided by the Europeans. There would, therefore, in addition to the astrometry to be done by the Fine Guidance Sensors, be four open slots for scientific instruments aboard the Space Telescope, and those would be filled through a competition organized by NASA.

The Science Working Group had ranked the Wide Field Camera and the Faint Object Spectrograph as the two most scientifically important instruments. When in the midst of the budget struggles of 1974 and 1975 the astronomers had defined the "minimum Large Space Telescope," these two had even been identified as "core instruments," instruments so central to the telescope's mission that it should not be launched without them (see Chapter 5).

The Wide Field Camera had been regarded by NASA and by the Phase B Science Working Group as so crucial to the Space Telescope's success that the detector for the camera had been designated a so-called facility-class detector. In other words, this detector would be the detector for the Wide Field Camera and would be furnished as part of the telescope, just as, say, the primary mirror would be. Such a plan was not unusual for NASA; the use of facility-class instruments was common for spacecraft missions to the planets. NASA, as we saw in Chapter 3, had designated the SEC Vidicon, being developed by a Princeton University Observatory/Westinghouse team, as a facility-class detector.

This detector was nevertheless not without drawbacks. A well-known weakness of the SEC Vidicon was that it showed poor response in the red part of the spectrum, a region of special importance to planetary scientists. Princeton astronomer Robert Danielson, leader of the Phase B High Resolution Camera Instrument Definition Team, was one of those particularly concerned by this failing.

OBSERVING PLANETS

Danielson had come to Princeton in 1958. There he became involved in the Stratoscope program for high-resolution photography from balloons. Danielson had an important role in modifying Stratoscope I so that the twelve-inch mirror and associated optics could be pointed to and guided on specific features on the sun's surface. His work on Stratoscope II, a telescope with a thirty-six-inch mirror, involved precise guiding, and between 1963 and 1971 the Stratoscope balloons produced photographs of galaxies and planets of very high resolution, nearly 0.2 arc second. Danielson thus brought to the Space Telescope program a wealth of experience in remotely controlled astronomical photography and an intense interest in both astrophysics and planetary science.

Astronomers and planetary scientists often tend to regard themselves as members of separate groups, each with its own goals, techniques, and programs. Danielson, however, moved freely among astronomers and planetary scientists, as well as spacecraft engineers, communicating well within each group and reconciling their scientific and instrumentation needs with what was technically feasible. Danielson was especially anxious that the Space Telescope be able to work well in the red part of the spectrum in order to observe distant clusters of galaxies, as well as planets. Some planetary scientists argued, for example, that the telescope would be extremely useful to them for long-term studies of planetary atmospheres. Unlike planetary flybys, which would take a spacecraft to a planet for only a relatively short amount of time, the telescope would make possible studies over several years. Planetary scientists also stressed the need for a detector with a good response in the red for such investigations.[49] Danielson therefore lobbied hard for inclusion of a "planetary camera" in addition to the Wide Field Camera, pressing his colleagues for serious consideration of such a camera, especially one with red-sensitive Charge Coupled Devices as its detectors.

In spite of Danielson's labors, the idea for a planetary camera found little favor with the Science Working Group. The members were wary of the novel technology represented by the Charge Coupled Devices and were generally far more interested in the science of stars and galaxies than of planets. One astronomer jokingly referred to the division between astronomers and planetary scientists as that between the "Lords of Creation," the stellar and galactic astronomers, and the "Grubbers after Facts," the planetary scientists. For the Lords of Creation there was little incentive in the mid-1970s to switch from the SEC Vidicon. Indeed, there was something of a feeling among space astronomers that planetary scientists had too long received the lion's share of NASA's space science funding. With the Space Telescope, it would become the turn of stellar and galactic astronomers, and this would very much be *their* telescope.[50]

Danielson died in mid-1976 after a lengthy period during which he knew he had an incurable illness. Although he was aware that he would not live to use it, Danielson continued to work on the telescope with determination, skill, and bravery to the very end.[51] Although long expected, Danielson's death was a professional and personal blow to the planetary scientists, who had regarded him as a spokesman for their cause on the Space Telescope program.

That loss was acute in the summer of 1976, when both the Space Telescope and the Jupiter Orbiter Probe were soon to be objects of vigorous selling campaigns on Capitol Hill (see Chapter 5). Noel Hinners, Nancy Roman, and some others in NASA Headquarters judged that the Space Telescope would stand a better chance of being approved by Congress if it were capable of high-quality planetary observations.[52] They reasoned that if that were so, it might help to head off criticism of the telescope by the advocates of the Jupiter Orbiter Probe, as well as help to build a wider and more solid base of support for the telescope. As far as Hinners saw,

> there was a good indication that the astronomy community didn't give a damn about planetary observations, and desired that the Telescope be devoted totally to the classical astronomy observations. And a number of us thought that getting the planetary capability into the Space Telescope was desirable from two points of view – one, that you could do some damn good planetary science with it, and two, that politically it would help to bridge that gap between the planetary and the astronomical community. That might be one of the selling points.[53]

That argument was hammered home by NASA staffers in a variety of arenas during Phase B, staffers who were anxious to extend the coalition that favored the telescope so as to make it more feasible politically.

The debate about detectors and the Space Telescope's ability to observe planets began to intensify. The aerospace industry's trade journal, *Aviation Week and Space Technology,* reported in August 1976 that NASA's inability "to secure more than one new planetary start in the last four fiscal years has increased the importance of the space telescope in terms of planetary observations." *Aviation Week* also noted that the National Academy of Sciences' Committee on Planetary and Lunar Exploration (COMPLEX) had recently stressed "that adequate instrumentation should be designed to enhance the instrument's usefulness in observing planets."[54] The article continued: "Some planetary researchers are concerned . . . that the heavy involvement in the space telescope program of scientists interested in observations outside of the solar system may prevent adequate planning for effective use of the instrument inside the solar system." Danielson's death was a cause for concern, because he had been "an avid supporter of planetary research via the space telescope."[55] Space Telescope program scientist Nancy Roman was cited as having said that discussions were being held with planetary researchers to assess the suitability of the Space Telescope's scientific instruments for planetary research.

Roman called such a discussion for August 5, 1976. Four planetary scientists attended, as well as NASA personnel prominent in the planning of the telescope's instruments. There was a long discussion on various detectors, and "the group expressed unhappiness with current performance specifications for the SEC Vidicon and urged consideration" of the Charge Coupled Device and other detectors for the Wide Field Camera. Certainly the perception was that,

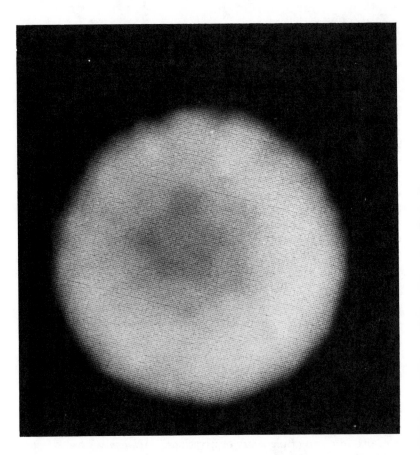

In 1976, Brad Smith, with the aid of
the Catalina telescope of the University of Arizona, secured the first CCD
images of an astronomical object,
Uranus, as seen at infrared wavelengths. The bright limb of Uranus
at these wavelengths was taken as
evidence for a hazy layer of ice crys-
tals high in the giant planet's atmosphere. (Courtesy of Lunar and Planetary Laboratory, University of
Arizona.)

as Danielson had argued, the potential performance of the Charge
Coupled Devices suited the interests of the planetary scientists much
better than did the SEC Vidicon.[56] The issue had been sharpened by
the fact that three months earlier, Brad Smith of the Lunar and
Planetary Laboratory of the University of Arizona had made the first
Charge Coupled Device observations with a telescope, obtaining several
dozen images of Uranus using a Charge Coupled Device camera provided by the Jet Propulsion Laboratory.[57] Smith, moreover, was one
of the planetary scientists who attended the meeting organized by
Roman at NASA Headquarters. The planetary scientists also "pleaded"
for NASA to allow a planetary camera to be proposed in the competition for the telescope's scientific instruments and to ensure that
there would be no bias against it in the selection process. Roman
was convinced, and she began to lobby for such a planetary camera
to be assured of a place aboard the telescope.[58]

Many planetary scientists had earlier shown relatively little enthusiasm for the telescope. But in 1976, with the planetary science
program faring relatively poorly in the NASA budget, they wanted
a piece of the telescope action.

At almost the same time as the meeting on the telescope and its usefulness for planetary science, the issue of a European instrument for the Space Telescope was coming to a head. As we noted in Chapter 4, NASA and the European Space Research Organization (soon to become the European Space Agency) had begun discussions on incorporating a European-built camera into the telescope's complement of instruments in 1974. At first those talks had encompassed several possible instruments, including a Faint Object Spectrograph. NASA, however, had balked at the idea of an international partner providing one of what it then regarded as the two most important scientific instruments for the Space Telescope. The issue had therefore quickly shifted to whether or not the Europeans would provide a Faint Object Camera based on a type of "Imaging Photon Counting System."[59]

Such a Faint Object Camera's main job would be to examine exceptionally faint objects, objects that typically would require many orbits of observation time to collect enough light for good images. A camera employing an Imaging Photon Counting System was attractive to the Europeans because, as one of the planning documents put it, "Europe has a certain lead in this area as University College, London has developed the only [such] Imaging system that is currently in routine use for optical astronomy."[60]

Following those earlier studies, the European Space Agency began the Faint Object Camera's official Phase A in late 1975. As with NASA studies, the bulk of the detailed design work was done by outside contractors. The Phase A study was therefore conducted by a European Space Agency team, with subcontracts to Dornier in Germany (the structure of the camera and thermal control), New Industrial Concepts Limited in the United Kingdom (the Imaging Photon Counting System), and Laboratoire d'Astrophysique Spatiale in France (the optical design). The design work was monitored by a Science Team composed of European Space Agency staff and outside astronomers.

NASA, meanwhile, before agreeing to any deal with the European Space Agency, wanted to be assured that the Europeans would be capable of providing a camera that was appropriately designed and constructed. ESA had refused to enter its planned instrument in the general competition for instruments, and so NASA wanted some other means of ensuring quality control. In consequence, NASA sent a high-powered group of U.S. engineers and astronomers to inspect laboratories and industrial plants in Europe and to assess the existing design for the Faint Object Camera.

The NASA review team concluded that the camera's Phase A design was too complicated.[61] It was uneasy that the camera included two possible light paths, as well as a spectrograph. These, the team argued, would involve too many mechanisms, thereby increasing the chances of something going wrong in orbit – a mechanism jamming, for example. As one of the leaders in the design of the camera has noted, at that very early stage of the design process, it "was a [Faint Object Camera] with bells and whistles attached."[62] The NASA review team also pressed for the design to better reflect the Faint Object Camera's main mission, that is, to secure the images of, and measure accurately the light emitted by, extremely faint objects.

The team's conclusion was that "prior to any NASA commitment, [the European Space Agency] Phase A design [must be] restructured to provide a greatly simplified instrument. . . . If the Phase A version of the [Faint Object Camera] is not so simplified, the Team recommends that it should not be part of the ST payload."[63] The review team, at the same time that it criticized the camera design, nevertheless drew an important conclusion. In its opinion, "the technology, the facilities, and the technical expertise required to design, build, test, and calibrate a Faint Object Camera . . . compatible with the Space Telescope meeting the scientific objectives established for this instrument currently exist in Europe." In the opinion of the high level review team NASA had assembled to address this issue, there was, then, no technical reason for NASA not to move forward with an international program with the European Space Agency as its partner.

AN OPEN COMPETITION?

The deliberations on the Faint Object Camera were taking place against the backdrop of a hard-fought debate within NASA on the approach to the "Announcement of Opportunity" for the telescope's scientific instruments. The announcement of opportunity would be a crucial document because it would give astronomers the specifications for the various kinds of scientific instruments that might be flown aboard the Space Telescope. One particularly contentious issue concerned the "core instruments." The Phase B Science Working Group had advanced that idea to indentify scientific instruments the group judged to be so central to the telescope's missions that the telescope should not be flown without them. As we noted earlier, the working group had picked two core instruments: the Wide Field Camera (which was planned to incorporate the SEC Vidicon detector) and the Faint Object Spectrograph.

The plan to designate core instruments did not meet with universal approval. It was opposed by both David S. Leckrone, the newly appointed instrument scientist at Goddard, and Nancy Roman, program scientist at NASA Headquarters. They argued that NASA should open up the process of selecting the scientific instruments and invite whatever type of proposal anyone wanted to make within a general set of guidelines. The other position was championed by O'Dell. He argued that the Phase B study had been thorough and expensive and had involved many competent people. Hence, its results should be adhered to closely. Because both Leckrone and Roman, as well as others at Goddard, had serious and growing doubts about the effectiveness of the SEC Vidicon and had been impressed by the promise of Charge Coupled Devices, the most important theme in the debate was whether or not to preselect the SEC Vidicon as the detector for the Wide Field Camera.[64] Charge Coupled Devices, as mentioned earlier in this chapter, were being strongly pushed by planetary scientists, and Leckrone and Roman, to a far greater extent than O'Dell, were sympathetic to the argument that the Space Telescope should be an effective tool for planetary science.

So in the fall of 1976, with the telescope likely to be a new start in the spring of 1977, the Science Working Group faced some tough decisions if it was to influence the writing of the announcement of opportunity. In fact, much of the last Phase B Science Working Group meeting was devoted to the issue of detectors, as had been its first meeting. Program scientist Roman reminded the group that the SEC Vidicon had been preselected for the Wide Field Camera. In Roman's opinion, however, "the development of the SEC Vidicon had not progressed as fast as expected and there is now doubt that it can meet the minimum performance specifications" established by the working group.[65] Roman wanted to reopen the question of which detector to select for the Wide Field Camera, and she announced that several groups and individuals had been invited to the meeting of the working group to present data on detectors, including the SEC Vidicon and Charge Coupled Devices. O'Dell noted that the performance specifications for the camera had been tailored to balance the ideal and the anticipated SEC Vidicon behavior, but it had begun to appear to him that it might not be possible to meet some of the specifications with the SEC Vidicon, such as adequate response in the red part of the spectrum and resolution over the entire field of view. After a break for lunch, at which more than food doubtless was chewed over, John Bahcall, who had come to regard the SEC Vidicon as "ancient technology" and who also had been impressed by Charge Coupled Devices,[66] offered a resolution that the Wide Field Camera, like the other U.S. instruments, be selected through the announcement of opportunity. The motion was carried unanimously. If NASA agreed, the SEC Vidicon would not necessarily be the detector for the Wide Field Camera, but would have to compete against what anybody else wanted to propose.[67] That, as one participant recalls, was a "very dramatic moment," because NASA and the astronomers had assumed for some years that the detector

247

for the Wide Field Camera would be the SEC Vidicon, and this announcement was "absolutely earth shattering."[68] To some others it was not so surprising, and Roman had maneuvered hard to bring about just such a result.

In his position as project scientist, O'Dell saw himself in many ways as the representative of the astronomical community within NASA, and so when that community spoke, in the form of the Science Working Group, he was able to change his public position on the Wide Field Camera and Charge Coupled Devices. Given the solid support of NASA Headquarters and Goddard, in the persons of Roman and Leckrone, for opening up the selection process to whatever type of proposal anyone wanted to make within the general guidelines, the issue of preselecting the Wide Field Camera dissolved. There would surely be a Wide Field Camera aboard the telescope, but the type of detector it would contain was still to be decided, although the lobbying of the planetary scientists had done a great deal to promote the claims of the CCDs.

FORMING TEAMS

Astronomers who wished to propose scientific instruments for the Space Telescope looked forward to the publication in March 1977 of the "Announcement of Opportunity for Space Telescope."[69] Interested astronomers had to assemble teams or join teams, and team leaders had to negotiate with contractors to decide which would construct their instruments if their proposals won out. The team leaders were also the prospective investigators, the people who would be responsible for the design and development of the scientific instruments. They would have to lead their teams in a scientific sense as well as managerial and engineering senses. As is characteristc of Big Science, the principal investigators would lead multidisciplinary teams and interact with other large teams composed of specialists from all sort of disciplines. In effect, they would be compelled to assume the role of hybrid scientist-engineer-manager.

For some of the people hoping to become principal investigators, their interest had by no means been obvious and had surfaced only late in the process. One such was Caltech astronomer James A. ("Jim") Westphal. Westphal was one of the planetary astronomers who had met with Nancy Roman in early August to urge that the telescope's instruments be capable of good planetary science. Westphal had also attended the October 1976 Science Working Group meeting as an expert on a type of detector known as the "SIT Vidicon," as well as to pitch for the planetary camera, but had not expected he would propose to be a principal investigator for the Space Telescope's Wide Field Camera.

Although always interested in astronomy, Westphal had studied

James A. Westphal in 1985. (Courtesy of Pearl Gerstel.)

geophysics in college and then begun work in the oil industry. After pursuing a career of developing and using geophysical instruments, Westphal went to Caltech in 1960. The idea was that he was to visit for a few months. Caltech, in fact, was soon to become his home. In making the physical move, he would also shift his research focus to constructing and using astronomical instruments.

It had nevertheless taken a lot of coaxing before Westphal became seriously involved in the Space Telescope program. Early in 1973, O'Dell, who had known Westphal for a decade, had tried to talk him into participating in the telescope's Phase B. O'Dell was especially impressed with Westphal's expertise with detectors and his uncanny knack of seeming to know just when a detector was ready to be taken to a telescope and put to astronomical work.[70] Westphal had declined, preferring to pursue his own ground-based projects.

However, one day late in October of 1976, Westphal recalls, his colleague James Gunn "came into my office, closed the door, stood by the side of the blackboard, looked me dead in the eye, and said 'we've got to build a wide field camera for the Space Telescope.' I said, 'you're out of your bloody mind.' " Gunn insisted that "if we don't, we're not going to be doing astronomy 10 years from now."[71] Westphal was still skeptical. He was heavily engaged in a range of other activities, but Gunn started to make calculations on Westphal's blackboard:

Now the calculations that he did were with the Charge Coupled Device performance that we knew about, and the quality of the images that we knew about. We knew the images were going to be 10 times better than the ground-based images, or better. We knew they were going to be a 10th of an arc second or better. The numbers were just mind boggling. A

249

factor of 10 improvement in the image size meant a factor of 100 in detectivity of faint objects, of faint stars, of faint point sources. Over the course of 45 minutes, he [went over] a list of 20 obvious things that you could do [with the Space Telescope] that you just couldn't do from the ground, no matter what. And it was extremely convincing. I finally said, "well, what do you want me to do?"[72]

Westphal and Gunn contacted several astronomers they wanted on their team, and to their surprise, all responded enthusiastically. The die was cast. Westphal had become a principal investigator for a Wide Field Camera proposal.

Westphal's team intended to exploit the Charge Coupled Device (CCD) as its detector, not the SEC Vidicon or any of the other detectors. When, in 1972, Westphal had first heard about CCDs he had not been at all impressed. At a NASA meeting on detectors in that year, "somebody from Bell Telephone came and described this wonderful new device called a CCD. I remember vividly at that time thinking, well, with that kind of performance, that thing is never going to be interesting to us."[73]

In the mid-1970s, however, he had become well aware of their improving performance, a consequence in part of development work at the Jet Propulsion Laboratory – itself a part of Caltech, Westphal's home institution – on the CCDs for use in planetary science missions. But in 1976 there was still a fundamental difficulty with the devices: their poor response in the ultraviolet, a region of the spectrum that provided one of the chief motivations for building and orbiting the Space Telescope in the first place. How to solve this problem?

At Goddard, Stanley Sobieski was leading a team that was preparing a proposal based on Intensified Charge Coupled Devices (ICCDs). As noted in Chapter 3, an ICCD is produced by placing a CCD inside a television tube. When ultraviolet light hits the front of the television tube, it is converted into a stream of electrons. The electrons then travel down the tube and are detected by the CCDs. In this fashion, the Goddard group hoped that by swapping electrons for photons, the poor ultraviolet performance of the CCD could be overcome.[74]

Two months earlier Westphal had told O'Dell he thought the ICCDs were the wave of the future. Now he sought a radically different solution. Clearly, a detector for the Wide Field Camera that did not work well in the ultraviolet would stand no chance of getting aboard the Space Telescope. Westphal, moreover, had negligible experience in working with the ultraviolet region of the spectrum. He therefore began to search for books and articles that might give some clue to a way out of his difficulty. When he checked under "ultraviolet" in the Caltech library's card catalog, he found an entry for James Samson's volume *Techniques of Vacuum Ultraviolet Spectros-*

250

copy. When he read Samson's book, he hit on an old technique in ultraviolet spectroscopy. It involved coating the detector with a material that fluoresces. Fluorescence is a well-known, but still remarkable, property of certain materials: Shine ultraviolet light onto quinine sulfate, for example, and it will emit blue light. When a material fluoresces, light of one wavelength is converted to light of a different wavelength. Westphal seized on fluorescence as the solution to the problem of the CCD's poor response to ultraviolet light. If he could coat the devices with a substance that would absorb ultraviolet light and then convert it to light of visible wavelength, the visible light could be detected by the device.

The comments in Samson's book on a coating substance called coronene particularly caught Westphal's eye.[75] Coronene, it seemed, could be deposited in vacuum and would leave a clear, thin film. "I thought about that," Westphal remembers, "and I realized that by golly there was a chance." If the coronene was really clear and was sensitive to ultraviolet light down to around 1,200 angstroms, "then it might work." The crucial test would be to coat CCD with coronene and examine its response in the ultraviolet. How to do this? Westphal, Gunn, and their team concluded that the best method was to take the coated CCD to the 200-inch Hale Telescope on Palomar Mountain and use it to observe a star.

By that stage, Westphal was receiving substantial backing from the Jet Propulsion Laboratory, and it was at the Jet Propulsion Laboratory that half of a CCD was coated with coronene. When it was placed on the Hale Telescope and directed toward a star, Westphal found that at a wavelength of 3,300 angstroms – that is, just in the ultraviolet – some 14 percent of the photons falling on the coronene-coated section of the CCD were detected. "And that just blew our minds . . . at that point we really knew we were home free."[76]

The coronene had given a good response in the ultraviolet. It was an innovation that was widely seen to strengthen significantly the claim of the CCD to be the detector for the Space Telescope's Wide Field Camera.

Westphal and his team had also hit on a method of splitting into four beams the light from the objects they were observing and placing that light onto one of two sets of CCDs (see Appendix 5 on the Wide Field/Planetary Camera). That design would permit the camera to have a wide-field mode, and another of higher resolution. So Westphal's team was proposing not only a Wide Field Camera but also what they termed the "Planetary Camera." Thus was born the Wide Field/Planetary Camera, although the term "planetary camera" was more of a selling tactic than a serious declaration regarding the kind of science to be done.

Princeton had meanwhile been preparing its Wide Field Camera proposal. Lyman Spitzer, Jr., led the team as the prospective principal investigator. Like the other Wide Field Camera groups, Spitzer's team contained astronomers from a wide range of institutions so as to secure a range of skills. Princeton worked closely with Ball Aerospace, its chosen industrial contractor, on the design of the camera. The Princeton proposal, though still founded largely on the SEC Vidicon, contained a CCD for its red response.

Because all the proposals for the Wide Field Camera incorporated CCDs, whichever one was chosen, CCDs would assuredly fly on the Space Telescope, despite the following facts: CCDs had barely entered astronomical service and still faced many development problems, and there was no thoroughly tested solution to the problem of their poor response in the spectral range (the ultraviolet) that provided one of the chief motivations for going into space to do astronomy in the first place. Why, then, was this switch made? We have seen that by 1976, the available SEC Vidicons had a long development history, but still had what were seen as serious technical problems. However, technical issues alone had not forced NASA to switch to a competition for the Wide Field Camera. The issue was much more complex than which detector would work "better" according to some set of technical criteria. The various detectors represented distinct individual and institutional commitments and investments in time and resources and held out different hopes for the two communities of potential users, the astronomers and the planetary scientists. As we have already seen, the strong pressure from planetary scientists to consider other detectors had been highly influential in swinging opinion in NASA against preselecting the SEC Vidicon.

The state of the art for CCDs had developed remarkably rapidly since discovery of the charge-couple concept in 1970. That advance, as we saw in Chapter 3, had been driven for the manufacturers by the potentially lucrative commercial and military markets. One central example was the attempt to replace Vidicon television tubes in commercial and home video cameras with CCDs. So even in the early 1970s, the development of CCDs was being fueled by a strong mix of "market pull" and "invention push." That is, the CCD innovators felt such confidence in the appeal of the devices that they believed CCDs would create a demand in themselves, but there were also clearly defined markets for the CCDs, and that market pull had played a significant part in the transformation of the devices into detectors suitable for astronomy. In contrast, the SEC Vidicon was based on a Westinghouse television tube that was attracting little attention. The tubes were thus being custom-made for the Princeton program and seemed to have no long-term future.

The historical process by which CCDs came to be aboard the Space Telescope cannot be explained in purely technical terms; rather, it was the product of a complex interplay of technical, scientific, commercial, institutional, and social reasons. In particular, the rapid

deployment of the devices as astronomical detectors was due to an
alliance of astronomers, planetary scientists, engineers, government,
and industry, made possible by industry's investment in the devices
for a range of possible commercial and military applications, as well
as more directed funding by NASA. Further, communication among
astronomers, planetary scientists, government, and industrial engineers – carried out by people who could move freely among the
several groups – was deliberately fostered, and to some extent controlled, by NASA program managers who sought to develop detectors for astronomical satellites and planetary spacecraft as well as a
clientele for those detectors.[77] The CCDs had also become a means
by which NASA managers had sought to reassure planetary scientists that the agency took their aspirations seriously, and so the CCDs
had played their part in coalition building.

Other choices

OTHER CHOICES

After much debate within NASA, the issue of core instruments had
finally been settled. Although the Wide Field Camera and the Faint
Object Spectrograph had not been designated as core instruments,
the compromise position set forth in the announcement of opportunity was that if good enough proposals were received for these instruments, they would both be flown. NASA saved another instrument slot for the European Space Agency's Faint Object Camera.
The other two available places were not assigned to particular instruments and so would be filled by a competition among whatever
instruments people wished to propose. For example, a High Speed
Photometer had not been specifically recommended by the Phase B
Science Working Group for the first flight of the Space Telescope,
but it was briefly described in the announcement of opportunity,
and reports on the Phase B work were available to interested astronomers.

One astronomer who proposed an instrument for an available berth
was Robert C. Bless of the University of Wisconsin–Madison. Bless,
as we saw in Chapter 1, was a long-time optical/ultraviolet space
astronomer whose involvement with flying astronomical instruments stretched back to the 1950s. He had taken his Ph.D. at the
University of Michigan, and in October 1957, he and the other
astronomy students there had been taken out by Leo Goldberg to
watch *Sputnik* pass overhead. Bless, however, had been aware of the
potential of astronomy from space for some years and had helped
develop a film from one of the Naval Research Laboratory's early V-
2 rocket shots.[78]

In 1958, Arthur D. Code was appointed director of the Wash-
burn Observatory at the University of Wisconsin–Madison, and he
soon founded a Space Astronomy Laboratory. Bless was one of his

Robert C. Bless. (Courtesy of Pearl
Gerstel.)

253

first appointments and was quickly immersed in space astronomy, flying instruments aboard rockets, balloons, aircraft, and satellites. In addition, Bless had been a member of Nancy Roman's Phase A working group, as well as the group chaired by O'Dell in Phase B. His attitude toward the Space Telescope was nevertheless somewhat ambivalent after the Phase B battles. There was even talk of closing down the Space Astronomy Laboratory at Madison, and Bless and Code, the laboratory's father figure, had begun to adjust to the idea of their space group dissolving.

However, when the announcement of opportunity for the Space Telescope was issued, Bless remembers, "we thought, well, there's still a group here that's been working together for a long time, and they want to continue . . . I thought as a matter of courtesy to the last people at [the Space Astronomy Laboratory], we should make an attempt at a proposal." But the area in which the laboratory had particular expertise was photometers, a type of instrument that had not received a high priority in Phase B. Bless still decided to propose a high-speed photometer, although "I really had a zero expectation that this would go, because it was not considered a core instrument." After a couple of meetings of the group to discuss the general approach, Bless sat down for several days with one of the Space Astronomy Laboratory's most experienced engineers, Don Michalski, to write the proposal. The photometer was to be a typical Wisconsin instrument:

It was going to be very small and simple. I get teased about this today because I characterized it as two thermos bottles and a shoe box. The thermos bottles were the detectors. They were image dissectors which go back to the mid-thirties. Image dissectors were the first successful television tubes. This was certainly not new technology. . . . And the shoe box was the electronics box, and the idea was that we would be some sort of tagalong on one of the prime instruments. We assumed that NASA would find a way of sticking the [High Speed Photometer] into the corner of an instrument using a pick-off mirror or something like that. We did not propose to be one of the major instruments since it just seemed probable from the Phase B studies that the photometer was not to be emphasized.[79]

MAKING THE SELECTION

The announcement of opportunity had been issued in March 1977. The responses from astronomers were due in the summer of the same year, but one issue was still of special concern to NASA: Which detectors to fly on the Wide Field Camera and the Faint Object Spectrograph? NASA therefore organized a special team to report on the status of the detectors. Chaired by Jeffrey D. Rosendhal of NASA Headquarters, the team's charge was to assess the strengths and weaknesses of the proposed detectors. To this end, during August

1977 the team visited all of the proposers for the Wide Field Camera and Faint Object Spectrograph. The team's findings and views were then passed on to the panels NASA had assembled to review the completed proposals for the scientific instruments. In their opinion, for example, the CCDs were a very promising and attractive technology.[80]

A total of thirteen teams proposed to build four scientific instruments for the Space Telescope. A fifth instrument had already been selected: the European Space Agency's Faint Object Camera, which was based on an Imaging Photon Counting System. With the issue of European participation in the building of the Space Telescope all but decided, the European Space Agency had established a project office for its Space Telescope activities. It had also appointed a project scientist,[81] assembled an Instrument Science Team under the chairmanship of the eminent astronomer Hendrik van de Hulst, and selected the major contractors for the Faint Object Camera (British Aerospace for the detector systems, and Dornier GmbH with Matra-Espace as co-contractor for the rest of the camera). The Instrument Science Team held its first meeting on February 16, 1977.

But how was NASA to choose the four U.S. instruments for the Space Telescope? NASA obviously would want to pick the best instruments. But what did "best" mean? A proposed instrument might promise superb science, but if the technology on which it was to be based was unlikely to be developed before the date of the telescope's launch, selecting that proposal would be risky. Those who reviewed the proposals had to take into account an instrument's cost too, for the complement of instruments would have to be built within a limited budget. Nor would it be sensible to choose an instrument if it was not compatible with the Space Telescope's expected performance.

The proposals for instruments from U.S. groups therefore went through a lengthy review process. After evaluations by panels of Goddard and Marshall engineers, NASA and outside astronomers, the proposals went before a "Synthesis Panel." From those deliberations emerged the proposed complement of instruments for the Space Telescope's first flight: Faint Object Camera, Faint Object Spectrograph, High Resolution Spectrograph, High Speed Photometer, and Wide Field Camera.[82] After an internal NASA review, Noel Hinners, head of the Office of Space Science, made the final selection. NASA, however, accepted the essential points of the synthesis panel's recommendations.

James Westphal's team won the competition for the Wide Field Camera. Westphal's solution to the CCD's poor ultraviolet response – the use of a coating of coronene – was seen by the review teams to be an elegant solution, even though it had never been tested in space. Further, CCDs were being developed for the *Galileo* mission to Jupiter (for which the Jet Propulsion Laboratory was the prime contractor, as it would be for Westphal's Wide Field/Planetary Camera). Because the *Galileo* mission's schedule would place it about a year ahead of the Space Telescope, NASA was optimistic that CCDs could indeed be ready in time for the telescope's launch. Although the CCDs were generally acknowledged to involve unproven technology, the choice, as a participant in the discussions recalls, was "one that was so exciting and so promising it was just irresistible. Because of that we put extra money and effort into the [Wide Field/Planetary Camera] Group." Because the *Galileo* mission would be launched before the Space Telescope, "we felt that they would solve all the problems and we would have the benefit."[83] (In fact, another type of CCD would be employed for *Galileo,* and so these hopes never came to fruition.)

The principal investigator for the Faint Object Spectrograph was the youngest of the proposers: Richard Harms, of the University of California, San Diego. His instrument team was paired with Martin-Marietta, a company that was already well-known to several of the San Diego space scientists. Harms, in fact, was a member of a group at San Diego that had received NASA funds for some years specifically to develop a type of one-dimensional detector known as a digicon. By 1977, the group had progressed to the stage that it had extensively tested digicons with ground-based telescopes.[84] The winning proposal for the High Resolution Spectrograph was the product of a team led by John C. Brandt of the Goddard Space Flight Center. Brandt was a well-known and experienced ground-based and space astronomer who had become head of the Solar Physics Branch at Goddard in 1967 and who in 1977 was chief of Goddard's Laboratory for Astronomy and Solar Physics. His team was paired with the Ball Aerospace Company for construction of the spectrograph.[85]

Both the Faint Object Spectrograph and High Resolution Spectrograph exploited digicons. In effect, devices for converting a beam of light into a stream of electrons and for focusing the electrons on a string of silicon diodes, digicons enable the different levels of light intensity in the original beam to be transformed into electrical signals.[86]

The competing proposals for the spectrographs featured two-dimensional detectors that held the promise of larger scientific returns, but the judgment of peer review and NASA groups was that they were a long way from being ready to fly in space. To have opted for any of these competing proposals, the argument went, would have introduced an unacceptable degree of risk.[87] Instead, all of the review panels judged digicons – wrongly, as it would turn out – to be relatively simple and well tested by the development work that several groups, including that at San Diego, had already conducted.

For the High Speed Photometer, NASA chose the instrument proposed by Robert Bless and his group from the University of Wisconsin. Bless would be the principal investigator, but some of the astronomers from a losing proposal were added to his team.[88] The photometer was to be built at the university.

It was not only the scientific instruments and the associated teams that NASA was selecting. The members of the Science Working Group, the main public forum in which the astronomers would be able to present their views on the issues of the design and development phase, were also being picked. The principal investigators for all of the scientific instruments, together with the project and program scientists, would automatically be members of the Science Working Group, but there were several other positions to be filled. Proposals had therefore been submitted by astronomers aspiring to be Science Working Group members by becoming one of the four "Interdisciplinary Scientists," two "Telescope Scientists," or the "Data and Operations Team Leader."

The synthesis panel had recommended that the interdisciplinary scientists be selected to provide scientific expertise in several broad areas, including planetary astronomy, galactic astronomy, extragalactic spectroscopy, and stellar astronomy. NASA agreed. Those astronomers picked – John Bahcall, John Caldwell, David Lambert, and Malcolm Longair – therefore represented a diverse range of interests, and Caldwell had been picked specifically because of his credentials as a planetary scientist. His presence was therefore a clear sign of the pressure the planetary scientists had been able to exert in the selling of the telescope. Two telescope scientists were selected: William G. Fastie and Daniel Schroeder, and Edward J. Groth was chosen as leader of the Data and Operations Team.[89] There was, in addition, to be an Instrument Definition Team for astrometry. That team was to be headed by William Jefferys, an expert in astrometry and celestial mechanics, who would also sit on the Science Working Group. The astrometry team, composed largely of astronomers from the University of Texas at Austin, would not lead the design and development of an instrument in the same manner as the other Investigation Definition Teams. Instead, the team would focus on the astrometric performance of the three Fine Guidance Sensors, which composed the telescope's Fine Guidance System. And the Fine Guidance System, through its intended ability to locate precisely the positions of the stars, was to be, in effect, a sixth instrument.

THE SHAPE OF THINGS TO COME

With the selection of the members of the Science Working Group and Investigation Definition Teams to build the scientific instruments, the major elements of the program design for the telescope

257

were complete. Four of the Phase B Science Working Group had survived into Phase C/D: John Bahcall, as interdisciplinary scientist, Robert Bless as principal investigator for the High Speed Photometer, Bob O'Dell as project scientist, and Nancy Roman as program scientist. The new members would soon be confronted with many of the same problems and issues that Bahcall, Bless, O'Dell, and Roman had lived with for years. Moreover, whether or not the astronomers had recognized it, with the new program design the power structure within the program had shifted dramatically from that operating in Phase B. In Phase B, the astronomers had been central to the planning for the telescope, and the working group had been able to exert a great deal of influence, despite being a purely advisory group. That state would not persist into Phase C/D. During Phase B and the selling of the telescope, it had been crucial for NASA managers to maintain the support of the astronomers: Without the astronomers the telescope could not be sold. In Phase C/D, that pressure no longer applied. Instead, with funds generally extremely tight, the working group and the scientists on the program became just another interest, and often a marginal interest at that. The Phase C/D Science Working Group was decidedly junior to the Phase B group — for example, such scientific and political heavyweights as Margaret Burbidge, Arthur Code, George Field, and Lyman Spitzer were no longer present — and only a few of the Phase C/D group had much experience in the practice of space astronomy.[90] Nor did the institutional device that the Phase B astronomers had created to give them a strong measure of scientific control, the Space Telescope Science Institute, come into being until well into the design and development phase.

But the astronomers, as we have seen in this chapter, were only one group in a complex program design. NASA had been the architect of that design, and agency managers now faced the task of managing the telescope's design and development by seeking to mesh together an extremely wide range of often competing interests within the framework of that program design.

8

Problems arise

I am old enough on the job to know . . . that estimates invariably are low and you run into cost overruns. Will you know when you complete Phase B what this project will cost?

Congressman George E. Shipley, 1975

We really thought that NASA and the contractors pretty well understood that project and what it involved. What went wrong?

Congressman Ronnie G. Flippo, 1981

We totally underscoped the job. I don't think it was done malevolently, with malice aforethought. I just think it was an honest misunderstanding or nonknowledge of just how difficult it is to build a 2.4 meter telescope that's stable to seven milliarcseconds that has to operate for fifteen years.

Edward J. Weiler, 1983

On Saturday, July 26, 1980, staff members from Marshall, Goddard, and NASA Headquarters met at Marshall to discuss the status of the Space Telescope program. From Marshall's perspective, the financial position was sufficiently bleak that the center's representatives were to suggest a series of drastic cost-cutting measures, including deletion of two of the scientific instruments from the telescope's first flight. It would be an explosive meeting and in project folklore would win the grim title of "Black Saturday." As we shall see, Black Saturday was only one of a string of events that had potentially crucial repercussions for the science to be conducted with the Space Telescope. Indeed, in 1980, and again in early 1983, the project passed through such troubled times that some participants feared the telescope might be canceled.

Why, less than three years after the start of detailed design and development work on the telescope, was the program in such a predicament? The telescope program, in fact, ran into severe financial and management problems almost from the start, but NASA seemed powerless to do much to shake these off until 1983, by which time the cost had more than doubled and the launch date had been delayed by almost three years. In part, the difficulties arose because the telescope was sold to Congress at too low a cost, but also there was a misunderstanding of how demanding it would be to design and build the telescope. Certainly the unfolding of events would disclose that because so many of the costs of the program had been underestimated, substantial cost increases were virtually built into the program from the start of Phase C/D. For example, NASA's estimated costs for operating, maintaining, and refurbishing the telescope in orbit were woefully short of the mark. The problems that were encountered by NASA, the astronomers, and contractors, however, were often not just technical problems that required for their solution more time and money than had been allocated. Rather, they were to a considerable degree consequences of the strains imposed on the program by the way the telescope had been sold, as well as the management approach and program design NASA had adopted in the mid-1970s, strains that quickly compounded the inherent difficulties of designing and building an extremely complicated spacecraft. In this chapter, then, we shall see how the telescope program was trapped by its own history.

CONTRACTS

At the start of Phase C/D, the telescope's detailed design and development phase, there was a widespread belief in NASA that the technical tasks generally were well defined. NASA managers accepted that taxing problems were bound to arise in a novel research and development enterprise of such large scale. For that reason, substantial reserves had been set aside, and the first major adjustment to the contracts had to be made within a matter of months of the project's start.

The contracts for the Space Telescope were of a type known as "cost plus fee." NASA would therefore pay whatever it cost to construct the telescope, but any changes to the contracts defined at the start of Phase C/D would have to be negotiated between the agency and the contractors, for such changes generally would imply extra costs, costs to be borne by NASA. The incentive for the aerospace contractors and optical houses to produce high-quality work within the budget and on schedule would come from the awarding of "incentive fees." NASA would assess a contractor's performance to decide how much incentive fee should be awarded in each contract period.[1]

But in February 1978, within a few months of the start of Phase C/D and due to various renegotiations of the contract for the Optical Telescope Assembly, Perkin-Elmer's contract had already been adjusted by $22.4 million to a total of $91.8 million. These changes included the addition of an external radiator to improve the performance of the Wide Field/Planetary Camera by allowing its CCDs to run at a lower temperature. The cost increases were portents of things to come.

During the program's early stages, NASA worked to establish schedules and define the milestones that would have to be accomplished to build the telescope, a task that required carefully dovetailing the work to be done by various contractors. The Optical Telescope Assembly, for instance, would eventually have to be incorporated into the Support Systems Module; in effect, the optical system would have to be integrated with its supporting spacecraft. The Support Systems Module was to be constructed by Lockheed, and so the mating was scheduled to occur after the Optical Telescope Assembly had been delivered to Lockheed. If the work at Lockheed and Perkin-Elmer was to proceed efficiently, it would have to be carefully coordinated.

NASA organized the building of the scientific instruments, the Optical Telescope Assembly, and the Support Systems Module into three major and partly overlapping stages: (1) design and development, (2) fabrication, and (3) assembly and verification. For example, design and development for the Optical Telescope Assembly was planned to last from late 1977 to mid-1980, and fabrication and assembly were to take from mid-1979 to late 1981. The completed Optical Telescope Assembly would be tested at Perkin-Elmer and delivered to Lockheed's Sunnyvale plant in mid-1982. There the assembly would be integrated with the Support Systems Module and the scientific instruments. After the entire Space Telescope had been tested, it would then be shipped to the Kennedy Space Center in Florida for launch aboard the Space Shuttle in late 1983.[2]

The order in which the telescope parts were to be built meant that Perkin-Elmer's activities usually were scheduled to be performed ahead of equivalent activities at Lockheed, and the various design reviews for the Optical Telescope Assembly were slated to precede those for the Support Systems Module. In effect, the telescope was to be built from the inside out. It was natural, then, that in the early years of Phase C/D, NASA's attention should be focused more on Perkin-Elmer's management and technical performance than on Lockheed's performance.

MONEY TO THE FORE

The overriding issue for the telescope program for both NASA Headquarters and Congress had long been its cost. As seen in Chapters 4 and 5, the funding battles of the mid-1970s had led NASA managers to develop a so-called low-cost approach to building the

Space Telescope, at the core of which were the capabilities of the Space Shuttle. The shuttle, NASA had contended, would offset some of the extra risks entailed by the low-cost approach, by enabling astronauts to replace components and subsystems in orbit, as well as return the telescope to the ground for maintenance and refurbishment. In line with the low-cost approach, Marshall had been conservative in its choice of safety factors for the various structures of the telescope. Marshall therefore hoped that it would not be necessary to do load testing on the structures (e.g., the sorts of tests that would indicate that the structures would survive the loads that would be imposed on them during the telescope's launch). The center could then reduce the amount of testing and the amount of hardware in the program. Mathematical analyses alone to "test" structures were expected to be less expensive; however, Marshall and the contractors ran into some problems with that approach with the Optical Telescope Assembly. The assembly was made of a composite material, graphite epoxy, and there were design problems with certain com-

The Optical Telescope Assembly's 200-inch-long graphite-epoxy metering truss. (Courtesy of Perkin-Elmer Corporation.)

ponents. As NASA chief engineer Jean Olivier recalls, there were difficulties with some

brackets and complicated joints. We'd design it. The design stress analysis would say this joint should carry so many thousand pounds. We would test it and it would carry half that much and fail. And so we had a big disparity in some cases between the tested strength and the design strength, and that undermined our confidence that the [standard] safety factor of 3 was adequate in the more complex aspects of the structure. So then we had to go back and build more component-level, special models of complex portions of the structure, to test and to see if they indeed were meeting the requirements. In some cases we found we had to re-design some of the [Optical Telescope Assembly] structure.[3]

The use of a different engineering approach, then, did not in that case reduce costs as much as NASA had reckoned. Other ways to save money that were debated in Phase B (e.g., to use many so-called standard components for the spacecraft and software that were "portable" between different organizations, as well as between different phases of the program) often were not, for a variety of reasons, put into practice. It was, in fact, one thing to discuss a low-cost approach theoretically, but quite another to carry it out in practice.

An early test of the low-cost approach had come in 1978 when the estimated cost for one of the scientific instruments, the Faint Object Spectrograph, reached $23.1 million, an increase of around $13 million in a few months. NASA project manager Keathley wanted to ensure that that did not set a trend. As he wrote to headquarters, "I need not more than mention that our decisions concerning their development [the instruments] will be closely watched by the Program participants; the outcome will demonstrate our seriousness relative to low-cost management."[4] The result was some tough negotiation — among NASA, Martin-Marietta, and the home institution of the spectrograph's principal investigators, the University of California, San Diego — in which the agreed cost for the spectrograph was hammered down to around $14 million (fiscal 1978 dollars).[5]

Some cost increases had nevertheless been incorporated into NASA's planning. The managers knew that as the contractors pushed ahead with their design and development work, the requirements for the telescope would become better understood, and the numerous interfaces between the contractors would become better defined. The refined and revised requirements would force alterations in the telescope's design. Costs would then rise to accommodate those changes. About half of the funding NASA had requested to build the telescope was in fact in reserve to cover such "expected" changes, as well as to meet unexpected difficulties.[6]

Despite the substantial reserves, within two years the financial pressure had begun to mount. In July 1979, Marshall compiled its "Program Operating Plan" for fiscal 1980, a plan for what money was to be spent where in the forthcoming fiscal year (October 1, 1979, to September 30, 1980), as well as in future years. The Program Operating Plan therefore detailed when money would be spent and for what. By that time, the calculations of both Lockheed and Perkin-Elmer regarding what it would cost to complete the work planned had overshot by many millions the amounts originally budgeted. To ensure that the program did not run out of money, Marshall, in what would be a familiar story for years to come, had decided to adjust the program's reserves, make changes in some planned activities, and reschedule some jobs to a time when money would be available.[7] By the summer of 1979, NASA Headquarters had also become uneasy about the project's cost increases and had asked Marshall for a complete cost review of the Space Telescope program.[8] Headquarters wanted it to begin in October 1979.

As the summer of 1979 moved into fall, the contractors produced their revised so-called cost-to-complete estimates, estimates of how much more it would take to complete their respective sections of the telescope. Those figures were to give rise to what later NASA briefing charts would highlight as "explosive cost growth since September 1979." The Lockheed cost-to-complete estimate, for example, was $145 million (1981 dollars). Even after allowing for the effect of inflation, that was a sizable jump from the $77 million contract negotiated no more than two years earlier.[9] Those increases, moreover, were occurring while the program was still in the detailed design phase. The bulk of the manufacturing work had not even started, and long experience had shown that it is the manufacturing stage during which shortcomings in the design process become apparent and costs really begin to increase in major aerospace projects.

NASA had made many claims during the selling of the telescope about readily accepting increased risks with less ground testing of the telescope and offsetting those with the use of the shuttle, but converting these claims into practice proved extremely difficult. At the start of Phase C/D, for example, there had been no plans for a so-called thermal-vacuum test, a rigorous operation of the systems of the Space Telescope within a chamber large enough to hold the completed spacecraft. The air would be pumped out of the chamber so as to create a vacuum and simulate the environment of space. Sets of sun-simulator lamps within the chamber would let engineers run tests in conditions mirroring the widely varying temperature conditions to be encountered in space. Deciding during the selling of the telescope one could do without such a test was one thing; it was quite another to accept it in the hard light of a highly complex design and development program.

Headquarters program engineer Arthur Reetz had joined the Space

264

Telescope program in late 1977. As a newcomer to the program and as someone who had not been involved in the compromises and negotiations of the previous years, the omission of the thermal-vacuum test had staggered him.[10] As the experienced Reetz knew well, the thermal-vacuum test usually is regarded as the most important test of all for a spacecraft. It allows engineers to put a completed spacecraft through its paces and test how all the different elements of the vehicle combine in a simulated space environment. A thermal-vacuum test lets engineers investigate a spacecraft's likely performance, as well as check for design, material, or workmanship defects.

The testing program for the telescope had instead initially emphasized a "modular approach." Writing to Lockheed in 1978, a Marshall engineer explained that the Support Systems Module and its sections would be tested by Lockheed, the Optical Telescope Assembly and its sections by Perkin-Elmer, and the scientific instruments by the principal investigators.[11] Marshall contended that testing in that fashion, in combination with analyses of mathematical models of how parts of the spacecraft should behave, and the use of the Space Shuttle – both central elements in the low-cost approach – would obviate the need to build a special thermal-vacuum chamber in which to perform the testing. What that meant, in effect, was that the various contractors would build separate parts of the spacecraft and test them; then the telescope would be assembled from the sections, placed into the shuttle, and launched.

Although Lockheed had not proposed a thermal-vacuum test in its contract bid, the company was soon pressing for its inclusion once the contract had been awarded. As a Lockheed manager stressed in May 1978, "we are concerned that the lack of ST integrated environmental testing" might lead to cost increases. NASA intended to retrieve the telescope with the shuttle after five years in orbit so as to perform maintenance and refurbishment of the telescope on the ground. Without more ground testing, it might be necessary to use more shuttle flights to perform repairs in orbit or to bring the telescope back to earth before the scheduled date.[12] As time went on, NASA would concede its earlier plans for testing were unrealistic; extensive ground testing of the completed Space Telescope was essential to ensure its success. The agency had therefore come full circle: The shuttle had earlier been exploited by NASA as a means to save money because it would enable the contractors to cut the amount of ground testing; now it was necessary to cut back the number of repair missions, and so expand the ground testing of the telescope because of the cost of shuttle flights.

A COST REVIEW

It was against this background of growing realization and acceptance, at NASA Headquarters as well as the field centers, that there were difficult technical problems to be overcome, as well as rising costs and pressure to include more items in the program, that in October 1979 T. Bland Norris led the cost review that NASA Headquarters had requested some months earlier. Norris, widely respected, and a long-time NASA engineer and manager, had recently retired after several years as deputy and then as head of the Astrophysics Division in NASA Headquarters. He was therefore quite familiar with the Space Telescope. Together with telescope scientist William Fastie, Norris was joined for the review by engineers and managers from Goddard, Jet Propulsion Laboratory, Marshall, and NASA Headquarters.

Norris's approach to leading the cost review was that "we'll just take every major subsystem and look at the critical path. We'll then try to look at the cost drivers, the risks in those critical paths, and then we as a group [will] put it on the blackboard and try to say, all right, what do we think the likelihood of this critical path making that schedule is? And if we don't, where do we think it's likely to run into trouble?"[13] In the light of that analysis, the team made changes to the existing schedules and estimated the costs that would be incurred if their new schedules were followed.

Grinding and polishing the Space Telescope's primary mirror began in May 1979. By December of that year it was undergoing its final grinding operations at the Perkin-Elmer plant in Danbury, Connecticut. (Courtesy of NASA.)

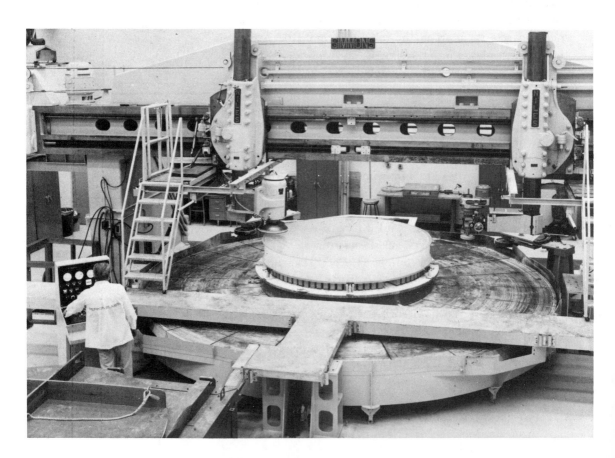

In that manner, the sixteen members of the cost review team arrived at a number of "Observations." Most important, they identified several critical paths that seemed to involve high risks, any of which might force postponement of the scheduled launch date of December 1983. Bundling together its findings, the review team concluded that the planned launch date of December 1983 was "in jeopardy. There is a 50% chance to contain the slip [in launch date] to $7\frac{1}{2}$ months (and a cost of $455 million in fiscal 82 dollars) and a 90–100% chance to contain the slip to $15\frac{1}{2}$ months (and a cost of $495 million in fiscal 82 dollars). Almost all of either slip is foreseen to be caused by the [Optical Telescope Assembly] with two scientific instruments running the [Optical Telescope Assembly] a close second."[14] In other words, the review team reckoned that the Space Telescope could be built with the current reserves, provided that no more major problems arose.[15] Of course, at that time, the major part of the detailed manufacturing work had not even started, and so that was an extremely big "if." On the other hand, the review team's assessment assumed that there were no ways to recover lost time in the schedules. The team therefore regarded its conclusions as highly conservative. The contractors had another three years to

A cost review

Lord Rosse's grinding and polishing machine, developed in the 1830s for use with thirty-six-inch-diameter telescope mirrors. Compare Rosse's device with Perkin-Elmer's technology of a hundred and fifty years later. (Courtesy of Director and Trustees, Science Museum, London.)

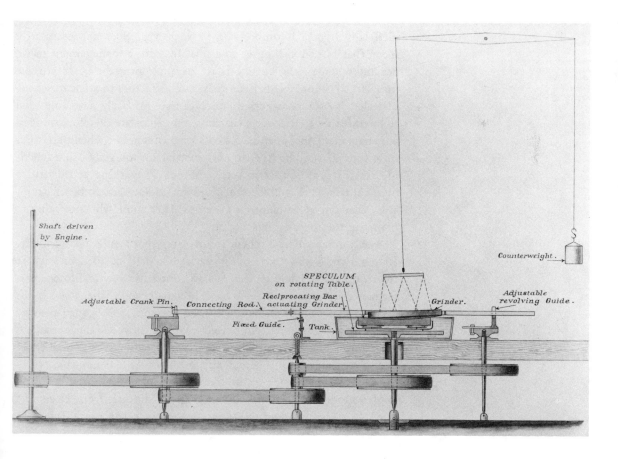

267

complete the telescope, and parts of the program originally thought to be essential might prove to be dispensable; if so, time and money could be saved.

Marshall, as the lead center for the Space Telescope, made a series of responses to the Norris team's "Observations." On most items there was little dispute. Marshall agreed that designing and building the Optical Telescope Assembly posed the biggest threat to meet the existing schedule, but, the center argued, "it's premature to project a launch slip at this time."

The crucial point about a slip in launch date is that, other things being equal, it drives up a program's total cost, as the contractors have to be paid for a period longer than originally planned. A postponed launch date is also a clear signal to Congress and the Office of Management and Budget that a program is not going as planned. Such slips tend to damage NASA's standing with the very people who provide the agency's funds. Also, if a program is delayed, costs rise, and so there is less money available to NASA to start new programs, and new programs are essential to ensure the future of the agency. The latter point will, as we shall see, be important in understanding the reaction of NASA Headquarters to the telescope's problems.

Even if a project's participants are generally agreed that a scheduled launch date is unlikely to be met, that does not necessarily mean that NASA will agree to a slip. Instead, a management tactic not unknown to the agency is to keep a contractor's "feet to the fire." In so doing, NASA compels its contractors to stick to schedules that NASA managers themselves may privately have conceded to be unrealistic. In some situations, schedules are therefore more of a management tool than reasoned statements about when particular milestones have to be reached. A potential advantage of using schedules in this way is that it keeps contractors working at full tilt; a disadvantage is that if the schedules are widely seen to be unattainable, that can be damaging to the morale of those who are trying to maintain the impossible schedules.

A slip in launch date, then, is not a step that NASA takes lightly. Rather than follow that thorny path, Marshall decided to stick to the road already laid out, but to watch carefully Perkin-Elmer's performance during fiscal year 1980. Marshall would then reassess its decision in the light of Perkin-Elmer's progress. So Marshall planned "no realignments of current project plans until we see those results." "We can," Marshall replied to the Norris review, "visualize problems that could cause a launch slip in the order of 6 months. Nothing short of a disaster could cause a slip of 15 months." Within twelve months the launch date was to be slipped, from December 1983 to a commitment to Congress to launch the Space Telescope

"DESCOPING"

Late 1979 and early 1980 saw several personnel changes on the Space
Telescope program. Two central changes involved severing the close
link between the program manager at NASA Headquarters and the
project manager at Marshall. In October 1979, NASA program
manager Warren Keller was promoted within NASA Headquarters,
and in February 1980, Marshall's project manager, William Keathley, left Marshall to assume a senior position at Goddard. The working day for Keathley and Keller usually had started with a lengthy
telephone conversation on the status of the program, but in subsequent years the relationship between project manager and program
manager would never again be so close.

Why was that important? On a program as complex as the Space
Telescope, it is impossible for any one manager to know fully what
is going on; certainly he or she cannot verify all the technical assessments presented. One of a manager's main jobs is thus to make
judgments on what he or she hears. But what is the manager to do
with the information received? Does it come from someone who
usually exaggerates problems, underestimates problems, is invariably sound on technical issues, is competent, is incompetent? In
other words, management can, to some extent, be boiled down to
personal relationships and experience. However, if new managers
are continually being introduced into leading positions in a program, the basis of experience a manager needs to function effectively
is regularly being lost. The arrival of a new manager can also lead
to the overturning of earlier decisions with which he or she dis-
agrees, one possible product of which is disruption to the program.
In the next three years, there would be four program managers in
NASA Headquarters (not counting acting program managers) and
four people in charge of the Office of Space Science in headquarters.
Not all such management changes, however, need have adverse effects. New faces can also mean new opportunities and new approaches, sometimes more realistic approaches. In fact, we shall find
that changes of project managers in 1980 and 1983 were closely
associated with major increases in funding for the telescope.

Keathley's replacement as project manager at Marshall was Fred
Speer. Throughout his childhood, Speer had been fascinated by rockets
and space flight. Although not one of the original group of German
engineers from Peenemünde who had joined Wernher von Braun in
the United States at the end of World War II, he had written to
von Braun from Germany to ask if there might be a position for him

in the United States. Speer, then an assistant professor of physics at the Technical University in Berlin, eventually joined von Braun at Huntsville in 1955.

When the U.S. Army Ballistic Missile Agency had been transformed into the Marshall Space Flight Center in 1960, Speer, along with many of the center's staff, began work on Saturn launch vehicles. In 1971, as the Apollo program was winding down, he became project manager for the High Energy Astronomy Observatory (HEAO) program. There he managed the design and development of three scientific spacecraft. Speer became known to astronomers as a man of high integrity and a strong manager who took pride in completing projects within cost and on schedule, even if, on occasion, that meant cutting back a spacecraft's scientific capabilities.[16]

In early 1980, the HEAO program was drawing to a close. Given his experience with scientific satellites, it was a natural step for Speer to transfer within Marshall to replace Keathley as a project manager for the Space Telescope, even if the reputation that preceded him made some of the scientists apprehensive.

Problems soon confronted Speer. Costs had continued to rise in early 1980. On February 21, 1980, for example, O'Dell noted that the estimate for the funds needed to design and develop the scientific

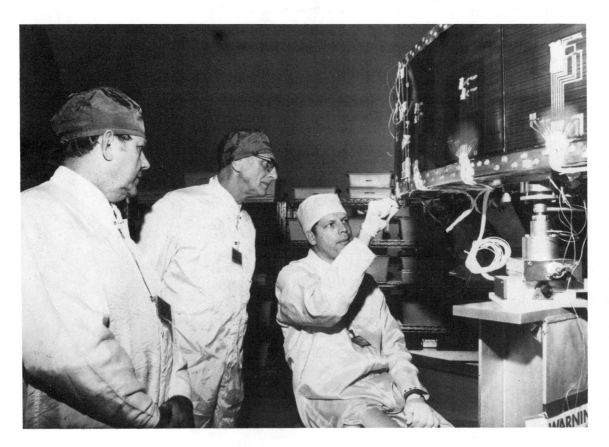

Marshall director William R. Lucas (left) and Space Telescope project manager Fred Speer (center) seen inspecting the installation of heaters on the primary mirror's main ring at the Perkin-Elmer plant in Danbury, Connecticut, in 1982. (Courtesy of NASA.)

instruments had jumped within only a few months from $89 million to $97.3 million (fiscal 1981 dollars).[17] Indeed, when in early 1980 Marshall prepared its latest Program Operating Plan, Speer and other Marshall managers calculated that they would, unless corrective measures were soon taken, run out of funds before the end of the current fiscal year, fiscal 1980. Speer had also decided by March 21, as he told O'Dell, that "we have already blown our launch date."[18]

At hearings on Capitol Hill four days later, Edward Boland, chairman of the House Appropriations Subcommittee dealing with NASA, wanted answers: "Is this project on time and will it really meet the December 1983 launch date?" The reply came from Thomas Mutch, head of the NASA Headquarters' Office of Space Science: "Yes, sir. The development of the Space Telescope is on schedule."[19] Headquarters, moreover, would not countenance a delay in the telescope's launch date for several more months.

In addition to problems in the fiscal 1980 budget, Marshall's own reserves were inadequate to meet all of the expected shortfall in the telescope's fiscal 1981 budget. Speer asked headquarters to release part of its reserve for fiscal years 1981 and 1982. But it was not simply a question for headquarters of moving funds from one budget line to another. The headquarters reserve is not a straightforward program reserve to be applied whenever there are shortages of funds, but a reserve to meet required adjustments in the program.[20] Instead of releasing its reserve, headquarters proposed another meeting to examine the Space Telescope's budget.[21]

Headquarters, in fact, wanted to keep its reserve available to back up an alternative plan Marshall was asked to fashion. One point was clear to Marshall, nevertheless: Extra money would not be readily forthcoming to solve problems. Marshall assured headquarters that it would "continue a very tight budget policy in all project elements." More significant was that at the same meeting, Marshall raised the issue of "descoping," that is, reducing the "scope" or performance of the Space Telescope in order to simplify the engineering task, to save time on the schedule, and to reduce costs. For the astronomers, descoping was a particularly unpalatable prospect, for it conjured up direct and serious threats to the telescope's capability to perform science.[22]

Speer also raised the possibility of descoping the scientific instruments, thus saving money by reducing their performance or delaying their development. That suggestion, according to one leading observer, made the headquarters managers in attendance "uncomfortable."[23] Despite that reaction, Marshall would soon make the suggestion again in a somewhat modified and more concrete form as the centerpiece in a cost-cutting plan, thus producing acrimony and a torrent of protests that were to have repercussions throughout the program.

FIGHTING OVER THE BONES

The Science Working Group met for two days in May 1980. With descoping in the air, the gloomy mood was captured by a cartoon sent to the group. It shows some cavemen picking disconsolately at a rather sorry collection of bones. The dialogue reads: "This meeting was called in order to discuss the meat. It has been pointed out that there is no more meat. A motion has been made to fight over the bones."

On the second day of the meeting, Speer explained the cost problems he faced. The planned and potential contract changes implied that unless some action were taken, the project would overspend in fiscal 1980 by some 6 percent. Speer cautioned that although the overall amount of money available appeared sufficient to handle long-term costs, the project reserves currently totaled only 5 percent of that amount, a small margin of safety indeed for a major research and development program still some years from completion. Speer therefore solicited "items from the Working Group which are potential cost increases but which have been overlooked. He suggested that we must all participate in going back to basic requirements." And, as Speer emphasized to the working group, he considered it impossible to request any more money from headquarters.[24]

Exactly how far Marshall could go or would be forced to go in its reevaluation was still unclear to the astronomers. What was obvious, however, was that there were likely to be some painful decisions ahead, and in the process of cutting costs, the telescope might emerge as a less powerful scientific tool. Such was demonstrated when the discussion by the Science Working Group turned to the question of flying infrared instruments aboard the telescope. The instruments for the first flight had, of course, already been chosen; an infrared instrument was not among them. The working group nevertheless very much wanted to ensure that no changes would be made to the spacecraft that might preclude the use of infrared instruments later. That issue was thrown into sharp relief because Marshall had identified a part of the telescope's design that could be changed to save money, but that change would mean that no infrared instruments could be employed on the Space Telescope, at least not until after the telescope had been returned to the ground by the Space Shuttle and modifications had been made.

The working group members also debated the issue of what to do if a scientific instrument should suffer severe cost and schedule problems. Should NASA press ahead and launch the telescope without that instrument? O'Dell thought the answer should be yes. Speer commented that although that was for him the least acceptable way of saving money, such a harsh solution would have to be considered if a truly difficult problem arose. The only people to back O'Dell to

any degree were John Bahcall, "who thought it a terrible alternative, but rational," and David Leckrone, Goddard's instrument scientist. Not surprisingly, the idea of leaving behind a scientific instrument met with no favor from any of the principal investigators.[25]

THE POSITION WORSENS

If the project's financial position looked pinched at the May Science
Working Group meeting, it was soon to become even worse. As
Marshall and Goddard continued to receive information from the
contractors and to revise their estimates of the costs for the next
fiscal year, fiscal 1981, the problems piled up. By June 27, Speer
was telling the project managers at Lockheed and Perkin-Elmer that
Marshall "is presently in the process of reviewing the total [Space
Telescope] budget and it is evident that an austerity program is
necessary for a successful completion of the [Space Telescope] project." He wanted both associate contractors to achieve better control
over costs by strengthening oversight procedures for their own subcontractors.[26] Marshall also assembled two cost review teams, one
run largely by Marshall, and the other chiefly the charge of Goddard
staffers, directed to examine the scientific instruments and science
operations.

At a July 18 meeting at Marshall on the "Space Telescope Budget
Situation," the project's financial position looked extremely serious
to Marshall managers. The latest cost estimates from the contractors
indicated that in fiscal 1981 the Support Systems Module would
require $14.4 million more than expected, the Optical Telescope
Assembly $9.8 million more, and the scientific instruments an extra
$9.8 million. The anticipated cost of operating the telescope once
it was in orbit had also grown. Goddard calculated that the money
to be spent on operations would meet the cost guidelines for fiscal
years 1980 and 1981, but by fiscal 1985, at which time the telescope would be in orbit, the expected outlays would have increased
$53.3 million over the guidelines. Similarly, in planning for main-
taining and refurbishing the telescope on the ground and in orbit,
Marshall's estimates were $7.5 million under the guidelines for fis-
cal 1982, but $11.1 million over budget for fiscal 1985.[27]

The message from Marshall was that somehow the books would
have to be balanced, and many members of the project were busy
scrutinizing all sorts of ways of cutting costs. They had gone back
to the drawing board, and by that stage even the previously unthinkable was being thought.

One item that came in for intense examination both within NASA
and at the contractors was the telescope's aperture door. Speer had
begun to wonder how necessary the aperture door was to the telescope's performance. From the time in 1974 when plans for an ex-

Problems arise

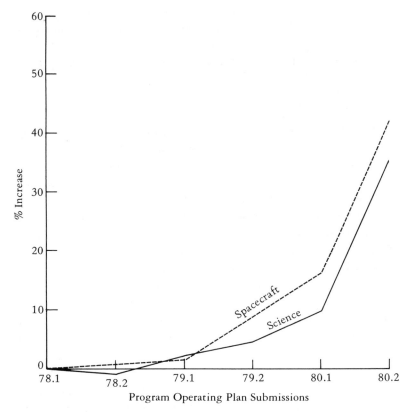

Changes in estimated costs for the Space Telescope. Graph prepared in mid-1980 by Marshall Space Flight Center. (Courtesy of Stansbury, Ronsaville, Wood Inc.)

tendable sunshade had been dropped for cost and technical reasons, the aperture door had been an integral feature of the Space Telescope. Designed by Lockheed as a movable cover over the telescope's light shield, it played two chief roles: first, to help prevent stray light from entering the telescope and degrading the quality of the images the telescope's optics produced; second, to keep contaminating particles from entering the spacecraft during the final assembly of the Space Telescope at Lockheed and during launch aboard the Space Shuttle.

In his position as project scientist — in effect, the representative of the scientific community on the project at Marshall — O'Dell was closely involved in the decision-making process on the aperture door. He warned Speer that if it were deleted, the science losses would be "major."[28]

Lockheed agreed. The company had studied the constraints that would be imposed on just where the telescope could point safely if the aperture door were removed. A central consideration was that a much smaller part of the sky would be available for observations at any one time (50 percent of the sky as opposed to 82 percent). Observations that required returning to the same object again and again (e.g., to monitor the changes in light output of Cepheid stars in distant galaxies in order to better determine the distance scale of the universe) would thereby be limited to 180 days each year, instead of 260. Lockheed nevertheless concluded that based solely on "engineering and cost considerations it is recommended that the aperture door be removed. That aspect which pertains to science must come from another source."[29]

Goddard, too, was strongly opposed to removing the aperture door. "The issue is not just that of reduced efficiency which would seriously impede the Telescope's use," Goddard argued, "but more importantly the irrecoverable loss of scientific opportunities whose timing is dictated by nature and not by operational planning. There are numerous examples, including observations of variable stars at key phases, supernovae, other cataclysmic variables, comets in the inner solar system, etc."

Goddard was also concerned that removing the aperture door might affect the well-being and safety of the scientific instruments and the Optical Telescope Assembly. The aperture door was planned to carry "sun-avoidance sensors." If the telescope were inadvertently aimed too near to the sun, the sensors would provide a warning to close the door before sunlight could stream in and damage the telescope's optics. With no sun sensors and no aperture door, a sudden loss of control over the telescope's pointing might mean that for a time the sun's light would fall directly onto the primary mirror and be reflected from there onto the secondary mirror. Such extremely bright light might damage irreparably the optical coating on the secondary, as well as some of the scientific instruments. These had, after all, been designed to observe incredibly faint objects, not the brilliant light of the unobscured sun.[30]

There was, then, a clear trade-off between cost and scientific performance: Removing the aperture door would mean major scientific losses and increase the risk of damage to the telescope. Nevertheless, the intense financial pressures led Marshall to halt all activities on the aperture door, the intention being to resume them the following May (1981). O'Dell was alarmed. He feared that once the work stopped, it might be impossible to start again. As he protested to Speer,

in effect, the aperture door has been deleted. If this proves to be the case, I have been party to a grievous error. I have observed that on the [Space Telescope] program things are only removed, not added, unless someone has goofed. In this case, we are potentially eliminating a key element of the [Space Telescope] design and potentially losing a major fraction of the [Space Telescope] science. . . . I urge you to ask that work on the aperture door not be discontinued, rather that it be rescheduled for a later time and that it be included in the [Space Telescope] budget.[31]

By June 18, Lockheed had estimated that deleting the aperture door would save a total of $580,000. Given the size of that sum, O'Dell complained that "I cannot support a tradeoff of a permanent loss (at least five years) of this much science for such a small fraction of our total budget, even the $300,000 in the critical year [fiscal 1981]."[32]

What was to be done to change the decision on the aperture door? A central point about any project as big as the telescope program is that large numbers of institutions, groups, and individuals are engaged. So even if the project manager makes or proposes to make a particular decision, that does not necessarily mean that it is cast in stone. And often the key to heading off or changing a decision is for those who oppose it to mobilize as much support as possible. Such support usually must be won by careful scientific and technical arguments. Indeed, for a person to be effective in such a large-scale project, he or she must be widely recognized to be technically competent and capable of building networks – pursuing the matter in question in official meetings as well as by energetic private lobbying – as well as tenacious and persistent. The informal network that sought to overturn the decision on the aperture door, though composed in part of people who also sat on the Science Working Group, acted independently of that group, despite the fact that the aperture door's removal had such important scientific consequences. Indeed, in Phase C/D the Science Working Group could only offer advice; certainly it could not dictate what would happen, and at that stage of the project, with money extremely tight, it had little clout.[33]

At that stage of the program, the influence of scientists, such as it was, tended to be manifest less through the Science Working Group than in the more specialized working groups that examined particular topics and most noticeably in the various informal networks that sprang up around specific issues, such as the removal of the aperture door.

CANCELLATION?

The debate surrounding the removal of the aperture door raises another fundamental question: Why was such a change even being seriously considered when everyone agreed – NASA, the scientists, and the contractors – that it would have a significant effect on the science that could be performed with the Space Telescope, but would save less than a million dollars, a small amount in the context of the cost of the telescope? The essential point is that, as one leading Marshall figure remembers, "we got to the point where the costs associated with the program were growing so much that we were told at the [Marshall] center that the Space Telescope program was at serious risk of being canceled."

That message was relayed to Marshall by Thomas Mutch, head of the Office of Space Science in NASA Headquarters.[34] "We thought," O'Dell recalls, "we were getting the message from the [NASA] administrator, Robert Frosch. We were told by our center director [William Lucas] that we must restructure the Space Telescope to come in at the present cost ceiling."[35] Hence, when the issue of removing the aperture door had arisen, Speer had emphasized that "Dr. Lucas has directed the Project to implement a broad range of reductions to keep the project on schedule and within previously agreed-to budget limits. It will be to our mutual advantage to minimize the science impact of these major reductions, although some serious impact may be inevitable. I encourage you to propose other reductions with no or minimal impact on science."[36]

Mutch had, in effect, made the point about staying within budget and on schedule shortly after the Norris cost review of late 1979 that we discussed earlier, for, as Mutch stressed in January 1980, "as you well know, our credibility as it relates to obtaining future missions is closely linked to our ability to accomplish the Space Telescope within our budgetary commitment to the Congress."[37] In other words, if the Space Telescope went over budget, it would damage the agency's chances of winning approval for future projects from the Office of Management and Budget and Congress. That crucial point had no doubt been brought home to Mutch in the course of his close involvement earlier in the decade with the Viking mission to Mars, for though in many ways it was widely judged a great technical and scientific success, Viking's cost overruns had damaged NASA's standing in the executive branch and on Capitol Hill. But without future projects, the agency would soon be out of business. The need for NASA managers to somehow strike a balance between current and future projects was a theme that was to be heard again and again in the Space Telescope's development.

The immediate problem for Marshall, then, was how to meet Lucas's direction to stay within budget and to keep to the existing schedule. For Marshall it would mean cutting costs, and not requesting any extra funds. Headquarters had, after all, strongly indicated that Marshall would not get any extra funds even if it asked, and Speer's request for the headquarters reserve had already been rejected.

The trouble with that approach was that despite strenuous efforts at Marshall and Goddard to engineer ways of saving money, there was little more that the centers thought could be easily cut. There was little, if any, "fat" left to be cut from the program. Marshall suspected that several million dollars might be saved on the Optical Telescope Assembly and Support Systems Module. But none of the identified and supposedly "painless" cuts was anywhere near large enough to solve the major problems the project was facing.

There was little, it seemed, to be saved on the scientific instruments. In late July, the Goddard study group that had been assembled to examine the status of the scientific instruments and the instruments' associated computer issued its findings. The Goddard group maintained that a strong case could in fact be made for increasing the instruments' budget. Instead of instruments ripe for pruning, they identified only $3.2 million that might be saved out of a total of over $90 million. That small amount, the study group reported, "reflected the fact that the [Space Telescope] Project Office had already conducted a number of cost reduction 'sweeps.' In general, we found that the program included almost no spare hardware and was already down to an absolute minimum of testing. . . . The hard reality is that most of the payload fabrication and assembly must be done during [fiscal year]1981 when funds are strictly limited, with no hope for relief." The instruments were already in what the leader of the Goddard study regarded as a "high-risk mode."[38]

If they were to bring costs under control, as NASA Headquarters and Lucas had dictated, Marshall and project manager Speer were facing some immensely tough decisions. It is in that context, in which even the most basic assumptions about the Space Telescope were being reexamined by Marshall, and in which talk of cancellation hung ominously in the air, that we should view the events of July 26, 1980, Black Saturday.

BLACK SATURDAY

On July 26, thirty-one representatives from Marshall, Goddard, and headquarters gathered for a special review of the Space Telescope's budget. Speer explained that to keep within the previously agreed levels in fiscal 1981, $28 million somehow needed to be saved. Marshall had decided that to reach that amount would require some descoping and a major restructuring of the telescope's development program. The potential items Speer listed for descoping were as follows:

Delete further spares:	$1.6 million
Transfer certain systems engineering tasks from the contractors to Marshall:	$2.2 million
Reduce testing:	$5.0 million
Delete some design features, increasing risk and making the telescope's operations less flexible:	$5.9 million
Certain items on transport of the telescope:	$2.5 million

The largest savings, however, came from changing some features that directly affected the science the Space Telescope would be able to perform. That included altering the telescope's design so that it would no longer be able to carry infrared instruments. Another proposed change was even more radical.

O'Dell was convinced that unless sizable economies could be achieved, the telescope might well be canceled. He thus felt himself

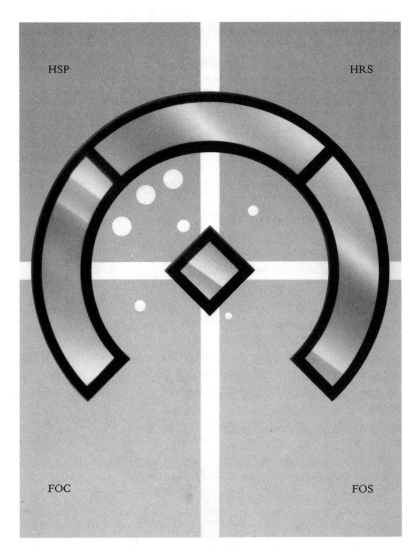

HSP   HRS

FOC   FOS

Space Telescope field of view. The light collected by the Space Telescope's primary mirror is reflected to the much smaller secondary mirror. From there it is directed back through the hole in the center of the primary. Part of this incoming light is deflected by small "pick-off" mirrors to the wide Field/Planetary Camera and the three Fine Guidance Sensors. The pick-off mirror positioned in the center of the beam reflects light into the radially mounted Wide Field/Planetary Camera. Three arc-shaped flat mirrors sited around the outside of the incoming beam direct light to the three radially mounted Fine Guidance Sensors. The light not interrupted by the pick-off mirror travels to the focal plane, where the four axially mounted scientific instruments are located. At the focal plane, the field of view is twenty-eight arc-minutes in angular diameter. The Wide Field/Planetary Camera, by use of its pick-off mirror, views a square region in the center of the field about three arc-minutes on a side. The rest of the field out to a radius of about nine arc-minutes is divided into quadrants, each of which is viewed by one of the axial instruments: the Faint Object Camera, the Faint Object Spectrograph, the High Resolution Spectrograph, and the High Speed Photometer. The outermost portion of the field of view is devoted to the three Fine Guidance Sensors. These are employed in pointing the Space Telescope as well as measuring precisely the positions of astronomical objects. The Fine Guidance Sensors, in effect, constitute a sixth scientific instrument.

under strong pressure to suggest ways of making substantial savings. O'Dell advanced a measure that he had first raised some months earlier. It would save a considerable sum of money – indeed, the bulk of the $15.4 million Marshall presented as the amount to be saved by "descoping the science": deferring the two spectrographs from the first flight.

What was the rationale behind the proposal to defer the two spectrographs? O'Dell reasoned that such a course was preferable to other cost-cutting measures. To economize on the science by deleting bits and pieces might leave the five dedicated instruments in place, but he thought that such a step would compromise the instruments' capacity to perform scientific observations over the long term. The Phase B Science Working Group had ranked the importance of the scientific instruments to be flown, and so in a way O'Dell was returning to that concept, although in this case he alone would be doing the ranking. He was also drawing on his experience with ground-based telescopes: They generally start their observing lives with a restricted set of instruments, usually a camera and photometer; then, once a telescope has operated for a few years, it is common for more scientific instruments to be added. Why not, given NASA's financial position, adopt that approach for the Space Telescope?[39] In the past, it had not been possible to make changes in the initial complement of instruments for a spacecraft once it had been launched. With the availability of the Space Shuttle, that need no longer be the case; the shuttle could rendezvous with the telescope after it had been in orbit for a time, and astronauts could then insert the spectrographs in the telescope.[40] NASA was already planning to fly "second-generation" scientific instruments, and so the concept of exchanging instruments later in the telescope's life was far from novel. Even if the spectrographs were deferred, O'Dell reasoned, the telescope would still be scientifically productive.

The notion of building up to a full set of scientific instruments had even been floated by NASA Headquarters in the mid-1970s during the funding battles with Congress. At that time the aim had been to drive down the telescope's development costs. However, the idea had been bluntly rejected by the Phase B astronomers, O'Dell included.

At a meeting on the telescope's ground system in January 1980, the attendees had spent much of the time trying to devise ways to save money. O'Dell had there suggested dropping the spectrographs.[41] At that time the proposal had caused few ripples and quickly sank from sight. But nine days before Black Saturday, O'Dell explained to Speer his idea for deferring both the High Resolution Spectrograph and the Faint Object Spectrograph from the first flight. Speer "liked it," and so it had been readily incorporated into Marshall's cost-cutting scheme.[42] The drawback to O'Dell's proposal was that it cut to the very heart of the science to be performed by the telescope, and it was of such a nature that it could command no support outside of Marshall. In fact, it provoked a storm.

Marshall had barely stopped short of some even more drastic proposals. From the perspective of the scientists, Black Saturday might have been even more somber had Marshall presented another set of measures that it held in reserve. A backup briefing chart prepared for the July 26 meeting, but apparently never used, had listed a further set of items for "additional descoping." Those included deleting more spares and tests from the planned program, removing the telescope's aperture door, and deferring yet another scientific instrument, the High Speed Photometer, until some time after launch.[43]

Marshall had not consulted with NASA Headquarters or the Science Working Group on its proposed program changes. Nor had the idea of delaying the two spectrographs been raised with the two principal investigators involved or with Goddard, the center responsible for the management of the scientific instruments. Goddard, Marshall had reasoned, would never accept it, would dig in, and thus perhaps would jeopardize the entire Space Telescope program.[44]

THE AFTERMATH

On the whole, Marshall's proposals got a hostile reception. NASA Headquarters was particularly unhappy with the suggestion to defer the spectrographs. The word spread quickly that when that recommendation had been made at the meeting at Marshall, it had blown "the roof off the whole place."[45] Marshall was told to rethink its approach. Headquarters directed that the scientific instruments would be the last items to go in any cost-cutting scheme. But, to begin with, at least, NASA Headquarters insisted that Marshall still had to remain within the existing budget guidelines for fiscal years 1981 and 1982 and was not to exceed the existing budget total of $575 million for the entire program.

Headquarters did allow some significant changes, however. The number of Marshall personnel who could work on the telescope was raised, thereby breaking the manpower cap imposed in 1977. The taboo on discussing a possible postponement of the launch date was also removed, and Marshall and Goddard were able to explore how the program could accommodate a delay of six, twelve, or eighteen months.[46]

About two weeks after Black Saturday, William Lucas, Marshall's director, wrote to NASA Headquarters with Marshall's latest assessment of the program's status. Lucas explained that in order to stay within the limits of $118 million for fiscal 1981 and $91 million for fiscal 1982, "we have identified $27 million of contractor work to be descoped without deletion of scientific instruments. The major reduction is for on-orbit replacement of scientific instruments during the first mission." In other words, when the telescope was carried aloft by the shuttle, the scientific instruments would remain in position until the telescope was returned to earth for its first major refurbishment (expected to be after five years). Only then could new instruments be switched for old.

Lucas also reported that Marshall was still investigating other ways to descope the program. The center had calculated that replanning would add $30 million to the telescope's total cost. Lucas added, however, that if more funds could be devoted to the telescope in fiscal 1982, that would reduce costs in later years and thus, in the long run, save money.[47] Because of the existing restrictions on the available funds, there would be a consequent delay in the delivery of much of the hardware, and so a delay in the telescope's launch date.

With such tight budgets, the replanning of the program would prove to be a lengthy, difficult, and intricate business. In a period when leading NASA managers were unsure what course events would take, it is hardly surprising that that period was bewildering for anyone not in the decision-making circles and unable to keep abreast of all of the proposals flowing around the program. The four U.S. principal investigators even formed a so-called gang of four. There was, as one of them recalls, "so much information and misinformation going around that [John Brandt, the principal investigator for the High Resolution Spectrograph] thought it would be useful if we could get together by phone occasionally . . . and discuss 'What have you heard lately?'"[48]

The principal investigators for the two spectrographs had not been told in advance of Marshall's plan to defer flying their instruments, a decision O'Dell was later to regret deeply and to view as a "tactical mistake of the highest order."[49] Not surprisingly, to those principal investigators and to most of the other members of the Science Working Group, Marshall's plan had been an extremely unpleasant bolt from the blue.

Meetings of the Science Working Group had never taken on the character of Sunday afternoon tea parties, and in September 1980, the first meeting after Black Saturday clearly demonstrated that. Bahcall remembers the discussion as "violent and heated," and one principal investigator termed it "lengthy, intense, and acrimonious."[50] O'Dell's defense of his proposal to defer the spectrographs was far from persuasive, and he was given a rough ride. In the opinion of one principal investigator, feelings were running so high that if the proposal to defer the spectrographs had been implemented, there most likely "would have been mass resignations . . . in the sense that the instrument people would have said, 'Sorry, we don't want to play under those circumstances.' "[51] It was widely assumed that if the spectrographs were left off the first flight, they might never fly. In the view of the astronomers, O'Dell had acted as a manager with his eye on the bottom line, not as an advocate for the science to be performed by the Space Telescope, and that undoubtedly damaged his standing with the Science Working Group.

But verbal warfare at a Science Working Group meeting should not be allowed to obscure the main issue. The root problem confronting the project's managers, as well as the astronomers who were anxious to protect the scientific capability of the telescope, was that the program was by that stage trapped in a financial box, and there seemed to be no ready escape. There were simply too many elements of the program jostling for insufficient funds. The constraints on

Marshall's possible courses of action were further tightened by the fact that the telescope's scientific performance could not be reduced too much. If it were, the telescope would lose its lead over ground-based telescopes, and the very reason for its construction would thereby be defeated.

A WIDER PICTURE

Here, if we are to comprehend better NASA's decisions regarding the Space Telescope, we need to take a step back from the telescope program in order to view NASA's overall financial position in early 1980. Early 1980 was a time when the Space Shuttle was still a year from its first launch. Thus far its design and development phase had not proceeded smoothly, in part because of changes in the planned funding for the shuttle mandated by Congress and the White House and the fact that it, like the telescope, had been sold at an unrealistic figure. Delays in the program had not been uncommon, and there were daunting technical problems still to be solved.

During 1979 there had been numerous NASA and congressional cost and management reviews of the shuttle program. At the end of that year, NASA had estimated that the shuttle's design, development, testing, and evaluation programs would cost $6.18 billion (in 1971 dollars), approximately $1 billion more than the $5.15 billion estimate NASA had been forced to accept at the start of the program.[52] The shuttle's schedule had also slipped by some years. The delays meant that the monetary requirements for developing the shuttle had not yet started to tail off, as NASA and the Office of Management and Budget had planned, and the so-called shuttle dividend that had been anticipated (see Chapter 5) was not materializing. Moreover, because it was, in the agency's view, far and away its most important program – the one that underpinned all its space activities – the shuttle took the bulk of the money the agency was receiving for design and development.

The fiscal 1980 funding problems for the shuttle were acute, and the amount allocated in NASA's budget fell far short of the actual spending. NASA had to seek a supplement of some $140 million from Congress, and an extra $300 million had to be diverted to the shuttle from within the agency's own budget. Such a request for supplemental funding is generally disliked at any period by federal agencies because it reflects unfavorably on their credibility and ability to manage.[53] NASA's request, moreover, was made at a time when President Jimmy Carter and the executive branch were toiling to reduce the federal budget deficit for fiscal 1981 and embarking on a policy initiative that required agencies to reduce the outlays originally proposed to Congress.

NASA, in fact, had been compelled to delete $214 million from its planned budget for fiscal 1981. During hearings on Capitol Hill, questioners had asked why part of that sum could not be sliced from money for the Space Telescope. As Edward P. Boland, the influential chairman of the House Appropriations Subcommittee dealing with NASA, and a long-time critic of the telescope whom we met in Chapters 4 and 5, succinctly put it in April 1980, "the Space Telescope garners the largest part of the Physics and Astronomy budget at $119,300,000 in 1981. Why didn't you take any of the reduction in the recent budget exercise out of this program? It doesn't require any specific launch window. It is scheduled to be launched in December 1983. If it doesn't meet that launch date, no science is lost."[54] It was under those difficult circumstances, then, that headquarters had to decide whether or not to provide extra funds for the Space Telescope, and if so, how much. There was, indeed, pressure to remove funds, not increase them.

To many of the scientists who had recently seen vivid examples of how cost overruns could and might affect the telescope's proposed scientific performance, matters looked grim. Writing on September 16, O'Dell told a correspondent that things "do look gloomy for [the Space Telescope] as a whole, but I think that much of what we fear will not come true in the end. The package we put together now (for [NASA administrator] Frosch) is what counts and we must be sure that the risks and trades are identified, then we'll see what the Administrator says."[55]

As part of their own decision-making process, headquarters managers were seeking detailed information on the program's status. One way they sought that was by establishing a "Space Telescope Cost Review Team." Chaired by former program manager Warren Keller (who, as noted earlier, had been promoted in late 1979 to another position in headquarters), the nine members were to examine the program and complete their review by early December 1980. Then, in the light of Marshall's revised program and the Keller team's findings, the NASA administrator would decide the telescope's funding.[56]

Marshall and Goddard, meanwhile, were continuing to investigate different ways of restructuring and revising the program. Although that process was incomplete, one point had become obvious: The new program would inevitably cost considerably more. As tasks were moved from fiscal years in which there would be no money to finish them, being rescheduled for future years when money would be available, inefficiencies were introduced that drove up the telescope's development cost. Whereas Marshall had previously been working to a $575 million limit, by October 22 the center's estimate of what it would cost to complete the telescope and pay for its first year of operations had risen to $612 million (fiscal 1982 dollars).[57]

284

A REVISED PROGRAM

Following numerous meetings, detailed technical studies, deliberations, and negotiations, a replanned program did emerge. After preparatory discussions between the directors of Marshall and Goddard,
as well as representatives of the Office of Space Science, the revised
program was presented to NASA administrator Frosch on November
5. There was no longer any question of deferring scientific instruments. In fact, Frosch emphasized that it was important to build
instruments correctly, as "the system must be appropriate for the
lifetime of the spacecraft."[58]

Marshall nevertheless proposed some significant changes to the
existing program. Perhaps most important was the drastic reduction
in the number of units that were explicitly designed to be capable
of being replaced while the telescope was in orbit, the so-called
orbital replacement units.[59] The number of orbital replacement units
was reduced by about a hundred, from 124 to 15 or 20, although
that number fluctuated. Consequently, much of the special equipment needed to carry the units in the shuttle payload bay, as well as
the tools required to insert the orbital replacement units into the
spacecraft, could be deleted, with further cost savings. On the debit side, there would be much less chance of being able to fix
something if it failed in orbit. Here, then, was yet another flip-flop
on the way NASA planned to use the shuttle to service the telescope
in orbit.

Marshall had also rearranged the activities on the program so that
the telescope would be scheduled for launch in October 1984, some
ten months later than originally planned. Both the Optical Telescope Assembly and the scientific instruments, however, were still
on what were termed "success-oriented schedules." That is, if the
October 1984 launch date was to be met, there could be no major
problems. Marshall also considered the schedule for testing the telescope to be very tight. To cover slips that might occur, Marshall
had therefore added contingencies of one month to the new schedules in case there was late delivery of the Optical Telescope Assembly or the scientific instruments, and two extra months for testing.
An extra two-month contingency was added to allow for additional,
currently unforeseen, problems that might arise. If all the contingency times had to be used, then the telescope would be launched
in March 1985.

Taking into account all of these changes in the program, Marshall
reckoned the telescope's total development cost to be $645 million.
However, the center argued that the most urgent need was for an
extra $25 million in fiscal 1982.[60] In Marshall's view, that would
be a crucial year for meeting and solving technical problems. If events
ran to plan, fiscal 1982 would see, among other important activi-

ties, the scientific instruments manufactured, assembled, and tested, the Fine Guidance Sensors assembled and tested, and a start made on assembling and aligning the Optical Telescope Assembly. Given the magnitude of those tasks, and the likelihood of having to solve unexpected problems, Marshall regarded the 20 percent reserve for fiscal 1982 as too low.

Frosch did not announce his decision at that meeting. He instead reserved judgment until the following month. By that time the Keller committee would have reported, and Marshall, too, would have finished its replanning. For the managers on the program, the goal became to present to Frosch a new plan that would have the support of all the institutions involved in designing and building the Space Telescope, in effect, to construct a consensus. In certain respects, the coalition that the telescope's advocates had fashioned to make the telescope politically feasible was once more being put into the field, this time to sustain the telescope program.

The backing of the Science Working Group was a key element in that process, and the group met on December 2 to offer its opinion on the replanning. Project manager Fred Speer again warned that although the total funding was adequate, there was some risk because the reserves allocated to fiscal years 1981 and 1982 were small. That, as we have already seen, had been and would continue to be a constant refrain in the program: Money may be short in the current fiscal year, but in a year or two, things will be fine. The continuing problem was that by the time those future fiscal years would arrive, the reserves would already have been eroded.

However, the new scheme entailed a substantial injection of funds. The recently appointed director of the Astrophysics Division in NASA Headquarters, Frank Martin, gave his support and asked for comments − in effect, approval − from the scientists. In contrast to their remarks at the September meeting, where they had been sharply critical of Marshall's lack of consultation earlier in the year, the members of the working group gave their strong backing to the plan. Though the astronomers might usually have relatively little influence on the program, there were still times, such as this, when their support was obligatory. The astronomers also asked O'Dell to write formally to Speer on their behalf, no doubt to provide ammunition for him at the forthcoming presentation to Frosch. "We unanimously agreed," O'Dell told Speer,

that you have been directly supportive of the scientific aspects of the Space Telescope program during the preparations of the new plan that will be submitted to Dr. Frosch. In particular, your reallocation of money [in fiscal years 1981 and 1982] to the Scientific Instruments will help guarantee the success of these instruments, and the subsequent success of the observatory.

. . . Everyone was pleased and impressed with your ability to balance the conflicting needs of the Project to produce a viable plan which we can all enthusiastically support. Please accept our thanks for this fine effort.[61]

The scientists mobilized other forces too. The prestige of the National Academy of Sciences' Space Sciences Board, whose promotion of the telescope had been so necessary for its selling in the 1970s, was brought behind the revised program. Writing to NASA administrator Frosch, the board stressed that the telescope "is the central element for astronomical research in space. We, therefore, urge that NASA take immediate steps to ensure that the effects of the present problems on the scientific potential of the Space Telescope will be minimal"[62]

In a climate of general agreement throughout the program on the new plan and the demonstration that a wide-ranging coalition still favored the telescope, the long-awaited encounter between leading project members and NASA administrator Frosch took place on the afternoon of December 9. Speer described the technical changes NASA had made since the November briefing to reduce the amount of risk in the program. He noted that the overall reserve had been raised to about 25 percent of the remaining cost, a decidedly healthier figure than the 5 percent Speer had mentioned to the Science Working Group as recently as May 1980.[63]

When Warren Keller's turn came to explain the work of his review team, he underscored that they had been conservative in treating costs and schedules. The Keller team had identified what it regarded as the major cost risks and potential problem areas for the Optical Telescope Assembly, the Support Systems Module, the scientific instruments, operations, and maintenance and refurbishment. It also listed the factors it believed to have led to the increases in costs, and whether or not those factors were still operating.

Many of those factors had to do with the way NASA had managed the program. For example, the use of associate contractors, the division of the project among many institutions through complex interfaces, and splitting the pointing and control system across the interface between Lockheed and Perkin-Elmer were all cited as having contributed to the rise in costs. The manpower cap at Marshall had played a role because it had impeded the center's ability to "penetrate" the contractors and manage them as tightly as Marshall would have liked. Nor, in the opinion of the Keller team, had the agency found it easy to break old habits and make the switch from NASA's traditional approach to building spacecraft to the new approach in which the ability to retrieve the telescope in orbit or make in-orbit repairs was intended to offset extra risks that might be built into the design.[64] Most important, however, the Keller team was in general agreement with Marshall's revised estimates on costs and schedules.

With that decisive stamp of approval – aided by the consensus among headquarters, Marshall, Goddard, contractors, and astronomers on the new plan – the NASA administrator agreed officially to increase both the funds and the time available to complete the Space Telescope. As one highly placed observer put it in 1984, the agency decided to "swallow hard" and "preserve the integrity of the program."[65]

The swallowing was difficult because they were accepting an increase in the total cost estimate from the previous range of $540 to $595 million to a range of $700 to $750 million (1981 dollars). There were substantial extra funds to come in future years, but not fiscal years 1981 and 1982. Some tasks originally allocated to those years would have to be postponed to a time when money would be available. The launch date, too, had been slipped, most likely to early 1985.

As in the selling campaigns of the 1970s, the coalition that favored the telescope had to be extended to the Congress, but the jump in cost and the delay in launch date caused surprise on Capitol Hill. The Capitol Hill rhetoric and the claims made during the selling, and sometimes overselling, of the Space Telescope a few years earlier provided a sharp contrast with the hard reality of what NASA now publicly conceded was a taxing design and development effort. During the selling campaigns, the argument had been made repeatedly by NASA, the contractors, and the astronomers that developing the technology for the telescope would not pose major problems. Yet, three years into the program, they faced a sizable cost increase, combined with an extra year to fifteen months more on the schedule, as well as warnings about the magnitude of the technical tasks. As the chairman of the House Subcommittee on Space Science and Applications asked in early 1981, "we really thought that NASA and the contractors pretty well understood that project and what it involved. What went wrong?"[66]

The Space Telescope was indeed proving much more exacting to design and develop than NASA, the contractors, and the astronomers had earlier claimed it would be. During Phase B, NASA had stressed that the telescope would be conservatively designed and would rely as much as possible on existing hardware, a typical selling point for a program, but one that usually is difficult to implement. The exacting technical demands of the program, however, meant that there would be less flight-proven hardware incorporated into the design than NASA had planned, and even the flight-proven hardware would have to be arranged and coordinated in new ways. As had been demonstrated repeatedly on the shuttle program, that can be just as challenging as developing new technology and proving it to be capable of operating in space. The Space Telescope would be a satellite that would, of necessity, push the state of the art very hard in many areas if it were to meet its scientific requirements.

In responding to congressional questions, William Wright, Lockheed's project manager, put in a nutshell the replies to the criticisms of NASA and the contractors: "It is a question of optimism at the outset, not fully appreciating the complexities of the program."[67] But optimism is surely too simple and easy an answer. The problems being encountered indicated that the Phase B studies had not been extensive enough, and not nearly enough work had been done to prove the concepts that were to be put into practice in Phase C/D, in part because NASA had not secured sufficient planning funds. The Fine Guidance Sensors, for example, had already been through one major redesign; yet their original design was one of the reasons Perkin-Elmer had been awarded the Optical Telescope Assembly contract. Nor was it only technical problems that had driven up costs. The Keller review team had emphasized that the agency had brought many of the difficulties on itself by its adoption of an unwieldy program design and a management approach that had not allowed the agency to penetrate its contractors. As events would show, the changes in the program in late 1980 did not really solve these structural problems. But to a degree, the problem was that the telescope program was, as Congressman George E. Shipley had predicted in 1975, proving to be in effect a buy-in. That is, the program had been approved at an unrealistic figure, and that was where to a significant degree the cost growth and schedule delays had their origins.

BACK INTO THE WOODS

Any hope that matters would go smoothly because of the extra time and money NASA Headquarters had inserted into the program was not to last long. On April 13, 1981, for example, Speer wrote to John D. Rehnberg, vice-president and general manager of the Space Science Division at Perkin-Elmer:

Our current objective should be to instill in our personnel the importance of schedule and cost and to elevate those to the same priority as technical performance. . . . I would like to emphasize the importance of getting the [Optical Telescope Assembly] back on schedule, staying within current cost plans, and extending your planning sufficiently far into the future, particularly through next year, so that any under-performance (deferred tasks) from this year can be accommodated within next year's planned budget by a realigning of priorities where necessary.[68]

A little over two weeks later, Speer was pressing hard again. Since October 1980, he complained, Perkin-Elmer had not met the planned completion dates for a "significant number" of events. He wanted prompt corrective measures to be taken and the assurance that the Optical Telescope Assembly's delivery date could be met within the existing budget. Perkin-Elmer should provide a detailed, carefully worked-out schedule, and then apply the necessary manpower to meet it. "I cannot overemphasize," Speer wrote, "the importance of accomplishing these two steps within the next two or three weeks. We must reestablish our credibility to meet the [Optical Telescope Assembly] delivery date" by drawing up appropriate intermediate milestones and meeting "these on plan."[69]

The costs for both the Lockheed and Perkin-Elmer contracts continued to rise, nevertheless. Marshall was again faced with extremely hard decisions on what work to defer and what work could be avoided altogether. This, remember, was a matter of months after NASA had made a major adjustment to the program.

So, by the summer of 1981, though the telescope was still scheduled for a January 1985 launch, the Support Systems Module was about three weeks behind schedule, the scientific instruments were about one month behind, and the Optical Telescope Assembly was some two and a half months behind schedule overall. More serious, significant sections of the Optical Telescope Assembly were lagging further still: fabricating the optics, five months; assembling the primary mirror, five months; building the Fine Guidance Sensors, four months. The shortage of money and time also meant that NASA managers, on occasion, had to accept additional risks rather than solve certain problems.

By August 1981, Perkin-Elmer had concluded that it would need a manpower increase of about forty, as well as six more months on the existing schedules, to finish the Optical Telescope Assembly. Marshall, however, would not provide the funds that would have entailed. NASA had sent its fiscal 1982 budget to the Office of Management and Budget the year before, and given the budget pressure the agency was facing, NASA did not seriously consider asking Congress directly for more money. As a headquarters manager recalls, that "was something that was unheard of. Nobody thought that we could get away with something like that."[70] Instead, there were negotiations that led in December 1981 to Marshall agreeing that Perkin-Elmer could delay the delivery of the Optical Telescope Assembly to Lockheed by two months. Marshall also added a few million dollars more to the Optical Telescope Assembly's budget for fiscal 1982.[71]

Once more a central theme for Marshall managers had become how to contain cost growth. One option Marshall examined concerned the primary mirror. The primary mirror was finished, although grinding and polishing it had taken Perkin-Elmer considerably longer than planned due to defects in the blank from Corning.

A close-up inspection of the Space
Telescope's 2.4-meter primary mirror
in 1981 prior to the mirror's coating.
(Courtesy of Perkin-Elmer.)

After rough grinding, Perkin-Elmer had begun the fine polishing of
the mirror's surface in August 1980. But for a time, Perkin-Elmer's
polishing work lagged behind Eastman Kodak's work on the backup
mirror. Marshall had debated the possibility of stopping the polishing of one of the mirrors well before it was completed.[72] That proposal had been, in part, a management tactic designed to spur performance at Eastman Kodak and Perkin-Elmer,[73] and it may have
worked, for Perkin-Elmer reevaluated its cost projections for polishing the mirror and "was able to reduce the funding requirements."
In any case, Perkin-Elmer had completed its fine polishing in April
1981.[74] Despite the delays, Perkin-Elmer had done what many people regarded as an extremely fine technical job, and the mirror was
generally agreed to be of exceptional quality. In the opinion of some
people, it was the Space Telescope program's crown jewel. Telescope
scientist Daniel Schroeder, for example, judged it "as good a mirror
as exists."[75] However, "the worsening situation at Perkin-Elmer,"

Speer wrote in August 1981, "requires that we give serious consideration to removing the primary mirror figure control actuators," the actuators that could alter the mirror's shape while the telescope was orbiting the earth.

The use of some sort of mechanical system to support the figure of a telescope mirror had been practiced by astronomers for over a hundred and fifty years. William Herschel had built the first large reflecting telescopes in the late eighteenth century. He had quickly been confronted by and had grappled with, sometimes unavailingly, the problem of "flexure," in which a telescope's mirror is bent out of shape because of the gravitational pull of the earth. In 1789, Herschel completed what at that time was potentially the most powerful telescope ever constructed, a reflector with a primary mirror forty-eight inches in diameter. It was not a great success, however. With hindsight, it is clear that Herschel pressed the technology too far. He had built an instrument that was very difficult to use and that, among other problems, suffered badly from flexure of its solid-metal mirror. Even astronomers after Herschel found it a formidable task to provide accurate support for such large mirrors.

There is no need to combat flexure of a mirror for the Space Telescope, because while in orbit the mirror is in a state of weightlessness. However, the primary mirror must be fabricated, ground, polished, and surface-coated while on earth. The switch from the gravitational pull on earth (1 g) to the zero gravitational pull of space might then result in some misshaping of the ninety-four-inch primary. Perkin-Elmer had taken enormous pains to ensure that the mirror would not undergo any distortion, but as a precaution, it had incorporated a system of "actuators" into its design. The actuators could be controlled from the earth to exert tiny pushes on the back of the mirror. Those forces could then correct small distortions in the mirror's figure even while it was in space.

In August 1981, Speer argued that removing the twenty-four figure-control actuators would save money and weight. "It appears to me," he told O'Dell, "that this is probably the only remaining capability that could be given up without direct and serious impact on science, although it does remove a desirable degree of freedom, a sort of insurance. To make this move will prove to [management] that we understand the seriousness of the budget problem and may help us get the necessary funds for the program later on."[76] To the disappointment of the astronomers, Marshall had already stopped the polishing of the backup mirror at a point short of the best that could have been achieved. As Eastman Kodak had taken each "lick" at the mirror, the results had been passed to Marshall. But at a time when Eastman Kodak was still improving the mirror's figure, Marshall had called a halt. To the astronomers, the amount of money that saved bore no relation to the potential loss in scientific perfor-

292

mance should the backup mirror have to be employed in the telescope.

In the case of the mirror actuators, it was difficult for the astronomers to point to any one specific area where removing them would
cause the telescope not to meet its performance specifications. However, as telescope scientist William Fastie maintained, though the
actuators were only insurance, they were "essential insurance."
"Without them," Fastie contended, and other astronomers reinforced, "we may run a risk that degraded performance will occur,
and there is nothing that can be done about it." Perkin-Elmer had
built a superb primary mirror, and it was "our responsibility to take
every reasonable step to guarantee that its potential will be realized."[77]

Replying to Speer's proposal to remove the actuators, O'Dell raised
a more general issue, one that probably echoed the sentiments of the
scientists on the program. "Over the last year," he protested,

the budget restrictions on the program have caused us to take steps that
have led to acceptance of a backup primary mirror of lower quality than
the Perkin-Elmer mirror, to acceptance of greater risk and lower performance of the Pointing and Control System, and now the consideration of
loss of our optical performance insurance. There appears to be no end to
the compromises we are asked to make and the incremental increases of
performance risk. In the final appraisal of the [Space Telescope] program
we shall all be measured by the performance of the observatory, from our
level through the head of the Agency. This message must be made clear to
the people in Headquarters who place such tight fiscal constraints on the
program. I am convinced that without the support of Headquarters we will
be asked to compromise on [the Space Telescope], perhaps to the point that
it will not be widely supported scientifically. . . . I believe that we have
already demonstrated our responsibility and need relief and help now – not
later.[78]

In the face of such opposition, again based in part in an informal
network, to removing the mirror actuators, Marshall eventually decided that they should stay in the program; once more the system of
checks and balances in the management system had been brought
into play to affect a decision. However, the trend of cost growth
among the contractors and builders of the scientific instruments continued, and it was clear to many project members that it would be
far from easy, if not impossible, to launch the telescope according
to the schedule and budget Marshall laid out in late 1980.

At congressional oversight hearings in May 1982, Frank Martin,
director of the Astrophysics Division at NASA Headquarters, sin-
gled out the delivery of the Optical Telescope Assembly from Perkin-
Elmer to Lockheed as the critical item in the schedule. "We believe," he cautioned, "that schedule is very tight. To deliver the
assembly in time, it is going to take some good, tough management

and progress from our associate [contractors], as well as our project team."[79]

MORE OF THE SAME

By the start of 1983, the schedule Martin had described to Congress only a few months earlier was in disarray. Once more the program was plunged into a crisis that, some participants feared, might mean cancellation for the Space Telescope. Although the contributing factors went far deeper than simply problems at one of the associate contractors, acceptance at NASA Headquarters that there was a crisis was forced by events at Perkin-Elmer.

NASA regarded the quality of Perkin-Elmer's technical work as generally extremely good. The agency looked on the company's optical group as perhaps the world leader in its field, but Marshall and headquarters managers had long been worried by what they saw as Perkin-Elmer's shaky management performance. Good schedules were essential for NASA to track Perkin-Elmer's progress accurately. Over the years, Marshall had pushed hard to ensure that Perkin-Elmer met NASA's demands. In April 1979, for example, project manager William Keathley had told the Marshall director that Perkin-Elmer "is well along in pulling the design and development schedules into an acceptable condition. A great deal of government pressure and direct involvement was applied." However,

it's my perception that an abnormal amount of government follow-up will be required to protect the new schedules. Over the past month, I've attempted to motivate [Perkin-Elmer] (from task managers to project director) to work the schedules, demonstrated the process, and insisted that they conduct weekly schedule *control* sessions. Unfortunately, I do not have a warm feeling that a contractor system will be in place to handle future schedule problems in real time. . . . I will continue to apply as much government pressure as time and manpower allows short of allowing other project elements to get in trouble from lack of attention.[80]

Over the years, NASA and Perkin-Elmer had tried various measures to meet NASA's needs. Another effort came in the summer of 1982. As two prominent Perkin-Elmer officials told NASA administrator James Beggs on July 1, there had been a shake-up and reorganization in their part of the Space Telescope program.[81] Also, a new project manager was soon to join the company to head the Space Telescope effort.

A few days after the meeting between Beggs and the Perkin-Elmer officials, Fred Speer was congratulating the company on its modification of a fixture for the Wide Field/Planetary Camera. "The effort," Speer wrote, "certainly demonstrates the ability of a dedicated [Optical Telescope Assembly] team to meet both technical and

schedule commitments. We look forward to this milestone as being
a turning point in the program."[82] Indeed, at a project-wide meeting on the telescope at Marshall in early August, the feeling among
most of the leading NASA managers was that Perkin-Elmer was
currently doing considerably better than before. NASA, indeed, should
start to look more toward Lockheed as the site of potential problems.
The point was that

Perkin-Elmer had just presented a set of schedules that showed them to be
a couple of weeks behind. Everybody was remarking it was the best they
had ever performed. For the four or five years preceding this they had never
performed that well, so they were getting slapped on the back [and NASA]
said, "you guys are doing a fabulous job, fantastic," and sent them
away. . . . Lockheed shows their data, and they show that they have
problems in developing a few of the electrical boxes . . . and a few problems here and a few problems there. And everybody starts to think, well,
Lockheed's going to become a problem. Perkin-Elmer is starting to get
straightened out.[83]

NASA project manager Speer was not convinced. Certainly Perkin-
Elmer was failing to maintain its schedule in several areas. Despite
a major revision and simplification of the design in 1979, Perkin-
Elmer still was far from completing the immensely demanding Fine
Guidance Sensors, and a number of people had grave doubts that
they would ever work to specification. Another pressing problem
had to do with "saddle bonding," that is, how the supports to hold
the primary mirror in its main ring were to be bonded to the back
of the mirror. Those were important joints, because the supports
had to be designed to flex and absorb structural stress before the
forces could be transmitted to, and therefore distort, the primary
mirror. In July, program engineer Arthur Reetz concluded that
Perkin-Elmer's planning on how it was to perform that immensely
delicate and difficult operation was inadequate and that problems
lay ahead.[84] Also, Perkin-Elmer had found that one of the essential
items of tooling needed to check that the primary mirror was properly installed in the main ring would not perform its job and so
would have to be rebuilt.[85] The tooling had been fabricated early in
the program, but the systems requirements had since been changed
by NASA. In fact, it would be more than six months before Perkin-
Elmer developed its procedures to bond the supports onto the mirror
and had appropriate equipment to insert the primary mirror into the
main ring.

The latches to hold the scientific instruments inside the Optical
Telescope Assembly were another problem area. The latches will be
discussed in more detail in Chapter 9, but here we should note that
on August 11, Speer was appalled to hear the latest estimate for
Perkin-Elmer's delivery date for the latches for the Wide Field/Plan-
etary Camera. "I believe this is a 6 week slip against your current

Problems arise

Installing the Space Telescope's primary mirror into its titanium main ring assembly was an especially significant technical milestone because it was through the main ring assembly that all else was attached: the secondary mirror, the metering truss, the optical baffling, the electronics boxes, and the Support Systems Module to the front and the focal-plane structure and scientific instruments package to the rear. (Courtesy of NASA.)

schedule," Speer told the Marshall manager who had direct responsibility for the Optical Telescope Assembly. "We *cannot accept* that and you need to review the reasons and make necessary changes in those 'plans.' " Speer wanted several extra measures to be investigated, including involving staff at Marshall on an emergency basis. "These continuing delays," Speer protested, "are just intolerable. We need to use unconventional methods to stem the tide. I am willing to call [Perkin-Elmer] management and pay a dollar price if you can find a way out. Pull the stops!"[86]

There was, nevertheless, little room to maneuver. Certainly Speer had to fit a number of extremely demanding extra items into the available budget. For example, Marshall was in the throes of revising the plan to prevent the telescope being contaminated. The weight of a large number of technical studies by the contractors and NASA had compelled the center to conclude that the existing plans were inadequate, and contamination, especially from hydrocarbons, might have an extremely serious effect on the scientific output of the telescope through, for example, a coating developing on the primary

296

mirror. NASA's plan at the start of Phase C/D had been to endeavor to assemble a very clean spacecraft. However, Marshall had recently decided that it would be necessary to "bake out" — that is, heat in a vacuum in order to drive off potential contaminants — the major components of the Optical Telescope Assembly and the Support Systems Module. That would be an expensive process, one that NASA had not judged at all likely earlier in the program.

The three rate gyroscopes, integral elements of the telescopes pointing and control system, were also causing problems. A central requirement for the telescope's builders was to ensure that it could be pointed to an accuracy of 0.01 second of arc and, once aimed accurately, would stay pointed on its astronomical target. But if the gyros (or any other mechanism on the spacecraft, for that matter) produced too much vibration, the Fine Guidance Sensors could lose their lock on their guide stars, and valuable time would be lost as the sensors searched again for guide stars, thus possibly destroying an observation. And the vibrations measured during the contractors' tests of the gyros suggested that the telescope might continually lose lock on its guide stars. In discussions with Lockheed, Marshall had become convinced that the gyros had to be improved by a technique known as "shrouding." So once again a solution to a technical problem would require more money.

As Marshall routed money to tackle the technical problems, so they ate into the funds available in the fiscal 1982 budget. By September of 1982, in fact, all of the budget was needed to cover the currently identified risks.[87]

BRIEFING THE ADMINISTRATOR

On September 21, NASA administrator James Beggs was briefed on the agency's science missions. The section on the Space Telescope was focused chiefly on the telescope's scientific capabilities. It was delivered by newly appointed program manager, Marc Bensimon. Bensimon, who had been in his job less than a month and was not completely familiar with the program, had been told to emphasize the science the telescope would be able to perform. Beggs, however, was disturbed by rumors he had heard of difficulties at Lockheed. So when Bensimon reached his one chart on the status of the program, Beggs "just went off into the ceiling. He says 'you guys aren't telling me what's going on. You've got problems at Lockheed like you don't understand. You're not telling me about them. . . .' So he gave us a hell of a tongue lashing, and said, 'come back to me with a detailed status review on Space Telescope.' "[88]

It was widely understood at Marshall and at headquarters that there was a problem with resources for the telescope; however, there was more in NASA's space science program than the Space Telescope. The Infrared Astronomy Satellite (IRAS), later to be widely judged a spectacular scientific success, was in deep trouble in early 1982. Many doubted that it would ever be launched, or, if it were, that it would work well in orbit. Headquarters managers were also endeavoring to find adequate funding for the recently approved Gamma Ray Observatory. So, as a leading headquarters manager remembered,

we were aware that Space Telescope had trouble. . . . But we were convinced that our best strategy was to put off worrying about Space Telescope's large problems until the 1985 budget, because we felt that we had a pretty clear course to get us through to the 1985 budget, and not draw attention on ourselves for those problems; but rather to use that budgetary year with the focus on getting the [Gamma Ray Observatory] on good ground, getting it off as a new start, getting it accepted and getting it moving.[89]

"We all realized," Bensimon recalls,

that we didn't have enough money to do the job [on the Space Telescope]. And we had had some discussions on how we were going to go about

Artist's conception of IRAS in space. In 1982, the IRAS program was widely regarded to be in deep trouble, and some people suspected the satellite might never get launched. NASA's Office of Space Science and Applications was therefore devoting much of its attention to IRAS. (Courtesy of NASA.)

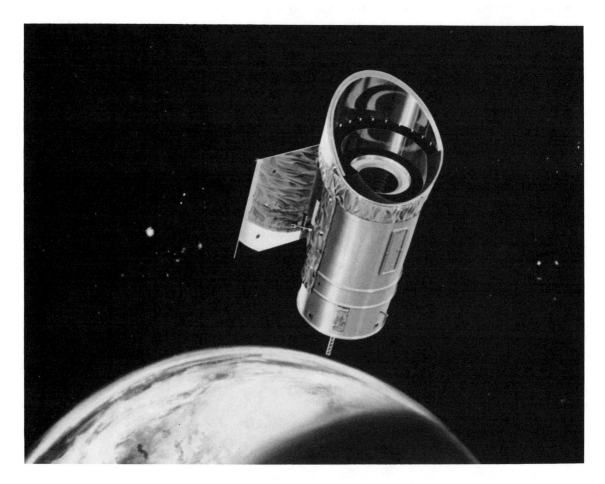

getting the necessary funding. We recognized we had to do some replanning of activities, particularly the integration activities, both at Perkin-Elmer and at Lockheed. We talked about how we would go about that and see if there was a painless way, without impairing the launch. The [fiscal] 1984 budget had already been submitted to OMB. So the question was, how do we get through [fiscal] 1983 and 1984 and increase our requirements in 1985 and possibly 1986?[90]

Meanwhile, Perkin-Elmer had been preparing for what was termed "Optical Test Number Two." In that test, various items would be bonded to the primary mirror, and its figure would be checked afterward to ensure that its quality had not been damaged. Program engineer Reetz had expressed doubt about Perkin-Elmer's preparations in July. Indeed, in preparing for Optical Test Number Two, Bensimon recalls, Perkin-Elmer

had pads to install on the bottom of the mirror that weren't machined properly. The bonding agent also came into question. They had to go back and do some tests on the bonding agent. The fixture that was supposed to hold it [the primary mirror] in turned out to be totally unsuitable. And then there were five or six other things in addition to that. Also, the roof of the facility, the clean room the mirror was stored in, had leaks. . . . So, every day you'd call up and try to find out "what's going on?" You would hit another problem. Just one after the other, bang, bang, bang, bang. And for a month we could show no progress. Every time you would look at something else there would be a problem, look at another thing and there would be another problem.[91]

By October, Marshall staff were in force at Perkin-Elmer trying to determine where things stood, poring over the Perkin-Elmer schedules.

It was against that background, then, that the briefing that NASA administrator Beggs had requested in September went ahead. It was delivered by Speer on November 3. He reported that Lockheed was making good progress: The basic design was complete, the telescope's requirements were being met, the subcontractors were delivering their hardware to Lockheed ahead of the dates on which it was needed, and the technical problems that might have affected the schedule were being resolved.[92]

In contrast, Speer painted a somewhat gloomy picture of conditions at Perkin-Elmer. He contended that the company had suffered from weak project management and poor schedule performance. Although it had advanced the state of the art in some areas, it had adopted what he termed a "hobby-shop" approach, with the focus on technical performance, and insufficient emphasis on schedule and cost. Speer described how NASA had pressed for various management changes to correct those faults. Marshall usually had three members of its staff in residence at Perkin-Elmer, but that number had been increasing, and by November a Marshall staff had been on site at Perkin-Elmer for some weeks to try to assess exactly where the schedules stood. That staff was also trying to find ways to recover time already lost on the schedules. Because Perkin-Elmer was short of some personnel with special skills, Marshall had assumed responsibility for designing and fabricating some of the equipment to install the scientific instruments in the Optical Telescope Assembly, in addition to performing selected tests at Marshall and Perkin-Elmer. For those witnessing Speer's briefing, however, the basic message, as a leading headquarters manager remembers, was that "we have some problems, it's going to be tight but we're going to make it."[93]

PROBLEMS PILE UP

Marshall was devoting a great deal of attention to Perkin-Elmer. Despite Beggs's earlier suspicions about Lockheed, the Optical Telescope Assembly was still regarded by Marshall as *the* item determining the pace of the program. On November 17, William Lucas, Marshall's director, traveled to Perkin-Elmer's plant in Danbury, Connecticut, to hear firsthand about the situation. Perkin-Elmer's project manager briefed Lucas and highlighted the saddle-bonding problems with which the company had been struggling for some months.[94] There was a more general discussion too. As Speer wrote to Perkin-Elmer a few days later, it was

agreed that we [Marshall] have not had the necessary visibility of the [Optical Telescope Assembly] program, particularly schedule and manpower requirements, to provide effective management and control. We also agreed this was a management problem and should be controlled as soon as possible. . . . This management activity of establishing a realistic program baseline, recognizing any real constraints such as available skills at Perkin-Elmer or resources at NASA, must receive highest priority.[95]

Shortly after the November 17 meeting, several NASA managers from headquarters and Goddard visited Perkin-Elmer to discuss the proposed Solar Optical Telescope, a new project with which Perkin-Elmer was involved. Not surprisingly, the conversation between the NASA and Perkin-Elmer representatives turned to the Space Telescope. One of those in the party was program engineer Arthur Reetz:

We were in an automobile riding from Danbury to Wilton, to go down to look at the coating chamber and some other optical facilities at Wilton. We were in the car with Jack Rehnberg, the manager for [Perkin-Elmer's] optical group, and [Don Fordyce] the Perkin-Elmer project manager. And it was Don Fordyce and Jack Rehnberg that said, for $5 million if we think we can put enough people on this . . . we can work our way out of it.[96]

There even seemed to be a relatively painless way of securing that extra amount. As Speer's deputy, James McCulloch, noted on December 13, $5 or $6 million was available in the Space Telescope's "maintenance and refurbishment" budget for fiscal 1983. With congressional approval, that could be switched into the development budget. McCulloch therefore recommended that "we immediately initiate this reprogramming activity so that we can maintain the momentum currently underway at [Perkin-Elmer]."[97] So, in mid-December, it seemed to Marshall and headquarters that they were facing a glitch in the program, not a major perturbation.

Perkin-Elmer, however, was still missing its milestones at an increasing rate. During 1982, NASA would later conclude, Perkin-Elmer's schedule had slipped approximately one week every two months between January and April, one week every month from April to August, and one week every two weeks in September and October.[98] Certainly during the fall and early winter, Marshall was pressing hard to receive Perkin-Elmer's revised schedules and was scrutinizing them as soon as they became available. But as Speer told the Marshall director on December 20, Marshall would not have all the information needed to compile a new overall schedule until early January; Perkin-Elmer's planning effort was late.[99]

The increasing flow of information out of Perkin-Elmer nevertheless meant that events were moving quickly. Just before Christmas, the picture began to darken. The news reached headquarters from Marshall that if Perkin-Elmer were to deliver the Optical Telescope Assembly to Lockheed on schedule, the extra price to be paid might be as much as $50 million in the current fiscal year, certainly far too much to be obtained by any simple reprogramming.

Christmas 1982 was neither restful nor festive for many of the Perkin-Elmer staff. Project manager Don Fordyce had his staff work over the vacation to better define the problems and schedules. By early January, Fordyce had seen enough. He concluded that matters were so bad that he had to go to Marshall and admit that "he had a situation he just couldn't possibly deal with, and that his intention was to come and see me with or without their approval."[100] The "me" in this case was Samuel ("Sam") Keller, deputy associate administrator in the Office of Space Science.[101] A big, jug-eared, avuncular-looking man with a reputation for blunt talking and a hard-nosed management style, Keller was a twenty-three-year NASA veteran who had joined the staff of the Goddard Space Flight Center in 1960. After holding a variety of management positions, he had moved to headquarters, becoming in 1981 deputy associate administrator in the Office of Space Science and Applications.

Keller had some familiarity with the telescope from his period at Goddard, but his new job soon brought responsibilities for the Space Telescope program. He therefore sought to become better acquainted with the telescope's status. The more Keller delved, the more concerned he became. Keller soon found that the scientists had become frustrated with their lack of influence on the program and

Samuel W. Keller, who was to play a crucial role in reshaping the Space Telescope program in early 1983. (Courtesy of NASA.)

deeply concerned by the telescope's difficulties. "Frankly, I don't think, in the late summer of '82," Keller recalls, "that half of the Principal Investigators would have bet they'd ever launch the Telescope. That may be a little strong, but there was developing a feeling of 'We can't get there from here.' "[102]

"The problems began to become more apparent," Keller recalled in late 1983,

at the Quarterly meetings, which are two or three day seminars held at Marshall with all the contractors, Principal Investigators, and advisors present. A series of reports are made by the various elements of the project, and when you would listen to the reports everything seemed fine, you'd go out in the hallway and talk to the presenter, and he would deny much of what he'd just said. . . . I think there was a very conscious effort at all levels of the project not to rock the boat in a public meeting. I understand that all the presentations that were made were sent to the project for review prior to these sessions. Some of these people have shown me copies of presentations submitted for clearance which came back marked up, with comments like, "This item is too controversial," or "don't bring this up in front of Headquarters." The reports were put together in such a fashion that they were not controversial.[103]

Nor was Keller the only one to regard the quarterly meetings as sterile; for the science teams, it was in part because of the insistence of Goddard's management at the time that Goddard review what they were going to say. "Part of the reason is perfectly OK," one principal investigator has argued:

Goddard did not want to be presented at a review, with Headquarters people there, with bad news that they had not heard about before. . . . But it went a bit further than that. They wanted to know what we were going to say and how we were going to say it and the impression this would give. They always wanted to — and this part is perfectly understandable — make sure that the impression would be given that they were on top of things, — "Yes, there was a problem there, but we're handling it." In fact, they knew about problems, but sometimes weren't handling them.[104]

Nor were the quarterlies seen by the participants as the forum in which to present new problems. Also, existing problems generally were described so that, to many observers, the emphasis was very much on accentuating the positive and downplaying or ignoring the negative. The result was sometimes a kind of fantasy world.

Marshall director William Lucas took a close interest in Marshall's major programs. He received regular notes and briefings on the telescope's status. When he offered a suggestion, he expected the Marshall project manager to jump to it. Although Lucas's style produced a sense of tight line management, it throttled public criticism, a point that was to be apparent during the Rogers Commission investigation after the explosion of the shuttle *Challenger* in January 1986. The result was that at the Space Telescope quarterlies, the impression given to many observers was of an almost endless round of good news, and most speakers basically seemed to be presenting what they thought Lucas wanted to hear. As O'Dell argued in 1985, "you see how you can get into a big problem – you have people being unwilling to surface problems. You can see how there would be an isolation from Headquarters, a lack of information at Headquarters, unless the Headquarters people were very competent and had sufficient time to look into things and detect them."[105]

On the Space Telescope program, the contacts between the NASA project manager and the contractor project managers were generally where the real information was passed. If Marshall judged that it needed to probe a problem or that a contractor needed to be taken to task, the center's style was to do that in private meetings, not in the relatively public forum of a quarterly meeting. Nevertheless, experienced Space Telescope hands usually could catch nuances in the public presentations at the quarterly meetings. They could thereby develop a reasonable grasp of the program's overall status, but to a beginner on the telescope, such as Sam Keller, there was an aura of schizophrenia to the program, a vast difference between what might be termed the public and private Space Telescope programs, a vast difference between what was going on in the plants and laboratories of the contractors and astronomers and the confident presentations at the quarterly meetings.

"OUT OF CONTROL"

Keller was not alone in having to get rapidly up to speed on the telescope. Another in the same position was Don Fordyce, the recently appointed Perkin-Elmer project manager. Fordyce, like Keller, had spent much of his earlier career at Goddard, and he knew Keller very well. During the late fall of 1982, these two had spoken privately about the telescope. Fordyce had voiced his increasing concern. Headquarters staffers, too, were reporting to Keller that the situation at Perkin-Elmer was deteriorating, although it is not clear how seriously those warnings were taken at the time. Matters came to a head at a meeting at NASA headquarters on January 14, 1983: Perkin-Elmer announced that it had enormous problems. In Fordyce's opinion, Perkin-Elmer would need a jump of about two hundred in its manpower (then at around 220) if the Optical Telescope Assembly was to be delivered to Lockheed on schedule in December 1983. It was obvious that Perkin-Elmer could not possibly get so many people to work so quickly and that the existing delivery date was impossible. Perkin-Elmer's bottom line, as Keller recalls, was that "we don't know what the schedule is, we don't know what the costs are going to be; we've got to go back and work it [out], but we have major problems and the program's out of control."[106]

The meeting was heated; Perkin-Elmer claimed that NASA had caused the problems because the agency had not given Perkin-Elmer money when it had been requested in 1981 and 1982. The NASA program manager replied that that claim missed the point; the agency had had to face hard budget realities. All budgets had to be submitted a year and a half in advance, and so it was impossible to make money available and to alter schedules on a month-to-month basis as Perkin-Elmer seemed to want. Even if additional funds were to be made available, what guarantee was there, given the company's past management performance, that Perkin-Elmer could properly use the extra money?[107]

More important, Keller had decided that the agency had "very serious problems," problems that had implications far beyond the Space Telescope program. In the previous few years, Congress had taken a strong interest in the agency's ability to manage large-scale programs; it had even requested a NASA review of the issue.[108] Keller had been intimately involved with that study. He therefore foresaw difficulties on Capitol Hill when the agency went for funds to bail out the telescope, for such action now seemed inevitable. He judged that "we are going to have to do something heroic to turn this around. We can't continue to do business as usual."[109] Early in February, Keller laid out his immediate plans.

Keller had concluded that the telescope could not be launched before the second half of 1985 and that a cost increase of "as much as" $100 million "may be incurred." Advanced programs inevitably meet unexpected technical problems; however, in Keller's opinion, "we should not permit a management situation such as has existed at [Perkin-Elmer] to surprise us as this one did." Keller proposed to introduce a new management group in NASA Headquarters to oversee the telescope's development. It would report directly to him. A review team should also be formed to investigate all elements of the program, not just Perkin-Elmer. Keller thought these actions necessary to secure a "detailed understanding of the [Space Telescope] situation as well as to emphasize to everybody in NASA the importance of the Telescope program and the need for a firm, realistic schedule with a cost plan that we are prepared to live with. It will also permit a time for assessment while we consider long-term changes that may be achieved in the project management structure."[110] Keller was pressing extremely hard to take control of the Space Telescope program, to be made, in effect, temporary program manager, but with decidedly enhanced powers and a free hand to do whatever he deemed needed to be done.

Marshall's preference was to work its way out of the problems by essentially doing "business as usual." Instead, Keller wanted to rethink the agency's entire approach and remove much of the risk that the compromises of the previous few years had built into the program. Keller's approach would be more costly in the short term, but might save money in the long run.

While Keller was urging a major policy shift, NASA staff members were busy informing the Office of Management and Budget, the Office of Science and Technology Policy, and key congressional staffers of the telescope's status. The aim, as a leading headquarters scientist has recalled, was to lead these important outsiders through the difficulties, explain what was happening, and "hope you can do enough stroking so that you get some understanding and didn't get into a big public confrontation when all this goes public at the time of the [congressional] hearings."[111] And go public it shortly would.

PROBING THE PROGRAM

The telescope's problems drew a lot of attention, from the executive branch, Congress, newspapers, and journals. In addition to the NASA review Keller had organized, the House Appropriations Committee assembled its own team of investigators. Members of the project were having to devote significant chunks of time to responding to inquiries rather than solving technical problems and managing the program.

The NASA review team, headed by James C. Welch, arrived at Marshall on February 9. Fred Speer told the Welch team that he judged the Fine Guidance Sensors to be the items determining the pace of the entire program. He further warned that the latches to hold the scientific instruments and Fine Guidance Sensors in place in the Optical Telescope Assembly might need a major redesign. If so, the rest of program would be substantially affected. Marshall's assistant director for policy and review, William Sneed, delivered a lengthy account of his own study of the telescope's status. He cited a range of factors that he believed had contributed to the delays and cost overruns. Those included what he termed general difficulties, such as, for example, the manpower cap headquarters had imposed on Marshall in 1977. Sneed contended, as had Warren Keller's review team in 1980, that the cap had prevented the center from managing the contractors as tightly as might otherwise have been the case. Sneed cited other problems, including the lack of engineering models of subsystems and, perhaps most important, the fact that Marshall had had to manage the telescope on an unrealistic budget.[112]

In addition to facing the special review teams, members of the program had to respond to the usual round of congressional hearings on the NASA budget. Not surprisingly, the telescope figured prominently in several of the hearings.

On February 28, 1983, members of the House Subcommittee on Space Science and Applications visited the Marshall Space Flight Center. Both Fred Speer and Marshall director William Lucas were quizzed at length on the Space Telescope. Speer explained that the Optical Telescope Assembly would not, as originally planned, be delivered from Perkin-Elmer to Lockheed in December 1983. In his opinion, delays had arisen because of a variety of technical problems. But he also argued that Perkin-Elmer's management had to be improved. Marshall, Speer pointed out, had already taken a number of actions to achieve that end:

We want to be closer to where the action is, so I have moved the entire optical telescope assembly project office to Perkin-Elmer to shorten the communication lines and to make sure that we can take corrective actions as required. . . . We certainly have not had the kind of forceful management that you need for this most difficult job. It is unfortunate that the most difficult job of this project was given to a contractor who was not strong enough from a management point of view. There was also simply an underestimation of the effort which, of course, is another indication of poor management. They did not have the ability to tell us early enough what is really required to do the various steps. . . . Now, Perkin-Elmer, I would have to say, has fully agreed with our assessment. They are taking very forceful actions. They have put in a new program manager [Don Fordyce], who has some experience at other programs in whom we have confidence. Corporate management has moved in very forcefully. The executive vice president of the corporation is taking charge of this optical group right now. . . . They have increased their manpower in order not to fall far behind schedule, and that manpower is now showing results. In addition we have given quite a bit of help from this center, forced Perkin-Elmer to do more detailed in-depth schedule penetration. In other words, we required them to look at each single drawing, each single operations sequence which is required to proceed.[113]

In summary, Speer concluded that "the Space Telescope project is making substantial progress in all areas in satisfying program requirements. Assembly, integration, and test operations are well underway, and all technical and scientific requirements are being met."[114]

Speer's judgment did not jibe with Sam Keller's assessment. When he saw the draft of Speer's testimony, Keller wrote in the margin: *"Not so."*[115] Keller's comment was indeed indicative of the quite distinct views of the telescope's status held at high levels in headquarters and at Marshall. It was almost as if Marshall and senior figures at headquarters such as Keller were talking about two different programs. Perhaps that was in part a reflection of the varying experiences of those engaged in working on the telescope and those investigating its status. Those who had lived with the Space Telescope for some years had grown accustomed to having to do business in a certain highly constrained manner because of the ever-present and massive financial pressures. They were therefore seeing the current situation in the light of what had gone before. In contrast, managers such as Sam Keller who had come relatively recently to the Space Telescope were inclined to be far more severe on what had gone before and to be much more purist in an engineering sense, to go back to basics and try to remove as much risk from the program as possible. They were also able to be more purist. As they were not responsible for, nor tied to, past decisions and were not enmeshed within webs of existing commitments, they were far freer to judge matters according to standards that had not been permissible for those managers who had actually had to manage the program on a day-to-day basis.

That point was underlined when on March 16 James Welch presented the findings of his review team to the NASA administrator. The briefing identified a string of problems and concerns, in addition to recommending solutions. The overall tone was extremely critical. There were two central arguments: first, "sound engineering management practices" were not being applied to all elements of the program, and, second, "technical performance [had been] subverted by program design and questionable decisions."[116]

One topic that was especially prominent in several sections of Welch's briefing was systems engineering. Lockheed was the contractor for systems engineering, but Welch argued that in practice, Lockheed had been able to perform only a coordination and system assembly function. The company had no direct authority over the other contractors and so could act only as advisor to Marshall. Lockheed had therefore been able to gain only a limited insight into what the associate contractors and science instrument contractors were doing.[117] The need to strengthen systems engineering had also been stressed in 1980 by Warren Keller's review team. Indeed, that was one of several of its recommendations that had not been fully implemented, and the same criticisms were being made in 1983 as in 1980, a point not lost on administrator James Beggs.

The report compiled by the House Appropriations Committee's surveys and investigations staff also was reproachful. One area it put under particular fire was NASA's management approach to building the telescope. Despite what the report agreed was the presence of a well-defined management apparatus, "serious management problems evolved early in the Space Telescope program. Many of these problems had a direct impact on technical requirements, [and] schedules, and resulted in additional funding far beyond that anticipated — even in a project where large scale reserves and/or contingency funding had already been plumbed into the program."[118]

Program scientist Edward J. Weiler (right) with Garth Illingworth and principal investigator Richard Harms (left) in 1985. (Courtesy of Pearl Gerstel.)

THE SCIENTISTS RESPOND

As the program's difficulties had become better known to him, the program scientist at headquarters, Edward J. (Ed) Weiler, had grown increasingly alarmed. After a spell at Goddard as part of the Princeton team concerned with the scientific operations of the *Copernicus* satellite, he had entered NASA Headquarters in 1978. In 1980, Weiler had succeeded Nancy Roman as the telescope's program scientist, and about the same time he had become chief of astronomy in NASA Headquarters. So by 1983 he was thoroughly versed in the scientific, technical, and management aspects of space astronomy. After attending Marshall's review of the telescope's status in November, he had judged that the program was not in serious difficulties. But by January, matters looked so bad to him that he was worried the telescope might even be canceled. "I invited directly each scientist in the Science Working Group," Weiler recalls,

to send me a letter which would be held in confidence, talking about anything they wanted to talk about. Tell me where the problems are. I don't
care if you name names or talk about management, or technical problems,
or scientific problems; just say whatever you want to say that is on your
mind. . . . And I did this in total independence of James Welch. I didn't
know that Jim Welch existed; that he had been commissioned to do a
similar type of review from an engineering and management point of view.[119]

After studying the replies from the scientists, Weiler, O'Dell,
operations scientist Albert Boggess, and instrument scientist David
Leckrone compiled their own report. They identified fifteen concerns
and proposed some actions; most significant, the astronomers had
independently corroborated many of the findings of the Welch review team. In particular, they had identified systems engineering as
their number-one concern, and their report painted a bleak picture
of the program's status.

As one astronomer put it,

in my opinion the most serious problem in the [Space Telescope] program,
across the board, is the lack of adequate systems engineering. A large part
of this problem appears to be a reticence on the project's part to do [systems
engineering] itself or allow the official [systems engineering] contractor to
do it. The problem varies over the different elements but ranges from severe to total. Although widely recognized nothing seems to happen. My
observation is that this critical function is often ignored or forgotten. There
are many cases where one is left with the clear impression that there is a
fear of systems engineering studies and audits – fear that they might "find
something" that would upset the schedule/cost. There's a pervasive attitude that "I don't want to hear of any more problems."[120]

Another astronomer commented on the available spares: "There aren't
any – anywhere." As we shall see at more length in Chapter 9, many
of the astronomers also had little confidence that the program of
testing that NASA had agreed to for the scientific instruments was
adequate to the task. Yet another astronomer commented: "I do not
believe that simply slipping schedule, or sending a Tiger Team of
managers from [Marshall] to [Perkin-Elmer] or pouring in money
will resolve the basic problem. I do believe that NASA Headquarters must take a position of much stronger leadership, because if ST
goes down the drain, NASA may go with it, which would be a
national catastrophe."

That Headquarters should play a more forceful role in managing
the telescope had been in Sam Keller's mind for some time. The
difficulties with the Space Telescope were especially troubling for
Keller and NASA administrator Beggs because they meant that the
NASA budget, submitted to Congress as recently as January, would
have to be revised. Some monies already allocated to other programs
would have to be transferred to the Space Telescope because of its
cost overruns.[121] If the size of the problems had been correctly as-

309

sessed and communicated to Beggs *before* the details of the NASA budget had been completed with the Office of Management and Budget, the administrator would not have been placed in such an embarrassing position. Now he had to ask Congress for budget adjustments almost as soon as the president's budget had reached Capitol Hill. The timing of the crisis, then, had added to its magnitude.

It was obvious that Beggs would have a lot of explaining to do on Capitol Hill. Equally clear, given the strength of the congressional reaction and the interest that had been stirred up in newspapers and journals, was that NASA would have to reorganize the program, if only as a signal to Congress that action was being taken. Beggs was also angry at what he saw as Marshall's management failure. The need to ask for extra funds for the Space Telescope hardly added to the agency's credibility at a time when Beggs was lobbying hard for approval to build a space station.

BEGGS'S VERSION

On March 22, Beggs presented his version of what had gone wrong to the House Appropriations Subcommittee. Facing its chairman, Edward Boland, Beggs conceded that "we have had a difficult time with this program at Marshall. I think I would characterize it not as a management breakdown but as perhaps a failure to take certain actions in the right time sequence. What we were trying to do was an exceptionally difficult task."[122] The House report had criticized communications between headquarters and Marshall. Beggs agreed, but pointed out that since the start of the telescope's detailed design and development in 1977, there had been a heavy turnover of personnel at headquarters. In those six years there had been five people in charge of the Office of Space Science, and five different program managers for the Space Telescope. In his opinion, "I think part of the communications breakdown was the fact that the people didn't know each other very well and that the Headquarters people were not able to ask the right questions and, therefore, they didn't get the right answers."[123]

Beggs also recounted for the House subcommittee how his unease with the telescope's status twice led him to ask for major reviews of the program in 1982. During the second of those reviews, in November, "we got a report that the program did have some difficulties but they were not severe. We were told we would not have to rebudget: the difficulties would be covered with program reserves." Beggs further argued that

if this program had been properly managed and communicated, without question, we would have been getting the severe rumblings from the contractor. The contractor was not coming clean to Marshall either. There is no question about that. They were covering over the problems in terms of both cost and schedule and they were not being transmitted to Marshall. But a good program management organization smells these problems out and we were not sniffing them out. [Nevertheless] we have not done all that bad either. I don't want to bare my breast here and beat us up too badly. We have made very, very significant progress on the program. We have done very well in bringing the pieces and parts of the job to this point. It is just that we have not been able to smell out the financial and schedule problems early enough, which is where I fault us. Technically, we have been doing a reasonably good job of progressing on it.

Beggs also announced that NASA would change its management of the telescope program. He wanted "a very strong man" on the Space Telescope at headquarters:

I need a mean, tough, program type . . . and then we can make more of a management impact from Washington. But we also need a strong program management structure at Marshall because that is where the detailed technical management takes place. We have not decided the names of the individuals but we are in the process of pinpointing the top flight people that we have at Marshall, and there are a lot of them, and we will put them on this program.[124]

Another point raised at the time by some headquarters managers was that headquarters had not understood the telescope's true status because there was a communications failure between Marshall and headquarters. Although such a claim was a useful means of distancing senior managers at headquarters from any blame, it does not easily mesh with the available evidence. Beggs argued that there had been a failure of management, but as another headquarters manager closely involved with the telescope recalled, "it was pretty well glossed over, the fact that Marshall was operating and doing what they were doing in very close collaboration with [Office of Space Science and Applications] management. . . . And I think there was no misunderstanding between [Office of Space Science and Applications management] and Marshall about what was happening. But the issue was raised as incompetence at Marshall and also as a cover up of information."[125] Marshall had not failed to transmit information to headquarters in late 1982, despite one source who claimed publicly that communications between headquarters and Marshall had been "horrible."[126] The information reaching Marshall *was* transferred to headquarters; the problem lay more in the difficulties Marshall had in extracting information from Perkin-Elmer, either because Perkin-Elmer did not understand its own position, because the schedules were in such a shambles, or, as Beggs had claimed, because Perkin-Elmer chose to hide the magnitude of its problems. However, given the nature of the quarterly meetings at Marshall, anyone who relied on them as a major source of information could easily have been misled as to the program's status.

A less strident, but sure more compelling and in some ways even more damning, account of the telescope's problems than Beggs's came in April, when Marshall director Lucas reviewed for Beggs the actions Marshall had taken on the telescope since September 1982 and the factors he thought had led to the program's current position. Those, he contended, had in many respects been rooted in the program design. "The Space Telescope," Lucas wrote,

has been a tremendous scientific, engineering, and management challenge. As a result, the program has encountered more than its share of problems, many of which are directly attributable to decisions made early in the program concerning the procurement and management strategy, and the protoflight development approach. In spite of our concerted efforts to establish a high-confidence program baseline in December 1980, we simply were not able to fully recover from the inherent problems introduced into the program as a result of those early decisions.

In Lucas's opinion,

I believe we have made considerable technical progress on the development of the Space Telescope. The extreme complexity and demanding requirements, coupled with the inherent problems associated with some early decisions, have made it extremely difficult to assess schedule progress or accurately predict cost requirements in a timely and effective manner. The inability to do this and the perceived necessity to remain under annual and budgetary commitments caused us to continuously understate our budgetary needs. This understatement of budgetary needs resulted in certain critical program decisions being made that, in retrospect, would be judged to have introduced too much risk into a project of such complexity and importance. They were, however, made with full knowledge of all parties at the time they were made. While I do not offer the above as an excuse, or justification, for the problems now confronting the Space Telescope, I do believe that appropriate consideration must be given them in assessing what went wrong, if for no other reason than to preclude similar decisions being made on future projects. [127]

Senior managers in headquarters nevertheless wanted heads to roll. Three days later, following pressure from administrator Beggs, [128] Marshall announced that Speer had been replaced by James B. Odom as the project manager for the Space Telescope. Goddard's recently appointed director, Noel Hinners, and many others on the program believed that Speer, who had managed the telescope according to the approach agreed between Marshall and headquarters, was made something of a sacrificial lamb for the program's difficulties. [129]

In addition to changes at Marshall, there were new people in place at headquarters. There was also a new management structure at NASA Headquarters. The "mean, tough, program type" Beggs told Congress he was hunting for turned out to be James C. ("Jim") Welch, the leader of the review team. As Keller has argued, "I think when you get a program in as much trouble as this was in, you have to

make changes. It's not a matter of trying to punish people or any- *Beggs's version*
thing like that because I think in the case of every individual here,
I could give you a set of mitigating circumstances why what happened happened. It was a combination of a lot of circumstances,
which we could blame on everybody and nobody, or you can blame
it on the system."[130] And there Keller surely hit on a central point,
for it was not a few individuals who had failed, but rather the system
within which they had had to work. To a significant degree the crisis
of late 1982 and early 1983 was the outgrowth of the overselling
and underfunding of the mid-1970s. The underfunding had also
been compounded by a range of other factors: managerial, technical,
scientific, social, and institutional. The crisis had nevertheless presented, and Keller and others had shaped it to present, an opportunity to escape to some extent from the program's history.

The managers on the program had always faced a difficult task
and had been compelled to manage according to the program design, schedule, and funds that existed – not what they might have
wanted or thought necessary. The new managers would have the
inestimable advantage of a large injection of funds into the project,
together with more time to finish building and testing the telescope, together with a somewhat revised program design. Exactly
how much extra time and money there would be, NASA had yet to
determine. The new managers, however, still would have to grapple
with extremely tough technical, scientific, and management challenges, challenges that raised doubts for some that the telescope
could ever work to its agreed specifications. Thus, there were many
new people in place, but no guarantee that an efficiently functioning
Space Telescope would result.

9

Closing in

We're gonna have to give up some niceties, since we've neither time nor money to make this perfect. We've eaten into reserves; this is not a sacred program.

James Beggs, 1984

I will have to admit that we are hanging on by our fingernails.

Burton Edelson, 1985

In November 1986, James B. Odom announced that he was to become associate director for science and engineering at Marshall Space Flight Center, following three and a half years as project manager for the Space Telescope. However, the atmosphere when he left the project was vastly different from that when he had joined. In early 1983, some people had harbored fears that the telescope might be canceled, and others had grave doubts that the telescope would ever perform to specifications.

For NASA and ESA staff, the contractors, and the astronomers, 1983 and the years that followed would be extremely demanding, years in which they faced an imposing array of problems. On several occasions the program seemed to teeter on the verge of another long delay, but whereas at the end of 1986 there were, for the program's participants, still major problems to be overcome, these problems were generally reckoned to be of a very different order from those in which the program had been mired in 1983.

The crucial point is that between 1983 and 1986, the program was able, to a large degree, to escape from its early history. The crisis of 1983 both prompted and allowed NASA managers to make significant changes in management structure, personnel, and program approach, as well as some of the power relationships between the different institutions, individuals, and groups that made up the program. The central themes in this chapter, then, are the ways in which NASA set about revising the program design, and how the changed relationships among NASA, ESA, the contractors, and the astronomers were played out in completing, bringing together, and testing the various sections of the telescope and its ground system, as well as the assembled Space Telescope.

There was, however, to be an unexpected coda to the telescope's design and development phase: another lengthy and jarring delay in the telescope's launch date. As we shall see toward the end of this chapter, the slip was, at least in the first instance, caused by events outside of the Space Telescope program: the grounding of the shuttle fleet because of the destruction in January 1986 of the Space Shuttle *Challenger*.

NEW STRUCTURES

The major scheduling and cost problems on the Space Telescope program that NASA confronted in late 1982 and early 1983 led, and in some ways allowed, the agency to assess the program afresh. But the crisis of early 1983 was fundamentally different from that of 1980. In an important way it was also more simple. In 1980, managers had had the option of trading cost and schedule for scientific and technical performance, as well as changing the telescope's design and omitting certain features. By 1983 the design work was mostly complete, and much of the hardware already manufactured. Thus, the issue for NASA managers essentially boiled down to securing extra funds. And, sure enough, following the NASA Headquarters campaign to present to the White House and Congress its version of why the program had gone awry, and to demonstrate that radical measures had been taken to correct matters, more funds had been forthcoming from Capitol Hill. That included an extra $45 million for fiscal 1983, $15 million of which was allocated to the troubled Fine Guidance Sensors. With the additional money in this and future years, NASA's estimate of how much it would spend on the telescope's design and development had jumped to $1,175 million (1983 dollars). After allowing for inflation, the telescope's cost had thus risen since 1977 by roughly a factor of two.

The launch date, too, had been slipped, with NASA, on this occasion, telling Congress it would launch the telescope in the second half of 1986. And there were other significant alterations to the program. According to one writer, the "two least credible sentences in the English language have been said to be 'The check is in the mail' and 'I'm from headquarters and I'm here to help you.' "[1] In 1983, in the wake of the Space Telescope's increasing difficulties, and whether or not the field centers liked it, NASA Headquarters had decided to take a much more visible and stronger role in managing the program.

315

Program manager James C. Welch (left), with Frank Carr (center) and Robert A. Brown (right). Before Welch was appointed, NASA administrator James Beggs told Congress he intended to appoint a "mean, tough" manager to head the program at NASA Headquarters. (Courtesy of Robert A. Brown.)

Until mid-1983, only a few people at NASA Headquarters had worked on the telescope: a program manager, a program engineer, a resource analyst, and a program scientist. At headquarters, Sam Keller told his superiors that such organization was insufficient. As was seen in Chapter 8, he wanted to overhaul the way headquarters managed the program. Over the objections of some other headquarters managers, his argument won out.

In order to strengthen the hand of NASA Headquarters with respect to the field centers, Keller – unable to assemble a NASA team in a short time because of civil service regulations – put into place a systems engineering group from the BDM Corporation, a corporation with no NASA experience, though it had worked extensively with the Department of Defense. NASA administrator Beggs had told Congress he needed a "mean, tough, program type" to head the Space Telescope program at headquarters. The manager appointed was James C. Welch. Welch had conducted the internal NASA investigation of the telescope program in 1983, and his earlier involvement with classified space programs that had similarities to the Space Telescope program helped to recommend him for the post. He was to head a group of around fifteen people, to be called the "Space Telescope Development Division," although Welch reported directly to Keller. Keller himself was devoting roughly half of his time to the telescope. In addition, NASA administrator James Beggs had emphasized that, after the shuttle, the Space Telescope was the agency's top-priority program.

In addition to the management changes and the announcement of increased emphasis on the telescope program at headquarters, NASA altered its decision-making process. So-called "Level 1" decisions, decisions affecting what had been defined as the Level 1 requirements of the telescope (which themselves had, amazingly, been explicitly stated only in late 1983, six years after the program had started), could be made only with the agreement of headquarters. It was a Level 1 requirement, for example, that the telescope "must have the capability to acquire a target and position it anywhere within any entrance aperture of any scientific instrument with an accuracy of 0.01 [second of arc]."[2] Thus, if any decision had to be made that might affect that requirement, it could be done only with the approval of NASA Headquarters.[3]

Such was indeed a notable shift in the relationship between headquarters and the field centers, although it was part of a more general process in which headquarters wrested back to itself some of the authority that had previously been delegated to Marshall and Goddard. By placing a systems engineering group at headquarters and by changing the manner of decision making, headquarters wanted to ensure it would be more closely engaged than ever before in the making of technical decisions, previously a jealously guarded prerogative of the field centers. Formerly, headquarters had acted essentially as advocate for the program; now the relationship between headquarters and the field centers (by deliberate design of headquarters) would take on adversarial overtones, with headquarters taking a more skeptical and questioning attitude toward the activities of Marshall and Goddard. Nevertheless, it would still be at the field centers that the overwhelming bulk of NASA's technical and managerial work on the telescope would be conducted, and that, in combination with the activities at the contractors, among the instrument teams, and at the Space Telescope Science Institute, would largely determine the telescope's success or failure.

POWER TO THE SCIENTISTS

In addition to redefining its relationship with the field centers, headquarters mediated other adjustments in the power structure of the program. One was to establish an institutional mechanism to allow the astronomers to exercise more power. By 1983, some, although not all, of the astronomers regarded the Science Working Group as toothless. For them, the group was just too large and diffuse a body to have much effect on overall program activities, particularly because it met only once every three months, and significant changes were being made in the program at a much more rapid rate. However, that is not to say that the scientists were without any say on program-wide decisions. Rather, the scientists tended to exert their influence through personal contacts with engineers and managers and through their participation in informal networks and official committees that sprang up on particular technical issues, as we saw in Chapter 8 on the decision to remove the aperture door, for example. Also, as the number of staff at the Space Telescope Science Institute increased, it grew in influence in many areas, as we shall discuss later.

Nevertheless, in early 1983 many of the astronomers had been deeply frustrated; the situation looked quite bleak, and they felt helpless in the face of the confusion surrounding the program (early 1983, remember, was before Congress had provided extra funds). What could they do? In March 1983, at a special meeting of the Science Working Group called to discuss the telescope's status, program scientist Edward Weiler proposed that a small executive committee of the astronomers be formed. He wanted the executive committee to meet more regularly than the working group and undertake regular reviews. In that process, the voice of the scientists would be heard more distinctly by the project's managers. Although the term was not used, the executive committee might become a kind of "war cabinet" for the scientists.[4]

The idea for such a committee was supported by James Welch, who saw two advantages in its formation: First, the committee would engage some of the members of the Science Working Group in the program more frequently than before. It would thereby give managers more chance to tap the expertise of the astronomers. Second, it would give the scientific community a sense of involvement

that it hadn't had before. If people feel isolated they tend not to offer as much advice as they would otherwise. . . . The science community had felt totally shut out, I don't think that's too strong a phrase . . . and when the money started getting tight, the project managers quite naturally felt that they could do without all the peripheral advice that was coming in to help them manage their program under those really severe constraints. So I can't be too critical of what the people in the project were doing. Given those same sets of conditions, I think any reasonable person might have done the same thing. But everything had conspired, all the circumstances had conspired, against detailed involvement in decision-making at the project level, and the science community . . . felt powerless to do anything about it.[5]

But what sort of group would best give the astronomers more voice? One complication for the astronomers was that the issue of the organization to which the committee would report became enmeshed in headquarters politics. The resolution of that issue took some jockeying. The astronomers, not surprisingly, wanted to report to headquarters, the highest level in the program. At a meeting in early May 1983, both Sam Keller and NASA's chief scientist disliked that notion. Reporting to headquarters, they contended, would set a bad precedent; Marshall would be cut out, and that would be

unfair to the new Marshall project manager, James Odom. However, agreement was reached that Robert Bless, a principal investigator with a wide range of experience in the nuts and bolts of space
astronomy, should chair the committee.[6] Despite its nebulous position within the structure of the program, what had come to be
called the "Space Telescope Science Coordinating Committee" held
its first meeting on May 26. Again, most of the members urged that
the committee report directly to headquarters. They also complained that, with nine members, it was too large. In what was a
novel proposal, given the project's earlier history, the members also
pressed for unlimited access to information at all levels of the project. Nor did they want their authority restricted to discussing only
scientific or technical matters.[7]

Still there was opposition in headquarters. On May 31, Frank
MacDonald, NASA's chief scientist, telephoned Bless to argue that
the committee should report to Marshall. Bless conceded that by
having it reported to headquarters, rather than Marshall, they were
not really giving Odom a chance. Bless nevertheless maintained that
there was no time for delay in solving the telescope's problems.
Headquarters was already heavily involved with management issues,
and it could not reflect badly on Odom if the committee reported to
headquarters. Yet if the committee began by reporting to Marshall
and then had to switch to headquarters because it was unhappy with
Marshall, that would indeed be bad. Five and a half years had already been spent in building the Space Telescope, so that "people
have become [used] to working in certain ways, had staked out their
turf," and Bless emphasized that the scientists had become frustrated. The new organization, responsible to headquarters, was ex-
pressly designed to give them a stronger say.[8] If MacDonald won
the argument, then the entire Science Working Group might become involved once more, and NASA might regard the committee
as little more than an executive committee of the working group
chaired by the project scientist. If so, the scientists feared that the
new committee would lose its punch, and the point of forming a
small group would be lost.

It is another indication of how power had shifted within the program that the views of the scientists on the committee prevailed.
The membership of the committee would be reduced. It would report to headquarters, and although it would report to program scientist Weiler, the members would have the opportunity to speak
directly to Sam Keller or Welch if they wished. The committee's
name was also to be changed, to the Space Telescope Observatory
Performance and Assessment Team (STOPAT). Most important, the
team had a license to probe into just about whatever issues it chose.
For the next two years, STOPAT would play a major role in the
telescope's development.

STOPAT members photographed at Lockheed in early 1985: left to right, James Westphal, Domenick Tenerelli (Lockheed), Robert Brown, C. R. O'Dell, Edward Weiler, Robert Bless, and Robert Jones (Perkin-Elmer). (Courtesy of Robert A. Brown.)

NEW PEOPLE AND NEW APPROACHES

Some headquarters managers had doubts about the formation of STOPAT, because its views might conflict with those of the Marshall project manager. That, however, proved to be a misplaced concern, at least in its early years.

In April 1983, James Odom had replaced Fred Speer as project manager. Odom had almost thirty years of experience with space hardware. He had joined the U.S. Army Ballistic Missile Agency in 1956, and like so many of the missile agency's staff, he transferred to NASA when the Marshall Space Flight Center was formed in 1960. After holding various positions at Marshall, in 1972 he became manager of the "External Tank Project" for the Space Shuttle. Odom realized that in contrast with the shuttle program, in which interactions had been essentially from one engineer to another, working on the Space Telescope meant a "completely different culture [where scientists] have a different set of motivations with totally different reporting chains. Whereas, we have very structural reporting chains within NASA or DOD."[9] Instead of treating the behavior of the scientists as a threat, and regarding STOPAT as an enemy, Odom actively sought the assistance of the astronomers and aided their efforts to examine various technical areas.

In addition to a new project manager, there were other important personnel changes. In 1982, Bob O'Dell, after an often grueling decade as the Space Telescope's project scientist, had decided to return to academic life to pursue astronomy full-time. His decision in 1972 to move to Marshall, with the administrative burdens his tasks

Project manager James B. Odom (right) in consultation with Goddard project manager Frank Carr (left) during a Science Working Group meeting in 1985. Jim Moore, who would succeed Carr in 1987, looks on. Odom brought a rich fund of experience to his work on the Space Telescope, including a spell as manger for the Space Shuttle's external tank. (Courtesy of Pearl Gerstel.)

at the center involved, had drastically curtailed the time he could spend pursuing research, something he had come to regret.[10] O'Dell had long been uncomfortable with the rather ambiguous position in which he found himself as a civil service scientist. By, in effect, aiming to represent the views of the outside community of astronomers within NASA as a NASA employee, he had felt "many many times that I wasn't either 'us' or 'them.' "[11] So in 1982, with the telescope then scheduled to be launched in less than three years, O'Dell concluded that if he was ever to return to academic life, that was the time. He therefore left Marshall to assume a position at Rice University in Houston.

O'Dell had continued initially as project scientist on a part-time basis. But in mid-1983, with the telescope beset by massive problems, Odom insisted on a full-time project scientist to be based at Marshall. When O'Dell had declined to return, Robert A. ("Bob") Brown, a member of the staff of the Space Telescope Science Institute, joined Marshall on a two-year leave of absence. A planetary scientist by training, Brown would prove to be an aggressive advocate of what he regarded as good management and scientific practices. As will be seen later, he would also press particularly hard to ensure that the Space Telescope truly would be capable of high-quality observations of planets. In so doing, Brown would rekindle a debate that had reached its previous high point in 1977 during the selling of the telescope (see Chapter 5). Then NASA had sought to win the support of planetary scientists with claims about the telescope's capacity for examining objects within the solar system. One of Brown's goals as project scientist, and as a member of STO-PAT, was to ensure that the telescope would live up to those claims.

It was not only the program's organization and staff that had been changed, but also the way NASA approached the program. NASA had assigned large teams to strengthen the systems engineering groups at both Lockheed and Marshall, a move that was a direct response to the criticisms of the various review teams of early 1983.

In 1983, as part of NASA's fundamental rethinking of the telescope project, the agency also decided that too much risk had been introduced into the program. With the recent injection of hundreds of millions of dollars into the program, NASA had a chance to remove some of that risk, to in effect escape from some of the program's earlier history. As James Welch has recalled, it "was too late to put back everything, of course, because when you get to a certain point in a program, you can't go back and do retrospective testing on some things that have already been built and integrated, but we did put back as much as we reasonably could."[12] More spares were added to the program, and more time was added to the schedule, particularly for testing.

When NASA reexamined the status of the program in 1983, Lockheed argued that the assembly and verification phase of the program — the phase in which the various sections of the telescope would be tested, assembled, and the entire spacecraft tested — was well defined. NASA, however, judged that Lockheed was relying too much on its experience in building military satellites, which are produced in the manner of an assembly line. NASA was sure that assembling and testing the Space Telescope would pose unique and major difficulties. "So," Welch recalls, "I had to turn a deaf ear . . . to Lockheed's claims of being on firm ground in the design for the Assembly and Verification phase of the program. We [NASA] redesigned it for them, put some things back in, stretched it out."[13]

In line with the decision to remove some of the risk from the program, the agency decided, in yet another flip-flop on this issue, to increase the number of units in the spacecraft designed to be capable of being removed in orbit. The advantage of such units was that if one became defective in space, it could be replaced by visiting astronauts from the Space Shuttle, with a replacement being inserted, and the spacecraft's operations continuing as before. Initially some 125 units had been so designed, but the number had been severely reduced to about a dozen when the program had been restructured in late 1980. In 1983, the number would be increased once more and would lead NASA to step up the training of the astronauts for in-orbit work on the telescope.

NASA's decision to provide funds for what would later come to be termed the "Wide Field/Planetary Camera Clone" was also part of the risk-reducing approach. During the building of the telescope, it had become widely accepted among the program's participants that though all of the scientific instruments might be equal, one was more equal than the others. That was the Wide Field/Planetary Camera. It was the most complex and the most costly, and because of the spectacular CCD images it was expected to produce, it was widely regarded as the most exciting instrument in terms of the telescope's appeal to the general public (and the Congress). The Wide Field/Planetary Camera was reckoned to be likely to produce about 90 percent, by volume, of the telescope's scientific data and often would be operated at the same time as other instruments. Images from the Wide Field/Planetary Camera could also be used to position astronomical targets on the entrance apertures of the telescope's spectrographs. Without an efficiently functioning Wide Field/Planetary Camera, the argument went, the spectrographs probably would not work as efficiently as intended.

If the Wide Field/Planetary Camera should stop returning data a short time after the telescope went into space, obviously that would be extremely damaging to the telescope program both scientifically and politically. Studies of the scientific instruments' expected failure rates had indicated that the camera was likely to fail sooner than any

Four shots of the Wide Field/Planetary Camera undergoing work at the Jet Propulsion Laboratory. (Courtesy of NASA.)

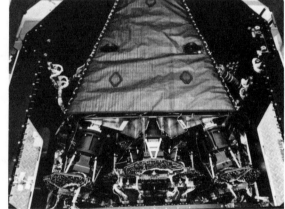

of the other instruments. During its development, the camera's builders had been compelled, for a variety of reasons, including cost, schedule, and weight, to remove much of the redundancy that had originally been designed into the camera. Although calculations of failure rates are regarded as notoriously unreliable by aerospace engineers, the prospect of early failure of the Wide Field/Planetary Camera was one that managers felt they had to take seriously. NASA managers were also concerned by a problem that, as we shall see later, had come to light with the camera's CCD chips and that threatened to have a significant effect on the science to be done with the camera: Quantum Efficiency Hysteresis. What was to be done? One option for NASA was to fund a backup Wide Field/Planetary Camera. Then, if the original camera failed or if its performance should degrade to a level deemed unacceptable, it could be switched, in orbit, with the backup. Even if it was too late to fix the QEH problem to the Wide Field/Planetary Camera, a fix might be found for its replacement.

The agency wanted to make a decision fairly quickly. If NASA delayed, the team at the Jet Propulsion Laboratory that had built the original camera would disperse. The institutional knowledge that underpinned its design and construction might quickly be lost. Even if construction on a backup were started right away, it would cost several tens of millions of dollars, but starting to build a backup camera from scratch in a few years time would surely be more expensive.

MAINTAINING THE SCIENTIFIC INSTRUMENTS

Program scientist Edward Weiler also wanted to push an issue wider than that of a replacement for a single scientific instrument: the planning for maintaining and refurbishing the Space Telescope. The telescope had been sold to the White House and Congress in the mid-1970s as a long-lived, serviceable observatory in space, one whose instruments would be regularly replaced and updated so as to keep it at the cutting edge of technology. Such aspirations, which, of course, meshed with NASA's strong institutional interest in emphasizing potential uses of the Space Shuttle, had slipped from sight over the years. In considerable part that was because of the financial problems the program had encountered, which had focused the available funds and the attention of managers on more immediate and pressing problems. But in Weiler's opinion, the existing schemes were certainly not adequate to maintain a scientifically productive observatory that could operate in space for well over a decade. He also wanted to strengthen his hand in negotiations on this issue by gaining the support of the Science Working Group.

Weiler took his case to the group in early 1984 and argued that

Using mock-ups in the neutral-buoyancy tank at the Marshall Space Flight Center to practice installation and removal of the Wide Field/Planetary Camera from the Space Telescope while the telescope is in the shuttle's payload bay. (Courtesy of NASA.)

it was essential to return to the concept of a serviceable observatory and for the agency to budget accordingly. What was the point of having a spacecraft in orbit that functioned well, but whose scientific instruments were ineffective or had failed? There were few spare parts for the instruments, and there was no detailed plan for personnel and facilities to receive and service a scientific instrument that might be returned from orbit.[14]

The issue of replacements for scientific instruments also arose in hearings before the House Subcommittee on Space Science and Applications. At the February 1984 Science Working Group meeting, John Bahcall noted that NASA had committed itself to supporting the Space Telescope for a ten-to-fifteen-year lifetime. The scientists therefore should all "drop hints" to various people in Congress to keep the issue alive.[15] Wherever the information had come from, by May 1984 the chairman of the subcommittee, Democrat Harold L. Volkmer of Missouri, had clearly been well briefed on the topic of maintenance and refurbishment, and this shows that the astronomers had been able to gain allies for their position on maintaining the scientific instruments even from Capitol Hill. Volkmer probed NASA's thinking on the topic during testimony by both Sam Keller and Goddard director Noel Hinners. When the questioning reached the backup Wide Field/Planetary Camera, Keller agreed that the camera probably was "our most significant instrument," but said that no decision had been made whether or not to build a backup.[16]

After detailed exchanges among headquarters, Goddard, and the Jet Propulsion Laboratory – the builders of the original Wide Field/Planetary Camera – headquarters decided that a backup should indeed be built. As James Welch told the director of the Jet Propulsion Laboratory, the

WF/PC [Wide Field/Planetary Camera] spare is a firm "starter" in the [Space Telescope program]. What remains to be done is to fit the JPL effort within the fiscal year fund constraints. While this longer term question is being worked, near term funding has been provided to JPL to retain key members of the WF/PC team. . . . Because of these fund constraints and the need to have the WF/PC spare capability available as soon as possible to assure support of the other Science instruments, the decision has been made to acquire a "clone" of the existing instrument, rather than to develop a camera incorporating additional capabilities. . . . It is our firm resolve that no additional requirements will be imposed on the WF/PC spare and that no changes in assumptions will be allowed. The funding and schedule realities make this the only possible course of action.[17]

Again NASA was removing some risk from the program by purchasing extra insurance, this time around an estimated $60 million worth.[18]

At the same time, NASA was changing its plans for maintaining and refurbishing the telescope in space. Rather than follow the previous scheme of returning the telescope to earth every five years or so, the agency decided to do what it could to avoid such a prospect. Not only would it be costly – in terms of shuttle flights and the need to maintain facilities to receive the telescope – but it might put the telescope at risk. Once down, the telescope might also have to compete with new missions for funds to get aloft again. Another of the aspirations on which the telescope had been sold – regular returns to earth by use of the shuttle – had failed to stand up to a careful study in the light of a taxing design and development program.

In addition to the new approach to building the telescope, NASA gave it a new name. After a committee had mulled over a variety of possible names, in 1983 the Space Telescope had officially become the Edwin P. Hubble Space Telescope, or Hubble Space Telescope.[19] The third name change in less than a decade, this designation won unanimous support from the selection committee. Given Hubble's central role in the development of cosmology (the main mission for the telescope) and Hubble's use of powerful telescopes in that enterprise (see Introduction), the selection was widely regarded as appropriate and was well received.

SOLVING PROBLEMS

Addressing his last Science Working Group meeting as project scientist in June 1983, Bob O'Dell warned that technical problems "abound."[20] To give some flavor of the sorts of problems that had to be tackled and the ways in which NASA, the contractors, and the astronomers went about doing so, two examples from among the many problems will be considered: the latch mechanisms for the

scientific instruments, the Fine Guidance Sensors, and some other
units designed to be replaced in orbit, and the "quantum efficiency
hysteresis" for the Wide Field/Planetary Camera.

The twenty-seven latch mechanisms to hold the scientific instruments and Fine Guidance Sensors in place in the focal-plane structure of the Optical Telescope Assembly had figured prominently in
the various investigations of the Space Telescope program in early
1983.[21] Each latch fitting consisted of one half attached to the focal-
plane structure and the other half attached to the scientific instrument or Fine Guidance Sensor. The latches had been designed so
that while the telescope was in orbit, a space-suited astronaut could
remove one scientific instrument or Fine Guidance Sensor and replace it with another. Yet, when the instrument or sensor was in
the focal-plane structure, it would have to be fixed in position to
very tight tolerances in order for the light from the secondary mirror
to reach it in focus.[22]

The latches also would act as thermal links between the focal-
plane structure and the scientific instruments and Fine Guidance
Sensors. They would therefore have to be designed so that they would
minimize the flow of heat between the instruments and the focal-
plane structure. Hence, as project manager James Odom put it,

to call those devices latches is a tremendous understatement and misnomer.
You are literally taking devices that are thermal insulators and that have
to hold phone booth size objects within one or two ten thousandths of an
inch through a total thermal gradient that you get in each orbit, as well as
accommodating the launch and ground handling loads. So, the actual thermal characteristics, not to even speak of the complexity of a device like
this, which can easily be opened and removed by an astronaut, but yet
buried deeply into the spacecraft, presented horrendous challenges to design, build, and qualify.[23]

During the phase B studies of the mid-1970s, NASA and the
contractors had put most of the emphasis on what they had then
identified as the most demanding technical challenges, including
producing an accurately shaped primary mirror, the Fine Guidance
Sensors, and the pointing and control system. Maintenance and refurbishment, as noted in the debate on a backup to the Wide Field/
Planetary Camera, had been studied only cursorily. However, as
Odom points out, "sooner or later you have to step up and recognize
that you have to operate and maintain [the spacecraft]."[24] Also, because of the early pace of the program, the systems requirements for
the latches were not fully developed until after work had started on
building the latches. There had been what was sometimes referred
to as a "rush to hardware." But building latches that would permit
precise replacement of new instruments and Fine Guidance Sensors
was a central, indeed essential, part of the maintenance and refur-

bishment strategy. It was one thing to make claims during the selling of the telescope, but quite another to ensure that the telescope could live up to those claims.

By 1982, the latches were ready for testing. There were two chief kinds of tests: dynamic and functional. In the most demanding kind of dynamic test, shocks and random vibrations were imposed on the latches to simulate the telescope's launch aboard the Space Shuttle. Functional tests involved checking that the latch fittings worked as they should and that the performance stipulated in the contracts for the latches had in fact been met. Because it was impossible to test the latches in space, Marshall instead tested them in its "Neutral Buoyancy Simulator," a 1.6-million-gallon water tank containing a full-scale mock-up of the telescope. The buoyant effect of the water closely simulated the weightlessness of space. The tank therefore enabled astronauts to practice removing and installing the scientific instruments in the telescope mock-up in conditions similar in some ways to those they would experience in orbit, and thereby check the operation of the latches.[25]

The testing, however, disclosed severe problems. Perhaps chief among them was that in dynamic tests, the latches tended to gall (chafe). Galling results when parts of joints slide against each other without being well lubricated, leading to localized buildup of heat and pressure, and eventually tiny spots where metal is pulled from one surface to another.

Marshall found that steel cylinders in the Space Telescope latches galled a steel surface and badly galled a titanium surface with which they came into contact. The tests on the latches implied that given the buffeting the telescope would experience as it was carried aloft in the shuttle orbiter's payload bay, there was a danger that parts of the latches would gall. If so, one consequence would be that the latches could not be closed exactly to their original positions. Were that to be the case, it would not be possible to regain the original alignments of the scientific instruments and Fine Guidance Sensors — a crucial requirement for the success of the telescope. For a time, one partly tongue-in-cheek suggestion making the rounds of the program was that the instruments would have to be welded into place, a solution that, of course, would have ruled out in-orbit replacement and undermined the very philosophy on which the telescope was being built and had been sold.

By early 1983, the astronomers had become troubled by the latches. Indeed, the latches were listed as item number 5 in the "Most Serious Concerns" of the "Space Telescope Scientists' View of Overall Space Telescope Program."[26]

In March 1983, a report compiled by the House Appropriations Committee's surveys and investigations staff emphasized that the latches, "critical to the fundamental success of the Space Telescope, are currently impacting cost, and threatening to unhinge the launch schedule. The original delivery date has slipped from 1980 to the fall of 1983, with serious outstanding questions as to the manufactured hardware's capacity to meet overall specifications."

The design and manufacture of the latches were the responsibilities of the Perkin-Elmer Corporation and its latches subcontractor, BEI, of Little Rock, Arkansas. However, as the House report argued, the "design and manufacture of these latches has proven extremely difficult." In support of that view, the Perkin-Elmer program manager, the telescope's chief engineer, and a senior NASA official were all listed as having characterized the latch problem as their number-one concern with the Optical Telescope Assembly.[27] At a Science Working Group meeting in March 1983, Fred Speer, the current project manager, had announced that work had begun on the design of alternative latches in case it became necessary to replace the existing ones.[28] But as the NASA administrator admitted shortly afterward, if it did become necessary to redesign the latches once more, that would have a "draconian effect on the program."[29]

However, a solution to the galling problem, it would turn out, was already at hand. Marshall engineers had hit on the idea of putting a tungsten carbide coating on the surfaces with which the steel cylinders came into contact. When the latches were dynamically tested with the new, harder coating, the problem dissolved. By February 1984, Odom was able to conclude that the latch problem "seems to be behind us."[30]

The latch problem was just one of the many technical issues that NASA, the contractors, and the astronomers were tackling. But it seemed that as soon as one problem was resolved, another would pop up. As Sam Keller pointed out in late 1983, "I think anybody who tells you that we're going to get from here to the launch without a lot of problems is kidding himself."[31] As he explained in congressional testimony in May 1984, "if we had . . . appeared before this committee 2 months ago, you would have had an equally impressive but different list of problems. I think you will continue to hear for the next 2 years, of a series of problems, and hopefully, we will continue to solve them."[32]

POINTING AT THE SUN

One such unexpected and troubling problem that arose had to do with "quantum efficiency hysteresis," involving the Charge Coupled Devices (CCDs) of the Wide Field/Planetary Camera, the instrument that had come to be widely regarded as the telescope's most important. Although it was a significant problem in itself, the way NASA, its contractors, and the astronomers hunted for and scrutinized possible solutions to the problem of quantum efficiency hysteresis tells us much about the program's response to unexpected obstacles. Most especially, however much the builders of the camera may have wanted to carve out and put into effect their own chosen plan, the central decisions rested not with them but with NASA managers. The process by which a solution was sought and agreed upon also engaged many more individuals, groups, and institutions than just those involved in designing and building the camera. The advocates of particular solutions also had to argue their cases before a variety of audiences. In addition, some of the astronomers had to labor hard to establish a consensus among their colleagues that a solution was strongly desirable. Earlier in the book we saw how the telescope itself was sold and approval for it won; here we shall see how the solution to a problem was assembled, sold, and approval for it won. As before, building a network of supporters will be key.

In the summer of 1984, the Wide Field/Planetary Camera was tested in conditions that simulated those it would experience in space, so-called thermal-vacuum tests. When the data from those tests were examined at the Jet Propulsion Laboratory, it became clear to the camera's engineers and astronomers that something was very wrong. In the corners of six of the eight CCD chips, as well as in other regions of the remaining two, the effect that would be termed quantum efficiency hysteresis (QEH) was found.

It was only by chance that the effect was discovered. Usually the data from the thermal-vacuum tests had been analyzed with the aid of a computer program. The program had been devised on a certain set of assumptions – assumptions that, it would turn out, in effect blotted out any evidence of quantum efficiency hysteresis. However, during one part of the test, a computer programmer had rigged up a plotting device to examine data directly from the camera, and it was the plotter that had helped to reveal the problem. Moreover, when the camera team scrutinized earlier test data for both the Wide Field/Planetary Camera and the ground-based version of the camera, which also used CCD chips, it found (now with the benefit of hindsight) the same phenomenon.

The problem was essentially that the sensitivity of a CCD to blue light depended on the area of the chip on which light fell and on the chip's exposure history. Each Wide Field/Planetary Camera picture is broken into 800 rows and 800 columns, but not all of those 640,000 (800 × 800) picture elements (pixels) have the same sensitivity to the light that falls on them. Their differences in sensitivity can nevertheless be relatively easily accounted for by electronic means. The final computer-processed images therefore appear as if all pixels had the same sensitivity. However, if the sensitivity of a pixel depends on what objects a CCD has previously been used to examine, then that straightforward solution is no longer available.

If a picture were taken of, say, a bright star, then later pictures would, for a time, have a region of higher sensitivity where the image of the star had fallen. If, then, the next picture were of a galaxy, superimposed on that would be a ghost image of the bright star. The problem became known as quantum efficiency hysteresis.

When NASA had selected the CCDs for the Wide Field/Planetary Camera back in 1977, it had been on the understanding that the CCD was a relatively unproven technology, a point that was about to receive new emphasis. If the camera were flown with no solution to the problem of quantum efficiency hysteresis, then there would be a significant, perhaps drastic, decrease in the quality of the science data it could provide.

As the builder of the camera, the Jet Propulsion Laboratory (JPL), and the camera's Investigation Definition Team probed the phenomenon, they soon hit on a measure that seemed to remove it completely, or at least for a reasonably lengthy period. The "fix" entailed shining bright ultraviolet light directly onto the CCDs, light of wavelengths between 1,800 and 2,900 angstroms. It would be a simple matter to carry out that solution for a ground-based instrument, in which it would be easy to arrange to direct a suitable lamp onto the devices; but what about in orbit, when the CCDs would be placed deep inside a scientific instrument that itself would be embedded in the telescope's focal-plane structure? Was it possible to generate sufficient ultraviolet light and ensure that it would reach the appropriate sections of the Wide Field/Planetary Camera?

In September 1984, James Westphal, the camera's principal investigator, submitted to NASA his science team's report for July and August. Attached to it were two cartoons, one of which carried the legend "Looking at the Sun – Try It – It'll grow on you."[33] Westphal was making what for many people was the dramatic proposal that by pointing the Space Telescope at the sun, more than enough ultraviolet light could be secured. His suggestion was that a hole about six inches in diameter be cut in the telescope's aperture door. A filter could then be inserted into the hole that would allow the passage of only ultraviolet light. After the telescope had been directed to the sun for a number of orbits, the quantum efficiency hysteresis effect would disappear. Although the expectation was that in time the effect would recur, the same procedure could be repeated.

The Space Telescope's designers had labored for years to ensure that the spacecraft would never be pointed toward the sun. They had devised sensors and software systems that would automatically close the aperture door if the telescope came within a certain angle of the sun. Not surprisingly, the idea of pointing at the sun drew startled and extremely skeptical reactions from some people. When it was raised at a public meeting in October 1984, there was, as one leading participant recalls, "certainly a swell of emotion, I'd say, with a lot of concern. . . . People felt there was a serious danger to the whole Telescope, not only the optics of the Optical Telescope Assembly, but the instruments as well because you'd be flooding the focal plane with sunlight."[34]

For many people, pointing at the sun raised obvious and disturbing problems. For example, if sunlight were directed onto the mirror, it might polymerize contaminants on the surface. The resulting coating would make the affected area of the mirror much less effective in reflecting ultraviolet light and so reduce the telescope's performance. Although it would be necessary to revise some of the spacecraft's software to enable it to be directed to the sun, none of the technical arguments against sun pointing was by itself conclusive for NASA managers.

For program scientist Edward Weiler and many others, the chief objection to the proposed fix was

the gut level feeling of everybody who's had any experience in space. People like me who have worked on Copernicus [*OAO-III*] to Frank Carr who worked on [the International Ultraviolet Explorer] and other programs, everybody in the program who had any space experience just has an internal relay that gets switched on when somebody tells you they're going to point the system at the Sun. Because every damned astronomical system we've ever developed has all sorts of safety mechanisms not to point at the Sun. . . . What we were worried about is the eventuality where about three or four single point failures occur at the same time, without the door closed, the filter in place or whatever. And it was so distasteful, we couldn't do it. Maybe I'm understating the technical arguments against it, but in my mind and a lot of other people's minds, we just didn't want to do that![35]

However, at the Space Telescope quarterly meeting at Marshall in October 1984, the Wide Field/Planetary Camera's project manager, David Rodgers, contended that the only feasible solution to quantum efficiency hysteresis was pointing at the sun. He argued that replacing the existing CCD chips with new ones would not work; the problem was generic.[36] If quantum efficiency hysteresis was to be fixed, ultraviolet photons would have to come from somewhere.

What about astronomical sources other than the sun? If the sun itself was too bright, what about sunlight reflected from the earth or the moon? Unfortunately, it seemed that the telescope would have to be directed at the earth for thousands of hours to remove the hysteresis effect, and for tens of thousands of hours for a full charge. Pointing at the moon was little better. For a full charge, that would take about 1,200 hours, a solution that was unacceptable if the telescope was to spend much of its time directed toward astronomical objects and being a productive astronomical observatory. Nor would flooding the CCD chips with an ultraviolet lamp just before launch be sufficient. That would be fine for a time, but the chips would need another flood within several months, and by then the telescope would be in orbit. Because he had already dismissed the idea of a special lamp placed inside the camera, Rodgers concluded that the flood of ultraviolet light, if such was to occur, had to be achieved in orbit.[37] However unappealing using the sun might be to many people, it seemed to be the only source that could produce a fix in a short time. But Rodgers's arguments did not win immediate acceptance from NASA, the contractors, or astronomers. Instead, the chase was on for an alternative solution, a chase that engaged many institutions, groups, and individuals.

In the pursuit of a fix they deemed acceptable for the problem of quantum efficiency hysteresis, NASA managers set several activities in motion, and Goddard assumed responsibility for developing a suitable plan of work. Another meeting to examine the problem and the effect it would have on the science to be performed by the telescope was scheduled for Goddard at the end of October 1984, a little over two weeks after Rodgers's presentation at Marshall. Prior to the meeting, however, Tom Harvey, head of the Systems Engineering Group at Lockheed, wrote an internal memorandum whose thrust was as follows: "Why not look into bringing the light into the Wide Field/Planetary Camera from the side of the Telescope?"[38]

If light were to enter the side of the telescope, instead of from the front, and if the ultraviolet light then were guided via a mirror inside the spacecraft and some kind of light path to the CCD chips in the Wide Field/Planetary Camera, the perils of sun pointing could be avoided altogether. But for some sort of light guide to be practicable, there would have to be enough room between the various items surrounding the Wide Field/Planetary Camera for a path to be found between the telescope's side and the camera. Was there indeed a gap? It was on that question that the debate on the "light pipe" centered.

The idea of the light pipe was raised again by Lockheed engineer Domenick Tenerelli at the Goddard meeting in late October. Marshall and Perkin-Elmer had already built a Plexiglas replica of the Wide Field/Planetary Camera to check its fit into the focal-plane structure. When engineers examined the replica, there appeared to be no room into which a light pipe might fit. That solution thus looked unrealistic.

The light pipe was but one of thirteen possible solutions Westphal described at the Goddard meeting. For those in attendance, the consensus was that the hole in the aperture door was the only possibility that was not an extremely long shot. The attendees therefore agreed that the ultraviolet flood was the only feasible way to fix the quantum efficiency hysteresis. If the light pipe would not work, the choice would be between the aperture-door fix and no fix at all.[39]

In October and early November, following a meeting between Westphal and Space Telescope Science Institute director Riccardo Giacconi, project scientist Robert Brown and his assistant John Clarke had drawn up a plan to examine the impact of quantum efficiency hysteresis on the science the Space Telescope would be able to perform.[40] The plan entailed a number of astronomers analyzing the effects of quantum efficiency hysteresis on specific astronomical problems. Each astronomer assigned to a particular problem was to consider the scientific consequences of *not* restoring the performance of the CCDs by use of an ultraviolet flood.[41] For example, James Gunn of Princeton studied the consequences of quantum efficiency hysteresis for investigations of Cepheid variables and the distance scale (see Chapter 1), Jim Westphal studied the imaging of the centers of galaxies, and William Baum studied the imaging of comets. The plan was intended by the astronomers to help them better understand QEH, as well as to help them convince NASA managers that a solution was necessary.

After the astronomers had completed their calculations, a two-day meeting was held in mid-November for Brown and others to chew over the data, as well as to craft the language of their final report. Brown, Albert Boggess, Goddard's project scientist for operations, Riccardo Giacconi, and Westphal — four of the program's most influential astronomers in terms of political clout — concluded that the chances of achieving many of the most important scientific objectives were significantly diminished by the problem of quantum efficiency hysteresis. Among those were some of what Brown termed the "Crown Jewels," objectives at the heart of the Space Telescope's mission. One of the crown jewels, for example, would be the determination of a more precise value for the distance scale of the universe. But James Gunn had reckoned that an important step in determining the distance scale, the use of Cepheids to measure the distance to the Virgo cluster, would be severely affected by quantum efficiency hysteresis, perhaps even rendered impossible.

In view of that and similar findings, Brown's group recommended that "we must fix the . . . problem. If the only fix is [an ultraviolet] flood through the aperture door and the risks are acceptable, then we should cut the hole and plan to use it." By involving so many leading astronomers in the analysis of the effects of quantum efficiency hysteresis, Brown and his colleagues had helped to strengthen the network of those who advocated solving the problem, and had also given the astronomers a powerful voice on the matter. By calling the astronomical questions that had been examined the "crown jewels," Brown and his colleagues had also elevated their status. After all, who would want to "steal" the crown jewels?

Matters had begun to move quickly, and meetings were being scheduled at a rapid rate. Another meeting was held at Lockheed in early December, for example. One body of opinion strongly advocated the use of a hole in the aperture door. As telescope scientist William Fastie argued after the Lockheed meeting, "in my view, the massive effort of the last few months has clearly established that the [quantum efficiency hysteresis] problem is one that requires a complete fix to insure the scientific integrity of [the Hubble Space Telescope]. It also appears to me that the filter in the door is a complete fix and that it is the only viable one. I believe that the work to implement the filter in the door fix should not be delayed."[42]

However, as one leading participant recalls, "by this time there had been just a tremendous swell or momentum against pointing the Telescope towards the Sun, coming from probably all sectors; engineers, scientists . . . and managers."[43] Marshall therefore asked Lockheed to examine in detail the use of a mercury lamp to generate the flood of ultraviolet light. But when Domenick Tenerelli and two other Lockheed engineers traveled to JPL to discuss the idea of the mercury lamp, they rapidly decided that the lamp would require so much energy for the flood that "we had a situation that the laws of physics would have to be violated to make this work."[44] The chief difficulty was that the lamp would produce too much heat for it to be easily dissipated. It seemed as if the meeting would be over quickly, but before the Lockheed staff left, Jim Westphal asked if they would like to check the light-path concept. When Westphal went to the blackboard and started to review what that would entail, it began to seem to the participants that a light channel might be feasible after all.

At a meeting on December 19 in NASA Headquarters, JPL proposed that it further check the use of a mercury lamp, but Lockheed protested that such a lamp would not work. Instead, Lockheed had analyzed the implications of cutting a hole in the aperture door and pointing the telescope at the sun. Westphal explained the light channel in detail and proposed that the concept be studied further. Program manager Welch also suggested an important revision to the light-channel idea: Put the mirror to divert sunlight to the camera on the rear end of the spacecraft, not in the side. In this way, the mirror could be made to be replaceable in orbit. It its surface became contaminated, a new mirror could be substituted by astronauts on a visit to the telescope. Although no decision was made to opt definitely for a light pipe, NASA managers dropped the option of the mercury lamp.[45] The light pipe became NASA's favored option, with the hole in the aperture door relegated to a backup role. There was still much to be done by the astronomers and contractors to convince themselves and the NASA managers and engineers that the light pipe was feasible, let alone that it should be adopted.

There was one other possibility that the managers had to consider seriously. As program manager Welch cautioned the Science Working Group in January 1985, his "overarching priority and concern . . . is the safety of the vehicle." Welch was chary of the sun-pointing option. He preferred not to override the several software safeguards that had been built into the flight and ground systems to prevent the telescope being pointed accidentally at the sun. Such manual overrides, Welch knew, had been tried on other NASA spacecraft in the past, and on at least two occasions had almost led to loss of the spacecraft. If he found the light pipe unacceptable, one option for Welch was therefore to do nothing to the Wide Field/Planetary Camera, but to press on with building the Wide Field/Planetary Camera clone and fix the quantum efficiency hysteresis problem for the clone. It that scheme was to be pursued, there would be a scientific price to be paid: the loss, for probably two and a half years, until the first maintenance visit of the shuttle to the telescope and the installation of the clone, of a considerable part of the science to be performed with the aid of the Wide Field/Planetary Camera.

But the day before Welch had spoken, engineers had measured the flight hardware to check the gap into which a light pipe might possibly be inserted. As part of a pitch to the Science Working Group on QEH and the importance of fixing the problem, Westphal announced that there really was a gap, one large enough to accommodate a light pipe. So Welch was able to respond that he was "very optimistic that we have a solution that's flyable and will work."[46]

After JPL had elaborated its design for the light pipe, a major meeting was held at JPL for NASA to select which, if any, of the possible fixes were to be adopted. A big advantage for the light pipe was that it could be designed to be modular. If installed, it would have little impact on the testing and assembling of the telescope still to be performed at Lockheed. Certainly NASA managers accepted that the light pipe, together with the changes it would force in other items of hardware, would cost several million dollars. However, weighing the cost against the significant loss of scientific data that would result if there was no fix, and after considering the other potential solution (pointing at the sun), Welch chose to adopt the light pipe.[47] But the program participants realized there still was no full explanation why quantum efficiency hysteresis had even occured in the first place. And so, though the light pipe solved the immediate problem by use of ultraviolet flooding, the hysteresis problem had emphasized that the CCDs involved new and relatively untried technology, with perhaps more surprises to come.

As the telescope's scheduled launch date of mid-1986 drew nearer, the issues of operating the Space Telescope in orbit and the kinds of observing programs it would undertake increasingly took center stage for NASA and the astronomers and contractors, particularly those working on the telescope's complex and costly ground system. At the same time, the Space Telescope Science Institute's position as something of an experimental kind of institution for NASA became obvious as the agency groped for the best manner to interact with the institute.

The institute would provide the "observatory staff" as well as become one of astronomy's major patrons and funding sources. The institute's plan was that an astronomer (or group of astronomers) who wished to observe with the telescope would send a proposal to the institute. If, after the proposal had been carefully reviewed by a committee of top astronomers, the proposal was accepted, the institute would channel NASA funds to that observer. In consequence, the institute was expected to distribute around $20 million each year for users of the Space Telescope, a sum that promised to have a major effect on the organization and practice of astronomy, and a significant fraction of the entire budget provided by the National Science Foundation for the support of all of ground-based astronomy in the United States. But, as we have seen throughout this book, there are distinct differences between operating a telescope on the ground and in space.

Because of the telescope's complexity and its position in space (about 590 kilometers from earth), drawing up the plans to operate it would be an intricate process for Goddard, the institute, and the numerous contractors involved. The usual constraints of near-earth orbits would come into play. One was that the telescope would pass through a full cycle of being illuminated by the sun, earth, and moon every 95 minutes or so. From its position in low earth orbit, there would be sections of the sky that the telescope could observe throughout an orbit, others that could not be observed at all (if they were too close to the sun, for example), and some sections that could be observed for certain periods.

One task of the Space Telescope Science Institute was to schedule the scientific observations of the telescope. Such schedules would then be passed to the operations center at Goddard for staff there to produce the commands to be sent to the spacecraft via the "Tracking and Data Relay Satellite System." The commands from Goddard would then direct the telescope's operations: for example, to which objects it was to point and for how long, and which scientific instruments were to be turned on and for what lengths of time.

The institute's scheduling of the scientific observations is perhaps

337

the most visible of its activities. But the institute is also charged with "providing scientific and technical support in evaluating and maintaining the instruments, planning the observations, handling, reducing and storing the data efficiently, and creating a stimulating atmosphere for the use of the Telescope."[48] For example, we have already noted that the telescope's Fine Guidance Sensors require sets of guide stars to lock on so that the telescope can be pointed accurately, and light directed precisely into the scientific instruments. When the institute came into being in 1981, no existing star catalog contained enough stars that it could be used as a source of guide stars for the Space Telescope. In consequence, the institute was quickly engaged in compiling such a catalog.[49]

The Guide Star Selection System was one of the items the institute had to deliver to NASA as part of the agency's contract with AURA, the university consortium responsible for running the institute. The institute and NASA therefore had to negotiate and settle on a schedule for completion of the Guide Star Selection System, and the institute had to secure the resources, in terms of manpower and machines, it would need for the job.

The construction of the Guide Star Selection System would illustrate two of the central aspects of Big Science. In particular, it would entail a team-oriented approach and a consequent division of labor, as well as intensive use of technology (in this case, computers for data analysis and machines for measuring the positions of stars on photographs of the sky). For the astronomers involved with the Guide Star Selection System, that generally meant adjusting to new ways of doing business. As one of them noted in 1984, "it was the first time in my life I'd ever done astronomy on a production line basis. Not only did we have to meet production schedules and fill in milestones and go to reviews and produce volumes of documentation, but we were designing a system that would do production line astrometry, which is something that's never been done. . . . Star positions on demand!"[50] In fact, it was an example of what might be termed "corporate astronomy," with the establishment of business methods for accelerating data analysis. Though the value of scientific goals could not be defined in production terms, the preconditions for important work (in this case, the development of an efficient Guide Star Selection System) could be.[51]

The comparison can be pressed further, for the institute exemplifies the transformation of the workplace between traditional ground-based astronomy and the large-scale space astronomy represented by the Space Telescope. Optical astronomers have long been used to working alone or in small groups, perhaps with the aid of a few engineers and graduate students, building apparatus for use on telescopes themselves, taking their own observations, and pursuing what we referred to in the Introduction as little science. Astronomers who

Computer consoles at the Space Telescope Science Institute illustrate the
transformation in the workplace of
the astronomer that has occurred in
the last four decades. (Courtesy of the
Space Telescope Science Institute.)

observe with the Space Telescope will be in a much different situation. They often will be members of teams of scientists and will be
employing scientific instruments aboard a telescope whose construction and operations they most likely will have had no part in. The
observer will have become physically and managerially distant from
direct control of the apparatus, and a wide variety of specialists will
be needed to supervise the scheduling of observations, the maintenance and calibration of the scientific data, and the operations and
well-being of the Space Telescope.

The new kind of workplace calls for astronomers to assume new
roles. An example of this transformation is provided by the career of
a senior member of the institute's staff, Ethan Schreier. Schreier had
trained at M.I.T. and in 1970 received a Ph.D. in theoretical high-
energy physics. To help support himself as a graduate student, he
had taken a part-time computing job at a local company, American
Science and Engineering. It happened that in the late 1960s American Science and Engineering had one of the leading groups in the
burgeoning field of x-ray astronomy. Shortly before American Science and Engineering launched its *Uhuru* satellite in 1970, Schreier
had decided to join the company full-time. He thereby became centrally involved in planning for the operations of *Uhuru*. After launch,
Schreier was charged with determining the satellite's precise position or "aspect," as it is also termed. That was far from being a
simple task, because various forces (including small variations in the
earth's magnetic field and the interaction of the earth's magnetic
field with the satellite) caused the axis around which the satellite
was spinning to drift continuously. Through his analyses, Schreier

Ethan Schreier (right) with Don Hall and Riccardo Giacconi (left) in Giacconi's office in 1982 at the fledgling institute (then based in temporary quarters in Rowland Hall at Johns Hopkins). (Courtesy of Jane Russell.)

became known to his colleagues as "Mr. Aspect."[52] In fact, operating spacecraft, together with the intimately related task of planning and devising data analysis systems, would become his specialty. Schreier entered x-ray astronomy without a background in laboratory physics, an unusual route for that time, but that did not matter in his area of expertise, scientific operations. He worked during the 1970s on a series of spacecraft that required increasingly sophisticated approaches in planning and executing their operations. Schreier was one of a new breed of astronomers who pursued their own scientific research but also devoted their energies to computing, not the construction of hardware. It was for Schreier's experience and talent in the operations of scientific spacecraft that he had been one of the institute's early hires in 1981.[53]

CONFLICT

The planning and preparations for the operations of the Space Telescope were enormously labor- and technology-intensive, but exactly how intensive would become a point of dispute between NASA and the institute. Soon after its establishment in 1981, the Space Telescope Science Institute and NASA had come to distinctly different positions on the appropriate size and scope of the institute's activities. Resources that the institute insisted were necessary for successful completion of its service functions sometimes were regarded by NASA as excessive. Into that sometimes intense debate were added complicated issues of the control of, and assignment of responsibilities for, various activities. The leading figure in this process of shaping exactly what the institute was to be was its director, Riccardo Giacconi.[54]

Giacconi was born in Genoa, Italy, in 1931. Trained at the University of Milan as a physicist, he came to the United States in 1956 as a Fulbright Fellow. In 1959 he made a move that would fundamentally shape his future career. Judging that his line of research on cosmic rays had little future, he switched his research field and physical location, joining American Science and Engineering in Cambridge, Massachusetts, to help start a program of space research. Soon he was engaged in x-ray astronomy, a subject that was pioneered by Herbert Friedman and a group at the Naval Research Laboratory near Washington, D.C., in the 1950s. But in 1959 relatively little was known for sure, except that the sun emitted x-rays.

Although the first two rocket flights with which he was involved failed, Giacconi and his colleagues soon succeeded in flying rockets to measure x-rays and gamma rays. Those radiations, however, were not from astronomical sources. Instead, the U.S. Air Force's Cambridge Research Laboratory had requested American Science and Engineering to undertake a crash program to investigate the effects of detonation of nuclear weapons at high altitudes. "The program," as Giacconi and a collaborator have written,

was to be completed by March 1962. Money was no object, but time was precious. AS&E accepted the project and began to prepare payloads to measure electrons, x-rays, and gamma rays produced by nuclear weapons explosions in the atmosphere. From the fall of 1961 to the summer of 1962, Giacconi's group expanded from half a dozen to seventy or eighty people. They designed, built, tested, integrated into vehicles, and launched twenty-four rocket payloads in an eight month period. They built and integrated six satellite payloads. In about 95 percent of the cases the payloads worked properly and the experiments were successful. This experience molded the AS&E group into a loyal, dedicated and highly skilled team. They had acquired a reputation and a confident "can-do" attitude; they were also aggressive in seeking funds to expand their research program.[55]

The American Science and Engineering group was quickly at the forefront of x-ray astronomy. In 1962, flying apparatus aboard a rocket funded by the air force, they detected the first x-rays from an object outside the solar system, and in 1970 they launched *Uhuru,* the first satellite devoted exclusively to x-ray astronomy.

In 1973, Giacconi joined the faculty of Harvard University. There he continued his pursuit of x-ray astronomy, aided by many of the former American Science and Engineering group. Those efforts led most notably to the Einstein Observatory satellite, launched in 1978, a project in which Giacconi was directly responsible to NASA for the scientific aspects. With the flying of rockets, as well as the satellite *Uhuru* and the Einstein Observatory, Giacconi had become widely recognized as a first-rank scientist, the winner of numerous prestigious scientific prizes, and a highly talented manager and leader of large-scale scientific enterprises. In addition, unlike nearly all of his astronomer colleagues, he had a rich fund of experience in private industry and was well equipped for managing the kind of "corporate astronomy" that would be central to the institute's mission.

It appears that his early years at American Science and Engineering building and flying experiments for the air force had a strong influence on Giacconi, for when he began to work with NASA, he found it extremely frustrating to have to battle so hard to establish x-ray astronomy and to play by the NASA rules of the game. What was most jarring was to be involved in projects in which the final decision-making authority lay not in the hands of scientists but in those of NASA project managers. Giacconi had quickly decided that astronomers should take responsibility for development and operation of the facilities they used, both managerially and scientifically (that step had already been taken by physicists working in high-energy physics, which had been Giacconi's background). Certainly that did not mean that he would not work productively with NASA. By the mid-1970s he had begun to sit on a number of NASA advisory committees, for example. Indeed, when in 1974 the position of head of the Office of Space Science had become vacant, Giacconi had been approached by one of the leading science managers in the agency to apply.

There was, nevertheless, something of a love–hate relationship between NASA and Giacconi. Giacconi needed NASA in order to pursue x-ray astronomy, and NASA was able to bask in the reflected glory of Giacconi's successes. Some in the agency, however, had never been happy with what they saw as his lack of willingness to compromise and his relentless pursuit of goals in which he believed. Hence, when in 1981 AURA, the university consortium charged with running the Space Telescope Science Institute, had conducted a search for a director for the institute, there had been decided unease for some in NASA when, from a field of some sixty candidates, AURA had selected Giacconi.

In the summer of 1981, AURA president John Teem had discussed Giacconi's candidacy with Andrew Stofan, the acting head of NASA's Office of Space Science. Teem, in his notes on the conversation, wrote against Giacconi's name, "Problem for NASA," and he underlined it twice. Nor had Stofan been the only one within NASA to raise objections to Giacconi.[56]

Immensely charming, Giacconi can also be an extremely tough negotiator. Commenting on his style in 1983, George Field, Giacconi's boss at Harvard for almost a decade, noted that he can be "quite aggressive even to the point of being difficult to deal with. He is demanding from the people he works for because he knows what he wants. His arguments are well reasoned from his point of view, but he can be very difficult. He was fond of pointing out the mistakes I made. But his record of success is such that, when he demands something, you sit up and take notice."[57] Sam Keller, the most senior NASA manager closely involved with the telescope, and sometime adversary, developed a high regard for Giacconi. Keller nevertheless ruefully reflected that "Riccardo's approach to a meeting is to first, roll a hand grenade through the door, and after he's gotten your attention he gets down to the subject matter."[58]

Giacconi was also known to have, in private, vigorously opposed the Large Space Telescope in the early part of the 1970s, arguing that an x-ray observatory should have higher priority. He and some other x-ray astronomers were particularly concerned that if the Space Telescope program went badly, then their own x-ray missions would be delayed. Although Giacconi had dropped his opposition well before 1981, it had not endeared him to a few in NASA and the astronomical community. The fact that an x-ray astronomer would be heading the Space Telescope Science Institute caused some surprise in the agency and among astronomers.

Yet what some in NASA disliked about Giacconi were seen by AURA as his great strengths. In particular, AURA believed that Giacconi's legendary energy and tenacity, as well as his long experience in dealing with NASA, would stand him in excellent stead in the difficult task of establishing the fledgling institute, a task that most likely would entail some hard battles, given the deeply entrenched opposition to the institute at Goddard. He would, in addition, bring with him a range of experience in space astronomy and large-scale projects that would be difficult to match.

Giacconi had long advocated institutes to run the scientific operations of space observatories, and he had been the principal driving force behind plans in the 1970s to establish an x-ray institute. Largely because of Giacconi's enthusiastic promotion and support, the Einstein Observatory had been organized along the lines of a national facility, wherein astronomers who had not been involved in designing and operating the spacecraft were able to propose observing programs. In fact, when NASA had intended to build the Einstein Observatory in the agency's usual manner – contracting with industry for the spacecraft and contracting with various research groups for the scientific instruments – the scientists, led by Giacconi, had balked. Instead, they formed their own consortium and, with Giacconi as principal investigator, proposed to build all the various scientific instruments under direct scientific control. It was with AURA's strong backing, then, that Giacconi had been appointed to head the institute in September 1981, replacing the acting director, Arthur D. Code.[59]

Giacconi and his staff faced some daunting challenges. The central one was that an institute would have to be fashioned that would be capable of performing its allotted work, and fashioned quickly. The two major groups at the institute would be the astronomers (about a fifth of the total staff) and the support contractors provided by Computer Sciences Corporation (CSC). For the institute managers, there would be the job of molding these groups together, as well as working closely with AURA, Goddard, other contractors involved with the ground system, the science teams, and the astronomical community.

By 1982, NASA was expressing some unease with the institute's projections for the resources it needed to complete its service functions. The pressures of preparing for the scheduled 1985 launch were also very high, and as the institute's astronomers strove to meet various deadlines they found little time for research. In a 1982 report drawn up by a senior NASA group[60] and chaired by former program manager Warren Keller, there were nevertheless some sharp words. The group contended that the institute was placing too much emphasis on scientific research, and not enough on the institute's service functions.[61]

That point went to the heart of a fundamental issue. For AURA and Giacconi, the best service to the ultimate users of the Space Telescope would be provided by an institute whose staff of astronomers were also top-flight researchers. But, they contended, to attract such people to the institute in the first place, it would be necessary to make it an agreeable place in which to perform research, and so opportunities to do so should be provided. As Giacconi pointed out in an interview in 1983, "there's a philosophical difference. I think NASA basically thought of the Institute as a data distribution system and a service institution. . . . But *I* conceive of it also as a first-rank research institute. I think that the very best scientists will give the very best service."[62] Giacconi's approach was in agreement with the line taken by astronomers who had championed the institute in the mid-1970s. As will be recalled from Chapter 6, the astronomers usually had seen a role for the institute considerably more extensive than the role anticipated by NASA. For the astronomers in the mid-1970s, the institute was to be in some ways a lead organization on the program. They envisaged that the institute would become closely engaged with all scientific issues, including the design of the telescope and its testing and calibration, not just its scientific operations. As the institute began to flex its muscle in these other areas, it ran into opposition, both from NASA and from some of the astronomers on the program.

GROWING PAINS

The institute's first two years were spent in a variety of temporary and generally cramped accommodations on The Johns Hopkins University campus, but in 1983 the institute staff moved into a new, specially designed building intended to house over 200 staff as well as visiting astronomers. Yet, even by the time of the building's

The Space Telescope Science Institute on the Johns Hopkins Homewood campus in 1983. Pressure for a larger facility was already being exerted by the institute and AURA. (Courtesy of the Space Telescope Science Institute.)

official dedication in June of that year, the institute's managers had become extremely uneasy about the building's ability to accommodate the staff judged to be necessary to accomplish all of the institute's tasks. That judgment had already developed into strong and concerted pressure from the institute and AURA for extra space, as well as more staff. What also exasperated NASA managers was that Giacconi wanted an institute rather different from that set out in the NASA/AURA contract. The winning AURA bid had laid stress on the extensive use of contractors to support the work of the institute's staff. Giacconi preferred an in-house staff.

NASA Headquarters, whose telescope managers lived every day with the severe financial constraints on the program, was disturbed because it believed the Space Telescope Science Institute was growing too rapidly. The dispute rumbled on until December 1983. Sam Keller and AURA then reached a temporary compromise. Instead of examining the Space Telescope program's money limits and deciding on the institute's manpower on that basis, in effect designing to cost, other questions should be asked: "What functions should the Institute fulfill, and how many people are needed for those functions?"[63] The issue was charged because, as NASA administrator Beggs had emphasized at a NASA meeting in January 1984, with funds very limited, "we've got to tell Riccardo [Giacconi] that there's a point beyond which we will not go."[64]

The widely differing views of NASA and the institute regarding the institute's size were presented at a meeting to discuss the issue at headquarters in February 1984. The institute pressed for more space and more manpower. Goddard, which monitored NASA's contract with the institute and whose own budget was tightly constrained, wanted, in effect, to "cap" the money going to the institute, as well as move some of the institute's staff to Goddard. Those proposals were resisted by the institute and AURA, which regarded the notion of transferring staff to Goddard as a crude threat to the institute's very nature.[65]

The NASA—institute dispute over resources reached its climax in early 1984. Giacconi had tried to end-run NASA by meeting with presidential science advisor George Keyworth on the form and scope of the institute. In seeking to gain allies for his position, Giacconi had discussed with Keyworth the results of the February 27 meeting at NASA Headquarters. But that set off alarm bells in headquarters, particularly as it was widely known that NASA administrator Beggs had become concerned by what he saw as an institute that was providing excellent scientific leadership, but was, he thought, expanding rapidly and exhibiting management weaknesses.[66] In particular, NASA wanted tighter management at higher levels in the institute. The agency also was anxious to avoid the impression of an institute at which charges of preferential treatment might be leveled, that is, the same kind of criticism that had been aimed at Goddard in the 1970s by non-NASA astronomers and had helped to create the institute in the first place.

At about the same time as Giacconi's end run to Keyworth, people at headquarters sought to isolate Giacconi and apply pressure on the institute by fostering and tapping doubts and anxieties about the institute among some astronomers. NASA staffers argued that there had been excessive growth of staff and funding at the institute, and appealed to the self-interest of the astronomers by contending that extra money devoted to the institute would have to be siphoned away from other projects in the NASA space science budget. Headquarters prompted a letter-writing campaign, one of the favored forms of politicking in Washington, D.C. Astronomers sympathetic to the headquarters position were invited to write directly to the NASA administrator to explain their views on the institute. There followed a volley of letters from at least fourteen astronomers, including three of the U.S. principal investigators on the Space Telescope.[67] O'Dell, one of the architects of the concept of an institute, sent a typical letter:

I am alarmed by the enormous growth that has occurred at the [institute] and expect that this is only the beginning. I believe that the best interests of the science community are served if immediate steps are taken. I encourage you to limit the [institute] to only its present building and to place a manpower ceiling on the staff size. Such actions will seem arbitrary, but only they can establish what the [institute] will be. Without such measures, I can only foresee continued bit by bit growth and expansion of responsibility, to the detriment of the space astronomy community.[68]

The letter writers did not dwell on the hundreds of millions of dollars siphoned from NASA's space science budget by the Space Telescope's cost overruns, and for the institute, such views were galling and revealed a lack of understanding of what the institute's tasks entailed. Some astronomers were unhappy that a campaign had been started against the institute. Writing to a highly placed official in NASA Headquarters, M.I.T. astronomer George Clark, a former colleague of Giacconi, complained of "misinformation, some of it originating from people at NASA Headquarters concerning projected costs at [the science institute] has roused the concern of many in the [scientific community] outside NASA."[69]

For a time, the relationship between NASA and the institute was extremely bad. Giacconi and others believed that some of the optical astronomers were still imbued with a romantic vision of what the telescope entailed, that somehow the Space Telescope would be operated in ways similar to ground-based telescopes and that the style of little science to which they were used could be maintained. But the unpopular point the institute was insisting on was that such a notion was wrong, that for the Space Telescope the rules, procedures, and ways of operating had to be very different. In that debate, Giacconi was working closely with AURA's president John Teem,[70] and though the going became rocky on occasion, and AURA had a reputation for not backing directors, the link between Giacconi and AURA remained intact throughout the battle with NASA.

In March 1984, Teem sent NASA a resolution passed by AURA's board of directors supporting the institute's arguments for extra space. Teem emphasized that both the board and the Space Telescope Science Institute Council (the group that oversaw the institute's activities for AURA) were concerned that NASA might impose constraints on the institute that could alter its fundamental character, as well as mar its efficiency and effectiveness.[71]

By July 1984, things had reached such a pass that the members of the Space Telescope Science Institute (STScI) Visiting Committee were reporting that they had been "made aware of an extreme state of tension – one might say polarization – that currently exists between the Space Telescope Science Institute and NASA on a number of important issues. The central question appears to involve the proper scope and function of the [institute] in the steady state era after launch."[72] In other words, how many staff should there be at the institute once the telescope had been launched, and what should they be doing?

The STScI Visiting Committee – composed largely of prominent astronomers charged with periodic reviews of the institute's activities – judged that the institute was overburdened with paperwork. Moreover, budget and policy decisions agreed between the institute and Goddard could be overturned elsewhere in NASA. Planning was therefore difficult. On the other hand, the STScI Visiting Committee concluded that the institute might be more flexible and conciliatory "recognizing that a large number of talented individuals are needed to make [the Space Telescope] operate successfully and that a spirit of teamwork and cooperation among all groups is needed if the various parts are to function effectively together."[73]

The points at issue, the STScI Visiting Committee stressed, often were quite abstruse and frequently involved the telescope's ability to perform specific functions. The committee noted that the discussions often revolved around whether or not a particular capability was needed to make effective use of the telescope and, if so, whether or not it had to be in place by the time the telescope was launched. If a given capability was not essential by launch, how soon after should it be added? "Decisions of this sort involve an extremely complex and delicate weighting of scientific priorities against cost," the committee noted. The consequence of the disagreement was "a noticeable effect on staff morale, imposing a special burden on certain departments and creating an atmosphere of uncertainty in which work schedules cannot be planned adequately in advance."[74]

By that time, following agreements with NASA, the institute had assumed more tasks than had been defined in the original contract. A number of the additional tasks involved the Science Operations Ground System (SOGS), which itself lay at the heart of the conflict between NASA and the institute over resources. SOGS consists of an extremely complex system of computer hardware and software whose main tasks are to enable the institute to plan, schedule, and perform observations with the scientific instruments aboard the telescope, as well as to analyze the data collected. The institute, as the link between the telescope and the observers, would therefore become the user of SOGS. Both before and at the time the SOGS contract had been let to TRW in 1981, many people and groups had urged NASA to hand management responsibility for SOGS to the institute immediately. That had been the assumption of the Hornig committee in 1976. In late 1980, the chairman of the Space Science Board's Committee on Space Astronomy and Astrophysics again appealed to NASA to ask "whether it would be more efficient, more administratively effective, and more scientifically productive for the Institute to assume responsibility for the SOGS functions."[75] NASA rejected those arguments. Its chief reason was that in 1981 the institute (itself delayed because of cost problems on the telescope program) was just getting established and should not be burdened with such an extra charge.[76] If the development of SOGS was delayed, it would not be ready for launch.

The $30 million contract let for SOGS and the development work conducted by TRW had therefore had relatively little scientific in-

put. Nor had NASA initiated any sort of Phase B study of the ground
system as a whole. TRW had proceeded to devise SOGS on the basis
of the requirements set out in its contract with NASA, but the
result was a SOGS that met NASA's requirements but that the institute vigorously protested (and NASA would later concede) was
grossly inadequate. SOGS seemed to be based on 1960s computing,
not that of the 1980s. For example, Ethan Schreier recalls that there
was "no concept of interactive data analysis, no concept of what it
means to have a command language, to have an environment where
the user interacts with the computer in real time and says, 'Now,
change this parameter and try it again.' "[77]

In a review of SOGS in May 1982, the institute, along with many
other members of the project, including the principal investigators
and their teams, listed some 750 items they had identified as potential failings or points they wanted to change in the system.[78] Some
of the proposed changes were major and pointed to fundamental
mistakes made in defining SOGS. One problem, for example, was
that SOGS did not permit the astronomers to analyze or calibrate
astrometric data. Nor, it would turn out, was the existing SOGS of
much use for planetary observations. The telescope's ability to observe the planets had been one of the chief selling points in the
1970s, when NASA managers had been anxious for the support of
planetary scientists and the telescope had been presented to them as
a powerful tool for their researches.

In the "Announcement of Opportunity" for the telescope issued
in 1977, the suggested items that the telescope might be used to
pursue had included determination of the compositions of the clouds
in the atmospheres of Jupiter, Saturn, Uranus, and Neptune, surface
mapping of the four Galilean satellites of Jupiter, and synoptic mapping of atmospheric features on Venus, Jupiter, Saturn, and Uranus.[79] However, when the telescope's Phase C/D began in 1977, the
tracking of planetary objects and planetary features had received a
low priority. As O'Dell has recalled, "we knew all along [the Space
Telescope] didn't have full capability" to do planetary tracking. "We
also knew that it would be very difficult and expensive to implement." So the plan had been to identify those times in the telescope's orbits when the features a planetary scientist might want to
study would be moving in a straight line, and it would be relatively
straightforward for the telescope to track that linear motion.[80] The
question of the telescope's decidedly limited capacity to track planets
within our solar system – together with features on their surfaces,
such as, for example, the great red spot on Jupiter, as well as the
satellites orbiting around them – had been heard for many years. In
1980, for example, a report to the Science Working Group by Edward Groth on how the telescope would acquire its targets stressed
that the tracking of moving targets (such as planets) would present

349

a major problem for the telescope, as it would be difficult to combine tracking and acquisition maneuvers.[81]

Planetary tracking had received little attention from NASA managers, in part because none of the astronomers had effectively hammered away at the issue or been able to mobilize much support. However, when planetary scientist Robert Brown became project scientist in 1983, he began to press hard, in private as well as public meetings, for the telescope to be made capable of tracking planets at all times, not just when the peculiarities of their motions presented the opportunity. One of the institute staff who had reviewed SOGS in 1982, he had already concluded that the existing system was flawed for planetary studies. As he repeatedly emphasized, "planets don't move in a straight line. They move in a very complicated fashion. And in order to be able to be at the right place at the right time, you need timing capabilities, you need to be able to describe that detailed track for that planet, you need to figure out where you want to point [the telescope]."[82] Rodger Doxsey of the institute's staff was widely regarded as something of a guru because of his expertise in the telescope's scientific operations, and in late 1982 Brown had worked with him to understand what was needed for the telescope to track planets. Through their analyses, Brown also knew that adding the planetary tracking capability would be an intricate and somewhat costly job. Not only were changes required in SOGS; the Lockheed flight software would also have to be altered.[83]

Planetary tracking, however, was only one of the crippling problems with SOGS in 1982 and 1983. Instead of a SOGS that could, in effect, be handed over to the institute, with the institute in place simply to operate and maintain the system, there was still a vast amount of development to be done, work the institute wanted to do and considered to be essential, but which would the hiring of more staff. What would constitute an acceptable SOGS at launch? If some things could not be achieved by the time of launch (e.g., because of cost and schedule), how much might be shifted to a later date? With funds for the program so tight, those were thorny issues for the NASA managers.

ENTER THE SPACE SCIENCE BOARD

SOGS was just one of the issues in dispute between NASA and the institute. However, by the time of the STScI Visiting Committee's report in July 1984, an answer to the immediate question of the size of the institute's staff was in hand. That had come by way of a committee organized by the Space Science Board of the National Academy of Sciences. Exactly who suggested such a committee to the Space Science Board — whether it was NASA alone, the Office of Science and Technology Policy in the White House, or astrono-

mers active in the National Academy of Sciences – is not clear.
Nevertheless, by May 1984, the Space Science Board had accepted
the charge of reviewing the institute's goals and objectives in the
light of the original plans and determining if those needed to be
changed.

The task group appointed by the Space Science Board got off to a
bumpy start when Arthur Code, the designated chairman, resigned
after the group's first set of meetings. The appointment of Code –
AURA's interim director of the institute before the appointment of
Giacconi – had been somewhat controversial. Whatever decisions
the task group arrived at, some critics complained, might be criticized because of Code's extensive private and professional involvement in the planning for and development of the institute. Code
himself had come to think that "I can not be completely objective
in my views and can not separate myself from the obvious conflict
that exists between NASA and the Institute and between individuals
involved."[84]

In his letter of resignation, Code noted that it had become clear
during the original contract negotiations between NASA and AURA
(in which he, as acting director of the institute in 1981, had played
a central part) that the amount of work the institute would have to
perform had been underestimated. Nor did Code think that the contract between NASA and AURA had properly reflected the fact that
AURA was a nonprofit organization; that is, the institute was not
like the aerospace contractors with which NASA was used to dealing. Despite those points, he judged that the institute should be
limited to the size Code and AURA had originally contemplated in
early 1981, that is, around 200 permanent staff members, not the
more than 300 the institute was seeking (about 50 of whom would
be astronomers).

When the Space Science Board's task group finally reported, it
essentially walked a diplomatic middle line between the positions of
the institute and NASA. As the group stressed, "neither the Hornig
Committee nor NASA nor AURA correctly anticipated the magnitude of the effort that would be required to carry out [the institute's]
functions." However, it emphasized that one of the points of the
Hornig committee was that the "Institute should be of sufficient
size, in facilities and staff, to carry out its functions, but should not
become so large as to absorb an inordinate fraction of the resources
devoted to astronomical research. The Institute we envision would
be comparable in budget and manpower to other national astronomical facilities." So the "Task Group recommends that the Institute
optimize the {Space Telescope's} science return within the available resources"
(emphasis in original). The task group also urged the overwhelming
importance of cooperation and mutual respect between the institute
and NASA for the telescope's success.[85]

351

By the time the task group's report appeared, much of the passion in the debate on the institute's size had subsided. That probably was largely because of NASA's decision to allow the institute to provide additional temporary space for the institute's staff pending a decision on another building or extensions to the existing building. So by the time the STScI Visiting Committee reported in June 1985, it was able to report that "relations with NASA, which were at a critically strained level a year ago, have greatly improved, but further improvement is still to be hoped for. Some constructive steps could be taken by the AURA Board."[86]

Despite some disagreements, NASA and the institute were able to establish what both sides would come to regard as a reasonable working relationship. Although a 1986 report prepared at NASA Headquarters was still critical,[87] the agency quickly disowned it. NASA came to concede the institute's claim that it required extensions to its existing building to accommodate all of the staff (also agreed by NASA to be around the 300 mark that the institute had consistently argued for with many of the hires driven by the problems with SOGS) it would need to run the scientific operations of the Space Telescope efficiently. Construction of the extensions began in 1987, providing a concrete sign of the resolution of a protracted and hard-fought struggle over resources.

ASSEMBLING THE PIECES

The debate over the size and scope of the institute was a significant theme in the development of the Space Telescope program in the mid-1980s. It was not, however, the central theme for the majority of the program's participants. Instead, that role was filled by the activities constituting "Assembly and Verification," the part of the program to which the Space Telescope's design and development phase had been geared. During assembly and verification (the "proof of the pudding") all of the numerous pieces of hardware that constituted the telescope would be collected together at Lockheed in Sunnyvale, California, and then assembled and tested to check that everything worked successfully, individually and in combination. In particular, the Optical Telescope Assembly and the Support Systems Module would be mated together at Lockheed, and the scientific instruments would be inserted into the spacecraft. The telescope's ground system would also be tested to ensure that it would be able to command the Space Telescope once the telescope was in orbit and that the spacecraft would indeed be capable of returning data to the ground.

The assembly and verification phase would not be the first time that the scientific instruments had been rigorously tested. In March 1983, the High Speed Photometer had been the first scientific in-

strument to be delivered to Goddard for its "Verification and Acceptance" program. The program involved a detailed series of tests in
which the instruments were first tested individually, and then in
combination with each other, very much as they might be operated
once the telescope was in orbit. The tests of the computer for the
scientific instruments began in April 1983, and the tests for the first
instrument in May. There then followed many months of activity
centered on Goddard and engaging over one hundred people from
thirteen organizations before the end of the verification and acceptance program in March 1984.[88]

At completion of that step, all of the instruments had to be returned to their contractors for any changes deemed to be necessary
by Goddard and the science teams. For example, some adjustments
had to be made in the alignment of the optics for the High Resolution Spectrograph, and both of the digicon detectors employed in
the Faint Object Spectrograph had to be replaced.[89] As Goddard's
project manager, Frank Carr, pointed out at a Science Working Group
meeting in February 1984, because of the complexity and amount
of work that had been required, the testing had been a "decided
struggle." He therefore expected that the much more extensive testing that would start at Lockheed during assembly and verification
would prove to be a long and difficult period. Lockheed, Carr contended, would need all the help it could get if the scientific instruments were to be fully tested with the rest of the spacecraft.[90]

Planning for the assembly and verification phase, however, had
started many years before, with the "Assembly and Verification
Working Group" providing the focus. One of a number of such
working groups that Marshall had fashioned to examine the often
difficult and subtle issues of the interfaces between numerous contractors and organizations, many people felt that by mid-1982 the
Assembly and Verification Working Group had become bogged down,
possessing responsibility but little authority. To some attendees it
seemed that issues debated at one meeting often reappeared at the
next, and the same old ground had to be plowed over once more.
Mid-1982 was also, it will be recalled, a time when the program
was acutely short of money. NASA managers were looking long and
hard at any proposed changes to the program at that time.

After attending the twelfth meeting of the working group in July
1982, Evan Richards, instrument manager for the High Speed Photometer, complained in his trip report that "future [Assembly and
Verification Working Group] meetings will be better than this one.
That is a safe prediction because it just isn't possible to get worse.
We achieved a new low with this meeting: the old issues remain
unresolved, ones thought to be resolved are not, new issues are arising, there is no evidence of any communication, and no apparent
mechanism for making decisions." The existing plans and sched-

ules, or "baseline" in NASA-ese, had been incorporated into a document known as SAV-1000, written by Marshall. The original intention was that it would be replaced by a new baseline detailed in a new document agreed on by all of the involved parties: Marshall, Goddard, the contractors, and the science teams. By July 1982, that idea had been dropped, and proposed changes to the existing baseline were strongly resisted by Marshall. Richards summarized matters: "So there it is folks, no runs, no hits, too many errors to count, nobody left on base. The score: [Assembly and Verification Working Group] 0, Speer's baseline 9."[91]

Following a meeting between the principal investigators and Goddard managers and engineers, Robert Bless spoke on behalf of all the principal investigators on the plans to test the scientific instruments at a meeting of the Science Working Group. Although it was in part a tutorial on scientific instrument testing for those members of the working group who knew little of such matters — Bless was a very experienced space hand — it was also a hard-hitting appraisal of where the program stood. "The Assembly and Verification (AV) program is not going well," Bless argued. "It is difficult for the [principal investigators] to make useful input to the AV Working Group and there are numerous problems to the present plan. The [principal investigators] consider the AV program to put the performance of their instruments at high risk. There will be an AV review the next week and this will be pivotal in shaping the future program."[92]

Bless was asked to appear at the assembly and verification meeting of November 1982. After he had delivered his pitch, he found that "most everybody agreed that what I was saying was correct, it's just that nobody had felt in a position to say this. . . . I was an outsider, I had not participated in the group. I had no particular axe to grind, and the axes were very sharp at that point."[93]

One aspect of planning with which the astronomers were unhappy was Goddard's stipulation on how the science teams were to communicate with Lockheed. Lockheed would run the tests on the scientific instruments during assembly and verification, but Goddard's plan was that the center's own instrument test engineer would be the primary link between the instrument teams and Lockheed. Lockheed reported to Marshall, and Goddard was also responsible to Marshall for the scientific instruments. Goddard's charge was to ensure that they were properly tested and ready for launch. "To perform that function," the Goddard deputy project manager had stressed in late 1982,

we must have control/cognizance of the requirements, and monitor/support the test implementation and data evaluation. We do plan to depend very heavily on our [science teams] and the instrument contractors to perform that function, but the project office at Goddard must have the central

354

leadership/coordination role. Consistent with this approach, we have tried to build technical teams, one for each instrument, including participation from all contractors, and have identified one instrument test engineer as the focal point for coordination.[94]

Goddard was especially concerned that if the members of the science teams talked directly with Lockheed engineers and managers, they would short-circuit the lines of management control and be able to devise new requirements for the tests, requirements that might increase costs. Given the exceedingly tight limits on the available funds, Goddard believed that would be disastrous. On the other hand, to those building the scientific instruments, the perception was that Goddard felt that its responsibility was to defend the interest of the instruments to the rest of the project, so that "if we all went off and made our own little deals with Lockheed we would be divided and conquered."[95]

For the scientists, the Goddard approach certainly smacked of an autocratic approach. At that time, Goddard was also insisting that presentations on the scientific instruments at the main quarterly meetings on the telescope's status had to be cleared with Goddard first.[96] Although the scientists did not expect to have the final word on decisions that would affect cost and schedule, they did expect to talk directly with the people who would test their instruments. Bless, who was leading the charge on that issue for the principal investigators, saw Goddard's plan as an extremely clumsy way to proceed, one that was almost sure to fail. He and others interpreted it as a means to increase Goddard's role and control at the expense of the input of the science teams. The Goddard approach to testing probably was one of the major reasons the scientists voiced extremely strong criticisms of the program's status in early 1983 (see Chapter 8).

Matters perhaps reached a low point when the instrument manager for the High Speed Photometer arranged a visit to Lockheed to discuss thermal-vacuum testing. As Richards recalls,

we set up this visit. When [Goddard] found out about it – and I didn't really try to keep it a secret, but I wasn't advertising it either, but I guess [the Goddard technical officer] wanted to know where I was going for a couple of days – I said "Well, I'm going out to Lockheed." "Oh? What are you going to do out at Lockheed?" "We're going to talk about thermal vacuum tests." Well! [the Goddard technical officer] told me that they were going to ask me that I not go. I said, "I understand your feelings, but I think it's important. I think we need to have this meeting." "Well, we're going to direct you not to go." I said, "I'm not sure you can do that." It got to be a very violent argument. We got a Rapifax, and were directed not to talk to Lockheed. Bless [the High Speed Photometer's principal investigator] got upset, and he was telling me to ignore those directives, that he would talk to [Goddard's project manager] and so on . . .

355

the day I was leaving for the airplane we got another much stronger Rapifax directing us not to go. They wouldn't fund it and on and on. But I went anyway.

This was at a time when we were doing some of our instrument tests out at Goddard. I had no sooner showed up at Lockheed when there was a message for me from [the Goddard technical officer], "Please call."

Well, they had just issued a directive and they were going to send it to all of the instrument people that they would refuse to put power on any instrument at Goddard for anything, unless the instrument manager was there in person, and that was going to be effective almost immediately.

So, what they were saying is, "You've got to be in Washington right now or we're going to shut down your testing."

It got to that level. It was really unfortunate.[97]

When the instrument teams' concerns about testing at Lockheed were brought to the attention of project manager Fred Speer, he was sympathetic.[98] Speer, however, was soon to be in no position to do anything about it, as he was replaced by James Odom as project manager in April of 1983. There had been earlier changes at Goddard too.

In 1982, Noel Hinners, a vigorous advocate of the Space Telescope in the 1970s and one of its most influential promoters during the selling campaigns, had been appointed director of Goddard. After reviewing the telescope program, he had soon decided that the situation on the Space Telescope at his center was bad:

I picked up enough in the way of grousing, complaints about Goddard, that I figured there must have been something to it. The relationship with Headquarters, with Marshall, with the Institute, and parts of Goddard had some strain to them.

How that really developed, I have yet to figure out. The roots of it have got to in some ways go back to the way the program management was structured, dividing it between Goddard and Marshall, throwing the Institute in, which Goddard didn't want and making Goddard responsible for it, making it work. You can go back and look and find all the good reasons, but for sure there's an attitude problem here. It didn't take long before I was convinced that I had to do something about it. . . . I could look at these other places and say, yes, they were behaving in some pretty gross ways too, nevertheless we had this problem at Goddard.

So I made up my mind pretty early that I had to do something to fix it. And the personalities became so ingrained that really the only way it was going to happen was to change out the project leaders here. So we removed the project manager and deputy project manager, and set up a new team.

The Space Telescope team here at Goddard had not really gotten the center's support. I don't want to put all the blame, if you will, on the project people here. The other directorates [at Goddard] were not giving them the support. It did not have a high sense center priority, pure and simple. I think when Goddard didn't get the role [on the Space Telescope program] it wanted, it said, well, we'll do the minimum — screw it. . . .

So, it was a bummer of a scene here, for all these different reasons. And to some degree, the top people of the project were the fall guys. It wasn't totally their fault, but just the whole past history.[99]

Hinners had therefore overhauled the Space Telescope Project at Goddard, elevating the project's status and visibility at the center and appointing some new staff. One of those was Frank Carr as project manager. Carr was a highly experienced spacecraft engineer and manager. He had joined Goddard in 1960 immediately after graduating from college and was soon working on *Explorer 12*, the first satellite designed, built, and tested at Goddard. After more than a decade of work on scientific and applications satellites, he had assumed a senior position on the International Ultraviolet Explorer project, a major astronomical satellite that was being developed by NASA, the European Space Agency, and the United Kingdom. It was that experience that was particularly appealing to Hinners in his hunt for a project manager for the Goddard portion of the extremely complex – in technical, scientific, and managerial senses – Space Telescope. Carr had also worked for some months at NASA Headquarters as program manager for the telescope, and so he would hit the ground running.

When Westphal, Bless and others discussed with Carr the concerns of the scientists and instrument managers about the planning for assembly and verification, Carr, a veteran of numerous test programs for scientific satellites, proved a ready listener.[100] Soon after taking over as Goddard project manager in late 1982, Carr had decided that over the years

the system-level test program had been continually scaled back and descoped to the point where it was virtually a "ship and shoot" philosophy, so to speak. The idea was to send everything out to Lockheed, bolt it all together, fire it up once or twice and ship it to the Cape and launch it. Now that's totally foreign to the way [Goddard has] done science spacecraft programs in the past, and the [principal investigators] were anxious to try to see what we could do to instill a notion on the part of the project that that was not the way to insure success.[101]

By May 1983, Carr had become convinced that the views of the astronomers, which meshed well with his own, had not been adequately represented in the planning for the telescope's assembly and verification. That failure, he judged, occurred because the existing body for discussing those issues, the Assembly and Verification Working Group, was ineffective. He advised project manager Odom that the group be reconstituted and its activities refocused.[102] The group was not reformed; instead, it simply slid into obscurity and became irrelevant as Lockheed began to assume more and more of the lead role for the testing.

However, even in 1984, NASA was concerned that Lockheed had

not fully grasped the size of the task it faced in assembling and verifying the Space Telescope. By April 1984, some people were detecting signs in the cost overruns and slipped milestones of the same sorts of difficulties that had precipitated the crisis at Perkin-Elmer in early 1983. Moreover, NASA administrator Beggs was lobbying the White House and Congress to drum-up support for the planned space station. He also worked hard to defend the telescope's budget at a time when the agency, like many government agencies, was being cut back, but Beggs had made it plain that he definitely was not going back to Congress to ask for more money for the telescope. He emphasized that point numerous times at a major project meeting in January 1984, even pounding the table on one occasion to ensure that the message struck home. Beggs had already gone to Congress for extra funds for the telescope in 1983 under embarrassing circumstances. If he had to return once more, it would reflect extremely badly on NASA's management abilities, thereby damaging the space station's chances for approval. [103]

The onus was therefore on the NASA managers to keep the telescope program on schedule and on budget. Writing to Lockheed two months after the meeting with Beggs, project manager Odom complained that the Support Systems Module program had had to be rescheduled three times in the past year. He conceded that those revisions had in part been driven by changes directed by NASA. The schedule had nevertheless been altered as recently as January because Lockheed was over budget and behind schedule. But already, within only a few months, Lockheed had fallen "significantly" behind. "We have discussed some of these activities that can be delayed until FY [fiscal year] 85 with little risk to the program," Odom noted. "What is very disconcerting about this situation is the increase in rate of missed milestones and the apparent inability of [Lockheed] management to arrest the schedule slips. Even more disturbing is the lack of planning for 'work-arounds' for the known milestones missed. . . . I am very concerned about the size of the total 'bow-wave' that will be pushed into FY 85 if effective corrective action is not taken immediately." Odom wanted Lockheed to complete a recovery plan by the end of the month. He concluded that "there have been elements of the [Lockheed] effort that have been and continue to be very productive. However, when one integrates the concerns I have identified above, it can only be concluded that timely and effective management actions are required at all levels. By the end of this year, virtually all of the Space Telescope flight hardware will be delivered to you for integration and test." [104]

After it had become accepted within NASA in 1983 that Perkin-Elmer would be late in delivering the Optical Telescope Assembly to Lockheed, the agency decided to insert into the schedule a series of tests of the scientific instruments at Lockheed known as "Pre-

A&V." In so doing, NASA hoped that the burden would be placed squarely on Lockheed to demonstrate the capabilities of the Spacecraft Automated Test System (SATS) to operate the Space Telescope's instruments safely and efficiently. That strategy meant delivering the scientific instruments to Lockheed as soon as possible. The instruments' very presence, agency managers felt, would apply pressure on Lockheed, however subtle.[105]

Speculating in March 1984 on the future course of the program, Goddard's project manager Frank Carr had contended that

the year and a half period beginning September [1984] is apt to be the most challenging period in the entire history of Space Telescope, in my view. Some people think we're over the hill and it's all downhill from here on, that we've overcome all our adversities and problems. No way. They're just beginning.

Now that we've got real hardware, we're going to have real problems. We're going to have failures. We're going to have things that have been forgotten, overlooked. I've never seen a *simple* spacecraft not have a monumental challenge during the systems integration and test phase.

As a matter of fact, I think if anybody did any surveys, we'd find that of our systems people, most divorces probably occur the year before launch . . . the systems test phase is an incredible, chaotic, high pressure, high stress, around-the-clock kind of an operation, and that's for a simple spacecraft.

For Space Telescope, I'm not sure what's going to happen, but I know it's not going to be easy.[106]

Three months after Carr's remarks, in May 1984, Robert Bless, as the chairman of the Space Telescope Observatory Performance and Assessment Team (STOPAT), addressed the telescope's status in congressional testimony. Bless told Congress that the test facilities and the mechanical and electrical support equipment to be used in the tests seemed to be in good shape. However, Bless reported STOPAT's opinion that the computer software for the test systems was not as mature as it should have been. The preliminary integration of the scientific instruments with the spacecraft was due to start in the third week of September, but STOPAT was worried that the target date might be missed (as indeed it was). In addition, "it appears to us that similar remarks could be made about other sections of the Lockheed test program, though with slightly reduced force since they are required somewhat later. We understand that steps are being taken by Lockheed to remedy this situation, and we applaud these efforts. For the moment, however, this aspect of the ST program remains a serious concern to us."[107]

One big hurdle in the program was successfully negotiated in September 1984: Perkin-Elmer's completion of the structural fabrication of the Optical Telescope Assembly. After the crisis of early 1983, Perkin-Elmer had greatly improved its performance, in NASA's

The completed Optical Telescope Assembly in its final assembly stand at the Perkin-Elmer plant in Danbury, Connecticut, in October 1984. (Courtesy of Perkin-Elmer.)

eyes. Although the Fine Guidance Sensors were late and still posed Perkin-Elmer some manufacturing problems, at least the difficulties were generally regarded as tractable, something that had not been so eighteen months earlier. The Fine Guidance Sensors, however, would not have to be delivered to Lockheed along with the Optical Telescope Assembly, and so they would not delay the assembly's delivery.

The transfer of the assembly from Perkin-Elmer's plant in Danbury, Connecticut, to Lockheed in Sunnyvale, California, was itself a major undertaking that had been meticulously planned. The move began on October 29, with the ten-foot-wide, thirty-foot-long Optical Telescope Assembly placed inside a special contamination-control unit and loaded on a flatbed truck. Together with its entourage of state police, the assembly began a thirty-mile journey to Stewart Air Force Base in Newburgh, New York. There it was loaded carefully into a cavernous "Super Guppy" transport plane. The assembly and shipping container were together so heavy that the Super Guppy could carry only enough fuel for short hops. The trip to California therefore had to be made in several steps.

On November 1, the assembly arrived at Moffett Field Naval Air Station at Mountain View, California.[108] The entire operation had gone flawlessly, and the assembly was at Lockheed's Sunnyvale plant a little ahead of schedule, ready for assembly and verification tests to get under way.

ASSEMBLY AND VERIFICATION

In October, Frank Carr had presented to Marshall the Goddard assessment of the status of Pre-A&V and SATS, the preliminary testing of the scientific instruments and the computerized test system. In Goddard's view, the telescope was facing "significant, but not insurmountable" threats to the 1986 launch date, the agreed budget, and the telescope's performance. The two main causes were (1) the immaturity of the Lockheed test system and (2) in Goddard's judgment that executing the tests would far exceed the available computer resources. "Solutions may exist," Carr noted, "but in my view they will not be simple nor straightforward, and will require substantial and dramatic actions."[109] Testing programs are, by their very nature, difficult and unpredictable times in which problems arise; that is why tests are performed. But with the Space Telescope, Carr argued, the difficulties were intensified because of the telescope's complexity, the numbers of people and institutions involved, the lack of readily applicable precedents, and the use of an automated system for testing the sections of the spacecraft. The software system for testing at Lockheed was completely different from that used at Goddard some months earlier for the verification and acceptance program. A much-discussed goal of the telescope program during Phase B, the use of common software for various activities in the telescope's development, had not survived into the practice of Phase C/D. Moreover, the software for actual operations in orbit would be different from both of the sets employed in the instrument tests at Goddard and Lockheed.

In addition to the problems of removing the bugs from Lockheed's automated test system, there were deep-seated institutional issues, particularly concerning the testing of the scientific instruments. The approach of Goddard and the astronomers, including project scientist Robert Brown, was to stress the execution of system-level tests, to operate a number of the scientific instruments together, and to exercise fully the spacecraft's systems. For some time, Lockheed as an institution (though not all its engineers) resisted that approach. Lockheed's institutional style, even in planning to test the entire spacecraft, was to emphasize tests at the "box levels," a level lower than that desired by Goddard and the astronomers,[110] but one that had been carefully worked out over the years.

Perhaps the major reason for the differences of opinion was that Lockheed was used to building military spacecraft, spacecraft that usually were built on an assembly line. The Space Telescope, however, was a one-of-a-kind spacecraft, and an extremely complicated one at that. Although Lockheed had built and tested hundreds of satellites, it had been little involved with scientific satellites; thus, in addition to the technical issues over testing, there was an institutional dimension: the Goddard/astronomer approach versus the approach of Lockheed.

Lockheed, moreover, was used to building its spacecraft for only one customer, the Department of Defense. The Department of Defense, as James Welch notes, was "represented in most cases by one person or a small office," and so if Lockheed got into trouble in military programs "or had a question about requirements or scheduling, money, then they would have one spot to go back to and coordinate the solution to it." With the Space Telescope, however, Lockheed was working with a heterogeneous group: Marshall, Goddard, the scientific community, and other contractors. That meant that it was exceedingly complex for Lockheed to design an integration and test program that would accommodate and knit together all of those various and at times competing factions.[111] By this point, the Space Telescope program had become the biggest kind of Big Science in the making, a series of activities engaging many large, multidisciplinary teams, activities that were focused on Lockheed's Sunnyvale plant, but that had links all over the United States and Europe. For the members of the science teams, it meant working in, and adjusting to, a very hierarchical system in which the rules of little science did not apply, and which was of a larger scale than any of them had encountered before.

Lockheed was facing enormous and far-reaching management challanges, but NASA managers wanted it to move more quickly in designing and running the test program. Writing to a highly placed Lockheed official in November, Burt Edelson, since 1981 the head of the Office of Space Science and Applications in NASA Headquarters, sought to keep the pressure on and underlined that it "is of critical importance that Lockheed management recognize that the progress accomplished over the last several months has been unsatisfactory and is unacceptable to NASA. Only your personal commitment and that of other Lockheed executives can restore this program to the projected schedules within the planned cost."[112]

Frank Carr's expectations about what the assembly and verification program would entail seemed to be borne out when the Pre-A&V activities, scheduled to last from November 20 to December 3, 1984, were not completed until February 1, 1985. That delay immediately used up fifty-three of the eighty-three slack days in the

Assembling the sections of the Space Telescope in the Vertical Assembly and Test Area (VATA) at Lockheed's plant in Sunnyvale, California, in 1985. The Support Systems Module's equipment section has already been mated to the Optical Telescope Assembly. (Courtesy of Lockheed.)

total schedule, and that was before assembly and verification proper had gotten under way. Though Pre-A&V was widely regarded by program participants as quite helpful, it had not, as had originally been hoped, seen the scientific instruments operated as they would be in orbit or established the automated test system as an effective tool.

The slow progress of the Pre-A&V test meant that the start on assembling the telescope was delayed. That nevertheless began on February 13 when the equipment section of the Support Systems Module was painstakingly lowered onto the Optical Telescope Assembly. After a short delay because of difficulty getting some bolts into their allotted holes, the operation was successfully finished.[113] The program to design and build the Space Telescope was definitely entering its final phase, but it was still a long way from completion. In fact, as Edelson told Congress in the spring of 1985, "I will have to admit that we are hanging on by our fingernails. We are using program reserves."[114]

THE RUN-IN

As the assembly and verification program got fully under way, the project's managers faced some tough challenges. The issues were succinctly put in a memorandum to the NASA comptroller from

Lockheed technicians work on the partially assembled Space Telescope inside the VATA at Lockheed in 1985. The open doors are part of the aft shroud. (Courtesy of Lockheed.)

one of the Space Telescope's analysts in NASA Headquarters. After attending an extensive review of the telescope's budget in early March 1985, he pointed out that the reserves in all areas of the program – development, operations, and maintenance and refurbishment – were

unusually low, especially when one considers the history of the Space Telescope. Without adequate reserves to buy one's way out of some of the schedule difficulties, I feel that the potential schedule problems will become more serious, leading to a greater chance of a slip in the launch and a [fiscal year] 87 budget impact. Even with an increase in funds to maintain necessary contractor manpower levels, currently unknown technical/test problems could very well force a schedule slip; added time may also be necessary.

Perhaps most important, headquarters had directed that the telescope's development budget – which would cover the assembly and verification program – would remain at $1,175 million. Schedule problems could not be solved by asking for extra funds. In fact, Marshall had already played the last card in its hand before it had to slip the schedule: It instructed Lockheed to gear up to a schedule of twenty-four hours per day, seven days per week, for most of the assembly and verification period. Only therein was there a chance to pick up some of the days already lost in Pre-A&V. An extra thirty days of slack could be earned by that move, thus providing a cushion of sixty-three vacant days in the assembly and verification schedule – time that, judging from the test program's progress thus far, almost certainly would be needed.[115]

As it turned out, 1985 was a punishing year for those members of the project closely associated with the testing and assembly of the telescope, including a number of Goddard staff who were now at Lockheed to help ride herd on the different groups engaged in testing the instruments. There was enormous pressure on Lockheed to keep on schedule, and that led to grueling demands on those conducting and preparing the tests.

The schedules were proving to be too optimistic, and Lockheed was continuing to slip behind. In December 1985, the leading figures on the Space Telescope from Marshall and Lockheed met to discuss why. The Marshall director, William Lucas, "asked Val Peline why [Lockheed] had missed so badly the October 21, 1984 baseline schedule," despite earlier Lockheed assurances to the contrary. Peline's answer was that it was because the Space Telescope was being built on the protoflight principle: The spacecraft being tested was the one that would be flown. In contrast, the many other spacecraft Lockheed had developed had followed the building of a prototype "to drive out and resolve development problems, thus minimizing schedule slippages on the flight article."[116] Marshall was particularly concerned that Lockheed be able to complete the telescope's thermal-vacuum testing on time. That was essential if Lockheed was to meet the June 21, 1986, date for delivery of the spacecraft to the Kennedy Space Center for launch aboard the Space Shuttle *Atlantis*.

Despite the earlier delays, Peline argued that the date could be met, the point being that the schedules had been revised to better define the work remaining to be done. Lockheed was so close to completing the testing that he was confident that not many unknowns were left in the program, and the facilities to be used for testing the telescope were almost fully operational. However, if there was any rebuilding of the spacecraft to be done, or any major test failures, then Peline conceded that the telescope might well be late.

For those engineers and technicians working directly with the

365

In March 1986, the assembled Hubble Space Telescope was removed from its assembly stand in the VATA for the first time. The forty-three-foot-long telescope is shown here being carefully lowered to a horizontal position so that tests can begin on its aperture door. (Courtesy of Lockheed.)

telescope and its components, the work remained at a high pitch throughout 1985. However, by the end of the year, the testing of the telescope was proceeding much more easily than it had earlier. Large numbers of tests had not been omitted because of schedule pressures, as some of the astronomers had earlier feared. The telescope's hardware was generally performing well, and the test system set in place by Lockheed was functioning far more smoothly than it had in 1984 and early 1985. It had taken quite a while for all of the disparate institutions, groups, and individuals to start to mesh with each other and the various test systems, but events were moving quickly toward what promised to be the most important test of all, the thermal-vacuum test, in which the telescope would be placed in a vacuum chamber to run simulations of its operations while in orbit.

The schedule had indeed slipped, but NASA had found some extra time by rescheduling the telescope's planned launch aboard the Space Shuttle. The earlier plan had been to aim for a lift-off on August 18, 1986, with the telescope stowed inside the cargo bay of

the shuttle *Atlantis*. But in January 1986, NASA announced officially that the telescope's flight had been rescheduled for October 27. In that way, another two months of slack had been added to the schedule for testing and for the telescope's planned sea voyage, via the Panama Canal, to Florida for loading into the shuttle. Whereas many people in the program had been downright skeptical of the official launch dates announced in previous years, often regarding them as little more than management devices to keep pressure on the contractors, it began to appear that the October date just might be achieved. A widespread view in early 1986 was that even if the October launch was missed, it would not be by much.

After nine difficult years of designing and building the Space Telescope, the astronomers were eagerly looking forward to receiving data at last. They were not the only ones. The December 1985 issue of *Life* magazine featured a cover proclaiming "Seeing Beyond the Stars. A Preview of America's Year in Space." The story announced that the Space Telescope and a teacher, Christa McAuliffe, would "head the cast of Space Show '86," and that "with a present

The telescope being moved in May 1986 from the VATA to another section of the same Lockheed building for acoustic tests. Air pads support the telescope during its transfer. (Courtesy of Lockheed.)

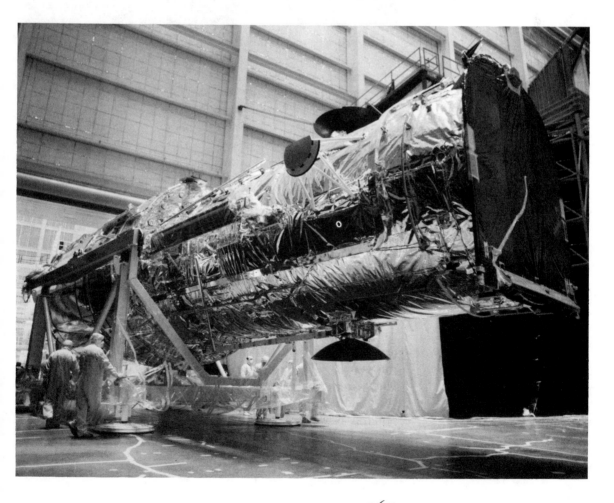

367

The Space Shuttle *Challenger* poised for lift-off on January 28, 1986. Following ignition of the solid-fuel rocket motor, dark smoke swirled between the right-hand motor and the external tank. Wisps of smoke are already visible near the critical joint of the rocket motor. (Courtesy of NASA.)

budget of over $7 billion, NASA is enjoying more clout with the administration and Congress than at any time since the Apollo moon program."[117] Also, commenting in December 1985 on ABC's Sunday morning news program "This Week with David Brinkley," columnist George F. Will contended that "I think the great event of this year is . . . going to be the launching of the Space Telescope. It is going to be the most important advance in astronomy in 400 years and is going to give us a quantum leap ahead in our understanding of who we are and how we got here and the nature of the universe . . . and that is even more important than who wins the Senate."[118] Will, however, could not foresee the events of a cold Florida morning late the following month.

At 11:38 a.m. on January 28, 1986, the Space Shuttle *Challenger* lifted slowly off its launch pad. Seventy-two seconds later, following a catastrophic series of events caused by the failure of a joint in the shuttle's right solid-fuel rocket booster, the booster crashed into the huge external tank that carried the shuttle's liquid oxygen. The tank ruptured, forcing the orbiter's main engines to shut down. *Challenger* spun wildly, and the orbiter was ripped apart. The crew compartment was thrown free, and the astronauts were hurled to their deaths.[119]

Some of the implications for the Space Telescope program were strikingly obvious almost immediately. Goddard project manager Frank Carr saw the launch on television. While "I was watching the screen still showing the smoke contrails from the Challenger, my own prediction was that we'd be delayed a year."[120] For those astronomers who had worked closely with the program, the delay due to the *Challenger* accident was a massive disappointment. As David Leckrone noted in March 1986, two months after the disaster, "it's been a very long haul. My own personal frustration with the possible delay this year is that I had hoped after a decade I could tie up all the loose ends of my own responsibilities and spend a year working with scientific data, and try to recapture my lost youth."[121]

What would the *Challenger* accident mean for the Space Telescope program? The program had gained enormous momentum by January 1986, and any schedule slips would be very costly, as hundreds of Lockheed and other contractor staff were at work on the telescope. The telescope was soon due to undergo its crucial thermal-vacuum tests: thirty days or more of quickly alternating heat and cold designed to simulate the conditions to be faced by the telescope in orbit, while all the time operating in a vacuum. As one of Lockheed's leading managers, Burt Bulkin, was quoted as saying, "that's the biggie. If we can get through that without any major problems – which I have every confidence in the world we will – then I think from there on, the problems will be minimal."[122] NASA and the contractors had spent years planning for that test and others still to be completed and had assembled highly trained groups of staff to conduct the tests. It therefore made no managerial or technical sense to NASA to postpone the tests or relax the schedules too much, particularly as it was unclear how long it would be before the shuttle fleet might be flying again. Hence, despite the *Challenger* accident, it was very much business as usual for those assembling and testing the telescope.

Hence on April 16, the telescope was rotated from its vertical position in Lockheed's "Vertical Assembly and Testing Area" and placed on the dolly that would transport it to the thermal-vacuum

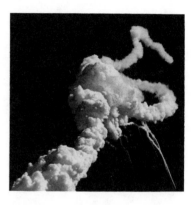

Seconds after the explosion of the *Challenger,* main-engine exhaust, solid-fuel rocket-booster plumes, and an expanding ball of gas were visible. (Courtesy of NASA.)

chamber. The telescope was moved into the chamber four days later.[123]

With numerous sensors attached to the telescope to monitor its test performance, the pumps that would remove almost all the air from the thermal-vacuum chamber began operating early on the morning of May 6. Thirteen days later the required pressure level had been reached, and tests could begin.[124]

The tests took longer to complete than had been planned. Among other things, there arose a variety of problems that caused sudden, sometimes alarming, upsurges in pressure in the thermal-vacuum chamber. If those increases in pressure had gone much higher, the consequences could have been disastrous for the telescope. Despite those heart-stopping moments, the testing was finished by July 1.

The early consensus among NASA and the contractors was that, on the whole, the telescope had performed well.[125] Some problems did surface fairly quickly, however. Perhaps the most serious of those had to do with the telescope's electrical power system. Careful audits of the power consumed during the thermal-vacuum testing led Lockheed and NASA to conclude that the spacecraft required some 100 to 300 watts more power than had been predicted. That meant there would be little margin between the power the spacecraft would be able to produce and the power it was expected to consume. That small amount was not expected by Lockheed to present difficulties early in the telescope's life in orbit, but as the cells on the telescope's solar arrays degraded and the spacecraft's batteries became less efficient, its safety margin would gradually be whittled away. Indeed, Marshall calculated that after a few years in space, the telescope's power needs would exceed its supply. Thus, a major effort by the contractors, NASA, and ESA was launched to investigate the problem and to examine possible solutions, including changing to more efficient kinds of solar cells and different kinds of batteries to power the telescope when the solar arrays were in darkness.

Other issues arose from the thermal-vacuum testing. None was major in itself, but when taken in combination with the power problem and the very immature state of the ground system, they, in the opinion of many participants, made it highly unlikely that the telescope could have come even close to meeting a launch date of late 1986. That question, however, had swiftly become irrelevant.

In the months immediately following the *Challenger* accident, it had seemed to some in NASA that the shuttles might require only a quick fix and that the telescope might soon be aloft. By July 1986 and the end of thermal-vacuum testing, it had become obvious to all that the shuttle fleet would be grounded for a time that would be measured in years. Thus, a new question confronted the telescope's managers: how to retain the staffs involved in developing and testing the spacecraft and its associated ground system, for it was their expertise that would be so necessary once the telescope was

A 1981 version of the solar array scheduled for use on the Space Telescope; one of two solar arrays is seen undergoing tests at British Aerospace's plant in Bristol, England. The wings were designed to provide about 4,000 watts of power; each was to be eighteen feet long and eight feet wide. (Courtesy of NASA.)

in space. With the telescope sitting in a vast clean room at Lockheed for some years, NASA expected that it would be difficult to maintain a large group of experienced people on the program, as they would be in demand elsewhere within their respective organizations. There would also be a considerable financial cost to be paid for the delay, on the order of $7 million per month. As it would turn out, there was still plenty of work to do on the telescope. For example, tests at Lockheed of the telescope's safing system, designed to keep the telescope operating even if contact with ground controllers was lost for a prolonged period, persuaded Lockheed and NASA that it was unreliable. The result was that Lockheed developed a modified safing system, one designed to keep the telescope's solar arrays directed toward the sun, and thus keep power on the spacecraft, even in the case of a major malfunction. Extensive work was also carried out on the scientific instruments. Even in the summer of 1988, the state of the ground system was such that it took three days of work by people and computers to schedule one day's worth of observations. With these and other problems, there was no real shortage of work to keep people busy. In fact, there was so much rework on the telescope that in 1987 there had been an intense debate on whether or not to have a second thermal-vacuum test (Marshall finally recommended against it).

By October 1986, the scheduled launch date before the *Challenger* accident, NASA and the European Space Agency had spent over $1.6 billion in designing and developing the telescope and preparing for its operations in orbit, as well as its maintenance and refurbishment. Marshall's expectation was that by the time it was launched, about $2 billion would have been spent on the Space Telescope, not counting the $250 million plus for its ride into space in the shuttle.[126] Although it had never been planned as such, the Space Telescope had become the most expensive scientific instrument ever built.

CODA

At the Science Working Group meeting of November 1986, project manager Odom announced that he was leaving the Space Telescope program to assume a new position at the Marshall Space Flight Center. The project scientist paid tribute to Odom, describing his "great managerial and technical skill" while doing a "magnificent job in a very difficult era." That won applause from all present, and further plaudits.[127]

It *had* been a difficult era. To a degree, agency managers had been able to escape from some of the telescope program's early history, but despite the injection in 1983 of hundreds of millions of dollars into the program, and extra time into the schedules, there had been moments when agency managers admitted they were "hanging on by our fingernails." These same managers had nevertheless set out to, and in part had been able to, administer the program in a fashion strikingly different from that prevalent earlier. They had, for example, approved a backup Wide Field/Planetary Camera; in 1980, Marshall had sought to remove the spectrographs from the first flight. They had increased the number of orbital replacement units; that number had been drastically cut in 1980. They had, at least for a time, actively assisted the operation of STOPAT, the scientists' "war cabinet"; before, the scientists had often been marginal to the wider process of decision making. Serious problems remained to be solved, but by late 1986, a Space Telescope that in some ways was complete had emerged.

IO

Reflections

The basic decisions that initiate and set the pattern for a large-scale endeavor are made by votes – within the administration, of members of Congress, and by the citizens on whom the other two depend. Votes determine whether an endeavor is to be started. Votes determine whether it is to continue and at what level, at what tempo, and for what changing or developing purposes.

James E. Webb, 1969

Prior to the disaster that destroyed the Space Shuttle *Challenger* in January 1986, the Hubble Space Telescope had been scheduled for flight at the end of 1986. For those who had labored so long to bring the telescope into being, the grounding of the shuttle fleet, as well as further problems with the telescope, meant a long delay in the launch date. The astronomers found themselves still looking forward to a time when the telescope would be in space and would return data.

Reflecting on his more than twenty-five years of building instruments for space use, Arthur D. Code argued that it is

an interesting kind of sequence, I think, that one goes through, in any major space experiment. . . . You learn about the opportunity to be able to propose to carry out certain science. In developing your proposal, there are the scientific objectives that initially capture your imagination. You design an instrument to carry out an investigation that never has been done before, and is the interesting and fun science you can do.

All right, the proposal is accepted, and you start to work on it, and I guess the first thing you encounter is, you have to build a contract, and get an organization in place to do it. And then you begin to, as the design progresses, get into engineering details, and the engineering, [the] electronics, [the] testing of the instruments becomes quite fascinating, and so you can be interested in man-made things the same way as in nature and try to analyze them.

Then when you get closer to launch, you've finished up much of the stuff and you begin to think about operations and launch, you realize: hey, this is a great device, look at the science you can do! You get excited about that, and have great expectations.[1]

With space astronomy, it is the design and development phase that determines the hardware and software to be built and the manner in which that will be done. It is during that same period, to a large degree, that one determines the science that can be pursued and the manner in which that must be carried out.

As has been emphasized throughout this book, the shaping of the telescope's scientific capabilities was an integral part of its political pilgrimage: the telescope's designs, the program to build it, and the claims made on its behalf continually had to be revised and refashioned as part of the effort to come up with a telescope that would be politically feasible. Eventually the telescope would become a concrete realization of the interests, hopes, and aims of those who built the coalition that won approval for the program. To help build and sustain that coalition, the telescope's scientific performance would have to be sufficiently impressive to enlist a substantial number of astronomers; if the telescope had not won the enthusiastic support of a large body of astronomers, it would never have gotten off the drawing board. That constraint was what had made the redesign of the telescope between 1974 and 1977 such an awkward and delicate task. NASA had to reduce costs; yet if the agency went too far in making cuts and thereby reduced scientific performance to a level the astronomers judged unacceptable, the telescope would be just as dead as if it had died at the hands of Congress.

Not until it has operated in space for several years will it be possible to render a final evaluation of the performance of the telescope. Although on many occasions, because of cost and scheduling problems, there was great pressure to agree to reductions in the telescope's scientific capabilities – witness the events of Black Saturday described in Chapter 8 – NASA held to the scientific specifications that were in effect at the start of Phase C/D. That, however, certainly does not mean that the problems encountered during its building will not influence the telescope's scientific output. For example, in the project's early years, the tests planned for the scientific instruments and the numbers of available spares were much reduced by NASA, and the use of engineering models was deleted from all of the programs to build the instruments, except for the High Speed Photometer (the photometer's builders had insisted, and NASA had agreed because in that case it was inexpensive). Thereby, NASA allowed more risk to be introduced into the instruments' construction than had originally been intended. In late 1982, newly appointed program manager Marc Bensimon attended a number of reviews on the scientific instruments, "and I was stunned. I couldn't believe that this was the most technologically advanced program in [the Office of Space Science and Applications], and that it was being done on a shoestring. . . . We weren't doing life testing on the mechanisms, and environmental testing was cut back drastically. You know, I came on this program, which is supposed to be in the big leagues, yet I knew we had done more testing on other little programs that weren't as important as this one. I was really shocked."[2] Also, the first of the second-generation scientific instruments probably will not be ready until some years after the telescope is in orbit, thereby raising the possibility that for certain periods the telescope may have to operate with a depleted complement of instruments. In that and numerous other ways, the problems encountered in building the telescope and its ground system will have been instrumental in shaping its eventual scientific output.

We have seen that the story of the development of the Space Telescope also illustrates a number of central themes in the history of post–World War II science and provides a rich case study on the making of U.S. science policy. In some ways, the history of science policy is the history of contemporary science. The telescope project, too, provides an example of a scientific enterprise of such large scale that it has been termed a "megaproject." The subjects of this chapter, then, are the major points, and perhaps some lessons, to be drawn from the story of the Space Telescope's design and development.

BIG SCIENCE: THE HUMAN EXPERIENCE

The building of the Space Telescope was intended to lead to the deployment of a powerful new scientific tool. But in late 1988, it still sat in the ultra-clean atmosphere inside the Vertical Assembly and Test Area at Lockheed in Sunnyvale, California, a mute giant stripped of a number of its scientific instruments and other sections that had been removed for more work.

By the time of its launch, over 12 years will have passed since the start of the telescope's construction, and over a quarter of a century since detailed planning for the telescope began. The time from inception to operation of a Big Science facility in the United States often is long, sometimes decades. Part of the reason is that Big Science generally is so expensive that any such project will come under intense and lengthy scrutiny within the sponsoring agency or agencies, as well as in Congress and the White House. In consequence, any proposed project must have advocates in the scientific community who are willing to work to mobilize its supporters, argue with doubters, and spend much time on advisory committees.

For the Space Telescope, the chief champion for many years was Lyman Spitzer, Jr. His 1946 report for RAND, "Astronomical Ad-

Some forty years after completing his report on the "Astronomical Advantages of an Extra-Terrestrial Observatory," Lyman Spitzer, Jr., stands in Lockheed's control center for the testing of the Space Telescope. The telescope can be seen over his left shoulder in Lockheed's VATA. (Courtesy of Lockheed.)

vantages of an Extra-Terrestrial Observatory," was a visionary look at the sorts of astronomical questions that could be addressed by a large telescope in space. But the primary claim that Spitzer made there, one that was to be repeated by Spitzer and his colleagues again and again over the years, was that such an instrument would provide a totally new capability: Its most important contributions to astronomy most likely would be "not to supplement our present ideas of the universe we live in, but rather to uncover new phenomena not yet imagined, and perhaps modify profoundly our basic concepts of space and time."

It was that belief, rooted in a vision of the history of astronomy and the effect of the introduction of other powerful telescopes, that did much to engage the interest and sustain the efforts of the astronomers. That was why, for example, Robert Danielson of Princeton, who knew he was to die shortly and would never have a chance to use the telescope, insisted on working on the plans to build it until the very end of his life in 1976. It was that level of commitment, shared by many people, that drove the program forward.

Accompanying that commitment was the realization that there was always a risk, however small, that the telescope would never produce scientific data. Moreover, because the telescope has been built on the protoflight principle, there is no replacement − with the Space Telescope there is only one chance.

During the 1977 campaign to sell the telescope, John Bahcall had an appointment on Capitol Hill with Congressman Max Baucus. When he entered Baucus's office, Bahcall found that Baucus was busy and in a hurry to leave. Baucus then "put me up on a stool, literally up on a stool, and he questioned, grilled me about Space Telescope science, what it would do. I was amazed. He really wanted to know what it would do. . . . He wanted to know all of the political aspects, and just like that, I mean really quick, and I had the feeling that if I couldn't answer in 15 seconds, then he was going to walk out." Then Baucus, who knew that no backup telescope was to be built, asked what would happen "if the Space Telescope just went into the drink." Remembering that in 1983, more than two years before the *Challenger* disaster, Bahcall emphasized that "I think what I said, I'm sure what I believed, was that it was so important that astronomers were willing to spend a decade of their lives, even if there was the danger it would all go into the ocean, because it was a historically important activity. What else could you say? That was the truth. We knew there was that danger, but we were spending our time because it was so important."[3]

Numerous astronomers spent a great deal of time planning for and then building the telescope. Those who worked continuously on the telescope from the start of Phase C/D in 1977 invested more than a decade of their careers. For others it was a much longer commitment. For Lyman Spitzer, Jr., involvement with the telescope meant four decades of proselytizing and detailed work. Laurence Fredrick, who began his studies of large telescopes in space in the early 1960s, was a member of Spitzer's National Academy of Sciences committee of the 1960s, as well as the Phase B Science Working Group, and later he became a member of the Phase C/D astrometry team. Arthur D. Code had been a leader in advocating and planning for the Space Telescope since the end of the 1950s. Robert Bless and Bob O'Dell had worked closely on the telescope since 1971, O'Dell as project scientist from 1972 to 1983. John Bahcall and several other astronomers had been associated with the telescope since 1973. Albert Boggess III, for example, was a team leader in Phase B and then a member of the team to build the Goddard High Resolution Spectrograph, before assuming a major role in Goddard's work on the telescope's scientific operations, then becoming project scientist in 1985. Ivan King had been a consultant in Phase B and in 1977 had joined the instrument team for the Faint Object Camera, in addition to becoming a member of the Science Working Group. Nancy Roman, at NASA Headquarters, worked on the Space Telescope in numerous capacities for almost two decades.

In 1984, Mike Disney examined his role in the Space Telescope program as a member of the science team for the Faint Object Camera:

I think the hardest thing . . . is hanging around for such a long time. It's very difficult. I think there's a sort of human time span of interest in a particular project. For example, when I first became interested in this whole field, I was working predominantly on active galaxies and nuclei and high energy objects for which Space Telescope is pretty appropriate. But over the course of years now, I've developed and worked in other fields, and most of my interest now is in low surface brightness galaxies, where the Space Telescope will make almost no contribution that I know of. And I suspect this is true for many of us. . . . There's no question, everybody's very excited about it, but a project like this, it can eat away at your time to do other things, and you sometimes get very fed up with it. I'm sure this is true of everyone who's been involved in the project, at times. "When is this thing going to happen?," you know. It's just, by the time it happens, I might either be dead or totally uninterested in astronomy or whatever it is. And I think that's a human thing. . . . for me, I think that's been the hardest thing, to believe it's going to happen. Intellectually I know it is, but emotionally I've waited too long.[4]

Principal investigator Richard Harms recalled that as an undergraduate at Stanford University he had visited one of the experimental high-energy physics groups at Berkeley. He had spoken to some of the graduate students who were working on enormous banks of electronics. They told Harms that they had already spent two or three years on the apparatus, and after another couple of years they would be able to take it to the accelerator to secure scientific data. "I thought, I'll never go into a field where it takes five years to get a project done like that. Being on Space Telescope now that seems ironic!" As O'Dell, who probably devoted the best years of his career to the telescope, put it before the *Challenger* disaster, "We'll soon know if we developers get the best science out of the observatory or if that distinction falls to those who stayed home doing good science instead of attending reviews for ten years."[5]

Some staff members at the contractors, as well as NASA engineers (although, significantly, not the leading NASA managers, who tended to stay two or three years at most), devoted large parts of their careers to the telescope. At Lockheed, for example, program manager Burt Bulkin started work on the telescope in 1972, and systems engineer Domenick Tenerelli in 1971. Marshall engineer Garvin Emanuel, along with numerous other Marshall and Goddard staff, worked for many years on the Space Telescope. In 1985, Emanuel was in the Marshall systems engineering office, and he noticed someone familiar. As he recalls, "I said, 'hey, there's Steve Treat. He came back on the project.' So I walked around the corner and he was gone. I waited a few minutes and here came a young individual and I said 'I'm waiting for Steve Treat.' He said, 'I'm Steve.' I said, 'Oh – oh. I worked with a Steve Treat back in 1971.' 'Oh, that's my dad, he works in another part of Lockheed. I'm Steve Jr.' " As Emanuel ruefully remarked, "those are the blessings of being on the project for a long time."[6]

For astronomers closely connected with the telescope, particularly the principal investigators, some institute and NASA staff, the program meant a reduction in time available for their own scientific research. It also meant working on a very large scale program in which the organizational dynamics were enormously different from those most of the astronomers had previously experienced. Although Big Science is characterized by enormous expense and often a huge program, cost and scale are not its only features. As we have seen, there is an important social dimension to Big Science. Little science is generally conceived and conducted by a principal investigator, perhaps leading a small laboratory or group. Big Science is characterized by large multidisciplinary teams, a division of labor, team commitments, agreements and negotiations on common purposes, and hierarchical organization, which on the telescope program are seen most vividly at the institute.

Optical astronomers usually have shown a strong preference for little science, working individually or in small groups. In the past, much of optical astronomy might even be said to have been almost a "cottage industry." But as astronomy has become a bigger science with larger-scale facilities, even ground-based optical astronomy has become more of a group activity and less personal in numerous ways. Optical and ultraviolet space astronomy has exhibited this trend even more clearly, and the development of the Space Telescope has meant the biggest sort of Big Science. Many scientists dislike Big Science because it can mean a loss of autonomy through a centralization of decision making. Often there was relatively little influence for the astronomers engaged in the Space Telescope program. Indeed, for a time during the early 1980s, with money extremely tight, they were almost irrelevant to the wider decision making on the project. In addition, although the Phase C/D Science Working Group demonstrated that it could be effective in addressing issues that had surfaced, it had at best a mixed record in anticipating problems. Nor did it have overarching powers to probe into matters. The working group's lack of influence, in fact, was to fuel the push in 1983 for STOPAT, a kind of war cabinet for the scientists that won approval from NASA managers in the wake of the crisis of late 1982 and early 1983.

The kind of Big Science exemplified by the telescope program means that astronomers are required to attend numerous meetings and prepare many documents to try to ensure the coordination of a complex program's vast range of activities. Because of the large number of institutions and groups engaged, one individual cannot possibly know exactly what is happening on all areas of the program. Unlike the situation for little science, it is often difficult, if not impossible, for a scientist acting alone to secure changes in, say, the design of a test program. For some astronomers, immersion in the highly ordered and hierarchical world of the Space Telescope was, on occasion, an uncomfortable and demoralizing experience, particularly as there sometimes were such deep and acrimonious divisions between different sections of the program. In addition, to be effective within the Space Telescope program, astronomers had to work diligently through the management system. For instance, in the debate in 1984 and 1985 over the appropriate solution to the quantum efficiency hysteresis problem, they had to fashion their arguments into forms appropriate to the decision-making processes of NASA managers and strive by careful technical arguments and active lobbying to build a consensus on technical and scientific issues they wanted to influence. The size of the Space Telescope program meant that anyone who was going to be effective would have to be able to address a variety of audiences, engineers and managers included. That kind of activity was a long way indeed from the usual rough-and-tumble debate in small-scale science, where the major interaction is between one scientist and another.

Since 1977, James Westphal had spent a tremendous amount of time leading the development of his own scientific instrument, the Wide Field/Planetary Camera, in addition to investigating wider issues on the telescope. In 1977, when he was preparing the proposal to build a scientific instrument for Phase C/D, Westphal realized he would have to work intimately with the aerospace industry and NASA and be placed in an engineering world "with its formalism and its structure and its massive amount of paper."[7]

Westphal even consulted his colleagues in the Planetary Science Division at Caltech on his decision to participate in the Space Telescope program. He was concerned that building the Wide Field/Planetary Camera would rule out his involvement in the day-to-day activities of his division:

In fact, I asked to have a meeting with the whole bunch of them in one room, and told them this was in my mind and this was the way I thought it was going to happen, and told them that although I knew I didn't have any obligation to do so, that I was in fact giving them a veto over this. . . . They discussed it a lot. . . . Most of them didn't believe me, that it was really going to be this bad, although I'm pretty sure Bruce Murray [director of JPL] was there . . . but certainly he had already had some experience. Dewey Muhleman had had a little bit of experience . . . and he said "Your judgement is probably optimistic."[8]

Another principal investigator, Robert Bless, had

often thought of [the Space Telescope program] as an enormous medicine ball, one of these things that's six feet high, that people form teams to push around. You could stand there and beat on it as hard as you could. After a while you'd make a little dent in the thing. And you'd come back a couple of hours later, the dent had disappeared and the medicine ball was completely unchanged. That was the feeling a lot of people had with [Space Telescope], and to an extent this is so even now [November 1986]. It's awfully hard to get anything changed, awfully hard, even when people at the top want to see some changes. It just doesn't happen easily.

Program scientist Edward Weiler likened the process to turning around a glacier.[9] Certainly the process is a long way from the relative autonomy of little science, in which it is the scientists who call the shots.

BIG SCIENCE: PROGRAM DESIGN

A central feature of the Space Telescope program has been its very size. The telescope surely represents Big Science of the biggest kind, as well as of a particular kind. From the first, NASA, the astronomers, and industry accepted that constructing and operating the telescope would involve many institutions, groups, and individuals. A key aspect of the telescope program was that there would be many technical, scientific, and management interfaces. If such a large program was to run smoothly, a high premium had to be placed on the way in which all the disparate institutions, groups, and individuals were arranged to fit and work together, in other words, the program's own design. Because the telescope was the charge of NASA, it was NASA that would be the program architect and the manager of the completed system. Despite the fact that a single agency was in charge, the result was a fragmented program design that was clumsy at best and that militated against designing and developing the Space Telescope as an observatory, the sum of which would be greater than its individual parts. Why was that?

NASA employed two associate contractors – Lockheed and Perkin-Elmer – to design and build the Support Systems Module and the Optical Telescope Assembly. That split responsibility, together with the manpower cap imposed on Marshall, made it difficult for NASA and Lockheed, the agency's chosen systems engineering contractor, to perform systems engineering. The systems engineering groups at Marshall and Lockheed were much strengthened in 1983, but that was six years after the program had started, and it was in direct response to the events of early 1983. During the planning stages of the program, the Hornig committee and some of the astronomers pressing for establishment of a Space Telescope Science Institute had intended the institute to play a kind of scientific systems engineering role from early in Phase C/D. That, however, did not work out in practice. First, the institute, for budget reasons, was started well after the telescope was in the design and development phase. Second, NASA very much intended the institute to be a service organization for scientific operations, not an organization to provide scientific systems engineering leadership.

NASA also selected two of its field centers to manage the telescope's design and development: Marshall had responsibility for the overall telescope system and for the Optical Telescope Assembly and the Support Systems Module, whereas Goddard was responsible for the scientific instruments and operations. The division often was effective, particularly in the later years, but it also created massive difficulties and complications. Rivalry between the two centers sometimes led to poor communications and to styles of working that on occasion were crudely confrontational. It was only in 1983, after James Odom at Marshall and Frank Carr at Goddard headed their respective Space Telescope organizations and strenuously set out to establish good working relations between the centers, that Marshall and Goddard began to work better together.

The NASA culture undoubtedly is composed of many disparate strands, but the program to design and build the telescope exemplifies the modern notion of technical change owing more to institutions and less to individuals, of invention being the province of corporations rather than of technical wizards, of human minds and knowledge making a difference in the way people live and work and die not through creative brilliance but through organization and control.[10] Whereas NASA always focused on the formal structures put in place to smooth the management of the program, the agency surely paid too little attention to the particular people involved and what was actually being achieved in the program's early years. Indeed, it probably was not until after the crisis of early 1983 that NASA Headquarters, Marshall, and Goddard each had, at the same time, what were generally agreed to be top-flight people leading their sections of the program. In fact, it took the crisis of 1983, and the consequent increase in visibility of the telescope program (and NASA managers had shaped and exploited the crisis to produce such an increase), to secure the program the priority in NASA that many of its participants believed it had long deserved. Only after 1983 did the agency's rhetoric about the telescope's potential as a scientific tool and as a superb piece of high technology begin to be matched by the reality of what was occurring on the telescope program. If it had been a simple spacecraft of undemanding requirements, the program design NASA adopted might not have mattered, but for one as complicated as the Space Telescope, the design did matter. However, it is essential to recognize that the program design was not devised in some ideal world, one in which the "best" means of managing the telescope's development was the only objective. Instead, the program design carried the cluttered baggage of NASA's institutional interests, competing and often intense science and industry interests, Department of Defense interests, and a number of compromises struck in order to initiate the program.

The program design was therefore strongly influenced by a medley of factors that had little to do with establishing a clear management structure. First, NASA's choice of Marshall as lead center was made in part to help fend off the thrusts of the Office of Management and Budget to close the center down, a move that was unacceptable to the agency because Marshall was essential to the agency's plans to build the Space Shuttle. Second, NASA might have joined with the European Space Agency to build the telescope even without pressure from Congress, but robust congressional criticism of the program's cost served to make such cooperation politically vital. Third, the manpower cap imposed on Marshall was at least in part the consequence of a NASA/Department of Defense agreement. In the early years of Phase C/D, the manpower cap compelled Marshall to manage the program in a manner that was far from the center's usual style. Instead of Marshall "penetrating" the contractors, the center was forced to some extent to watch the proceedings, and that from a rather obscured seat. It should nevertheless be added that even after the increase in manpower in 1980, NASA still failed, in the NASA/Department of Defense phrase, to "penetrate the contractors," as the events of late 1982 and early 1983 would show.

The original manpower cap was at least in part the result of the design and construction of the telescope by contractors who had established a tradition of building photographic intelligence satellites and whose experience was largely outside that of optical space astronomy. The history of space astronomy in optical and ultraviolet wavelengths was in many ways almost irrelevant to the construction of the Space Telescope. Instead, the dominant technical heritage of the telescope derived from the building of reconnaissance satellites, the key technology in the history of the U.S. space program. Perhaps those institutional and technical links gave NASA and its contractors what proved to be a false sense of confidence and fed a consequent lack of sensitivity to the unique requirements of an orbiting optical telescope for astronomy. If similar spacecraft had already been built, why should the telescope prove difficult?

Where the early history of space astronomy was important for the Space Telescope, however, was in forming a core of advocates and potential users of the telescope. That group pressed for the telescope to be built, helped in its design and in framing the broad questions it should tackle, was involved in designing possible scientific instruments, pressed for the institute, and played a part in making the telescope politically feasible.

BIG SCIENCE: FUNDING AND POLITICS

Before World War II, most of the funding for U.S. science came from industry, private foundations, and the states. During the war, that pattern changed radically, and the federal government became the chief sponsor of the scientific enterprise, a shift that was swiftly institutionalized in the years following 1945. That shift brought with it not only more money but also new obligations and a new political framework within which scientists had to work.

In the first few years after the war, many astronomers had been chary of accepting government money for fear that the source might soon dry up. By the 1960s, this fear had disappeared. And even in the early 1960s it was clear to its advocates that the price of the telescope put it well beyond the reach of private or industrial patronage. It was therefore agreed by everyone that if it was to be built at all, it would, like most of the major science facilities since the war, have to be built with the support of the federal government. James Webb was NASA's second administrator and the person who stamped his leadership on the agency in the decade the United States reached the moon. In his opinion, in

our pluralistic society any major public undertaking requires, for success, a working consensus among diverse individuals, groups, and interests. A decision to do a large, complex job cannot simply be reached "at the top" and then carried through. Only through an intricate process can a major undertaking be gotten underway, and only through a continuation of that process can it be kept going. The basic decisions that initiate and set the pattern for a large-scale endeavor are made by votes – within the administration, of members of Congress, and by the citizens on whom the other two depend. Votes determine whether an endeavor is to be started. Votes determine whether it is to continue and at what level, at what tempo, and for what changing or developing purposes.[11]

Hence, because of the need to win government patronage, winning approval for the telescope was an intricate process. Also, because of the particular set of circumstances in which the telescope was initiated, it was fashioned to a significant degree by extrascientific pressures, pressures that introduced severe tensions into the program design.

It was not only in the program design that those circumstances made an impression. Recall, for example, the changes made between 1973 and 1977, between the Large Space Telescope and the Space Telescope. The telescope designs of 1973, and the program to turn those designs into reality, were very different from the designs and programs that emerged in 1977. The 1973 design for the Large Space Telescope had a primary mirror three meters in diameter; the 1977 telescope had a 2.4-meter mirror. In 1973 there were seven scientific instruments; in 1977, five. In 1973 the Support Systems Module sat behind the primary mirror; in 1977 it surrounded the primary. In 1973 the telescope was fifty-five feet long; in 1977, just over forty-three feet long. In 1973 there was an extendable light shield; in 1977, only an aperture door. In 1973 NASA planned to build an engineering model and was considering flights of a precursor telescope; by 1977 the decision had been made to use the protoflight concept. Even the telescope's name had been altered, from the Large Space Telescope of 1973 to the Space Telescope of 1977.

Why were those changes made? A major goal behind all of them was to lower the telescope's cost to the level which NASA's top administrators thought the "market" – in the shape of the White House and Congress – would accept. Given NASA's pessimistic view of how Congress and the White House would respond to the prospect of funding the Large Space Telescope, the low cost was regarded by the agency as essential. Thus, we have a vivid illustration of how the telescope's dependence on government patronage and the political context in which it was initiated, including the existing set of NASA's own institutional interests, were exceptionally important in shaping the design of the program as well as the telescope itself. For example, if Congress and the White House had readily approved the early 1974 version of the telescope, a three-meter Large Space Telescope would surely have been built.[12]

Nor was the transition to a politically feasible Space Telescope program – that is, one that the White House and Congress would fund – a particularly orderly or rational process. That was hardly surprising. Although the telescope was a joint NASA/ESA program, NASA was the dominant partner, and the telescope was chiefly an American creation. The coalition-building, bargaining, and compromise entailed in winning approval for the Space Telescope were woven into the very fabric of American society, and policy-making in the U.S. form of democracy can at times be complex and apparently disordered.[13] But with the Space Telescope, policy-making did not simply seem to be disordered – it was disordered. In Chapter 5, the process was termed ad-hocracy. The question "Should the Space Telescope be built?" was in fact continually being reframed as the issues surrounding the decision were reshuffled, repackaged, and hammered out in a variety of arenas according to ever shifting rules and players.

Nor did the struggle to win political support for the telescope begin and end with the hearings in Congress in the mid-1970s. Rather, the efforts began in earnest in the 1960s, and the common theme running through them all was coalition-building.

Coalition-building was very much a goal of the committee, assembled in the late 1960s and led by Lyman Spitzer, that reported to the National Academy of Sciences on the scientific uses of the Large Space Telescope. The committee's members chose not to examine the telescope's possible scientific programs in any great detail. Such, when they knew the telescope was years from construction, would not have been a fruitful activity. Instead, they saw their role in a missionary light, to go out among the wider community of astronomers and explain what would be the scientific potential of such a powerful new instrument. In so doing, the committee members sought to help to defuse the criticisms of opponents, as well as to establish a strong base of support for constructing what was then called the Large Space Telescope.

In certain ways, the Phase A and Phase B Science Working Groups played a role similar to that of the Spitzer committee. Their discussions and debates went far beyond technical and scientific issues. Indeed, one of the reasons for the existence of those groups was to provide a sizable number of supporters for the telescope. After all, if the people who were planning for the telescope and who fully understood its objectives were not prepared to defend it if it ran into trouble, who would?

Coalition-building was also one aim of NASA Headquarters in pressing for the planned scientific instruments to be capable of performing high-quality planetary science, often in the face of a diffident, if not hostile, Phase B Science Working Group. The Space Telescope Science Institute, too, was created by NASA in part to widen the telescope's political base.

As seen in Chapter 6, one of NASA's aims was to fashion an institution that would be attractive to ground-based astronomers, a group that until the mid-1970s had never been excited about the telescope. Support was in fact built for an institute because astronomers outside of NASA saw the institute as a means for them to secure control of the telescope's science operations. That was a consequence of three assumptions shared by many of the astronomers: (1) Once the telescope was in orbit, scientific operations might not receive the priority within NASA they deserved and thus not receive the appropriate funding. (2) The best scientific operations would be secured by a staff functioning in a university-like environment. (3) Similar types of establishments had been commonplace in the structure of postwar U.S. science. Certainly the institute came into being in part because those other organizations – especially those devised to link the users with high-energy physics facilities and ground-based astronomical observatories – were taken by NASA and the astronomers as models. In many respects, the institute was perceived by the astronomers as the "natural" way to proceed, the sort of organization with the best chance of ensuring the telescope's scientific success. Whereas the Hornig committee had argued that "the institute approach is the one most likely to provide an optimal scientific return for a given dollar investment in the [Space Telescope] program," from the perspective of the 1970s, NASA did not see any overwhelming technical reason that there had to be an institute, though the agency did see that there would be political benefits from strengthening the coalition in favor of the telescope.[14]

The selling campaigns undertaken by the astronomers between 1974 and 1977 played a leading part in winning approval for the Space Telescope. Bahcall, O'Dell, and Spitzer, together with George

Field and others, were following in the steps of George Ellery Hale *Funding and politics* and other astronomer-entrepreneurs of the late nineteenth and early twentieth centuries who had financed the building of new telescopes by persuading their patrons of the telescopes' worth. In the 1970s, the patrons were not rich individuals, but the White House and Congress. The selling campaigns for the Space Telescope were also much different in character from those conducted by Hale and other leaders of the astronomical community of his time. In fact, the essence of the selling campaigns launched by the astronomer advocates of the Space Telescope was the large number of their colleagues who participated and who, by 1977, had been mobilized into an energetic lobbying group.

It would, however, be a mistake to think that the astronomers acted in isolation; rather, they joined with managers and professional lobbyists employed by the contractors, as well as anybody else who could help, in a loosely coordinated coalition, one that also had strong NASA links. The contractors, in addition to mounting their own lobbying campaigns and suggesting arguments, particularly regarding the jobs created in the building of the telescope, were able to help educate the astronomers to the ways of Capitol Hill and what selling the telescope would actually entail.

NASA, the White House, and Congress in fact gave their approval for the telescope for reasons much more diverse than simply the quality of the science the telescope's supporters promised – for example, to promote international ties, to strengthen the scientific/technological base of the United States, to help maintain the capability of the Marshall Space Flight Center, to provide employment in the districts and states of many congressmen and senators, and as part of a major initiative in the Ford presidency to promote the development of basic research.

The telescope's advocates were fortunate in some ways in the timing of the plans to initiate the telescope program. In the decentralized, pluralistic structure of post–World War II science, any sort of overall centralized planning of federal funding had generally proved to be impossible. That usually meant that policy definition had been initiated at low levels in the collection of executive agencies (such as NASA) and that higher policymakers, including, for example, the President's Science Advisory Committee, had played only a coordinating role at the end of the process.[15]

Starting in 1975, with the coming of the Ford presidency, federal funding for basic research, after having declined for a few years, resumed the upward trend that has been characteristic of U.S. science policy since World War II. That boost in funding sat excellently with NASA's proposal to build the Space Telescope, and the telescope was seen in the Ford administration as a fine example of the kind of science program that was worthy of support. By 1975, support for the telescope had already been established at the lower end of the hierarchy of the executive branch, but then support began to arrive from the upper direction too. The telescope therefore became tightly meshed with policy definition within the White House. With a different White House and another set of policies, the telescope program might never have gotten into gear.

UNDERESTIMATIONS

The process of making the telescope politically feasible involved its advocates in coalition-building. It also meant negotiation and compromise on the telescope's design and the planned program to build it. The result was to be a telescope whose construction was to be a formidable engineering task. That task was further complicated by the program design, a program design that, as argued earlier, can be interpreted in considerable part as a product of pork-barrel politics. The road to a completed telescope that would perform to its specifications would be littered with technical obstacles and cost growth.

In addition to the problem of political pressure making the engineering job more difficult, unrealistic technical expectations had been fostered when the telescope was being sold. The claims by NASA of the mid-1970s that the technology to build the telescope was well understood and was within the state of the art crumbled under the pressure of a taxing design and development effort. Much of the technology was indeed within the state of the art; however, it had to be arranged in novel ways in a very complicated spacecraft, one, for example, with an extremely large number of ways of taking observations. Again it took some years for the reality to catch up with the rhetoric of the early selling campaigns, perhaps not until the crisis of 1983.

Another illustration of the chronic public underestimation of how difficult it would be to build the telescope is seen in the winning bids of Lockheed and Perkin-Elmer for the design and development of the Support Systems Module and the Optical Telescope Assembly. Both of the bids proved to be unrealistic. The Keller review of 1980, led by former program manager Warren Keller and composed of senior NASA staffers (see Chapter 8), termed both bids "buyins," but in 1977, without the benefit of hindsight, NASA had calculated that they were only slightly too low. Nor were Lockheed and Perkin-Elmer the only ones to underestimate costs. The early estimates regarding the costs and complexity of the scientific instruments, the Science Institute, and the ground system to operate the telescope in orbit, as well as the costs and problems presented by the telescope's maintenance and refurbishment program, would prove to be significantly too low.

The telescope was highly complex, and its overall design was pioneering in several areas. It was, for example, the first scientific spacecraft explicitly designed to be serviced in orbit. Building the spacecraft to meet that goal had been far more demanding than anticipated in Phase B. The latches to hold the scientific instruments in place and to ensure that they would maintain their precise alignment, as seen in Chapter 9, proved to be difficult to design and build, for instance. Not surprisingly, it was the aspects of the telescope's design that at the time of Phase B NASA expected to be most difficult — perhaps most notably the Fine Guidance Sensors and the primary mirror — to which the agency and its contractors devoted the most attention. The latches, in contrast, received relatively little study.

MONEY

A number of issues — the telescope's technical complexity, the lack of systems engineering at the beginning of the program, the agency's failure to penetrate the contractors, the contractors' low bids, and the poor program design — contributed to the root problem facing the NASA managers throughout the telescope's detailed design and development phase: lack of money to accomplish all of the tasks they had defined at the start of the program in 1977, as well as the extra tasks added in later years. The NASA Headquarters attitude toward the telescope's difficulties exacerbated matters. As James Welch, head of the Space Telescope Development Division from 1983 to 1987, has contended, when

it became clear that the earlier estimates of costs, for what it would take to achieve the perceived performance requirements, were seriously in error, there was not the inclination among the managing divisions here in Headquarters to go back and ask the Comptroller for more money, that is, during the budget development. So whenever the project came to Headquarters to ask for more money to accomplish what was only their original set of goals, then almost invariably Headquarters's response was, "No, live within your budget."

And in some cases the response was not only "No," but "We're going to cut your budget." And it created an absolute disaster. The things that started disappearing from the program — and necessarily so, because the Project Manager had no choice, when the money went away — was to start slowing down, re-phasing it, even stopping work at some subcontractors in some cases. What this meant was that the best talent migrated away to other large programs that the contractors and subcontractors had. This meant that things got cut out, like reliability processes and practices which you would ordinarily want to keep in a space program. Spares were cut out, tests, vital tests in some cases, were cut, because of false economies. . . . So whenever these penny-wise, pound foolish judgements were made, Space Telescope kept getting deeper and deeper into trouble. Combine that

389

with a company [Perkin-Elmer] that had serious amounts of difficulty in planning and scheduling and performing work, combining that with a general lack of understanding of what performance requirements really were going to be needed in the subsystems and modules, and combine that with a lack of understanding, penetration, and oversight from the highest levels of management: it was a recipe for disaster. In fact, that is what happened [in 1983].[16]

Given the commitments to Congress in the mid-1970s and the intensity of the selling campaigns, campaigns that had not emphasized potential difficulties, it was inevitable that the telescope program would be fiercely criticized when it ran into trouble.

The criticisms that it was easy to predict would be leveled at the program and the space agency, particularly after the addition of more money and extra time on the schedules in 1980, together with the agency's always tight budget and continuing problems with the shuttle, help in part to explain why NASA Headquarters was reticent about asking the White House and Congress for another substantial injection of funds. There had, in fact, been pressure from the House Appropriations Subcommittee to cut funds on at least two occasions: in 1978, when the subcommittee had tried, unsuccessfully, to remove $15 million from the telescope's budget to be devoted to a Space Shuttle contingency fund,[17] and during the fiscal 1981 budget hearings. A former senior Goddard manager, later to be a Perkin-Elmer manager, remembered that "there was a great fear that if you really dug down too deep in the hole, the program might be canceled. People were a little unsure if you tried to go back for another 100 million dollars or something, it might just kill the program. The Shuttle was in trouble. There were a lot of people who had a fear of that. The community, once they start something this big, can talk themselves into fear, just as easily as they can be talked out of it."[18]

A program that was oversold and underfunded, that was beset by tough technical problems, that was committed to overambitious schedules and at best a fragmented program design, that suffered from extreme reluctance at Marshall and NASA Headquarters to ask for additional money and time, and that had a management system that fundamentally boiled down to trusting the contractors to do the job — all that indicated a program that was headed for derailment. When in early 1983 headquarters decided that the program had indeed left the tracks, the agency had no choice but to plead for more money. But that move had been made only after what was interpreted at a high level in headquarters as a disaster.

The initial underfunding of the telescope, when combined with a poor management system and technical difficulties, led to imposing problems. Like many of the issues that surfaced in the design and development phase, the underfunding could in large part be traced back to the selling process for the telescope. Money

As John Logsdon pointed out, the "problem with overselling a program is that advocates may later be expected to deliver on the promises made to gain approval and may find it difficult to back off their public commitments. Further, the expectations created by program advocates influence the policy framework by which program success will be judged. Unrealistic expectations obviously lead to later policy failures."[19]

"Usually, when a project is completed within cost, on time, and in a manner that leaves the user of the product satisfied," writes policy analyst W. Henry Lambright, "it is called a success. One symptom of 'bad' management is the cost-overrun, but, even here, appearances are deceiving. Cost-overruns may not be overruns at all but realistic costs of a program whose original estimates were unduly optimistic." Lambright further notes that

Apollo shares with Polaris the glory of being a "well-run," large-scale program in terms of avoiding an overrun. But how did NASA avoid such an overrun in a $24 billion program? By some miraculous management techniques? To be sure, NASA pioneered many management innovations and willingly offered itself for study as a showcase of "project management" at its optimum.

NASA, because of the high priority of Apollo, was able to avoid an overrun by estimating *realistically*. Hence, when Apollo came in at a figure which was approximately that suggested almost a decade before, NASA was heralded for its "good management." Good management, in this case, was merely cost realism in an environment in which costs were not a very important consideration in the decision to initiate Project Apollo. The pace of the program was far more important to President Kennedy than was the cost.[20]

That does not mean that Polaris and Apollo were not well run. Lambright's point is that "standards for successful management are elusive and . . . the political environment may determine far more than internal administrative genius, the conditions under which it is possible to be realistic as to costs and thus win a reputation for managerial efficiency. Projects with modest backing would have great difficulty in getting initiated, much less implemented, if their proponents were entirely 'realistic' about their costs."

Such, surely, was the case for the telescope. As Don Fordyce put it, "it was a matter of what Congress would buy at the time. It was a matter of how people tried to de-scope a program, and talk themselves into doing something that they thought they could do."[21] As chief engineer Jean Olivier recalls, the project had started out as puritanical in an engineering sense. But the financial pressures were extremely high, and you're told, "hey, you want this project or not? We can all go home and not even have a project, because it ain't going to get sold, because you guys have to get the cost down." And that's an incentive to try to get the cost down, right. And so you begin to work out how you can do it, and so you rationalize: "I can do without this and this and this, and therefore, I can cut out enough money to maybe shoehorn it in under some kind of a magic number somewhere." A lot of that went on, wave after wave of it, until finally it was shaken down to the point where it was economically acceptable, and technically acceptable. . . . But with the size and complexity of [the Space Telescope] there were too many hidden problems in this to keep it totally in bound. I guess it got squeezed a little too much.[22]

As O'Dell remembers,

of course, the cost of ST had always been hokey, depending on the situation and the individuals. Throughout Phase B, I felt that the contractors and NASA at the working level, were all working to produce artificial numbers. All honest people involved, but the system was such that it led to the generation of false numbers. [Headquarters] was continuously cutting this goal figure, and very naturally, the project and contract managers were coming back and saying, "yes, we can do it, we can shave this off but we can still do it."

Shaving this off didn't mean a performance reduction, rather, "Well, we'll use this approach instead of that approach, after all, we can accept 80% probability of success because we've always got the Shuttle there to take care of it." . . . I think Marshall, and also the people at headquarters, were much more naive [than the contractors] about what it was going to end up costing. After all, they didn't have the experience that the contractors had. So I think that both Headquarters and Marshall were honestly surprised by the events that then occurred, so it wasn't a case of duplicity, but rather of genuine ignorance, incompetence is too strong a word, but failings.[23]

But NASA had not really wanted to hear realistic cost estimates from the contractors. When managers such as George Low urged that "unless we [have] a very much less expensive telescope, we won't have any," the result can hardly be surprising.

The telescope program as approved was in effect a buy-in. It (and the shuttle program too) was initiated under circumstances very different from those surrounding Apollo. NASA's tight budgets of the 1970s bore little relation to those of the buoyant years of the early 1960s. To simplify somewhat, to 1969, NASA designed to get the job done; after 1969, NASA designed to cost. And as with the Space Shuttle in the early 1970s and the space station in the mid-1980s, the Space Telescope ran into political problems very early. The pressure to keep the telescope's costs down (largely because of sometimes strident criticism from the House subcommittee on appropriations) forced NASA to accept that the program's measure of success would be completion of the telescope's design and development on schedule and within the cost negotiated during the selling process. Yet the selling process had ultimately produced what would prove to be an unrealistic cost figure and overly optimistic schedules. The result was that there was no hope of producing a success within the terms set by Congress.

THE SHUTTLE APPROACH

One of the central points about the Space Telescope is that it was a creature of its parent government agency. The telescope program was susceptible to the shifting sands of NASA's changing fortunes and circumstances, in particular the other projects the agency was funding or hoping to fund. In the case of the Space Telescope, it was the Space Shuttle, NASA's key program for the 1970s and 1980s, that most influenced its development. The availability and capabilities of NASA's launch vehicles have always played a central role in space science. At the most basic level, without the means to get an experiment into space, there can be no space science, as the shuttle accident graphically demonstrated.

The Space Shuttle had been the telescope's planned launch vehicle since 1972, and once that decision was made, there was no worthwhile debate on other alternatives. The choice of the shuttle meant much more than simply imposing certain design constraints on the telescope (e.g., that it could be no longer than the orbiter's payload bay). Since the early 1960s, some of the astronomers who had advocated the building of the Large Space Telescope had argued that it would be unwise to spend so much on the telescope and then, through lack of repairs, have it fail after a few years in space. They had therefore pressed for it to be repairable and maintainable in orbit. The advent of the Space Shuttle had reinforced that drive and also led to plans for returning the telescope to earth. Indeed, the shuttle, at least on the surface, had to a considerable degree merged the interests of the astronomers and NASA – the shuttle provided the astronomers the capability to service the telescope in orbit, and for the agency the telescope provided a solid justification for, and added some much needed scientific legitimacy to, the shuttle. But there nevertheless had been much unease with the shuttle among astronomers. That unease, and sometimes hostility, as we saw in Chapter 5, helped to fuel the astronomers' anger at the exclusion of the telescope from NASA's budget in fiscal year 1977.

During the 1970s, NASA packaged the telescope and shuttle together and devised an engineering approach to the telescope that placed the Space Shuttle firmly in center stage. The argument ran that by exploiting the shuttle, the contractors could reduce the amount of testing that would be needed for the spacecraft and its components, as well as the number of spares required, thereby saving money

393

New route. (Copyright 1981 by Herblock in *The Washington Post*.)

in the design and development phase. That approach would increase the risk in the program. It probably would mean that there would be more failures of parts of the telescope in orbit than might have been the case if strenuous efforts had been made to produce more reliable systems and subsystems. The extra risk could, nevertheless, be offset by use of the shuttle to effect repairs in orbit or to bring the telescope back to earth. NASA, however, never pursued that approach with much conviction. Anyway, the selling of the telescope produced such a set of claims and expectations that by the time of the 1983 crisis, the risk of a major failure of the telescope had become politically unacceptable to NASA Headquarters.

But the shuttle approach had begun to be undermined even earlier, for in the late 1970s the costs of potential shuttle launches had risen rapidly. That cost growth came at a time when the budget for the telescope's design and development phase had been fixed. By the mid-1980s, NASA had decided that it made little sense to return the telescope to earth. A return to earth would involve technical risks, in terms of damage and possible contamination of the telescope, as well as the high cost of maintaining a facility to receive the telescope. There was also the risk that once down, it might never get aloft again. If grounded, the telescope might be compelled to compete with new space science missions for perhaps hundreds of millions of dollars worth of funding, and so might be kept earthbound for economic, if not technical, reasons. By planning to keep the telescope in orbit, NASA could exploit the telescope once again to help sell another program – not the shuttle this time, but the space station. In congressional hearings in 1985, for example, the agency touted the benefits of docking the telescope at a space station so that it could be refurbished and maintained by astronauts.

By 1986, NASA had calculated the cost of operating, maintaining, and refurbishing the telescope in orbit to be around $190 million each year. That was a much higher figure than those quoted in the late 1970s, which had been based on relatively few studies (now clearly known to have been deficient). In fact, the emphasis in Phase B had been on the cost of the telescope's design and development, and that was the ground over which NASA and the telescope's lobbyists, on one side, and Congress, on the other, had fought. Despite persistent questioning, NASA generally had shied away from discussing in detail the costs of operating the telescope in orbit, or the costs of maintaining and refurbishing the telescope in space, most likely because such things had never been done and the agency's figures could hardly be reliable. Nevertheless, during congressional testimony in 1977, NASA administrator Fletcher had cited a cost of $10–15 million each year to operate the telescope in orbit, and Noel Hinners had suggested a figure of around $10 million to bring the spacecraft to the ground and refurbish some major parts of the telescope.[24]

Even after allowing for inflation, those figures now seem hopelessly inadequate. In fact, in 1986, NASA estimated that the telescope would cost around $90 million per year to operate in orbit and would cost around $100 million each year for maintenance and refurbishment. The sizes of those figures raise fundamental questions. In particular, did the availability of the shuttle lead NASA to push too hard and too soon the approach of repairing and maintaining a large and complex spacecraft in orbit? If the Space Shuttle had been able to provide the cheap and regular access to low earth orbit that in the 1970s NASA asserted it would, then the approach of trading risks against costs would have made obvious sense. The shuttle, however, has not been able to live up to NASA's earlier claims; it is not, nor is it ever likely to be, the "space truck" that was promised in the earlier Space Telescope planning. The explosion of the *Challenger* was a clear demonstration of the remarkable frailty of the U.S. space program. In one accident, a quarter of the launch fleet was lost, and that caused a delay of at least two and a half years in major scientific, commercial, and military programs.

The use of the shuttle as the telescope's launch vehicle, together with the shuttle-related plans for maintenance and refurbishment, presented a mix of opportunities and problems. But it eventually led to an absurd situation. In the 1970s, NASA had justified the telescopes high cost in part by the long life promised for it by the shuttle. Yet for the cost of a few shuttle missions to maintain and repair the telescope, an entire second telescope could have been built.

WHITHER BIG SPACE SCIENCE?

The costs of operating, maintaining, and refurbishing the Space Telescope, as well as its cost overruns, made some astronomers critical of the telescope program. One reason was that NASA was driven to siphon money away from potential new space science programs to shore up the Space Telescope's budget. One mission to suffer was the Advanced X-Ray Astronomy Facility (AXAF), a large x-ray observatory planned for launch in the 1990s.[25] AXAF had been close to initiation since the early 1980s, but it appeared in the president's budget as a new start only in 1988.

The Space Telescope's difficulties, together with those of the shuttle program, raise the question whether or not it makes sense to plan to maintain and refurbish AXAF and the proposed infrared observatory, SIRTF, in the same manner as the Space Telescope.[26] In addition, how many observatories in space can the United States, even with international partners, afford to maintain? More pertinent, how

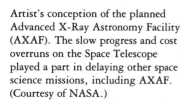

Artist's conception of the planned Advanced X-Ray Astronomy Facility (AXAF). The slow progress and cost overruns on the Space Telescope played a part in delaying other space science missions, including AXAF. (Courtesy of NASA.)

many space observatories will Congress and the White House be prepared to finance at a construction cost, if the Space Telescope is much of a guide, of over $1 billion each and an annual running cost of around $200 million each in an era of constraints on the federal budget? Also, are there more appropriate institutional arrangements than those employed for the building of the Space Telescope, new institutional ways of doing business that might improve the quality of the science to be produced and save money?

These questions highlight a larger issue: At what point does it become prohibitively expensive to expend the resources necessary to build and operate the instruments of Big Science, the point at which its gains become less cost-effective than the gains that can be achieved with a large number of small-scale projects? The trend in particle physics and astronomy is toward ever larger and more sophisticated scientific tools to probe smaller and more distant regions, and scientists have secured impressive results through the access provided by Big Science tools to previously inaccessible phenomena. NASA, as fundamentally a research and development organization, has shown a predilection for large-scale, complex projects. The agency's institutional interests have thus helped to promote the trend toward bigger and more complicated instruments. But does this development always make scientific sense? The eminent mathematical physicist Freeman Dyson has been an eloquent critic on this point. In his opinion, the Space Telescope "is a basket with too many eggs riding in it. It would have been much better for astronomy if we had had several one-meter space telescopes to test the instrumentation and see how the sky looks at one-tenth-of-a-second-of-arc resolution, instead of being stuck with a single one-shot 2.4 meter telescope for the rest of the century." Writing before the *Challenger* accident,

Dyson conceded that the telescope would be a splendid instrument that "would extend massively the boundaries of optical astronomy," which "can hardly fail to make big discoveries when it comes into service in the 1980s. But I have a sneaking fear that it may end up in the 1990s . . . a glorious technical success but scientifically twenty years behind the times."[27] Those miss, nevertheless, the political realities of the 1970s. NASA and the astronomers, as seen in Chapters 2 and 3, did toy with such an evolutionary program, but it faded from sight under the glares of a skeptical Congress and doubtful NASA managers. It did prove possible to sell the very large scale Space Telescope program because it contained so many different interests; a series of smaller-scale missions, such as the ASTRA program that we discussed in Chapter 2, could not be sold.

But Dyson's remarks raise another related and more general question: Should Big Science be allowed to grow at the expense of little science? Indeed, has Big Science, when it reaches the cost levels of such megaprojects as the Space Telescope and the proposed multibillion-dollar "Superconducting Super Collider," reached the limits of what society can afford, no matter how great the potential scientific rewards, and however hard its advocates might strain to justify it in terms of short-term social benefits such as new insights into energy production (in the middle and late 1970s) or industrial competitiveness (in the late 1980s)? The Office of Technology Assessment of the U.S. Congress studied that problem in 1982 in an extensive evaluation of the nation's space science program. It concluded that

large science projects are necessary to sustain scientific progress, but tend to crowd out smaller scale projects, and, given current budget constraints, they have been mounted less often than required to maintain space science teams. In the present situation of level overall funding divided among fewer, but generally more expensive activities, an increasingly heterogeneous space science community has been forced into a mode of divisive competition for available resources. The diversity of the community, set in the context of constrained funding, makes consensus on priorities set by means of broad-based peer review especially difficult to achieve. Thus, officials at NASA, whose responsibility it is to make these decisions, face growing difficulties. A good solution to this complex set of problems is not yet evident.[28]

The history of post–World War II science is certainly to a significant degree the history of Big Science. That history reveals that Big Science projects often are enormously demanding, both organizationally and technically. A number of Big Science projects have also proved to be beyond the talents, capabilities, and money of their advocates and builders, at least the talents, capabilities, and money defined to be required in the planning stages. Big Science projects have even had to be aborted after large sums of money had been spent on them. Examples here would be the ill-fated Mohole project to drill an unprecedentedly deep hole in the ocean floor, and "Isabelle," a high-energy physics accelerator that was never completed because of technical and budgetary problems.[29] Nor is it unknown for Big Science projects to require large injections of personnel and resources to get them back on track after a badly planned and executed start or after encountering major, unexpected problems. Examples are the attempts in the late 1940s to build the first nuclear reactor at the Brookhaven National Laboratory and the construction of the national radio astronomy facility at Green Bank, West Virginia.[30] The Space Telescope belongs in this category of Big Science projects, and with the telescope NASA both created — through its chosen program design and methods of managing the program — and confronted massive problems.

All of these Big Science enterprises demonstrate how essential it is to define the program well — scientifically, organizationally, managerially, and technically — from the outset. In the case of the telescope, the particular set of circumstances in which it was initiated and the adopted management system militated against this. At the least, the history of the Space Telescope surely argues that scientists, engineers, managers, legislators, and policymakers urgently need to consider new instruments of scientific leadership — new strategies and perhaps even new institutions — for the successful advance of science and high technology in future years.

Artist's conception of Isabelle (the main ring of which is shown in the upper section of the illustration), a high-energy physics accelerator that was never completed, although many millions of dollars had been spent on it. The lower part of the illustration shows the Alternating Gradient Synchrotron. (Courtesy of Brookhaven National Laboratory.)

Appendixes:
The Space Telescope

APPENDIX 1: INTRODUCTION

The appendixes gather together materials that the reader might wish to consult to better understand the narrative. The technical information on the Hubble Space Telescope and its associated systems is drawn from many sources and reflects the state of the system in early 1987. Readers seeking more detailed information on the telescope and its scientific objectives should consult the list of selected items.

APPENDIX 2: SOURCES

Oral history interviews

The following oral history interviews were taken during the course of research on the history of the Hubble Space Telescope.

Interviewee	Date	Interviewer
Adams, R.	01/09/86	JNT
Aucremanne, M. J.	11/21/83	RWS & AN
Bahcall, J. N.	11/03/83	PH
Bahcall, J. N.	12/20/83	PH
Bahcall, J. N.	03/22/84	PH
Bahcall, J. N.	03/28/84	PH
Bahcall, N.	03/27/85	RWS
Baum, W. A.	06/22/86	RWS
Belton, M.	10/11/84	JNT
Belton, M.	12/12/84	JNT
Belton, M.	12/13/84	JNT
Bensimon, M.	10/05/83	RWS
Bensimon, M.	12/05/83	RWS
Bless, R. C.	11/03/83	RWS
Bless, R. C.	02/21/84	RWS
Bless, R. C.	11/11/86	RWS
Boeshaar, G.	05/20/86	RWS
Boggess, A., III	04/20/84	RWS
Brown, R.	04/03/84	JNT
Brown, R.	06/07/84	JNT
Bulkin, B.	01/11/85	RWS
Burbidge, E. M.	11/29/84	RWS
Caldwell, J. J.	10/01/84	JNT
Carr, F.	03/07/84	RWS
Carr, F.	03/14/84	RWS
Carr, F.	05/29/86	RWS
Chapman, C. R.	10/10/84	JNT
Clark, J.	05/15/87	RWS
Code, A. D.	02/22/84	RWS
Danielson, E. G.	09/27/85	JNT
Davidsen, A.	01/29/84	RWS
Davis, M.	12/12/85	JNT
Disney, M.	05/01/84	RWS
Downey, J. A.	01/18/84	RWS
Doxsey, R.	07/22/87	RWS
Edmondson, F. K.	06/08/84	JNT & RWS
Elliott, J. L.	11/21/84	JNT
Emanuel, G.	12/02/85	RWS
Fastie, W.	06/04/86	RWS
Fastie, W.	06/06/86	RWS
Field, G.	03/10/86	RWS
Fordyce, D.	10/31/83	RWS & PH
Fredrick, L.	05/02/86	RWS
Giacconi, R.	01/25/84	RWS & PH
Goldberg, A.	05/02/85	RWS & JNT
Goldberg, A.	10/21/85	RWS
Groth, E.	03/15/84	PH

399

Guha, A. K.	07/26/83	RWS	Rosendhal, J.	10/01/84	JNT
Guha, A. K.	08/01/83	RWS	Rosendhal, J.	12/23/85	RWS
Hall, D.	05/03/84	RWS	Russell, J.	12/18/84	RWS
Harms, R.	05/26/87	RWS	Schreier, E.	07/06/87	RWS
Henry, R. C.	11/29/83	RWS & PH	Schroeder, D. J.	10/07/85	RWS
Hinners, N.	10/17/84	RWS & JNT	Sherrill, T.	09/25/84	JNT
Hinners, N.	11/16/84	RWS & JNT	Sherrill, T.	12/07/86	JNT
Hinners, N.	12/14/84	RWS & JNT	Simmons, F. P.	09/10/84	RWS & JNT
Hunten, D. M.	12/14/84	JNT	Simmons, F. P.	09/11/84	JNT
Keathley, W.	12/05/84	RWS	Sobieski, S. S.	02/08/85	RWS
Keller, S.	12/15/83	RWS & PH	Speer, F.	02/26/85	RWS
Keller, S.	02/11/85	RWS	Spitzer, L.	10/27/83	PH
Keller, W.	11/28/84	RWS	Spitzer, L.	03/07/84	PH
Keller, W.	12/09/85	RWS	Stockman, P.	02/15/85	RWS
King, I.	05/02/84	RWS	Stuhlinger, E.	04/02/84	JNT
Lane, A. L.	10/09/84	JNT	Stuhlinger, E.	04/05/84	JNT
Lane, A. L.	11/10/84	JNT	Teem, J.	09/26/84	RWS
Lasker, B.	12/08/83	RWS	Teem, J.	10/04/84	RWS
Laurance, R.	05/26/83	RWS	Tenerelli, D.	01/11/85	RWS
Leckrone, D.	08/14/84	RWS	Tenerelli, D.	12/19/85	RWS
Longair, M.	06/14/84	RWS	Tenerelli, D.	12/23/85	RWS
Lowrance, J.	03/28/85	RWS	Tenerelli, D.	12/27/85	JNT & RWS
Macchetto, D.	04/10/84	RWS	Tenerelli, D.	01/16/86	RWS
McCandless, B.	01/08/86	JNT	Tifft, W. G.	12/13/84	JNT
Meserve, K.	11/01/85	RWS & PH	van de Hulst, H. C.	05/27/83	RWS
Mitchell, J.	12/10/84	RWS	Weiler, E.	10/20/83	RWS
Nein, M.	06/05/84	RWS	Weiler, E.	02/11/86	RWS
Noah, D.	01/08/86	JNT	Weiler, E.	03/17/86	RWS
Norman, M.	09/06/84	RWS	Weiler, E.	03/09/87	RWS & JNT
Norman, M.	09/12/84	RWS	Welch, J.	08/20/84	RWS & PH
Norris, B.	04/19/84	PH	Welch, J.	02/12/87	RWS
O'Dell, C. R.	04/12/82	DHD	Westphal, J. A.	08/09/82	DHD
O'Dell, C. R.*	05/21/85	RWS	Westphal, J. A.	08/12/82	DHD
Odom, J.	02/26/85	RWS	Westphal, J. A.	09/14/82	DHD
Odom, J.	01/07/87	RWS	Westphal, J. A.	09/28/85	JNT
Olivier, J.	01/18/84	RWS	Westphal, J. A.	05/15/87	RWS
Olivier, J.	02/27/85	RWS	White, R. L.	04/06/84	RWS
Olivier, J.	01/08/87	RWS	Zedekar, R.	01/08/86	JNT
Pellerin, C. J.	08/01/83	RWS			
Reetz, A.	06/27/83	RWS			
Reetz, A.	09/21/83	RWS			
Rehnberg, J.	11/01/83	RWS			
Richards, E.	02/20/84	RWS			
Roman, N.	02/03/84	RWS			
Rose, J.	07/24/87	RWS			
Rosendhal, J.	09/12/84	JNT			

*These interviews are part of the collection of the Center for History of Physics, American Institute of Physics.

Key to interviewers: DHD, David H. DeVorkin; PH, Paul Hanle; AN, Allan Needell; RWS, Robert W. Smith; JNT, Joseph N. Tatarewicz.

The space telescope history project resource files

The Space Telescope History Project Resource Files were assembled from 1982 to 1987 by the staff of the Space Telescope History Project. They include documentary materials, including published and internal reports, meeting minutes, copies of correspondence, and so forth, collected from all levels and segments of the Hubble Space Telescope Project. Also included is a collection of approximately one thousand images on various media, including photographic prints, slides, and transparencies.

Further reading

Those seeking further information on the telescope and its scientific objectives may begin with the following selected list of publications.

Bahcall, J. N., and C. R. O'Dell, "The Space Telescope Observatory," in *Scientific Research with the Space Telescope*, NASA CP-2111, edited by M. S. Longair and J. W. Warner (Washington, D.C.: NASA, 1979), pp. 5–46.

Bahcall, John N., and Lyman Spitzer, Jr., "The Space Telescope," *Scientific American* (July 1982):40–51.

Belton, M. J. S., "Planetary Astronomy with the Space Telescope," in *Scientific Research with the Space Telescope*, NASA CP-2111, edited by M. S. Longair and J. W. Warner (Washington, D.C.: NASA, 1979), pp. 47–76.

Bless, R. C., and HSP Investigation Team, "The High Speed Photometer for the Space Telescope," in *The Space Telescope Observatory*, NASA CP-2244, edited by D. N. B. Hall (Washington, D.C.: NASA, 1982), pp. 106–13.

Boeing Company, Aerospace Group, "A System Study of a Manned Orbital Telescope," report no. D2-84042-1, October 1965.

Bok, Bart J., "The Promise of the Space Telescope," *Mercury* (May–June 1983):66–75.

Brandt, John C., and HRS Investigation Definition and Experiment Development Teams, "The High Resolution Spectrograph for the Space Telescope," in *The Space Telescope Observatory*, NASA CP-2244, edited by D. N. B. Hall (Washington, D.C.: NASA, 1982), pp. 76–105.

Caldwell, John, "Uranus Science with the Space Telescope," in *Uranus and the Outer Planets*, edited by G. Hunt (Cambridge University Press, 1982), pp. 259–74.

Cowens, M. W., M. M. Blouke, T. Fairchild, and J. A. Westphal, "Coronene and Liumogen as VUV Sensitive Coatings for SI CCD Images. A Comparison," *Applied Optics* 19(1980):3727–8.

Davies, J. K., "Space Telescope: Eye on the Universe," *Spaceflight* 24(1982):434–41.

Dougherty, H., A. Nakashima, J. Machnick, J. Henry, and K. Tompetrini, "Magnetic Control Systems for Space Telescope," *Journal of Astronautical Sciences* 30(1982):229–50.

Dougherty, H., C. Rodoni, K. Tompetrini, and A. Nakashima, "Space Telescope Pointing Control," in *Automatic Control in Space*, edited by P. T. L. M. van Woerkom (Oxford: Pergamon Press, 1982), pp. 15–24.

Doyle, Robert A. (ed.), *A Long Range Program in Space Astronomy*, NASA SP-213. (Washington, D.C.: NASA, 1969).

Duncombe, R. L., G. F. Benedict, P. D. Hemenway, W. H. Jefferys, and P. D. Shelus, "Astrometric Observations with the Space Telescope," in *The Space Telescope Observatory*, NASA CP-2244, edited by D. N. B. Hall (Washington, D.C.: NASA, 1982), pp. 114–20.

Dyson, Freeman J., "Science and Space," in *The First Twenty-Five Years in Space: A Symposium*, edited by Allan A. Needell (Washington, D.C.: Smithsonian Institution Press, 1983), pp. 90–106.

Elliot, J. L., "Direct Imaging of Extra-Solar Planets with Stationary Occultations Viewed by a Space Telescope," *Icarus* 35(1978):156–64.

Field, George B., "The Space Telescope," *Astronomy* (November 1976):6–15.

Friend, David, "Science Beyond the Stars," *Life* (December 1985):29–42.

Giacconi, Riccardo. "Science Operations with Space Telescope," in *The Space Telescope Observatory*, NASA CP-2244, edited by D. N. B. Hall (Washington, D.C.: NASA, 1982), pp. 1–15.

Hanle, Paul A., "Astronomers, Congress, and the Large Space Telescope," *Sky and Telescope* 69(1985):300–5.

Harms, R. J., and FOS Science and Engineering Team, "Astronomical Capabilities of the Faint Object Spectrograph on Space Telescope," in *The Space Telescope Observatory*, NASA CP-2244, edited by D. N. B. Hall (Washington, D.C.: NASA, 1982), pp. 55–75.

Hinners, N. W., "NASA In-House Astronomers – Verse and Converse," in *Scientific Research with the Space Telescope*, NASA CP-2111, edited by M. S. Longair and J. W. Warner (Washington, D.C.: NASA, 1979), pp. 321–7.

Hunt, Garry E., and Vivien Moore, "Space Telescope Studies of Jupiter and Saturn Simulated by Voyager Observations," in *Astronomy from Space, Advances in Space Research*, vol. 5, no. 3, edited by G. G. Fazio, J. A. M. Bleeker, P. A. J. de Korte, and J. J. Caldwell (Oxford: Pergamon Press, 1985), pp. 181–8.

Jefferys, W. H., "Astrometry with the Space Telescope," *Celestial Mechanics* 22(1980):175–81.

Jones, C. O., "Space Telescope Optics," *Optical Engineering* 18(1979):273–80.

Large Space Telescope – A New Tool for Science, CP-101. American Institute of Aeronautics and Astronautics, 1974.

Leckrone, D. S., "The Space Telescope Scientific Instruments," *Publications of the Astronomical Society of the Pacific* 92(1980):5–21.

Leckrone, D. S., "Spectroscopic Equipment for the Space Telescope," *Philosophical Transactions of the Royal Society of London* 307(1982):549–61.

Lightman, Alan P., "First Light: The Space Telescope," in *Infinite Vistas: New Tools for Astronomy*, edited by James Cornell and John Carr (New York: Charles Scribner's Sons, 1986), pp. 3–20.

Longair, M. S., "The Space Telescope and Its Opportunities," *Quarterly Journal of the Royal Astronomical Society* 20(1979):5–28.

Longair, M. S., "Reflections on the Space Telescope," in *The Space Telescope Observatory*, NASA CP-2244, edited by D. N. B. Hall (Washington, D.C.: NASA, 1982), pp. 121–34.

Longair, Malcolm, "The Scientific Challenges of Space Telescope," *Sky and Telescope* 69(1985):306–11.

McCarthy, D. J., and T. A. Facey, "The 2.4 Meter Space Telescope Program," *Perkin-Elmer Technical News* (September 1981):12–22.

Macchetto, F., "European Astronomy and the Space Telescope: The Space Telescope European Coordinating Facility," in *The Space Telescope Observatory*, NASA CP-2244, edited by D. N. B. Hall (Washington, D.C.: NASA, 1982), pp. 16–19.

Macchetto, F., and FOC Instrument Science Team, "The Faint Object Camera," in *The Space Telescope Observatory*, NASA CP-2244, edited by D. N. B. Hall (Washington, D.C.: NASA, 1982), pp. 40–54.

Macchetto, F., H. C. van de Hulst, S. di Serego Alighieri, and M. A. C. Perryman, *Faint Object Camera for the Space Telescope*, ESA SP-1028 (Paris: ESA, 1980).

McRoberts, Joseph J. *Space Telescope*, NASA EP-166 (Washington, D.C.: NASA, 1982).

Morrison, D. "Investigation of Small Solar System Objects with the Space Telescope," in *Scientific Research with the Space Telescope*, NASA

CP-2111, edited by M. S. Longair and J. W. Warner (Washington, D.C.: NASA, 1979), pp. 77–98.

Murphy, J. T., and C. R. Darwin, "Lower Payload Costs Through Refurbishment and Module Replacement," *Astronautics and Aeronautics* 11(May 1973):40–7.

Oberth, Hermann, *Die Rakete zu den Planetenraumen* (Muchen: R. Oldenburg, 1923), pp. 81–9.

O'Dell, C. R., "The Large Space Telescope Program," *Sky and Telescope* 44(1972):369–72.

O'Dell, C. R., "The Space Telescope Observatory," in *The Space Telescope Observatory,* NASA CP-2244, edited by D. N. B. Hall (Washington, D.C.: NASA, 1982), pp. 20–7.

Oemler, A., Jr., "Space Telescope Observations of Normal Galaxies," in *Scientific Research with the Space Telescope,* NASA CP-2111, edited by M. S. Longair and J. W. Warner (Washington, D.C.: NASA, 1979), pp. 165–80.

Pipher, J. L., S. P. Willner, and G. G. Fazio, "Infrared Astronomy on the Hubble Space Telescope," In *Astronomy from Space, Advances in Space Research,* vol. 5, no. 3, edited by G. G. Fazio, J. A. M. Bleeker, P. A. J. de Korte, and J. J. Caldwell (Oxford: Pergamon Press, 1985), pp. 173–9.

Roman, N. "Space Telescope: A Versatile New Instrument," *Celestial Mechanics* 22(1980): 165–74.

Schroeder, D. J., "Science at the Performance Limits of the Hubble Space Telescope," in *Astronomy from Space, Advances in Space Research,* vol. 5, no. 3, edited by G. G. Fazio, J. A. M. Bleeker, P. A. J. de Korte, and J. J. Caldwell (Oxford: Pergamon Press, 1985), pp. 157–67.

Smith, Robert W., and Joseph N. Tatarewicz, "Replacing a Technology: The Large Space Telescope and CCDs," *Proceedings of the Institute of Electrical and Electronic Engineers* 73(1985):1221–37.

Spitzer, Lyman, Jr., "The Beginnings and Future of Space Astronomy," *American Scientist* 50(1962):473–84.

Spitzer, Lyman, Jr., "History of the Space Telescope," *Quarterly Journal of the Royal Astronomical Society* 20(1979):29–36.

Tammann, G. A., "Cosmology with the Space Telescope," in *ESA/ESO Workshop on Astronomical Uses of the Space Telescope,* edited by F. Macchetto, F. Pacini, and M. Tarenghi (Paris: European Space Agency/European Southern Observatory, 1979), pp. 329–43.

Tinsley, B. M., "Galactic Evolution with the Space Telescope," in *Scientific Research with the Space Telescope,* NASA CP-2111, edited by M. S. Longair and J. W. Warner (Washington, D.C.: NASA, 1979), pp. 181–96.

Tucker, Wallace, and Karen Tucker, "Where the Stars Don't Twinkle: The Space Telescope," in *The Cosmic Inquirers* (Cambridge, Mass.: Harvard University Press, 1986), pp. 171–215.

U.S. General Accounting Office, "Space Telescope Project," PSAD-76-66, January 1976.

U.S. General Accounting Office, "Status and Issues Pertaining to the Proposed Development of the Space Telescope Project," PSAD-77-98, May 4, 1977.

U.S. General Accounting Office, "NASA Should Provide the Congress Complete Cost Information on the Space Telescope Program," PSAD-80-15, January 3, 1980.

Westphal, James A., and WF/PC Investigation Definition Team, "The Wide-Field/Planetary Camera," in *The Space Telescope Observatory,* NASA CP-2244, edited by D. N. B. Hall (Washington, D.C.: NASA, 1982), pp. 28–39.

Winter, Frank H., "Observations in Space, 1920s Style," *Griffith Observer* (June 1982):2–8.

Hubble Space Telescope Personnel History

	1977	1978	1979	1980	1981	1982	1983	1984	1985	1986	1987

ORGANIZATION POSITION

NASA Headquarters

Administrator — J. Fletcher / R. Frosch ---------- A. Lovelace ---------- (Acting) J. Beggs----------------------------W. Graham ---------(Acting) J. Fletcher

Assoc. Adm. for Space Sci. — N. Hinners A. Stofan---T. Mutch A. Stofan ----B. Edelson --L. Fisk
(Acting) (Acting)

Deputy AA for Sp. Sci. — A. Lovelace-A. Stofan -------------------------------S. Keller -------------------

Astrophysics Div. Director — B. Norris---------------------F. Martin ------------------------C. Pellerin ---

ST Program Mgr — W. Keller -----------D. Burrowbridge-------------F. Carr-A. Reetz (Acting) J. Welch---------------------------------- M. Bensimon

ST Program Engr. — A. Reetz --H. Estes------------------------

ST Program Sci. — N. Roman -------------------------------E. Weiler --

Marshall Space Flight Center

Center Director — W. Lucas ---J. Thompson

ST Project Mgr. — W. Keathley----------------------F. Speer --------------------------J. Odom --------------------J. Richardson

ST Deputy Proj. Mgr. — J. McCulloch---S. Wojtalik

ST Mgr., OTA — M. Rosenthal--------J. Richardson ----------------------M. Rosenthal

ST Mgr., SSM — E. L. Field-J. Harlow --

ST Project Scientist — C. R. O'Dell---R. Brown--------------A. Boggess

ST Chief Engr. — J. R. Olivier --

Goddard Space Flight Center

Center Director — R. Cooper ------------R. Smylie-T. Young -----------N. Hinners------------------------ (Acting)

ST Project Mgr. — G. Levin ------------------------G. Burdett -------------------F. Carr-----------------------------

ST Instrument Scientist — D. Leckrone---

ST Operations Scientist — R. Hobbs ------------------- (Vacant) A. Boggess----------------------

Perkin-Elmer Corp. Optical Division

ST Project Mgr. — D. McCarthy--------------------------K. Meserve ------D. Fordyce----------------W. Raiford ----------------------

ST Deputy Proj. Mgr. — C. Bryant---R. Jones -------P. Brickmeier (OTA)-------------------------------- L. Brisco (FGS)-------------------------

Lockheed Missiles & Space Company

ST Program Mgr. — W. Wright---B. Bulkin -------------------------------------P. A. Coffman

ST Program Dep. Mgr. — M. W. Hunter----B. Bulkin --------------------------T. Harvey-----------M. A. Stuart-P. A. Coffman

Note: All dates are approximate.
Certain position titles were changed, created, or abolished as a result of reorganizations.

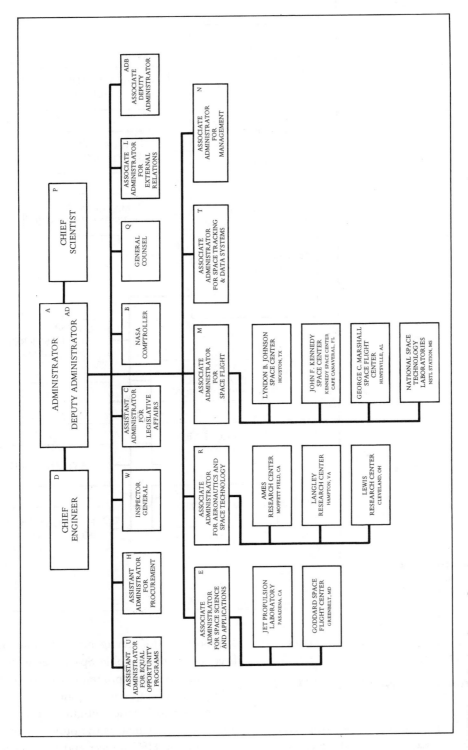

NASA organization chart, 1983.

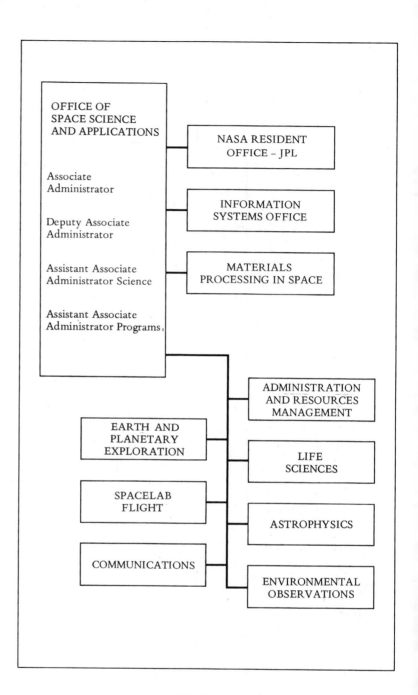

OFFICE OF
SPACE SCIENCE
AND APPLICATIONS

Associate
Administrator

Deputy Associate
Administrator

Assistant Associate
Administrator Science

Assistant Associate
Administrator Programs

NASA RESIDENT
OFFICE – JPL

INFORMATION
SYSTEMS OFFICE

MATERIALS
PROCESSING IN SPACE

EARTH AND
PLANETARY
EXPLORATION

SPACELAB
FLIGHT

COMMUNICATIONS

ADMINISTRATION
AND RESOURCES
MANAGEMENT

LIFE
SCIENCES

ASTROPHYSICS

ENVIRONMENTAL
OBSERVATIONS

NASA Office of Space Science and Applications organization chart, 1981. (Courtesy Stansbury, Ronsaville, Wood Inc.)

APPENDIX 4: MAJOR CONTRACTORS
AND INSTITUTIONS

Major institutions involved in the HST (as of 1977)

The Office of Space Science and Applications (OSSA), NASA Headquarters, Washington, D.C., is responsible for the agency-wide planning and direction of the Hubble Space Telescope (HST) program. The HST program scientist is responsible for the overall science policy.

The Marshall Space Flight Center (MSFC), Huntsville, Alabama, is the project management center for the HST project and is responsible for meeting cost, schedule, and technical performance goals. The HST project manager heads a Project Office that is responsible for directing all NASA and contractors' efforts, for establishing and maintaining effective project management, and for preparing and maintaining the detailed technical specifications for all elements of the project.

The Goddard Space Flight Center (GSFC), Greenbelt, Maryland, is responsible to the HST project for the scientific instruments (SIs), the SI control and data handling (SI C&DH) subsystem, the total HST ground system, the Space Telescope Science Institute, and the conduct of mission and science operations, including operations planning.

The Johnson Space Center (JSC), Houston, Texas, is the lead center for the Space Shuttle, which will be used to place the HST into orbit and, during maintenance missions, to retrieve, repair and refurbish, and reboost. JSC is responsible for defining and establishing all crew system design and training requirements.

The Kennedy Space Center (KSC), Cape Canaveral, Florida, is responsible for launch activities.

The Lockheed Missiles and Space Company, Inc. (LMSC), Sunnyvale, California, is the associate contractor for the Support Systems Module (SSM). The contract includes design, development, fabrication, assembly, and verification of the SSM, integration of systems engineering and analysis for the overall HST, and support to NASA for planning and conducting ground, flight, and orbital operations.

The Perkin-Elmer Corporation (P-E), Danbury, Connecticut, is the associate contractor for the Optical Telescope Assembly (OTA). The OTA contract includes design, development, fabrication, assembly, and verification of the OTA and support of HST integration and development operations.

The European Space Agency (ESA) provides the solar arrays (SA), the Faint Object Camera (FOC), as well as personnel to participate in science operations activities.

HST investigation definition teams as of 1977

Astrometry science team

W. H. Jefferys, University of Texas, principal investigator
G. F. Benedict, University of Texas
R. L. Duncombe, University of Texas
O. G. Franz, Lowell Observatory
L. W. Fredrick, University of Virginia
P. D. Hemenway, University of Texas
P. J. Shelus, University of Texas
W. F. Van Altena, Yale University

Data and operations team leader

Edward Groth, Princeton University

Faint object camera instrument science team

H. C. van de Hulst, Sterrewacht, Leiden, chairman
R. Albrecht, University Observatory, Vienna
C. Barbieri, Instituto di Astronomia, Padua
A. Boksenberg, University College, London
P. Crane, European Southern Observatory, München
J. M. Deharveng, Laboratoire d'Astronomie Spatiale, Marseille
M. J. Disney, University College, Cardiff
T. M. Kamperman, Sterrekundig Instituut, Utrecht
I. King, University of California, Berkeley
C. D. Mackay, Institute of Astronomy, Cambridge
F. Macchetto, European Space Agency, Noordwijk
G. Weigelt, Physikalisches Institut, Erlangen-Nürnberg
R. N. Wilson, European Southern Observatory, München

Faint object spectrograph

Richard Harms, University of California, San Diego, principal investigator

The Space Telescope Science Working Group photographed during a meeting in January 1985 at the Lockheed Missiles and Space Company in Sunnyvale, California. Front row, left to right: Richard Harms, James Westphal, David Leckrone, Edward Groth, Robert Brown, William Fastie, David Lambert, and Duccio Macchetto. Back row, left to right: John Caldwell, Edward Weiler, John Brandt, Robert Bless, William Jefferys, C. Robert O'Dell, Daniel Schroeder, Peter Jacobsen, and Garth Illingworth. Members absent were John Bahcall, Philippe Crane, Riccardo Giacconi, Ivan King, and Malcolm Longair. (Courtesy of Robert A. Brown.)

R. Angel, University of Arizona

Frank Bartko, Martin-Marietta Corp.

Edward Beaver, University of California, San Diego

E. M. Burbidge, University of California, San Diego

A. Davidsen, The Johns Hopkins University

Holland Ford, University of California, Los Angeles

Bruce Morgan, University of California, Los Angeles

High resolution spectrograph

John Brandt, NASA Goddard Space Flight Center, principal investigator

E. A. Beaver, University of California, San Diego

A. Boggess, NASA Goddard Space Flight Center

S. R. Heap, NASA Goddard Space Flight Center

J. P. Hutchings, Dominion Astrophysical Observatory, Victoria, British Columbia

M. A. Jura, University of California, Los Angeles

J. L. Linsky, University of Colorado, Joint Institute for Laboratory Astrophysics

S. P. Maran, NASA Goddard Space Flight Center

B. D. Savage, University of Wisconsin

A. M. Smith, NASA Goddard Space Flight Center

L. M. Trafton, University of Texas

R. J. Weymann, University of Arizona

High speed photometer/polarimeter

R. C. Bless, University of Wisconsin, principal investigator

G. W. Van Citters, University of Texas

A. D. Code, University of Wisconsin

J. L. Elliot, Cornell University

E. L. Robinson, University of Texas

Interdisciplinary scientists

J. N. Bahcall, Institute for Advanced Study, Princeton

J. Caldwell, University of Calgary

D. L. Lambert, University of Texas

M. Longair, Cambridge University

Telescope scientists

William Fastie, The Johns Hopkins University

Daniel Schroeder, Beloit College

Wide field/planetary camera

James Westphal, California Institute of Technology, principal investigator

W. A. Baum, Lowell Observatory

A. D. Code, University of Wisconsin

D. G. Currie, University of Maryland

G. E. Danielson, California Institute of Technology

J. E. Gunn, California Institute of Technology

J. A. Kristian, Carnegie Institution of Washington (D.C.)

C. R. Lynds, Kitt Peak National Observatory

P. K. Seidelmann, U.S. Naval Observatory

B. A. Smith, University of Arizona

HST major subcontractors

Actuator Control Electronics	Perkin-Elmer (P-E)
Aft Latch, Solar Array	Lockheed Missiles and Space Co. (LMSC)
Antenna Pointing System	Sperry
Battery	Eagle Pitcher/ General Electric
Charge Current Controller	LMSC
Circulator Switch	Electromagnetics
Coarse Sun Sensor	LMSC
Computer	Rockwell Autonetics
Data Interface Unit	LMSC
Data Management Unit	LMSC
Deployment Control Electronics	European Space Agency (ESA)
Dish and Feed for High Gain Antenna	General Electric
Electric Power/Thermal Control Electronics	P-E

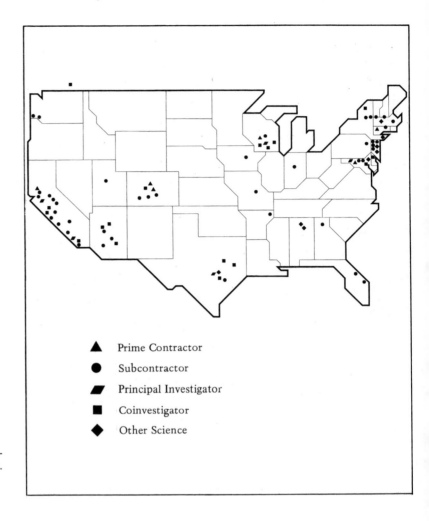

United States geographical distribution of Space Telescope participants. (Courtesy of Stansbury, Ronsaville, Wood Inc.)

Fixed Head Star Tracker Light Shade	Bendix	Image Dissector Camera Assembly	P-E
Faint Object Camera	Dornier	Instrument Control Unit	LMSC
Faint Object Spectrograph	Martin-Marietta	Interconnect Cables	LMSC/P-E et al.
Fine Guidance Electronics	Harris	Latch, Aperture Door	LMSC
Fine Guidance Sensor	P-E	Latch, High Gain Antenna	LMSC
Fixed Head Star Tracker	Ball Aerospace/ Bendix	Low Gain Antenna	LMSC
		MA Transponder	Motorola
Focal Plane Assembly	P-E	Magnetic Torquer	Ithaco/Bendix
Forward Latch, Solar Array	LMSC	Magnetic Sensing System	Schoenstadt/ Bendix
High Resolution Spectrograph	Ball Aerospace		
		Mechanism Control Unit	LMSC
High Speed Photometer	University of Wisconsin	Metal Matrix Mast	LMSC
		Multilayer Insulation	LMSC/P-E
Hinge, Aperture Door	LMSC	Off Load Device	ESA
Hinge, High Gain Antenna	LMSC	Optical Telescope Assembly	P-E

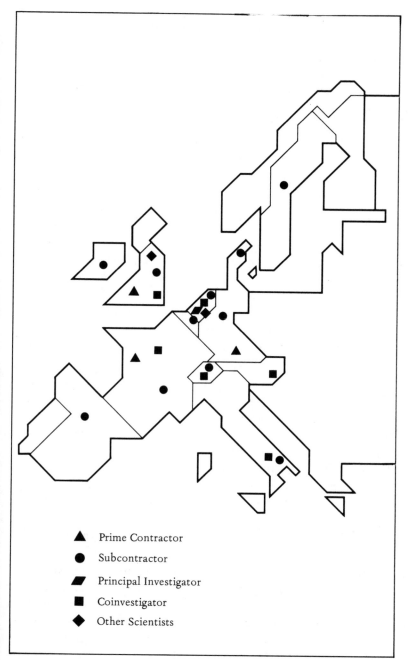

▲ Prime Contractor

● Subcontractor

◢ Principal Investigator

■ Coinvestigator

◆ Other Scientists

European geographical distribution of Space Telescope participants. (Courtesy of Stansbury, Ronsaville, Wood Inc.)

Optical Control Electronics	P-E	Power Control Unit	LMSC
Oscillator	Frequency Electronics	Power Distribution Unit	LMSC
		Primary Deployment Mechanism	ESA
Photomultiplier Tube Electronics	P-E		
		Primary Mirror Assembly	P-E
Pointing Safemode Electronics Assembly	Bendix	RF Multiplexer	Wavecom
		RF Switch	Transco

RF Transfer Switch	Transco
Rate Gyro Assembly	Bendix
Reaction Wheel Assembly	Sperry
Retrieval Mode Assembly	Northrop/ Bendix
Rotary Drive	Schaeffer
SAD Adapter	ESA
SI C&DH	Fairchild/IBM
SSA Transmitter	Cubic
Science/Engineering Tape Recorder	Odetics
Secondary Deployment Mechanism	ESA
Secondary Mirror Assembly	P-E
Sensor Electronics Assembly	P-E
Solar Array Blanket	ESA
Solar Array Drive	ESA
Solar Array Drive Electronics	ESA
Star Selector Servo	BEI Motion Systems Company
Temperature Sensor	LMSC/P-E
Thermostat/Heater	LMSC/P-E
Umbilical Drive Unit	Sperry
Waveguide	LMSC
Wide Field Camera	Jet Propulsion Laboratory

APPENDIX 5: THE HUBBLE SPACE TELESCOPE

The HST is basically a high-quality optical telescope inside a spacecraft designed to operate semi-autonomously in low earth orbit. The **Optical Telescope Assembly (OTA)** contains a 2.4-meter $f/24$ Ritchey-Chrétien Cassegrain telescope, support structures, baffling to reduce stray light, a fine-guidance system, and attachments for scientific instruments. Its purpose is to deliver high-quality images to the instruments. The OTA is surrounded by the spacecraft, called the **Support Systems Module (SSM)**. The SSM provides structural support, thermal control, electrical power (from the solar arrays), communications, data handling, and pointing control for the telescope.

Support systems module

The SSM is the means by which all of the HST hardware is supported. Structurally, the SSM consists of a **light shield** onto which is attached the **aperture door**, a **forward shell** supporting two high-gain antennas and two solar arrays, a toroidal **SSM equipment section** containing electronics and reaction wheels for pointing the spacecraft, and an **aft shroud** enclosing the scientific instruments.

When closed and latched, the *aperture door* prevents contaminants from entering the telescope. It can be closed in one minute from its fully open position of 105 degrees, and it will automatically close to prevent direct sunlight from entering the telescope if the sun comes within 35 degrees of the telescope's main axis.

The *light shield* contains internal baffles to suppress stray light, latches for the solar arrays and high-gain antennas in their stowed positions, coarse sun sensors, magnetometers, and a low-gain antenna.

The *forward shell* surrounds part of the OTA. It is the main attachment point for the solar-array wings, the high-gain antennas, and four magnetic torquers. It also has one remote manipulator-system grapple fixture for deployment/retrieval and two forward trunnions for latching the spacecraft in the Space Shuttle orbiter payload bay.

The *SSM equipment section* is divided into ten equipment bays containing the vast majority of the control, communications, power, and data-management electronics. Most of this equipment is in the form of "orbital replaceable units" (ORUs) and can be replaced by a space-suited astronaut by opening hinged doors or by removing panels.

The *instrumentation and communications subsystem* connects the two low-gain antennas and the two high-gain antennas to the rest of the HST systems via the data-management subsystem (DMS). The multiple-access transponders can communicate with the Tracking and Data Relay Satellite (TDRS) through either the high-gain or low-gain antennas. Because of the very high data rates (1.024 Mbps), scientific data can be sent only from the single-access transmitter through the two high-gain antennas. When TDRS is unavailable, the data will be stored on the science tape recorders for later transmission.

The *aft shroud* surrounds the scientific instruments and focal-plane structure and contains the access doors and equipment such as Extra Vehicular Activity handholds and lights to support replacement of instruments by astronauts. Facility for a continuous dry-nitrogen purge of the instruments (from the time they are installed at Lockheed until the orbiter payload bay doors are closed before launch) is part of the aft shroud. The end of the aft shroud is the **aft bulkhead,** which has three pins for docking the HST to the flight-support structure in the orbiter payload bay. The aft bulkhead also has quick-disconnect fittings for the dry nitrogen, electrical umbilical connectors, vents, a low-gain antenna, and a docking target for retrieval by the orbiter.

Optical telescope assembly

The OTA supplies high-quality images to the focal plane, where pickoff mirrors for the scientific instruments and Fine Guidance Sensors (FGSs) intercept the light.

The *primary mirror assembly* contains the 2.4-meter (94-inch) primary mirror inside the main ring. The **main baffle** is attached to the main ring and suppresses stray light. The slightly conical **central baffle** surrounds and extends from the central hole in the primary mirror. Twenty-four **figure-control actuators** can react against the **reaction-plate assembly** to adjust the figure of the mirror in orbit if required.

The *primary mirror* is an ultralight sandwich construction of Corning ultra-low-expansion glass. The thin front and rear faceplates were fused to an egg-crate-construction inner core. The mirror was then ground and polished to a precise figure and vacuum-coated with aluminum; then a thin coating of magnesium fluoride was evaporated onto the surface to enhance ultraviolet reflectivity.

The *secondary mirror* is a 0.31-meter (12-inch) convex Zerodur glass mirror, coated with aluminum and magnesium fluoride like the primary mirror. The secondary mirror focuses the light reflected from the primary mirror and sends it through the central baffle and central hole in the primary mirror to the focal plane. The secondary mirror is mounted inside a mounting ring and has its own baffling and six control actuators for fine adjustments and focusing in orbit.

The *metering truss structure* is a cylindrical lattice of graphite-epoxy tubes that keeps the primary and secondary mirrors in precise alignment. The metering truss assembly keeps the two mirrors separated within 118 microinches, their centers aligned within 394 microinches, and parallel within 2 arc seconds as the HST goes from direct sun to shade every forty-five minutes.

The *focal-plane structure* (FPS) is also made of graphite epoxy. Assisted by heaters, the FPS maintains the precise alignment of the scientific instruments and FGSs with respect to the light path through the OTA. The focal-plane structure attaches to the primary-mirror main ring and is enclosed by the aft shroud.

Alignment of the scientific instruments (SIs) and FGSs with respect to the focal plane is crucial to the scientific performance of the telescope. The FPS must allow the SIs and FGSs to be installed and removed in precise alignment by space-suited astronauts. Guide rails, registration fittings, and retention fittings (latches) make this possible. The FPS also supports three fixed-head star trackers for coarse pointing and three rate-sensor units for perceiving the motion of the telescope about its center of mass.

The *OTA equipment section* is a 150-degree toroidal structure that does for the OTA what the SSM equipment section does for the SSM. The nine equipment bays house ORUs containing most of the electronics for the OTA. A data-interface unit (DIU) connects the electrical systems of the OTA to those of the SSM. The electrical-power/thermal-control electronics control all electrical power to the OTA units, as well as heaters located throughout the OTA. Control of the figure of the primary mirror and control of the fine tuning of the secondary mirror are under the control of the actuator-control electronics. The fine-guidance electronics receive data from the FGSs and transmit those data via the data-interface unit to the pointing-control system in the SSM.

HST optical system and focal plane

The HST OTA produces images at a focal plane that lies just behind the 94-inch primary mirror. The OTA FPS provides mounting for four axial and four radial instruments. Three of the radial slots are occupied by the FGSs, leaving one slot for a radial camera. In the initial complement of instruments, this slot will be occupied by the Wide Field/Planetary Camera (WF/PC). In the initial complement, the axial slots will be occupied by a Faint Object Camera, two spectrographs, and a photometer. Only the radial camera intercepts light from the on-axis beam. The FGSs receive light from three "pickles" at the edge of the focal plane, 10–14 arc minutes off-axis. The axial scientific instruments pick off their light from intermediate locations.

The $f/24$ Ritchey-Chrétien optics of the HST provide diffraction-limited images from 1,150 angstroms to 1 millimeter. Each of the SIs is able to use a portion of that spectral range to advantage. Future second-generation instruments will take advantage of the near-infrared region.

Scientific instruments

Wide field/planetary camera

The WF/PC is a single SI that can be operated in two different modes. Functioning as the **Wide Field Camera** at $f/12.9$, the WF/PC provides a field of view 2.57 arc minutes square, for imaging extended faint objects, such as galaxies, clusters of galaxies, galactic nebulae, and so forth. As the **Planetary Camera** at $f/30$, the WF/PC provides a field of view 77 arc seconds square, to cover bright extended objects, such as the planets, with short exposures.

The central beam from the OTA at the focal plane is intercepted by the flat pickoff mirror and sent through a shutter and a choice of selectable filter wheel, polarizer, or dispersing element. The light then reflects off one of two (WF or PC) four-faceted pyramids. The pyramid selected precisely divides the image into four segments, each of which is then relayed and focused via a Cassegrain Ritchey-Chrétien repeater onto an 800- × 800-pixel Charge Coupled Device (CCD). The four images can then be assembled (on the ground) into one 1,600- × 1,600-pixel image.

The CCDs are cooled by an external radiator to −95 °C to reduce the level of background electrical activity in the detectors. They are coated with an organic phosphor, coronene, in order to enhance their response in the ultraviolet. This gives the WF/PC the widest spectral range of all the instruments (1,150 Å to 1.1 μm) and a very wide dynamic range (visual magnitude 8–28).

In 1984, an effect known as Quantum Efficiency Hysteresis (QEH) was found to present a serious problem by leaving residual images in the CCD arrays. The WF/PC was modified to include a "light pipe," consisting of optics and baffles to allow periodic back-illumination of the CCDs with ultraviolet light from the sun to remove the residual images.

The WF/PC is expected to be the most-used instrument in the telescope. In addition to its use as a primary instrument, it frequently will be used in parallel with another instrument to obtain wide-field images of the surrounding region.

Faint object camera

Provided by the European Space Agency, the Faint Object Camera (FOC) complements the WF/PC. The spectral responses of the two cameras are comparable in the middle of the range, with the FOC more sensitive in the ultraviolet (UV), and the WF/PC more sensitive in the red. With a smaller field of view, the FOC was designed to use the full optical performance of the OTA, recording the faintest objects possible at the highest angular resolution. The FOC contains two independent optical systems, operating at $f/96$ and $f/48$, as well as a variety of filters and special optical components.

Each optical camera system in the FOC focuses the incoming light onto a three-stage, magnetically focused image-intensifier tube. The output of this tube is scanned by a high-gain, high-resolution, electrostatically focused television tube.

The *f/96* system contains a coronagraphic occulting disk that can be centered on a bright object to suppress the light. The FOC is expected to be able to image a faint object 1 arc second away from another object with a magnitude difference of 16.7.

The *f/48* system contains a relay mirror and grating for recording spectra of extended objects. With a 20- × 0.1-arc-second square slit, a spatial resolution of 0.1 arc second, with spectral resolution for bright objects comparable to that of the FOS, can be obtained.

Faint object spectrograph

The Faint Object Spectrograph (FOS) can obtain moderate- to low-resolution spectra in the visible and ultraviolet of very faint objects. In addition, it can perform spectropolarimetry, as well as time-resolved spectroscopy.

Two identical optical paths conduct the light from the entrance port to the detectors via the optional polarizing assembly, a grazing incidence mirror, an order-blocking filter, and carousel-selectable gratings.

The wide spectral response of the FOS is achieved by using two single-stage, 512-channel, photon-counting digicon tubes, one optimized for blue response and one for red response by the choice of faceplate and photocathode.

Goddard high resolution spectrograph

The Goddard High Resolution Spectrograph (GHRS) is a photon-counting ultraviolet spectrograph with resolution comparable to that of the largest ground-based coudé instruments.

The GHRS is the only HST instrument designed to operate solely in the ultraviolet. It is deliberately designed to be insensitive to visible light, to allow observation of faint ultraviolet emissions from stars that produce intense visible flux.

The GHRS presents two entrance slits to the OTA focal plane. The larger slit (2 arc seconds) is used for target acquisition, and the smaller (0.25 arc second) is used for actual observations. The slits are selected by orienting the entire telescope. The incoming light is reflected and collimated by a mirror, dispersed by a carousel-selectable grating, and focused onto one of two single-stage, 512-channel, photon-counting digicon tubes, each optimized for a different portion of the ultraviolet range of the spectrum.

For time-resolved spectroscopy, the GHRS can take a single frame in as little as 50 milliseconds and reset for another frame in 2 milliseconds.

High speed photometer

The High Speed Photometer (HSP) is by deliberate design the simplest of all the HST instruments and contains no moving parts. The pointing accuracy of the HST itself is used to select one of approximately 100 filter/aperture combinations. The HSP can resolve fluctuations in brightness separated in time by 10 microseconds with a photometric accuracy of 0.2 percent. Light enters the instrument through a focal-plane filter/entrance-aperture assembly containing a variety of filters, apertures, and polarizing filters. The detectors used are four magnetically focused image dissectors for the visible and ultraviolet and one red-sensitive photomultiplier tube.

Fine guidance sensors

The HST contains four radial instrument bays located just behind the focal plane, occupied by the WF/PC and three FGSs. Whereas the WF/PC pickoff mirror protrudes into the center of the focal plane, each FGS can receive light from a 90-degree annulus ("pickle") at the edge of the field, giving a field of view 69 arc minutes square. Two FGSs are required for pointing the HST, and the third is available for astrometry.

Light from any star within the FGS field of view

can be directed into its pair of interferometers through a series of rotating mirrors. The encoder readings of the mirrors provide the relative position of the star in the field of view, and the interferometers generate a fine error signal. The FGS can obtain positional information on objects that are either stationary or moving in the field, or it can allow the object to track across the field of view for calibration.

The FGS, using various filters, can obtain positions for objects that range in brightness from visual magnitude 4 to 18, with a relative positional accuracy of 0.003 arc second.

Light from the OTA focal-plane "pickle" is sent via a pickoff mirror into the FGS collimator. The collimator magnifies the apparent scale of the field of view by 57.25, making tiny displacements large in the FGS. The light then passes through two star selectors in turn. Each star selector is a high-precision servomotor whose hollow shaft contains either mirrors and lenses (first) or a double-rhomb prism (second). By positioning each shaft, the 5-× 5-arc-second instantaneous field of view can be placed anywhere in the 69-arc-minute square "pickle" to select a particular star.

A combination beam-splitter/polarizer sends the x or y component of the object's light independently to one of two Koester prisms. Each Koester prism then further divides its beam, sending each component to a photomultiplier tube. For each component (x or y), the difference in intensities of the signals arriving at the pair of photomultiplier tubes is an indication of the degree to which the star is off axis. This error signal is then used by the pointing-control system to steer the spacecraft or is analyzed for astrometry.

Space support equipment

The HST is the first major astronomical spacecraft designed to be a semipermanent facility. As such, with an expected lifetime of fifteen years, it is designed to be serviced by space-suited astronauts using the space transportation system (Space Shuttle) and eventually the space station. Special provisions are being made in the Space Shuttle orbiter. The special racks and carriers for replacement scientific instruments, orbital replaceable units (ORUs), and tools are all grouped under the heading of **space support equipment (SSE)**.

The *SSE maintenance platform* is a specially modified version of the "Multimission Modular Spacecraft Flight Support Structure." It latches the HST at three points on the aft bulkhead, provides electrical power and monitoring umbilical connections, and allows the entire HST to be rolled and tilted into positions convenient for astronaut work. The SSE maintenance platform is also used to attach the HST to the orbiter for the periodic reboost mission to correct for the decay of the HST orbit.

The *orbital replaceable unit (ORU) carrier* is a modified Spacelab pallet. It can carry various combinations of SIs, FGSs, ORUs, tools, and support equipment. It also provides various crew aids, such as tether attachments, work lights, and so forth.

APPENDIX 6: GROUND SYSTEM — COMMAND AND DATA FLOW

The overall HST ground system consists of all the elements needed to operate a national facility in orbit for scientific research. It must integrate all of the steps from the initial proposal by an independent scientist or team to the release of the acquired data.

In simplest form, the procedure for a typical scientific observation is as follows. An investigator or team submits a proposal to the Space Telescope Science Institute. The proposal contains the objects to be observed, the HST scientific instruments and special modes and filters to be used, and a scientific justification. The proposal is reviewed for scientific merit and technical feasibility and is checked to see if it duplicates other proposals or already accomplished observations. If selected, the individual observations from the proposal are interleaved with those from other proposals into an overall observing schedule, taking into account the overall status of the HST, the availability of schedule time on the communications networks, and so forth. That schedule is then translated into sequences of individual commands, interleaved with commands required for engineering and housekeeping requirements. The command sequences are then transmitted from the Goddard Space Flight Center via domestic communications satellite to the Ground Station at White Sands, New Mexico. From there, the signal is transmitted to the HST via the Tracking and Data Relay Satellite (TDRS). Computers on board the HST store and then execute the commands and send the data back to Goddard via the reverse path. The investigator then has exclusive rights to those data for a period of one year, after which the data are made available to others.

APPENDIX 7: HST MISSION OPERATIONS GROUND SYSTEM

All of the elements required to routinely operate the HST are organized into the Mission Operations Ground System. In general, scientific matters are the responsibility of the **Space Telescope Science Institute (STScI)** located on The Johns Hopkins University Homewood campus, Baltimore, Maryland, and operated for NASA by the Association of Universities for Research in Astronomy. Engineering and operational activities are the responsibility of the **Space Telescope Operations Control Center (STOCC)** and other support elements located at the NASA Goddard Space Flight Center in Greenbelt, Maryland.

The STOCC is "mission control" for the HST. It is there that controllers will issue commands to and receive data from the HST. Controllers at the STOCC monitor the overall health and safety of the spacecraft, analyze engineering data, and schedule observations, communications, maneuvers, housekeeping, and maintenance. They maintain the archive of engineering data and sep-

arate out the science data for transmission to the STScI. Real-time control and continuous monitoring of the spacecraft take place in the **Payload Operations Control Center (POCC)**. Science data are "captured" in the **Data Capture Facility (DCF)**, where the information representing the results of the science observations is stripped out of the data stream coming into the STOCC from the HST via the NASA communications network. Those data are then sent to the STScI.

Also located in the STOCC is the **Science Support Center (SSC)**. In the SSC, representatives of the STScI and the STOCC work together to translate the observing schedules generated at the STScI into detailed overall control schedules for the HST. Incoming science data ("quick look") are monitored for quality, the status of the HST scientific instruments is monitored and investigated, and the performance of the HST in acquiring and tracking targets is monitored. For special observations that require real-time control, investigators will work with observational astronomers in the SSC during the observation.

Operating the HST requires other services that are provided to all satellites, including the Space Shuttle, by Goddard. These include scheduling the NASA Space Flight Tracking and Data Network, scheduling coverage on the TDRS system, analyzing the elements of the HST orbit and predicting the orbit's future behavior so that communications links and observations can be scheduled, and so forth.

The STScI serves as the primary point of contact with the HST for the scientific community. The STScI establishes policy, solicits and selects proposals for observations, plans scientific observations, and maintains the archive for HST data. Similar activities are performed for European astronomers by the European Coordinating Facility (ECF). STScI personnel staff the SSC and other positions at the STOCC.

The STScI also develops and maintains the data bases of information on the scientific instruments, including data required for calibration of observations, long-term trends in the health of the SIs, and information needed for planning the periodic maintenance and refurbishment of the HST and its instruments.

Astronomers and other specialists at the STScI provide detailed information and support to the "General Observers" and "Archival Researchers" concerning the abilities and limits of the scientific instruments, planning observations, and interpreting data. The STScI develops and maintains, and in some cases distributes, calibration and analysis software that researchers can use at their home institutions.

Short essay on sources

The common belief that we gain "historical perspective" with increasing distance seems to me to utterly to misrepresent the actual situation. What we obtain is merely confidence in generalizations which we could never dare make if we had access to the real wealth of contemporary evidence.

Otto Neugebauer, 1957

As this is written, the Space Telescope program is far from complete. Analyzing the early stages of the telescope's history therefore presents challenges different from those historians usually face. Although there are the usual difficulties of finding, selecting, and interpreting published and unpublished sources, these are complicated by the fact that many of the unpublished sources are "active" and have not yet been retired to an archive. In consequence, I have spent a considerable amount of time acting as my own archivist. However, the complications have their pluses as well as their minuses, because of the chance to study materials that might be destroyed shortly or thrown in the trash can. Also, by studying a program as it is developing, one has the opportunity to take oral history interviews with the historical actors when some of the events they are describing are fresh in memory. Such interviews also help to correct for some of the problems historians of modern science face in that so many communications are now made by telephone and by electronic mail instead of by letter. In addition, Joe Tatarewicz and I have been permitted to observe many meetings on the Space Telescope.

It is, however, on the examination of documents that a study of this kind must be grounded. And as traditional forms of documentation such as letters have become less central to communication between scientists and engineers, newer types of materials have become key to documenting the history of science and technology.

In this book I have employed three main types of sources: first, published material (technical papers, congressional hearings, and so on); second, unpublished correspondence, memoranda, diaries, and so on; third, interviews.

PUBLISHED MATERIALS

One of the most bewildering aspects of in-depth research on the history of a large program such as the Space Telescope is the peculiar language in which its participants often talk and write. Learning to translate this language is a major undertaking. To do so means coming to grips with a prodigious number of acronyms, for example. One way to become at least partly proficient in "space speak" and knowledgeable about some of the technical and scientific aspects of the Space Tele-

419

scope is to immerse oneself in the published literature.

There are already available several technical and scientific expositions on the Space Telescope. One of the leading ones is C. R. O'Dell's "The Space Telescope Observatory," in *Telescopes for the 1980s* (Palo Alto: Annual Reviews, 1981), pp. 129–93. In addition to providing a lucid account of the telescope's technical and scientific aspects, it touches on the selling of the telescope. *The Space Telescope Observatory*, NASA CP-2244, edited by D. Hall, is a handbook on the telescope and its scientific instruments (as understood in 1982) for working astronomers. An excellent account of how the Large Space Telescope was perceived in 1974, and the scientific problems it was expected to tackle, is provided by the papers in *Large Space Telescope – A New Tool for Science* (AIAA, 1974). A similar but slightly later work is *The Space Telescope*, NASA SP-392 (Washington, D.C.: NASA, 1976). This contains the authors' summaries of papers presented at a special session of the American Astronomical Society in August 1975. For an account of the scientific problems that astronomers in 1979 expected the telescope to be able to tackle, see *Scientific Research with the Space Telescope*, NASA CP-2111 (Washington, D.C.: NASA, 1979), edited by M. Longair and J. Warner. This also includes an important essay by Noel Hinners on the origins of the Space Telescope Science Institute.

There are, in addition, hundreds of scientific papers describing observations that the Space Telescope might undertake. For information on these items, the reader is referred to the "Space Telescope Bibliography" prepared by the Space Telescope History Project. This bibliography also details many technical papers and reports that concern aspects of the telescope's design.

There are many other kinds of published materials besides scientific papers that deal with the Space Telescope. The records of NASA's congressional hearings often are surprisingly good means of gaining a view of the Space Telescope program at any one time, although a reader needs to be aware that that view often was outdated even at the time it was presented, and congressional hearings often are set piece affairs in which the proceedings have been carefully arranged beforehand. Also, the emphasis usually is on problems solved and issues resolved, rather than current matters. Nevertheless, when read in conjunction with other sources, the hearings record can often be illuminating, and the questions submitted for written responses often elicit useful information on such matters as costs and official NASA policies and positions. The transcript of the June 1983 hearings on the Space Telescope conducted by the House Subcommittee on Space Science and Applications, for example, is crammed full of material on the problems the program encountered in early 1983.

Reviews of the program conducted by groups of investigators from Congress are also helpful. The General Accounting Office has conducted three examinations of the Space Telescope program: "Space Telescope Project," PSAD-76-66 (Washington, D.C.: GAO, 1976), "Status and Issues Pertaining to the Proposed Development of the Space Telescope Project," PSAD-77-98 (Washington, D.C.: GAO, 1977), and "NASA Should Provide the Congress with Complete Cost Information on the Space Telescope Program," PSAD-80-15 (Washington, D.C.: GAO, 1980). When the telescope ran into major difficulties in 1983, the House surveys and investigations staff conducted three examinations of the program: "A Report to the Committee on Appropriations, U.S. House of Representatives, on the NASA Space Telescope Program," prepared by House surveys and investigations staff, March 1983, "The NASA Space Telescope Program," report prepared by House surveys and investigations staff, June 1983, and "The NASA Space Telescope (ST) Program," report prepared by the House surveys and investigations staff, February 1984.

Articles from aerospace and scientific journals and newspapers sometimes help to provide a broad picture. But most merely repeat basic information, and because reporters rarely disclose their sources, the accuracy of their comments on controversies is hard to judge. A researcher also needs to be aware that articles on the telescope often have been "planted" by advocates of the program or advocates of particular interests within the program. For example, John Walsh's article, "Astronomers Go Into Orbit," *Science* 191(1976):544–5, was, as is clear from the papers of John Bahcall, very much shaped by information fed to Walsh by Bahcall.

UNPUBLISHED SOURCES

In this book I have made extensive use of unpublished sources from a wide variety of institutions, groups, and individuals, and as part of its activities the Space Telescope History Project has compiled a "Guide to Space Telescope Archives." In consequence, what follows is only a sketch of the unpublished sources. For a fuller account, the reader is referred to the guide, much of which is based on a folder-level listing of items. The incredible abundance of material might surprise those historians who seem to presume that all information in late-twentieth-century science is conveyed at meetings or over the telephone. As Peter Galison has found in his studies of high-energy physics, if one looks beyond notebooks and letters, one can find a great wealth of material: minutes of meetings, project proposals, progress reports, photographs, film records, conference proceedings, blueprints, technical memoranda, artifacts, and so on. Planning, designing, and building the Space Telescope have involved large multidisciplinary teams, and one consequence of this teamwork is a huge amount of paperwork to keep the members of a team, as well as sometimes outsiders, informed on the team's activities.

The central institution for the history of the Space Telescope is NASA, and the best starting point for those interested in using NASA records is *History at NASA* (Washington, D.C.: NASA Headquarters, 1986). This includes brief descriptions of the holdings at NASA Headquarters as well as the various field centers. It is impossible to generalize about the quality and quantity of NASA records even on such a focused topic as the Space Telescope. Much depends on who retired

the records, how many changes there have been in the particular office that generated the records, whether or not a project office has changed buildings during its lifetime, and so on.

The richest collections of NASA material for study of the Space Telescope are those at the Marshall Space Flight Center and the Goddard Space Flight Center. The project scientist papers at Marshall are particularly noteworthy and form the single best collection for those seeking to understand the role of science and scientists in the decision making on the Space Telescope program. The project manager's papers at Marshall are also well ordered and are essential for a study of activities on the telescope at the center. The Goddard project manager's papers are especially good for the Phase B period. Unfortunately, much of the Goddard material for the late 1970s and early 1980s tends to be merely routine requests and responses to those requests. Although, in fact, a considerable fraction of the NASA material is of this kind, by no means all is so. Memoranda for the record and the minutes of meetings often are most helpful to the historian in shedding light on the decision-making process. But even here there are awkward issues of interpretation. Because NASA is a large government bureaucracy, NASA letters, memoranda, and briefing charts often are phrased exceedingly cautiously and are focused on the results of the agency's decision-making processes, not its dynamics. The minutes of the Science Working Group, for example, are written in a flat, bureaucratic language that eschews controversy. I have attended the meetings of the Science Working Group for over five years and have interviewed people who have been members for far longer, it is clear to them that although the minutes certainly are important sources, they are a pale reflection of the intensity of the group's discussions. The George M. Low papers in the archives of the Rensselaer Polytechnic Institute, Troy, New York, are also one of the best sources for gaining a view of the ways in which high-level NASA officials regarded the telescope program in the early and middle 1970s.

When reading NASA documents, it is also

crucial to bear in mind that many agency letters and memoranda were signed by people who did not originate them. Rather, a letter or memorandum may have been passed to someone simply because that person sat in a particular place in the management hierarchy. Similarly, letters and memoranda sent to a particular person in the agency were not necessarily dealt with by that person, but often were passed to other levels in the management chain for a response.

A central part of the NASA culture involves meetings with contractors at which various speakers present a series of viewgraphs. In major meetings, it seems, almost no one is willing to go to a blackboard for his or her presentation. Often the viewgraphs are the only records of those meetings, but even the viewgraphs, though in many ways disappointing for historical purposes, serve as a guide to what was agreed policy at any time, which problems were being actively investigated, and who was working on them. Viewgraphs must also be treated skeptically. For example, at one meeting in 1985, I witnessed a presenter claim the performance of a system he was describing "exceeded specifications." The viewgraph reflected this. When, however, it was pointed out by a puzzled questioner that it had *failed* to meet its specifications, the presenter confidently asserted, "Yes, it exceeds it in the negative direction."

More Space Telescope records reside in other government files. There is material at OMB that is important for understanding NASA/OMB interactions in the 1970s, particularly fiscal year 1977, when the telescope was omitted at a late stage from the NASA budget. A number of documents in the Gerald R. Ford Library at Ann Arbor, Michigan, also help to place the Space Telescope in the wider picture of the NASA budget and the budget for federal research and development generally.

The private papers of astronomers have proved invaluable in tracing the development of the telescope and the way scientists thought about it and organized to secure its approval. The collections of John Bahcall and Lyman Spitzer, Jr., are ex-

tremely rewarding in all of these regards, as is the collection of George Field's papers in the archives of the Smithsonian Institution. The language in correspondence and memoranda exchanged between astronomers is decidedly less inhibited than that in government documents. C. R. O'Dell was project scientist for the Space Telescope from 1972 to 1983, and during that period he kept a collection of short daily notes. These have helped me on many occasions to untangle a confused chronology, as well as to better understand the evolution of a number of issues. Arthur D. Code's papers at the University of Wisconsin–Madison are a good resource on the development of space astronomy, as well as the history of the Space Telescope.

The Data Center at the Lockheed Missiles and Space Company in Sunnyvale, California, contains an exceptionally helpful collection of correspondence, engineering memoranda, and studies, as well as chronologically arranged sets of presentations and reviews. Unfortunately, I was not granted access to any materials at Perkin-Elmer.

INTERVIEWS

The Space Telescope History Project has been fortunate in that nearly all of the people approached by us agreed to be interviewed. The majority of interviews were taped and transcribed, and a listing appears in the appendixes. The National Air and Space Museum's "Space Astronomy Oral History Collection" also contains a number of interviews of interest to a researcher on the history of the Space Telescope. Catalogs of this collection are available from the museum.

Such oral history interviews, of course, raise profound issues of reliability. But in an age when telephone- and computer-generated ephemera have so often replaced correspondence and other sorts of records, the testimony recorded in such interviews, despite the newer kinds of documentation that have flourished, has assumed special importance for historians. On occasion during my researches and those of my colleagues, interviews have proved to be a means of securing access to documents, as participants have been stimulated to check points in their files, files that they had sometimes previously regarded as uninteresting. A number of interviews were also taken at the start of the history project to provide an orientation and to help identify what participants viewed as key issues. Another tactic was to interview some participants more than once so as to gain an insight into their changing views on the program.

As to reliability, interviews, like all sources, must be treated skeptically. In particular, a user needs to be aware that interviews can be systematically unreliable. It has been noted in many places that this is not necessarily due to dishonesty or even a failing memory, but rather the shifting contexts of meaning and use within which the events are retrospectively set, even by those with reliable memories. I therefore attempted to examine any particular topic in several interviews, as well as in extremely close combination with the documentary record.

Essay on sources

Notes

INTRODUCTION

1 Oral history interview (OHI) (see Appendix 2), J. Bahcall, with PH, November 3, 1984, p. 44. On Spitzer's career, see OHI, L. Spitzer, Jr., with DHD, American Institute of Physics, April 8, 1977, and May 10, 1978, and OHI, L. Spitzer, Jr., with DHD, June 17, 1982, and with PH, October 27, 1983, and March 7, 1984. For a description of one of Spitzer's activities in World War II, his role as director of the Sonar Analysis Group organized as an advisory staff to the U.S. Navy, see Lincoln R. Thiesmeyer and John R. Burchard, *Combat Scientists* (Boston: Little, Brown, 1947), pp. 120–2.

2 H. N. Russell, "Where Astronomers Go When They Die," *Scientific American* 149(1933):112–13. I am grateful to David DeVorkin for bringing this reference to my attention.

3 See, for example, Joan Lisa Bromberg, *Fusion: Science, Politics, and the Invention of a New Energy Source* (Cambridge, Mass.: M.I.T. Press, 1982), p. 1.

4 See Richard S. Westfall, *Never at Rest: A Biography of Isaac Newton* (Cambridge University Press, 1980), pp. 232–7, and H. W. Turnbull (ed.), *The Correspondence of Isaac Newton: Vol. I. 1661–1675* (Cambridge University Press, 1959), pp. 3–9. See also Anthony J. Turner, "The Prehistory, Origins, and Development of the Reflecting Telescope," in Centro Internazionale A. Beltrame Di Storia Dello Spazio E Del Tempo, Bollettino No. 3–4 (1984), as well as A. A. Mills and J. Turvey, "Newton's Telescope: An Examination of the Reflecting Telescope Attributed to Sir Isaac Newton in the Possession of the Royal Society," *Notes and Records of the Royal Society* 33(1979):133–55. Mills and Turvey have tentatively suggested that the telescope usually identified as Newton's first was in fact his second.

5 On this, see Albert Van Helden, "The Invention of the Telescope," *Transactions of the American Philosophical Society* 67(pt. 4, 1977):5–67.

6 This point has been made by, among others, Richard S. Westfall in "Science and Patronage: Galileo and the Telescope," *Isis* 76(1985):11–30. See also Albert Van Helden, "The Invention of the Telescope," p. 27, and Edward Rosen, "Stillman Drake's *Discoveries and Opinions of Galileo*," *Journal of the History of Ideas* 18(1957):446.

7 His early observations are described in "The Starry Messenger," in Stillman Drake (ed.), *Discoveries and Opinions of Galileo* (New York: Doubleday Anchor, 1957), pp. 27–58.

8 I. B. Cohen, *Revolution in Science* (Cambridge, Mass.: Harvard University Press, 1985), p. 9.

9 Albert Van Helden, "Building Large Telescopes, 1900–1950," in O. Gingerich (ed.), *The General History of Astronomy*, vol. 4, pt. A (Cambridge University Press, 1984), pp. 134–52. See also Henry C. King's classic work, *The History of the Telescope* (New York: Dover, 1979).

10 C. C. Gillespie (ed.), *Dictionary of Scientific Biography*, s.v. "Hubble, Edwin Powell," by G. J. Whitrow, p. 529. On Hubble, see also N. U. Mayall, "Edwin Powell Hubble," *Biographical Memoirs of the National Academy of Sciences* 41(1970):175–214.

11 On these discoveries and the context in which they were made, see Robert W. Smith, *The Expanding Universe: Astronomy's "Great Debate" 1900–1931* (Cambridge University Press, 1982).

12 A Russian telescope has been built that is larger, 236 inches, but it has been plagued by technical problems that distort the shape of its primary mirror.

13 Westfall, *Never at Rest*, pp. 157–61, 212–16.

14 C. R. O'Dell, "The Space Telescope," in G. Burbidge and A. Hewitt (eds.), *Telescopes for the 1980s*

(Palo Alto: Annual Reviews, Inc., 1981), p. 132.

15 The figure of o.1 second of arc might prove to be a conservative estimate.

16 See Malcolm S. Longair, "The Space Telescope and Its Opportunities," *Quarterly Journal of the Royal Astronomical Society* 20(1979):5–28.

17 OHI, L. Fredrick, with RWS, May 2, 1986, p. 54. Philosopher of science Ian Hacking has argued that experimentation "sometimes pursues a life of its own": Ian Hacking, *Representing and Intervening: Introductory Topics in the Philosophy of Science* (Cambridge University Press, 1983), p. 215. The dominant assumption among astronomers that larger telescopes mean more discoveries has surely also led to the conception that telescopes in some sense possess lives of their own. This point has also been made by Peter Galison with respect to scientific instruments in general: Peter Galison, "Bubble Chambers and the Experimental Workplace," in Peter Achenstein and Owen Hannaway (eds.), *Observation, Experiment, and Hypothesis in Modern Physical Science* (Cambridge, Mass.: M.I.T. Press, 1985), p. 355.

18 See H. Miller, *Dollars for Research: Science and Its Patrons in Nineteenth Century America* (Seattle: University of Washington Press, 1970).

19 Leo Goldberg, "Solar Observatories in Space: Getting Started," in *National Air and Space Museum: Research Report 1986* (Washington, D.C.: Smithsonian Institution Press, 1986), p. 49.

20 On the rise and use of federal funding for science after World War II, see, among others, Harvey A. Averch, *A Strategic Analysis of Science and Technology Policy* (Baltimore: Johns Hopkins University Press, 1985), Harvey Brooks, *The Government of Science* (Cambridge, Mass.: M.I.T. Press, 1968), J. Merton England, *A Patron for Pure Science: The National Science Foundation's Formative Years, 1945–1957* (Washington, D.C.: National Science Foundation, 1982), James Everett Katz, *Presidential Politics and Science Policy* (New York: Praeger, 1978), Daniel Kevles, "The National Science Foundation and the Debate Over Postwar Research Policy, 1942–45," *Isis* 68(1977):5–26, W. Henry Lambright, *Governing Science and Technology* (New York: Oxford University Press, 1976), Don K. Price, *The Scientific Estate* (Cambridge, Mass.: Harvard University Press, 1965), Nathan Reingold, "Vannevar Bush's New Deal for Research: or the Triumph of the Old Order," *Historical Studies in the Physical and Biological Sciences* 17(1987):299–344, Michael D. Reagan, *Science and the Federal Patron* (New York: Oxford University Press, 1969), Margaret Rossiter, "Science and Public Policy Since

World War II," *Osiris* 1(1985):273–94, and Jeffrey K. Stine's "A History of Science Policy in the United States, 1940–1985," being a report prepared for the Task Force on Science Policy, Committee on Science and Technology, House of Representatives, Ninety-ninth Congress, Second Session, Serial R, September 1986. The classic introduction to the federal funding of science before 1940 is A. Hunter Dupree, *Science in the Federal Government* (Baltimore: Johns Hopkins University Press, 1986).

21 See Alvin M. Weinberg, "Impact of Large-Scale Science on the United States," *Science* 134 (1961):161–4, and *Reflections on Big Science* (Cambridge, Mass.: M.I.T. Press, 1967), p. 39. A classic work on Big Science is Derek de Solla Price's *Little Science, Big Science* (New York: Columbia University Press, 1969). On Big Science, see also, among others, Philip H. Abelson, "Instrumentation and Computers," *American Scientist* 74(1986):182–92, Bromberg, *Fusion*, David Edge and Michael Mulkay, *Astronomy Transformed: The Emergence of Radio Astronomy in Britain* (New York: Wiley, 1976), Paul Forman, "Behind Quantum Electronics: National Security as Basis for Physical Research in the United States, 1940–1960," *Historical Studies in the Physical and Biological Sciences* 18(1987):149–229, Galison, "Bubble Chambers and the Experimental Workplace," Daniel S. Greenberg, *The Politics of Pure Science* (New York: New American Library, 1967), J. L. Heilbron, R. W. Seidel, and B. R. Wheaton, "Lawrence and His Laboratory: Nuclear Science at Berkeley," *LBL News Magazine,* 6(1981), special issue, Armin Hermann, John Krige, Ulrike Mersits, and Dominique Pestre, *History of CERN I* (Amsterdam: North Holland, 1987), Chunglin Kwa, "Representations of Nature Mediating Between Ecology and Science Policy: The Case of the International Biological Programme," *Social Studies of Science* 17(1987)413–42, Spencer Klaw, *The New Brahmins* (New York: Morrow, 1968), Ralph E. Lapp, *The New Priesthood: The Scientific Elite and the Uses of Power* (New York: Harper & Row, 1965), Allan Needell, "Nuclear Reactors and the Founding of the Brookhaven National Laboratory," *Historical Studies in the Physical and Biological Sciences* 18(1986):89–110, and "Berkner, Tuve and the Federal Role in Radio Astronomy," *Osiris* 3(1987):261–88, and Robert Seidel's "Accelerating Science: The Postwar Transformation of the Lawrence Radiation Laboratory," *Historical Studies in the Physical Sciences* 13(1983):375–400, and "A Home for Big Sci-

ence: The Atomic Energy Commission's Laboratory System," *Historical Studies in the Physical and Biological Sciences* 16(1986):135–75.

22 Quoted in Forman, "Behind Quantum Electronics," p. 218.

23 Price, *Little Science, Big Science*, p. 92.

24 Daniel J. Kevles, *The Physicists: The History of a Scientific Community in Modern America* (New York: Vintage Books, 1979), p. 395.

I. DREAMS OF TELESCOPES

1 G. Kuiper to O. Struve, May 29, 1953, copy in F. K. Edmondson's active files, Indiana University, Bloomington, Indiana. The STHP is grateful to Professor Edmondson for permission to quote from this and several other letters in his active files.

2 R. S. Richardson, "Luna Observatory No. 1," *Astounding Science Fiction*, February 1940, pp. 113–23.

3 Gregory P. Kennedy, *Vengeance Weapon 2: The V-2 Guided Missile* (Washington, D.C.: Smithsonian Institution Press, 1983), p. 38. See also Frederick I. Ordway III and Mitchell R. Sharpe, *The Rocket Team* (New York: Thomas Y. Crowell, 1979), and Rowland F. Pocock, *German Guided Missiles of the Second World War* (New York: ARCO Publishing, 1967), for an account of the German work on missiles.

4 OHI, Lyman Spitzer, Jr., with PH, October 27, 1983, p. 3. See also Leo Goldberg, "Solar Physics," in Paul Hanle and Von Del Chamberlain (eds.), *Space Science Comes of Age* (Washington, D.C.: Smithsonian Institution Press, 1981), p. 16.

5 There was no independent U.S. Air Force until 1947. The Department of the Air Force was then created by the National Security Act and began operations as an autonomous military service in September of that year.

6 Douglas, however, was soon to drop its support. U.S. Air Force backing, together with financial assistance from Wells Fargo and the Ford Foundation, was required to place RAND on firm footing as the RAND Corporation. See Bruce L. R. Smith, *The RAND Corporation: Case Study of a Nonprofit Advisory Corporation* (Cambridge, Mass.: Harvard University Press, 1966), pp. 30–65; also see Walter A. McDougall, *The Heavens and the Earth: A Political History of the Space Age* (New York: Basic Books, 1985), pp. 89–91, and Fred Kaplan, *The Wizards of Armageddon* (New York: Simon & Schuster, 1983).

7 This, in fact, was not Spitzer's first work for RAND; he had already written for them on the atmosphere above 300 kilometers.

8 OHI, L. Goldberg, with DHD, February 22, 1983, p. 13.

9 Leo Goldberg, "Solar Physics," p. 17. See also L. Goldberg and L. Spitzer, Jr., "Proposal for High Altitude Spectroscopic Project Under the Auspices of the Office of Research and Inventions of the Navy Department," July 15, 1946, Spitzer papers, Princeton University, Princeton, New Jersey, and Lyman Spitzer, Jr., "Memorandum for File: Notes on Washington Trip – July 14–17 [1946]," Spitzer papers. At that time, Spitzer probably was unaware of the earlier writings of Hermann Oberth and Hermann Noordung on the advantages of placing telescopes in space: Hermann Oberth, *Die Rakete zu den Planetraumen* (Munich: R. Oldenbourg, 1923), p. 85; Hermann Noordung, *Das Probleme der Befahrung des Weltraums der Rakete Motor* (Berlin: Richard Carl Schmidt and Co., 1929), pp. 144–6.

10 L. Spitzer, Jr., "Astronomical Advantages of an Extra-Terrestrial Observatory," Douglas Aircraft Company, September 1, 1946. A draft of "Astronomical Advantages of an Extra-Terrestrial Observatory," dated July 30, 1946, is in the Spitzer papers. Interviewed in 1981, Spitzer still predicted that "the most spectacular results with the Space Telescope probably will be those we haven't thought of"; Dave Dooling, "A Telescope in Space," *Huntsville Times*, December 10, 1981, p. A6.

11 "Preliminary Design of an Experimental World-Orbiting Spaceship," Douglas Aircraft Company, Santa Monica Plant Engineering Division, May 2, 1946, report SM-11827, pp. II and VIII (Spitzer's report was Appendix 5).

12 R. Cargill Hall, "Early U.S. Satellite Proposals," *Technology and Culture* 4(1963):412, and McDougall, *The Heavens and the Earth*, pp. 102–3.

13 Hall, "Early U.S. Satellite Proposals," p. 419.

14 OHI, Lyman Spitzer, Jr., with PH, October 27, 1983, p. 3.

15 G. Kuiper to O. Struve, May 29, 1953, copy in F. K. Edmondson's active files.

16 See Clarence G. Lasby, *Project Paperclip: German Scientists and the Cold War* (New York: Atheneum, 1971), pp. 252–7. As Lasby points out, a V-2 was also launched from the aircraft carrier USS *Midway* (see p. 256). A more recent account is Tom Bower, *The Paperclip Conspiracy: The Battle for the Spoils and Secrets of Nazi Germany* (London: Michael Joseph, 1987). Bower argues that the security records of some of the V-2 engineers, in-

cluding von Braun, were systematically sanitized after the war by the U.S. Army. The aim was to ensure that these engineers could be granted U.S. citizenship and so continue their rocket work.

17 On the use of V-2s for scientific research, see David DeVorkin's "Organizing for Space Research: The V-2 Rocket Panel," *Historical Studies in the Physical Sciences* 18(1987):1–24. DeVorkin's forthcoming study of the origins of space science also analyzes this and other issues.

18 On the Naval Research Laboratory's use of V-2s, see Bruce Hevly, "Basic Research Within a Military Context: The Naval Research Laboratory and the Foundations of Extreme Ultraviolet and X-ray Astronomy, 1923–1960" (Ph.D. dissertation, The Johns Hopkins University, 1987).

19 Herbert Friedman, "Rocket Astronomy – An Overview," in Hanle and Chamberlain (eds.), *Space Science Comes of Age*, p. 32. The paper describing the results is W. A. Baum, F. S. Johnson, J. J. Oberly, C. C. Rockwood, C. V. Strain, and R. Tousey, "Solar Ultraviolet Spectrum to 88 Kilometers," *Physical Review* 70(1946):781–2.

20 Of course, even at the highest altitudes reached by the V-2s, there is a tiny amount of atmosphere.

21 E. T. Byram, T. A. Chubb, H. Friedman, and J. E. Kupperian, Jr., "Rocket Observations of Extra-Terrestrial Far UV Radiation," *Astronomical Journal* 62(1957):9.

22 Occasionally researchers used both; for example, James Van Allen and Herbert Friedman flew the composite rocket and balloon known as a "rockoon."

23 M. Schwarzschild, J. B. Rogerson, Jr., and J. W. Evans, "Solar Photographs from 80,000 Feet," *Astronomical Journal* 63(1958):313.

24 Hall, "Early U.S. Satellite Proposals," p. 425.

25 McDougall, *The Heavens and the Earth*, pp. 97–134.

26 Homer E. Newell, *Beyond the Atmosphere: Early Years of Space Science*, NASA SP-4211 (Washington, D.C., 1980), p. 52. James A. Van Allen, *Origins of Magnetospheric Physics* (Washington, D.C.: Smithsonian Institution Press, 1983), 33–42.

27 L. G. deBey, "Systems Design Considerations for Satellite Instrumentation," in James A. Van Allen (ed.), *Scientific Uses of Earth Satellites* (London: Chapman & Hall, 1956), p. 49.

28 Philip Morrison, "Concluding Remarks," in Allan A. Needell (ed.), *The First 25 Years in Space* (Washington, D.C.: Smithsonian Institution Press, 1983), 135. Walter McDougall has argued per-

suasively that the support of the U.S. government for the U.S. satellite program in the IGY was partly founded on the desire to establish the principle of satellite overflight of the Soviet Union. The overflight of a scientific satellite would then pave the way for overflight by reconnaissance satellites. See McDougall, *The Heavens and the Earth*, pp. 112–40. On the IGY, see also J. Merton England, *A Patron for Pure Science: The National Science Foundation's Formative Years, 1945–57* (Washington, D.C.: National Science Foundation, 1982), pp. 297–304.

29 Newell, *Beyond the Atmosphere*, p. 31.

30 Van Allen, *Origins of Magnetospheric Physics*, p. 46. See also OHI, J. Van Allen, with DHD and AN, July 16, 1981, p. 254. During the interview, Professor Van Allen referred to his notebooks of the period. On reactions to *Sputnik*, see, for example, Paul A. Hanle, "The Beeping Ball That Started a Dash Into Outer Space," *Smithsonian* 13(1982):150, and James R. Killian, Jr., *Sputnik, Scientists, and Eisenhower: A Memoir of the First Special Assistant to the President for Science and Technology* (Cambridge, Mass.: M.I.T. Press, 1977).

31 John Foster Dulles, "The Role of Negotiation," *Department of State Bulletin* 38(1958):159.

32 Van Allen, *Origins of Magnetospheric Physics*, p. 46. See also OHI, J. Van Allen, with DHD and AN, July 16, 1981, p. 254.

33 McDougall, *The Heavens and the Earth*, pp. 141–56, and Stephen E. Ambrose, *Eisenhower: The President* (New York: Simon & Schuster, 1984), pp. 423–6.

34 Ambrose, *Eisenhower*, p. 458.

35 John M. Logsdon, "U.S. Organizational Infrastructure for Space Programs: Strengths and Weaknesses," in Uri Ra'anan and Robert L. Pfaltzgraff, Jr. (eds.), *International Security Dimensions of Space* (Hamden, Conn.: Archon Books, 1984), p. 166.

36 This point is made by, for example, William H. McNeill in *The Pursuit of Power* (University of Chicago Press, 1982), p. 369.

37 T. Keith Glennan, in a talk at the National Air and Space Museum, Smithsonian Institution, January 28, 1982; transcript in NASM collections.

38 The suggestions sent directly to NASA were prompted by telegrams and letters sent by Nancy Roman, the recently appointed chief of NASA's astronomy programs.

39 George B. Kistiakowsky, *A Scientist at the White House: The Private Diary of President Eisenhower's*

Special Assistant for Science and Technology (Cambridge, Mass.: Harvard University Press, 1976), p. 385.

40 For one analysis of some of the difficulties involved in deciding on priorities for astronomical funding, see Martin Harwit, *Cosmic Discovery: The Search, Scope, and Heritage of Astronomy* (New York: Basic Books, 1981).

41 Homer E. Newell, "Space Astronomy Program of the National Aeronautics and Space Administration," *Astronomy in Space,* NASA SP-127 (Washington, D.C.: NASA, 1967), p. 6.

42 On this issue, see Edward Coyne, "Genesis of a Space Astronomy Program in NASA" (M.A. thesis, Case Institute of Technology, 1965). This thesis is also HHN-43 in the series of the NASA History Office.

43 For an account of the Goddard Space Flight Center's early years, see Alfred Rosenthal, *Venture into Space: Early Years of Goddard Space Flight Center,* NASA SP-4301 (Washington, D.C.: NASA, 1968).

44 It was originally planned that a series of small ultraviolet telescopes would be provided by the Smithsonian Astrophysical Observatory, but as the development of the detectors for the telescopes was late, the Smithsonian's experiment package was delayed to a later flight. The alternative payload thus had to be assembled very quickly.

45 OHI, C. R. O'Dell, with DHD, April 12, 1982, p. 6.

46 L. Spitzer, Jr., "Ultraviolet Spectra of the Stars," in Hanle and Chamberlain (eds.), *Space Science Comes of Age,* pp. 3–4. Rocket-borne instruments capable of pointing at the sun were developed in 1952 by the University of Colorado, These, however, were what are known as two-axis stabilized rockets. In other words, although they could be directed to the sun, they could not be fixed to any one point; rather, they were free to rotate about the line of sight to the sun.

47 In a major review of space science in 1962 (which will be examined later) the point was made that "it has been said that most of the satellites now in orbit but inoperative could be made to function again if a man with a screwdriver could get at them": *A Review of Space Research,* publication 1079 (Washington, D.C.: National Academy of Sciences/National Research Council, 1962), p. 2.20.

48 On the OAOs, see John B. Rogerson, Jr., "The OAOs," *Space Science Review* 2(1963):621–52, and Joseph Purcell, "The OAO Series of Space Telescopes," in *Optical Telescope Technology,* NASA SP-233 (Washington, D.C.: NASA, 1970), pp. 41–54.

49 This point is considered at length in Chapter 6.

50 Memorandum from A. D. Code, L. Goldberg, L. Spitzer, and F. L. Whipple to Nancy G. Roman, June 15, 1960, Leo Goldberg papers. Copy provided to STHP by Professor Goldberg.

51 Homer E. Newell to Leo Goldberg, April 18, 1960, Leo Goldberg papers. See also a conference report, "Orbiting Astronomical Observatories Program, Held at NASA Headquarters, 17 June 1960," Leo Goldberg papers. Copies of these materials were provided to the STHP by Professor Goldberg.

52 Newell, *Beyond the Atmosphere,* p. 97. On the Office of Space Science and Applications' use of launch vehicles to 1965, see Edgar M. Cortright, "Space Science and Applications: Where We Stand Today," *Astronautics And Aeronautics* 4(1966):42–3.

53 See Hans Mark and Arnold Levine, *The Management of Research Institutions: A Look at Government Laboratories,* NASA SP-481 (Washington, D.C.: NASA, 1984), pp. 127–31.

54 Now named, once again, Cape Canaveral; it was named Cape Kennedy during 1963–73.

55 "At Last, UW Has Its Big Moment," *State Journal* (undated, but in the April 1966 collections of Space Astronomy Laboratory, University of Wisconsin–Madison). See also OHI, R. C. Bless, with RWS, November 3, 1983, p. 9.

56 This account is based on the recollections of R. C. Bless and Arthur D. Code, as well as the "Summary Report of Review of the Orbiting Astronomical Observatory, OAO I, Failure by the Observatory-Class Spacecraft Projects Review Board," October 1966, Goddard Space Flight Center papers (hereafter cited as GSFC papers), 255-80-859, box 2 (file "Garbarini Committee, 1966–7"), Federal Record Centers, Suitland, Maryland.

57 OHI, R. C. Bless, with RWS, November 3, 1983, p. 8.

58 On the review board, see material in GSFC papers, 255-80-859, Box 2 (file "Garbarini Committee, 1966–7").

59 See Arthur D. Code (ed.), *The Scientific Results from the Orbiting Astronomical Observatory (OAO-2),* NASA (SP-310 Washington, D.C.: NASA, 1972).

60 For a brief account of the planned satellite, see "Another Orbiting Astronomical Observatory," *Sky and Telescope,* December 1970, pp. 349–50. See also OHI, A. Boggess III, with RWS, April 20, 1984, pp. 31–5.

61 OHI, A. Boggess III, with RWS, April 20, 1984, p. 34.

62 A small x-ray telescope designed by a group of University College, London, was the subsidiary experiment.

63 This was a considerable advance on *OAO-II's* pointing accuracy of thirty seconds of arc and pointing stability of three seconds of arc, an increase driven by the type of observations to be secured by *Copernicus:* Raymond N. Watts, Jr., "An Astronomy Satellite Named Copernicus," *Sky and Telescope,* October 1972, pp. 231–2, 235, and Lyman Spitzer, Jr., *Searching Between the Stars* (New Haven: Yale University Press, 1982), pp. 57–67.

64 For a dissenting view on *Copernicus,* see Freeman J. Dyson, "Science in Space" in Allan A. Needell (ed.), *The First 25 Years in Space* (Washington, D.C.: Smithsonian Institution Press, 1983), pp. 90–106. In Dyson's opinion, the delays in instituting the *Copernicus* program meant that *Copernicus* "was not the instrument astronomers would have chosen to answer the exciting scientific questions of the 1970s" (p. 93).

65 Goldberg, "Solar Physics," p. 26.

66 See, for example, "Research, Design, and Construction Leading to a Large Aperture Orbiting Telescope," Kitt Peak National Observatory, June 1961, copy provided to STHP by F. K. Edmondson.

67 *AURA: The First Twenty Five Years 1957–1982* (Tucson: AURA, 1983), pp. 8–9, and James E. Kloeppel, *Realm of the Long Eyes: A Brief History of Kitt Peak National Observatory* (San Diego: Univelt, 1983), pp. 75–83.

68 See, for example, OHI, R. C. Bless, with RWS, November 3, 1983, pp. 10–11; a similar point is made by Fred Hoyle in *Man in the Universe* (New York: Columbia University Press, 1966), p. 6.

69 Hoyle, *Man in the Universe,* p. 7.

70 See also Fred Hoyle, "The Big Bang in Astronomy," in Nigel Henbest (ed.), *Observing the Universe* (Oxford: Basil Blackwell, 1984), p. 15. In 1966, Richard Berendzen sent a questionnaire on career development and personal background to a large sample of U.S. astronomers. Some two-thirds replied, and Berendzen found that "the item about NASA evoked the greatest emotionalism of the questionnaire. For unspecified reasons, some scientists harbored derogatory feelings about NASA." Richard E. Berendzen, "On the Career Development of Astronomers in the United States" (Ph.D. dissertation, Harvard University, 1968), vol. 1, sect. 4.8.2.

71 For an excellent account of why the United States went to the moon, see J. Logsdon, *The Decision to Go to the Moon* (Cambridge, Mass.: M.I.T. Press, 1970).

72 James Webb to J. F. Kennedy, November 30, 1962, quoted in McDougall, *The Heavens and the Earth,* p. 380.

73 L. Goldberg to T. Paine, June 5, 1970, Newell papers, 255-79-0649, box 25 (file "AA Reading Files"), NASA Headquarters papers, Federal Record Center, Suitland, Maryland.

74 Max Planck, *Scientific Autobiography and Other Papers,* translated by Frank Gaynor (New York: Philosophical Library, 1949), p. 33.

75 R. C. Bless to L. Meredith, December 12, 1969, Goddard active files (file "Post-Launch Operation"). See also OHI, R. C. Bless, with RWS, November 3, 1983, pp. 18–19.

76 NAS/NRC, *A Review of Space Research,* p. 2.1. Plans for observations with the X-15 were being made by a group at the University of Wisconsin-Madison led by Arthur D. Code.

77 NAS/NRC, *A Review of Space Research,* p. 2.11.

78 Lyman Spitzer, Jr., "The Beginnings and Future of Space Astronomy," *American Scientist* 50 (1962):473–84. This contains the text of a talk Spitzer delivered in 1962 that included consideration of a 400-inch telescope.

79 NAS/NRC, *A Review of Space Research,* p. 2.12. For notes taken during many of the sessions on astronomy, see F. K. Edmondson's notebook, now in the collection of the National Air and Space Museum. On the Iowa meetings, see also OHI, J. Van Allen, with DHD and AN, July 21, 1981, pp. 317–21.

80 NAS/NRC, *A Review of Space Research,* p. 2.13. There was at least one vote against the majority view and one abstention: OHI, American Institute of Physics, L. Goldberg, with Spencer Weart, May 17, 1978, p. 88. Goldberg recalls that A. J. Deutsch, of Mount Wilson Observatory, and in effect representing the West Coast astronomers, voted no, and Richard Tousey of the Naval Research Laboratory abstained.

81 Boeing document D2-84042-1: *A System Study of a Manned Orbital Telescope* (Seattle: Boeing Company, 1965). The chief astronomical advisor for this study was Zdenek Kopal. See Kopal's *Of Stars and Men: Reminiscences of an Astronomer* (Boston: Adam Hilger, 1986), p. 291.

82 See also "Final Report Feasibility Study of a 120-inch Orbiting Astronomical Telescope," AE-1148, J. W. Fecker Division, American Optical Company (undated, but 1963).

83 See the NASA report "Manned Space Astronomy. Section I – Early Efforts. Astronomy Subcommittee" (undated, but ca. 1966), copy in STHP files.

84 In Chapter 6, this point is considered in more detail.

85 For a brief account of the National Astronomical Space Observatory within the context of NASA's space science and applications programs, see *Objectives and Goals in Space Science and Applications 1968*, NASA SP-162 (Washington, D.C.: NASA, 1968), p. 4/9.

86 *Space Research: Directions for the Future*, part 2 (Washington, D.C.: Space Science Board, National Academy of Sciences/National Research Council, 1966), p. 1.

87 On June 3, 1965, astronaut Ed White had walked in space during his flight aboard *Gemini IV*.

88 NAS/NRC, *Space Research: Directions for the Future*, part 2, p. 3. For a brief survey of the major recommendations made at Woods Hole, see Harold M. Schmeck, Jr., "Orbiting of Huge Astronomical Telescopes Urged," *New York Times*, February 1, 1966, p. L6.

89 See, for example, R. J. Davis, F. L. Whipple, and C. A. Whitney, "An Astronomical Telescope in Space," *Astronautical Sciences Review*, January–March 1959, pp. 9–12, 14.

90 Homer E. Newell, "Space Astronomy Program of the National Aeronautics and Space Administration," p. 7.

91 N. U. Mayall and Aden Meinel, however, both stepped down.

92 OHI, N. Roman, with DHD, August 19, 1980, p. 48. The NASA staffer was John Clark, later to become director of the Goddard Space Flight Center.

93 OHI, N. Roman, with RWS, March 1984, p. 14. See also OHI, R. C. Bless, with RWS, November 3, 1983, pp. 16–17, where a similar point is made.

94 Memorandum, Edward R. Dyer, Jr., to Space Science Board, "Status Report on the SSB Large Space Telescope Committee," October 19, 1967, Aucremanne papers (file "LST 1966–1969"), and memorandum, Edward R. Dyer, Jr., to Space Science Board, on "Status Report on the SSB Large Space Telescope Committee," December 10, 1966, Aucremanne papers (file "LST 1966–1969").

95 L. Spitzer, Jr., "Astronomical Research with the Large Space Telescope," *Science* 161(1968):225–9. See also L. Spitzer, Jr., "Astronomical Research with a Large Orbiting Telescope," in S. Fred Singer (ed.), *Manned Laboratories in Space* (New York: Springer-Verlag, 1969), pp. 88–98.

96 L. Spitzer, chairman, *Scientific Uses of the Large Space Telescope, Report of the Space Science Board ad hoc Committee on the Large Space Telescope* (Washington, D.C.: National Academy of Sciences, 1969).

97 For example, Tifft was a finalist in the selection of astronaut/scientists in 1965 and 1967. See also W. G. Tifft, "The Overall Spectrum of Astronomical Research Possibilities Utilizing Manned Earth Orbital Spacecraft," in *Science Experiments for Manned Orbital Flight* (Tarzana, Calif.: American Astronautical Society, 1966), pp. 283–92.

98 OHI, W. Tifft, with JT, December 13, 1984, pp. 61–2. See also W. Tifft, "Astronomy, Space, and the Moon," *Astronautics and Aeronautics* 4(1966):40–53.

99 See also OHI, L. Spitzer, Jr., with DHD, July 17, 1982, pp. 48–9, and L. Spitzer, Jr., with PH, October 27, 1983, pp. 10–11.

100 OHI, L. Spitzer, Jr., with DHD, June 17, 1982, pp. 48–9, and W. Tifft with JT, December 13, 1984, pp. 61–2.

101 Spitzer, *Scientific Uses of the Large Space Telescope*, p. 1.

102 Spitzer, *Scientific Uses of the Large Space Telescope*, p. 2.

2. BUILDING A PROGRAM

1 In 1970, for example, there were nine NASA field centers: Manned Spacecraft Center (Houston, Texas), Marshall Space Flight Center (Huntsville, Alabama), Kennedy Space Center (Cocoa Beach, Florida), Ames Research Center (Moffett Field, California), Flight Research Center, (Edwards, California), Langley Research Center (Hampton, Virginia), Lewis Research Center (Cleveland, Ohio), and Goddard Space Flight Center (Greenbelt, Maryland). Whereas the Jet Propulsion Laboratory was officially part of the California Institute of Technology, it worked in effect as a NASA center.

The early NASA owed much to the existing structure of NACA, the National Advisory Committee for Aeronautics, which the new agency absorbed on its inception: Alex Roland, *Model Research: The National Advisory Committee for Aeronautics, 1915–1958*, 2 vols. NASA SP-4103 (Washington, D.C.: NASA, 1985), and James R. Hansen, *Engineer in Charge: A History of Langley Aeronautical Laboratory, 1917–1958*, NASA SP-4305 (Washington, D.C.: NASA, 1987). On the development of NASA in its early years, see Richard Hallion, *On the Frontier: Flight Research at Dryden, 1946–1981,* NASA SP-4303 (Washington, D.C.: NASA, 1984), Richard Hirsch and Joseph Trento, *The National Aeronautics and Space Administration* (New York: Praeger, 1973), James R. Killian, *Sputnik, Scientists, and Eisenhower: A Memoir of the First Special Assistant for Science and Technology* (Cambridge, Mass.: M.I.T. Press, 1977), George B. Kistiakowsky, *A Scientist at the White House* (Cambridge, Mass.: Harvard University Press, 1976), Clayton R. Koppes, *JPL and the American Space Program: A History of the Jet Propulsion Laboratory* (New Haven: Yale University Press, 1982), John Logsdon, *The Decision to Go to the Moon: Project Apollo and the National Interest* (Cambridge, Mass.: M.I.T. Press, 1970), Walter McDougall, *The Heavens and the Earth: A Political History of the Space Age* (New York: Basic Books, 1986), Elizabeth Muenger, *Searching the Horizon: A History of Ames Research, 1940–1976,* NASA SP-4304 (Washington, D.C.: NASA, 1985), Homer Newell, *Beyond the Atmosphere: Early Years of Space Science,* NASA SP-4211 (Washington, D.C.: NASA, 1980), Robert A. Rosholt, *An Administrative History of NASA, 1958–63,* NASA SP-4101 (Washington, D.C.: NASA, 1966), Enid C. B. Schoettle, "The Establishment of NASA," in Sandford A. Lakoff (ed.), *Knowledge and Power: Essays on Science and Government* (New York: Free Press, 1966), pp. 162–270, and Herbert York, *Making Weapons, Talking Peace: A Physicist's Odyssey from Hiroshima to Geneva* (New York: Basic Books, 1987), pp. 128–76.

Several of the works cited earlier were sponsored by NASA. A full list of NASA history publications is provided in *History at NASA. The NASA History Office,* HHR-50 (Washington, D.C.: NASA, 1986). A brief but very helpful review of literature on space in general is that by Richard Hallion, "The Next Assignment: The State of the Literature on Space," in Alex Roland (ed.), *A Spacefaring People: Perspectives on Early Space Flight,* NASA SP-4405 (Washington, D.C.: NASA, 1985), pp. 61–7. A much more extensive work is that of Katherine M. Dickson (ed.), *History of Aeronautics and Astronautics: A Preliminary Bibliography,* NASA HHR-29 (Washington, D.C.: NASA, 1968).

2 C. H. Danhof, *Government Contracting and Technological Change* (Washington, D.C.: Brookings Institution, 1968), R. J. Fox, *Arming America: How the U.S. Buys Weapons* (Cambridge, Mass.: Harvard University Press, 1974), H. L. Nieburg, *In the Name of Science* (Chicago: Quadrangle, 1966), Merritt Roe Smith, *Harper's Ferry Arsenal and the New Technology: The Challenge of Change* (Ithaca: Cornell University Press, 1977). The use of arsenals versus private suppliers is a theme in Alex Roland's review, "Science and War," *Osiris* 1(1985):247–72.

3 Arnold S. Levine, *Managing NASA in the Apollo Era,* NASA SP-4102 (Washington, D.C.: NASA, 1982), p. 66.

4 There is a rich and rapidly growing literature on this topic. Good starting points are E. Bijker, Thomas P. Hughes, and Trevor J. Pinch (eds.), *The Social Construction of Technological Systems: New Directions in the Sociology and History of Technology* (Cambridge, Mass.: M.I.T. Press, 1987), Melvin Kranzberg, "Technology and History: 'Kranzberg's Laws,'" *Technology and Culture* 27(1986): 544–60, and Donald Mackenzie and Judy Wajcman (eds.), *The Social Shaping of Technology: How the Refrigerator Got Its Hum* (Philadelphia: Open University Press, 1985). An outstanding case study of how nontechnical factors shaped the development of a technological system, in this case networks of electrical power, is provided by Thomas P. Hughes, *Networks of Power: Electrification in Western Society, 1880–1930* (Baltimore: The Johns Hopkins University Press, 1983).

As noted in the Introduction, in this book I focus on the notion of coalition-building, and I argue that the coalition that was constructed to make the telescope politically feasible did much to shape the technology of the telescope that was finally produced. Although my analytical tools derive in part from political science, I believe my approach to be compatible in some ways with that of Hughes. He emphasizes, in the words of John Law, that "those who build artifacts do not concern themselves with artifacts alone but must also consider the way in which the artifacts relate to social, economic, political, and scientific factors. *All* these factors are interrelated, and all are potentially malleable. The argument, in other words, is that innovators are best seen as system builders." John Law, "Technology and Heterogeneous Engineering: The Case of Portuguese Expansion," in Bijker, Hughes, and Pinch (eds.), *The Social Construction of Technological Systems,* p. 112. In a way, we can regard some of the advocates of the telescope as system builders, for making possible the construction of the Space Telescope would require the manipulation and juxtaposition of a range of scientific, social, technical, economic, and political variables. Hughes's notion of reverse salients could also be applied to the problems the telescope's advocates had in persuading the Congress (as we shall see in Chapters 4 and 5).

5 In 1970, Goddard advanced a concept called STAR ("Space Telescopes for Astronomical Research"). STAR pictured a variety of telescopes in space, some of them weather satellites and other earth-applications satellites. The Large Space Telescope was, nevertheless, a central feature of the plan.

6 Joseph Purcell, "The OAO Series of Space Telescopes," in *Optical Telescope Technology*, NASA SP-233 (Washington, D.C.: NASA, 1970), p. 53. See also memorandum, "SGT/Program Manager, Advanced Programs and Technology, to The Files," on "GSFC/LaRC," October 21, 1969, Aucremanne papers, NASA Headquarters, Washington, D.C. (file "LST: 1970–71"). This memorandum describes a meeting at which Goddard presented its OAO "follow-on" study.

    An example of the differing cultures of Marshall and Goddard is that artists' conceptions used to illustrate Goddard's plans for space telescopes rarely showed astronauts, whereas the Marshall (a manned space flight center) illustrations almost always depicted astronauts, as well as the telescope(s).

7 Briefing, "Systems Study of the Large Space Telescope (3 Meter Observatory)," Grumman Space Astronomy (undated, but 1969), P. Simmons file in STHP files.

8 Levine, *Managing NASA*, pp. 172–3.

9 Marshall, for instance, built production models of the first stage of the Saturn I launch vehicles. See Roger E. Bilstein, *Stages to Saturn: A Technological History of the Apollo/Saturn Launch Vehicles*, NASA SP-4206 (Washington, D.C.: NASA, 1980), pp. 70–2. On the origins of Marshall, see Bilstein, *Stages to Saturn*, pp. 38–42, and York, *Making Weapons, Talking Peace*, pp. 173–6.

10 These space station concepts included, for example, the "Manned Orbiting Research Laboratory"; for example, see volumes 1 and 2 of Laurence W. Fredrick (ed.), "Final Report for Applications in Astronomy Suitable for Study by Means of Manned Orbiting Observatories and Related Instruments and Operational Requirements," studies done for Langley Research Center, October 1963.

11 Wernher von Braun, "Crossing the Last Frontier," *Collier's*, March 22, 1952, pp. 25–59, 72–4.

12 Von Braun, "Crossing the Last Frontier," p. 72. Von Braun was also a promoter of space stations for military uses. On early plans for space sta-

tions, see also John M. Logsdon, *Space Stations: A Policy History*, prepared for Johnson Space Center, contract NAS9-164661.

13 Some of these, given that they were producing merely feasibility studies, received considerable sums of money. The "Optical Laser Technology Program," for instance, spent a total of some $15 million (in fiscal 1982 dollars) between fiscal years 1962 and 1966. See briefing charts for "Advanced Optical Systems Technology Study," presentation by Max Nein, undated, Nein file, STHP files. The Marshall-managed *Skylab* program also involved a variety of telescopes. See. W. David Compton and Charles D. Benson, *Living and Working in Space: A History of Skylab*, NASA SP-4208 (Washington, D.C.: NASA, 1983).

14 See Lockheed document "Large Space Telescope Support Systems Module: Phase B Definition Study Proposal," July 8, 1974, LMSC-D 4222750, folder 1, vol. 2, pp. 1–3.

15 See *A Long Range Program in Space Astronomy: Position Paper of the Astronomy Missions Board*, NASA SP-213 (Washington, D.C.: NASA, 1969), pp. 46–76.

16 The name was chosen because it is Latin for "star," although in NASA's insatiable quest for acronyms, it would sometimes be referred to as "Astronomical Space Telescope Research Assembly."

17 NASA document, "Report of the ASTRA Advisory Committee," June 1968, NASA Headquarters papers, 255-73A-368, box 1 (file "Report of the ASTRA Advisory Committee").

18 NASA document, "Summary," Headquarters papers, 255-76-600, box 2, p. 2 [file "Astronomical Space Telescope Research Assembly (ASTRA) Correspondence 11/13/67–3/28/69"]. See also R. O. Doyle (ed.), *A Long Range Program in Space Astronomy: Position Paper of the Astronomy Missions Board July 1969*, NASA SP-213 (Washington, D.C.: NASA, p. 52).

19 F. G. Allen to M. J. Aucremanne, June 13, 1968, Headquarters papers, 255-73A-368, box 1 [file "Man-Maintained OAO (ASTRA) 1967–1969"]. The conversion factor used to take account of inflation (3.178) is taken from *Space Science Research in the United States: A Technical Memorandum* (Washington, D.C.: Office of Technology Assessment, 1982), p. 42.

20 Of course, NASA is involved in far more than building spacecraft. We shall here refer to a spacecraft for the sake of simplicity.

21 Levine, *Managing NASA*, pp. 158–61, and C. R. O'Dell, "The Space Telescope," in *Telescopes for*

the 1980s (Palo Alto: Annual Reviews, 1981), p. 137.

22 OHI, J. Mitchell, with RWS, December 10, 1984, p. 74.

23 See memorandum, "Astronomy Technology Programs in OART," attached to the memorandum, M. Aucremanne to "The Files," on "Telescope Technology Program," October 28, 1968, Aucremanne papers (file "Optical Telescope Technology Workshop"), and memorandum, J. Beggs to J. Naugle, on "Telescope Technology Program," October 2, 1968, Aucremanne papers (file "Optical Telescope Technology Workshop").

24 On the planning of this workshop, see Aucremanne papers [file "Optical Telescope Technology Workshop (OTTW)"].

25 *Optical Telescope Technology*, NASA SP-233 (Washington, D.C.: NASA, 1970). See also OHI, G. Emanuel, with RWS, December 12, 1985, p. 4, and D. Schroeder, with RWS, October 7, 1985, pp. 12–14.

26 R. Curtin to NASA centers, November 15, 1969, Aucremanne papers (file "Optical Facilities").

27 N. Roman, "NASA Goals and Objectives," in *Optical Telescope Technology*, pp. 8–9.

28 For a brief account of the family of Titan launch vehicles, see J. G. Davies, "The Titan Launch Vehicle Family," *Spaceflight* 25(1982):36–8.

29 Bilstein, *Stages to Saturn*, pp. 348–51.

30 NASA document, "Astronomy Position Paper," Astronomy Planning Panel of Planning Steering Group, September 1, 1969, p. 114, copy in STHP files.

31 H. Anderton to W. von Braun, September 25, 1969, Aucremanne papers (file "LST 1966–69").

32 Memorandum, Wernher von Braun to H. L. Anderton, November 18, 1969, copy in STHP files.

33 Memorandum, J. Mitchell to associate administrator for space science and applications, on "Perkin-Elmer Contracts in Active Optics," July 9, 1970, Aucremanne papers (file "LST 1970–71"). See also OHI, Garvin Emanuel, with RWS, December 12, 1985, p. 8.

34 See *Astronomy Programs Review* (Office of Space Science and Applications, NASA Headquarters) (Washington, D.C.: NASA, 1970), p. 81, and OHI, M. J. Aucremanne, with RWS and AN, November 21, 1983, pp. 1–5.

35 Memorandum, M. J. Aucremanne, on "Large Space Telescope; Rationale for the 2-meter Telescope," August 31, 1970, Aucremanne papers (file "LST 1970–71").

36 N. Roman, "Stellar Astronomy," in *Astronomy Programs Review*, p. 26.

37 The official decision was announced at a press conference held by President Nixon, in which NASA administrator James Fletcher participated: NASA news release no. 72-05, January 5, 1972. For the background to this decision, see J. Logsdon, "The Decision to Develop the Space Shuttle," *Space Policy* 2(1986):103–19. The different shuttle designs are traced in J. Guilmartin and John Mauer's forthcoming chronological history of the shuttle.

38 M. Aucremanne, "Space Telescopes," *Astronomy Programs Review*, p. 63. These possibilities had also been quickly grasped by the manager of the Orbiting Astronomical Observatories. He, too, thought that the shuttle could lower the cost of building spacecraft because designers need no longer be overly conservative to ensure against a crippling failure in orbit. The chance to exchange subsystems in orbit would permit new approaches: memorandum, "Projects Directorate, OAO Project, to Shuttle Project Manager, Mr. Robert Thompson," on "Space Shuttle," June 22, 1970, Space Shuttle chronology papers, Johnson Space Center, location 006-13.

39 "OAO/LST Shuttle Economics Study," Grumman Aerospace Corporation, September 1970, contract NAS5-17149. The figures would have to be drastically revised later in the shuttle program. As it would turn out, the staggeringly optimistic, not to say absurd, assumption in the study was that each shuttle flight would cost only $5 million. With more realistic shuttle costs, the study's conclusions would have been undermined. By 1985, the cost of a shuttle flight had risen to around $250 million.

40 W. R. Lucas to E. Rees, September 22, 1970; copy in STHP files. Draft memorandum, M. Aucremanne, to Associate Administrator (Homer Newell) for Administrator, on "Large Space Telescope," September 1970, Aucremanne papers (file "LST 1970–71").

41 Draft memorandum, M. Aucremanne, to Associate Administrator (Homer Newell) for Administrator, on "Large Space Telescope," September 1970, Aucremanne papers (file "LST 1970–71"). See also memorandum, W. Lucas to Dr. Rees, September 22, 1970, no title, where Lucas reported that "Jesse [Mitchell] is thinking in terms of [Marshall] having overall responsibility and [Goddard] being responsible for the telescope, instruments, and operations"; copy in STHP files. See also briefing charts, "Large Space Telescope Program," October 29, 1970, and "Comments on LST Vugraphs for the Administrator's Presen-

tation," draft by M. J. Aucremanne, October 19, 1970, Aucremanne papers (file "LST 1970–71").

42 Memorandum, G. Chandler, Jr., on "Memorandum of Meeting Held Thursday, October 29, 1970, [on] Large Space Telescope," November 2, 1970, Space Shuttle chronology papers, Johnson Space Center, location 006-26.

43 Memorandum, J. Mitchell to W. Lucas, on "Transfer of Surplus Optical Equipment," November 13, 1970, Aucremanne papers (file "LST 1970–71").

44 A brief biographical sketch is included in *Astronomy Programs Review,* June 2–3, 1970 (Office of Space Science and Applications, NASA Headquarters), p. 80.

45 Memorandum, J. Mitchell to W. Lucas, on "Transfer of Surplus Optical Equipment," November 13, 1970, Aucremanne papers (file "LST 1970–71").

46 OHI, J. Mitchell, with RWS, December 10, 1984, pp. 77–80.

47 Large Space Telescope Task Team, "Minutes of the First Meeting," February 25, 1971, Aucremanne papers (file "LST Task Team").

48 See, for example, OHI, C. R. O'Dell, with DHD, April 12, 1982, p. 25.

49 "Summary Minutes," Large Space Telescope Steering Group, July 14–15, 1971, and also Ernst Stuhlinger, "Memo for the Record," July 19, 1971, both in O'Dell papers, Rice University (file "NASA-LST Science Steering Group, 1971–72[1]").

50 Memorandum, J. Mitchell to J. Naugle, on "The 2.5 Meter Diffraction-Limited Telescope – on the Ground and in Space," June 24, 1971, Aucremanne papers (file "LST 1970–71").

51 In November 1971, Itek had been awarded a twelve-month contract worth $400,000 to define the OTA, see Marshall release 71-219.

52 See, for example, a paper on Martin's study: Carl L. Kober, "Future Orbital Observatory Modules for Stellar and Galactic Astronomy," in *Astronomy from a Space Platform* (Tarzana, Calif.: American Astronautical Society, 1972), pp. 115–32.

53 This was the "Scientific Instrument Package" (SIP), defined by the Kollsman Instrument Company in collaboration with Goddard. See *"Large Space Telescope* (LST) Preliminary Study. Volume 1: Executive Summary," February 25, 1972, Marshall active files (file "Program Development").

54 See, for example, "Summary Minutes, Large Space Telescope Steering Group," July 14–15, 1971, p. 5, O'Dell papers (file "NASA–LST Science Steering Group 1971–72[1]").

55 Memorandum, J. Mitchell to J. Clark, on "Large Space Telescope," January 17, 1972, and memorandum, J. Mitchell to E. Rees, on "Large Space Telescope," January 17, 1972, both in Aucremanne papers (file "LST 1972").

56 The figures are taken from Bilstein, *Stages to Saturn,* p. 450.

57 Richard S. Lewis, *The Voyages of Columbia: The First True Spaceship* (New York: Columbia University Press, 1984), p. 30, and John Logsdon, "The Space Shuttle Program: A Policy Failure," *Science,* May 30, 1986, pp. 1099–105.

58 OHI, J. Mitchell, with RWS, December 10, 1984, p. 51.

59 Simmons's recollection is in OHI, P. Simmons, with RWS and JT, September 10, 1984, pp. 42–3. Goddard's management problems are discussed in personal notes, #67, March 26, 1972, box 69, folder 5, George M. Low papers, Rensselaer Polytechnic Institute Archives, Troy, New York.

60 Memorandum, J. Downey III to E. Stuhlinger, October 5, 1965, Program Development papers, Marshall active files [file "(P)6 Optical Astronomy – Kollsman (Wells)"].

61 OHI, R. C. Bless, with RWS, November 3, 1983, p. 38. See, too, George Low's description of a visit to Marshall: personal notes, #28, August 8, 1970, box 70, folder 4, George M. Low papers, Rensselaer Polytechnic Institute Archives, Troy, New York.

62 OHI, J. Mitchell, with RWS, December 10, 1984, p. 48.

63 Memorandum, J. Fletcher to R. Frosch, "Problems and Opportunities at NASA," May 9, 1977, Fletcher papers, NASA History Office (file "Correspondence 1977").

64 Memorandum, J. Naugle to G. Low, on "Project Management Center for the Large Space Telescope (LST)," April 25, 1972, Aucremanne papers (file "LST 1970–71 [sic]"). As a manned space flight center, Marshall reported to the Office of Manned Space Flight at NASA Headquarters. The decision therefore had to be approved by the head of that office.

65 J. Naugle to E. Rees, May 5, 1972, on "Assignment of Marshall Space Fight Center (MSFC) as Project Management Center for the Large Space Telescope (LST)," May 5, 1972, Aucremanne papers (file "LST 1972"). See also "MSFC Given Management of Large Space Telescope," *Marshall Star,* May 10, 1972, p. 1.

66 Memorandum, E. Rees to J. Murphy, on "LST," May 15, 1972, copy in STHP files. This includes

a transcript of a telephone conference between Rees and Clark on May 11, 1972.

3. ASTRONOMERS, INDUSTRY, AND MONEY

1 Memorandum, E. Rees to J. Naugle, on "[LST] Program Data," March 21, 1972, Aucremanne papers (file "LST 1972").
2 Memorandum, J. Naugle to E. Rees, on "Large Space Telescope," June 5, 1972, Aucremanne papers (file "LST 1972").
3 Allan A. Needell, "Lloyd Berkner, Merle Tuve, and the Federal Role in Radio Astronomy," *Osiris* 3(1987):261.
4 Memorandum for the record, J. Downey III, on "Enclosure A – LST Meeting in Headquarters with Drs. Fletcher and Low, December 21, 1972," January 9, 1973, Marshall presentation files (file "Briefing to Dr. Low at Headquarters 12/21/1972").
5 Another company (Eastman Kodak) would emerge a couple of years later.
6 J. Downey III to M. Aucremanne, January 18, 1973, Aucremanne papers [file "Large Space Telescope (LST) 1973–74"].
7 On these points, see "LST Procurement Strategy Rationale for Pairing Phase B Contractors," Aucremanne papers (file "LST 1972"), memorandum for the record, J. Downey III, on "Enclosure A – LST Meeting in Headquarters with Drs. Fletcher and Low, December 21, 1972," January 9, 1973, Marshall presentation files (file "Briefing to Dr. Low at Headquarters, 12/21/1972"), and Marshall presentation files (file "OSS Definition & Supporting Research Activities Meeting with Dr. Naugle and Mr. Mitchell [March 1973]"). On the meeting, see also memorandum, George M. Levin, on "MSFC Presentation on LST to Drs. Fletcher and Low, 12/21/72," December 22, 1972, GSFC papers, 255-81-305, box 1 of 4 (file "April 1971–April 1973").
8 This figure excludes civil service and support contractor costs, as well as any sustained operations cost.
9 Memorandum for the record, J. Downey III, on "Enclosure B – Follow-up Meeting with Mr. Mitchell and Mr. Aucremanne, December 21, 1972," January 9, 1973, Marshall presentation files (file "Briefing to Dr. Low at Headquarters, 12/21/1972").
10 The protoflight concept was already being adopted elsewhere at Marshall. The program to construct the so-called High Energy Astronomy Observatories had in fact been suspended in January 1973, and Marshall's switch to the protoflight concept was just one of the measures taken to revive it by drastically lowering its costs: Wallace H. Tucker, *The Star-Splitters: The High Energy Astronomy Observatories,* NASA SP-446 (Washington, D.C.: NASA, 1984), pp. 23–8.
11 For a discussion of the protoflight concept by one of those closely involved with its introduction, see OHI, Fred A. Speer, with RWS, February 26, 1985, pp. 7–10.
12 Charles R. Pellegrino and Joshua Staff, *Chariots for Apollo: The Making of the Lunar Module* (New York: Atheneum, 1985), p. 8.
13 OHI, L. Spitzer, Jr., with DHD, June 17, 1982, p. 53, and L. Spitzer, Jr., with PH, October 27, 1983, pp. 17–18.
14 OHI, C. R. O'Dell, with DHD, April 12, 1982, p. 22; also E. Stuhlinger to J. Tatarewicz, private communication.
15 OHI, C. R. O'Dell, with DHD, April 12, 1982, pp. 1–13.
16 OHI, C. R. O'Dell, with DHD, April 12, 1982, pp. 20–1.
17 O'Dell daily notes, September 5, 1972, O'Dell papers, Rice University, Houston, Texas.
18 OHI, R. C. Bless, with RWS, November 13, 1983, p. 29, and D. Schroeder, with RWS, October 7, 1985, p. 23.
19 This astronomical satellite was to be launched later in the 1970s.
20 OHI, C. R. O'Dell, with RWS, American Institute of Physics, May 21–23, 1985, session II, p. 38.
21 C. R. O'Dell, "The Science Management Plan for the Large Space Telescope," November 3, 1972, project scientist papers (file "Science Community Involvement in LST Project").
22 See "Instrumentation Package for a Large Space Telescope (LST)," design report for Goddard Space Flight Center, November 1970, X-670-70-480.
23 Memorandum, G. Pieper to J. Naugle, on "GSFC comments on the MSFC/O'Dell Plan to Involve the Scientific Community in the LST," November 1, 1972, Aucremanne papers (file "LST 1972").
24 For two accounts of this meeting, see memorandum, C. R. O'Dell to J. Murphy, on "Meeting with G. Pieper and L. Meredith, 3 November," November 6, 1972, O'Dell papers, and G. Pieper to C. R. O'Dell, on "Marshall/Goddard Agreements on the LST," November 7, 1972, Aucremanne papers (file "LST 1972").

25 Memorandum, G. Levin to Robert S. Cooper, November 16, 1976, on "Problems with Marshall During Phase B," GSFC active files (file "GSFC/MSFC Intercenter Agreement").

26 The Science Working Group was officially titled the "Large Space Telescope Operations and Management Working Group" throughout Phase B. In Phase C/D this became the "Space Telescope Science Working Group." To avoid confusion, we shall adopt "Science Working Group" throughout the text.

27 See O'Dell appointment books, O'Dell papers, and OHI, C. R. O'Dell, with RWS, May 21–23, 1985, pp. 13–14, and J. Olivier, with RWS, February 27, 1985, addendum, p. 8.

28 These teams were as follows: Imaging Optics, High Resolution Spectrograph, Low Resolution Spectrograph, Infra-Red Devices, Astrometry, and Data Handling and Operations.

29 "Selection of Phase B Instrument Definition Teams and Working Group Members," particularly documents by Nancy G. Roman, "Instrument Definition Teams and Scientific Working Group for the Large Space Telescope (LST)," being "Recommendations to the Space Science and Applications Steering Committee" (undated), O'Dell papers. See also "Opportunities for Participation in the Definition of Scientific Instruments for the Large Space Telescope," memo change #55, NHB 8030.1 A, December 1, 1972, signed by John E. Naugle, copy in Aucremanne papers (file "LST 1972").

30 For a Lockheed view of the company's history, see *Lockheed Horizons* 12(1983).

31 See, for example, "Large Space Telescope Experiment Program (LTEP)" (Final Technical Report, Volume 2: Technical Proposal), January 5, 1970, Lockheed papers A958175, support of NAS 8-21497, p. iii.

32 OHI, D. Tenerelli, with RWS, January 11, 1985, p. 2.

33 See "Program Review, Large Space Telescope (LST) Support Systems Module (SSM)," October 29, 1972, box 12, Lockheed papers sorted by RWS.

34 See the Lockheed document "Large Space Telescope Support Systems Module Definition Study" (Folder 1, Technical Proposal, Volume 2: Engineering and Design Capability, Organization, and Management), July 8, 1974, D422750, pp. II-7/8.

35 Marshall news release 72-142, November 3, 1972.

36 Memorandum for record, J. Downey III, February 8, 1973, Marshall presentation files (file "LST Briefing to Dr. Petrone, 2/7/73").

37 Marshall presentation files (file "LST Planning Presentation to Dr. Low at Headquarters, 4/13/1973").

38 See Marshall presentation files (file "LST Planning Presentation to Dr. Low at Headquarters, 4/13/1973"), especially memorandum, J. Downey III to Mr. Murphy, April 9, 1973, memorandum for the record, James A. Downey III, on "LST Meeting with Dr. Low – April 13, 1973," April 26, 1973, and memorandum for the record, M. Aucremanne, on "LST Planning Meeting with Dr. Low," April 23, 1973. See also memorandum, G. Low to J. Fletcher, on "Activities during week of April 8–14, 1973," box 35, folder 2, George M. Low papers, Rensselaer Polytechnic Institute, Troy, New York.

39 Minutes, Operations and Management Working Group, June 11, 1973, Project Scientist papers.

40 One was to investigate and develop designs for an infrared instrument; another was for an astrometry instrument, that is, an instrument for accurately measuring the positions of astronomical objects. A third team was selected to provide advice on data management and operating the instruments and telescopes in orbit.

41 OHI, N. Roman, with RWS, February 3, 1984, p. 43.

42 OHI, D. Schroeder, with RWS, October 7, 1985, p. 20.

43 See, for example, "GSFC Systems Level Designs of the LST Scientific Instrument Ensemble," prepared by G. Levin, December 4, 1973, Goddard Study Office, copy enclosed with minutes to Operations and Management Working Group, December 13–14, 1973, project scientist papers.

44 OHI, I. King, with RWS, May 2, 1984, pp. 6–7, and I. King, "Content and Structure of Galaxies as Observed with LST," in *Large Space Telescope – A New Tool for Science* (Washington, D.C.: American Institute of Aeronautics and Astronautics, 1974), p. 28. But see also J. Bahcall to C. R. O'Dell, December 5, 1973, being enclosure 7 to minutes of Operations and Management Working Group meeting, December 13–14, 1973, project scientist papers. Here John Bahcall and Neta Bahcall called for a camera to work in conjunction with the telescope's spectroscopic and photometric observations.

45 The "radial" instruments, too, would be modular.

46 See enclosure 7, "LST Working Group Recom-

mendations and Consensus Statements," enclosed with minutes to Operations and Management Working Group, April 16–17, 1974, project scientist papers.

47 See, for example, OHI, A. D. Code, with RWS, February 21, 1984, p. 14.

48 Michael Disney, *The Hidden Universe* (New York: Macmillan, 1984), p. 229.

49 OHI, N. Roman, with RWS, February 3, 1984, p. 43. For sympathetic views of using astronauts to perform astronomy, see W. G. Tifft, "The Overall Spectrum of Astronomical Research Possibilities Utilizing Manned Earth Orbital Spacecraft," in *Scientific Experiments for Manned Orbital Flight* (Tarzana, Calif.: American Astronautical Society, 1966), pp. 283–92, and K. G. Henize, "Manned Versus Unmanned Space Based Astronomy," *Astronomy from a Space Platform* (Tarzana, Calif.: American Astronautical Society, 1972), pp. 385–91.

50 Summary minutes, Large Space Telescope Science Steering Group, July 14–15, 1971, p. 2, O'Dell papers (file "NASA LST Science Steering Group 1971–72[1]"). There was only one dissenting voice to this opinion.

51 OHI, L. Spitzer, Jr., with DHD, June 17, 1982, p. 45.

52 OHI, C. R. O'Dell, with DHD, April 12, 1982, p. 33. See also OHI, N. Roman, with RWS, February 3, 1984, p. 43.

53 Ground-based astronomers were interested in similar devices. In the 1970s, for example, Leo Goldberg instituted a "Panoramic Detector Program" at the Kitt Peak National Observatory. Led by Roger Lynds, it gave rise to the development of the so-called Video-Camera based on a "Charge Induced Device," or CID, private communication, Leo Goldberg to RWS.

54 W. S. Boyle and G. E. Smith, "Charge Coupled Semiconductor Devices," *Bell System Technical Journal* 49(1970):587–92, and G. F. Aurelio, M. F. Tompsett, and G. E. Smith, "The Experimental Verification of the Charge Coupled Device Concept," *Bell System Technical Journal* 49(1970):593–600. On the operation of CCDs, see also *CCD: The Solid State Imaging Technology* (Palo Alto: Fairchild, 1976), and C. H. Sequin, "Image Recording Using Charge-Coupled Devices," in *Advanced Electro-Optical Imaging Techniques*, NASA SP-338 (Washington, D.C.: NASA, 1972), pp. 51–68.

55 See J. T. Williams, "The Intensified Charge Coupled Device as a Photon Counting Imager," in *The Space Telescope*, NASA SP-392 (Washington,

D.C.: NASA, 1976), pp. 88–9, and OHI, S. Sobieski, with RWS, February 8, 1985, pp. 16–21.

56 House Committee on Science and Astronautics, *Statement of John E. Naugle, Associate Administrator for Space Science, NASA Large Space Telescope: Hearings before the Subcommittee on Space Science and Applications*, 93rd Cong., 2nd sess., March 7, 1974, part 3, p. 421.

57 Minutes, LST Operations and Management Working Group, February 7–8, 1974, pp. 3–4, project scientist papers.

58 Minutes, LST Operations and Management Working Group, May 30–31, 1974, project scientist papers.

59 OHI, S. Sobieski, with RWS, February 8, 1985, pp. 18–19.

60 See *CCD: The Solid State Imaging Technology*, p. 122, and, for example, *Aviation Week and Space Technology*, February 13, 1984, p. 174, where it was noted that the Hughes Aircraft Company was developing a CCD system for guiding the U.S. Navy's Trident submarine-launched ballistic missile. The CCD camera sights on a star and electronically verifies the trajectory in the missile's inertial guidance system.

61 On "pull" and "push," see F. R. Jevons, *Science Observed: Science as a Social and Intellectual Activity* (New York: George Allen & Unwin, 1973), pp. 102–3.

62 Private communication, S. Faber to RWS, March 6, 1986.

63 "Itek–Perkin-Elmer Set to Receive Phase B Studies for Telescope," *Aviation Week and Space Technology*, July 2, 1973, pp. 51–2. See also Marshall release 73-110, August 9, 1973, and *Defense/Space Daily*, August 13, 1973, p. 223.

64 See the report of James A. Downey's comments in minutes, LST Operations and Management Working Group, July 31 and August 1, 1973, project scientist papers.

65 It was, in the jargon of mirror makers, accurate over 90 percent of its surface to lambda/50 rms. See viewgraphs of J. Olivier for meeting of LST Operations and Management Working Group, June 11, 1973, project scientist papers.

66 The support structure for the Apollo Telescope Mount used on *Skylab*, for example, used three control moment gyroscopes, each with a 53-cm rotor, weighing 65.5 kg, and spinning at a constant 9,000 rpm: W. David Compton and Charles D. Benson, *Living and Working in Space: A History of Skylab*, NASA SP-4208 (Washington, D.C.: NASA, 1983), p. 170.

67 As it would turn out, they were right.
68 See Jean R. Olivier, "LST Design Considera-
    tions," in *Large Space Telescope – A New Tool for
    Science* (Washington, D.C.: AIAA, 1974), pp. 53–
    7.
69 Lockheed document, "Large Space Telescope
    Support Systems Module Definition Study," July
    8, 1974, LMSC-D422750, folder 1, vol. 2, and
    OHI, D. Tenerelli, with RWS, January 11, 1985,
    p. 15.
70 OHI, D. Tenerelli, with RWS, January 11, 1985,
    p. 15.
71 A. D. Houston, L. W. Hodge, Jr., and T. J.
    Kertesz, "An Analytical and Experimental Eval-
    uation of Actuator Vibration on Space Telescope
    Image Distortion," in *The Space Telescope,* NASA
    SP-392 (Washington, D.C.: NASA, 1976), pp.
    146–50.
72 R. L. Gates, D. H. Wine, R. W. Seifert, and
    N. A. Osborne, "Development of a Large-Inertia
    Fine Pointing and Dimensional Stability Simu-
    lator," and W. W. Emsley, T. D. Fehr, D. C.
    Foster, and D. L. Knobbs, "Three Axis Simula-
    tion of the Pointing Control Subsystem – A Mul-
    tidiscipline Activity," in *The Space Telescope,* being
    pp. 151–2 and pp. 174–6, respectively.

4. SELLING THE LARGE SPACE
TELESCOPE

1 This "Voyager" is not to be confused with the
   later NASA missions of the same name that were
   directed at the outer planets.
2 E. C. Ezell and L. N. Ezell, *On Mars: Exploration
   of the Red Planet, 1958–1978,* NASA SP-4212
   (Washington, D.C.: NASA, 1984), pp. 113–18,
   and J. Tatarewicz, "Where Are the People Who
   Know What They Are Doing? Space Technology
   and Planetary Astronomy" (Ph.D. dissertation,
   Indiana University, 1984), pp. 190–318 for the
   context of this decision.
3 For brief accounts, see Carl M. York, "Federal
   R&D Programs: Who Picks 'Em?" *Science Policy
   Reviews* 5(1972):3–6, and Albert H. Teich, "Coordination of United States Research Programs:
   Executive and Congressional Roles," *Science and
   Technology Studies* 4(1986):29–36. See also Richard Barke, *Science, Technology, and Public Policy*
   (Washington, D.C.: Congressional Quarterly, Inc.,
   1986).
4 OHI, J. Mitchell, with RWS, December 10,
   1984, pp. 66–74, and N. Hinners, with RWS
   and JT, October 17, 1984, p. 28. W. Henry

Lambright, *Governing Science and Technology* (New
York: Oxford University Press, 1976).
5 OHI, J. Mitchell, with RWS, December 10,
   1984, pp. 90–1.
6 House Committee on Science and Astronautics,
   Statement by Dale D. Myers, 92nd Cong., 1st sess.,
   March 4, 1971, H.R.3981 (superseded by
   H.R.7109), p. 136.
7 House Committee on Science and Astronautics,
   *Supplemental Statement for the Record by John E.
   Naugle: Hearings Before the House Subcommittee on
   Space Science and Applications,* 92nd Cong., 2nd
   sess., February 22, 1972, H.R.12824 (superseded
   by H.R.14070), p. 68.
8 In 1984, Mitchell recalled that "I think all of us
   were involved with that. I was never one to set
   down on paper that sort of thing. I tend to run
   those things through my head and integrate, integrate the effect," OHI, J. Mitchell, with RWS,
   December 10, 1984, p. 91.
9 Minutes, Management and Operations Working
   Group in Shuttle Astronomy, October 22–23,
   1974, p. 8, project scientist papers (file "Management and Operations Working Group in
   Shuttle Astronomy: 22–23 October 1974").
10 OHI, P. Simmons, with RWS and JT, Septem-
    ber 10, 1984, pp. 3–5.
11 J. Naugle, "Epilogue," in *Large Space Telescope –
    A New Tool for Science* (Washington, D.C.: AIAA,
    1974), p. 124.
12 See "Space Science Research in the United States:
    A Technical Memorandum," Office of Technol-
    ogy Assessment, September 1982, especially Ap-
    pendix A, "Trends in the Space Budget," and
    "Funding Trends in NASA's Space Science Pro-
    gram," Office of Science and Technology Policy,
    1984.
13 The crisis was precipitated by the oil embargo
    begun by the Organization of Petroleum Export-
    ing Countries in October 1973. On the energy
    crisis, see Martin V. Melossi, *Coping with Abun-
    dance: Energy and Environment in Industrial America*
    (Philadelphia: Temple University Press, 1985).
14 For example, in early February the committee had
    visited the Martin-Marietta Company in Denver,
    Colorado, and had seen some of the supporting
    research and technology work that had been ac-
    complished on the LST program: " Large Space
    Telescope Program Briefing to House Space Sci-
    ence and Applications Committee," Martin-
    Marietta, Denver, Colorado, February 11, 1974,
    Aucremanne papers (file "LST 1973–74").
15 House Committee on Appropriations, *HUD–Space–
    Science–Veterans Appropriations for 1975: Hearings*

Before the Subcommittee on HUD–Space–Science–Veterans, pt. 3, NASA, 93rd Cong., 2nd sess., March 26–27, 1974, pp. 248–9, 327–30. See also J. Naugle to N. Hinners, August 9, 1974, Naugle papers, NASA History Office, box 1 (chronology file 3/19–12/31/1974).

16 House Committee on Appropriations, *HUD–Space–Science–Veterans Appropriations for 1975: Hearings Before the Subcommittee on HUD–Space–Science–Veterans, pt. 3, NASA*, 93rd Cong., 2nd sess., March 26–27, 1974, pp. 248–9, 327–30.

17 House Committee on Appropriations, *Report No. 93-1139 of the Subcommittee on HUD–Space–Science–Veterans on the Department of Housing and Urban Development, Space, Science, Veterans, and Certain Other Independent Agencies Appropriations Bill, 1975*, June 21, 1974, p. 16.

18 *Congressional Record*, 93rd Cong., 2nd sess., June 26, 1974, H5746-73.

19 OHI, C. R. O'Dell, with RWS, May 21–23, 1985, p. 17. Also, as one well-placed observer put it at the time, "Nancy Roman was as surprised as anyone else at this turn of events and had no prior clue as to what would happen," H. M. Gurin to B. J. Bok, July 8, 1974, F. K. Edmondson active papers.

20 On resolving House–Senate differences, see Walter Oleszek, *Congressional Procedures and the Policy Process*, 2nd ed. (Washington, D.C.: CQ Press, 1984), pp. 201–24.

21 In the early 1950s, he had been an organizer of VOTE, a lobby group that proposed changes in New York State's election laws on voter eligibility. See "Resident Honored for Space Astronomy Work," *Tupper Lake Free Press*, July 2, 1980, copy in F. Pete Simmons file, STHP files.

22 OHI, P. Simmons, with RWS and JT, September 10, 1984, pp. 71–2. On Simmons's calls to O'Dell, see, for example, O'Dell daily notes, June 25, 1974, and June 28, 1974, O'Dell papers.

23 See material in University of Wisconsin–Madison Space Astronomy Laboratory archives, box 101 (file "LST and Congress"), especially memorandum, D. M. Green to distribution, on "Results of 22 June Meeting with Congressman Jim Symington held in his Clayton Office re LST Funding Problem," June 22, 1974.

24 O'Dell daily notes, July 5, 1974, O'Dell papers. For one view of the lobbying activities of defense contractors, see Gordon Adams, *The Politics of Defense Contracting* (New Brunswick: Transaction Books, 1982). This book contains profiles of three of the contractors who hoped to win the Phase B

Support Systems Module contract: Boeing, Lockheed, and McDonnell Douglas.

25 See H. M. Gurin to B. J. Bok, July 8, 1974, F. K. Edmondson active papers.

26 OHI, J. Bahcall, with PH, November 3, 1983, p. 25.

27 Bahcall, for example, recalls that Spitzer "was widely recognized as the scientific inspiration of the Telescope," OHI, J. Bahcall, with PH, November 3, 1983, p. 23.

28 See Wallace Tucker and Karen Tucker, *The Cosmic Inquirers: Modern Telescopes and Their Makers* (Cambridge, Mass.: Harvard University Press, 1986), p. 184.

29 OHI, J. Bahcall, with PH, November 3, 1983, p. 4.

30 OHI, J. Bahcall, with PH, November 3, 1983, p. 9.

31 OHI, J. Bahcall, with PH, November 3, 1983, p. 10.

32 OHI, C. R. O'Dell, with RWS, American Institute of Physics, May 21–23, 1985, session I, p. 15.

33 O'Dell's appointment book for 1974, O'Dell papers.

34 OHI, C. R. O'Dell, with RWS, American Institute of Physics, May 21–23, 1985, session I, pp. 30–1, 157–9.

35 See Alice K. Smith, *A Peril and a Hope: The Scientists' Movement in America, 1945–47* (University of Chicago Press, 1965).

36 OHI, J. Bahcall, with PH, November 3, 1983, p. 28.

37 *Astronomy and Astrophysics for the 1970s* (Washington, D.C.: NAS, 1972), vol. 1, p. 8.

38 *Astronomy and Astrophysics for the 1970s*, p. 105. A similar recommendation had been made in a 1971 study by the Space Science Board. See *Priorities for Space Research 1971–80: Report of a Study on Space Science and Earth Observations Conducted by the Space Science Board, National Research Council* (Washington, D.C.: NAS, 1971), p. 10. The board had also concluded that it should not be considered as a project for the 1970s because of cost.

39 Personal communication, J. Greenstein to P. Hanle, October 23, 1984. See also C. R. O'Dell to "Dear Colleague," June 26, 1974, project scientist papers (file "Reading File 1974").

40 See OHI, C. R. O'Dell, with RWS, American Institute of Physics, May 21–23, 1985, session I, pp. 4–6.

41 OHI, G. Field, with RWS, March 10, 1986, pp. 2–3.

42 *Opportunities and Choices in Space Science, 1974* (Washington, D.C.: NAS, 1975), p. 3.

43 House Committee on Appropriations, HUD–Space–Science–Veterans Appropriations for 1975: Hearings Before the Subcommittee on HUD–Space–Science–Veterans, pt. 3, NASA, 93rd Cong., 2nd sess., March 27, 1974, pp. 742–5.

44 This was the Senate Subcommittee for HUD, Space, Science, Veterans, and Certain Other Independent Agencies.

45 J. Bahcall to W. Proxmire, July 1, 1974, Bahcall papers, Institute for Advanced Study, Princeton, New Jersey.

46 OHI, Lyman Spitzer, Jr., with DHD, June 17, 1982, p. 62.

47 OHI, J. Bahcall, with PH, November 3, 1983, p. 43. As we shall see later, the building of the Space Telescope did lead to delays and cuts for other projects.

48 OHI, J. Bahcall, with PH, December 20, 1983, p. 52.

49 Emphasis in original, J. Bahcall and L. Spitzer, Jr., to Burt L. Talcott, August 6, 1974, Bahcall papers.

50 *Congressional Record*, 93rd Cong., 2nd sess., July 2, 1974, H.22088-9. This statement is consistent with the views Greenstein had expressed in 1972 in a letter to O'Dell: J. L. Greenstein to C. R. O'Dell, December 18, 1972, project scientist papers (file "Reading File 1972"). There Greenstein had reported that "I am delighted that the advent of the Shuttle project has pushed the LST into a much earlier launch date than we were told by NASA when we started the astronomy survey." On his concern that the telescope was sometimes being sold at the expense of ground-based astronomy, see Paul Hanle, "Astronomers, Congress, and the Large Space Telescope," *Sky and Telescope* (1985):300–5.

51 *Opportunities and Choices in Space Science, 1974* (Washington, D.C.: NAS, 1974), p. 40.

52 Senate Committee on Appropriations, *Senate Hearings Before the Committee on Appropriations, Department of Housing and Urban Development, Space, Science, Veterans and Certain Other Independent Agencies Appropriations*, 93rd Cong., 2nd sess., April 2 1974, H.R.15572, pp. 504, 706.

53 *Congressional Record*, 93rd Cong., 2nd sess., August 16, 1974, S.15161-5.

54 "Transcript of President Ford's Address to Joint Session of Congress and the Nation," *New York Times*, August 13, 1974, p. 20A.

55 *Congressional Record*, 93rd Cong., 2nd sess., August 21, 1974, H.29693. See also C. R. O'Dell

to "Dear Colleague," August 22, 1974, project scientist papers (file "Reading File 1974").

56 J. Fletcher to J. Naugle, August 6, 1974, Naugle papers, NASA History Office, box 1 (chronology file 3/19–12/31/74).

57 National Aeronautics and Space Act of 1958, as amended, section 102(b)(7). A central reference for this section is John M. Logsdon, "U.S.–European Cooperation in Space Science: A 25 Year Perspective," *Science* 223(1984):11–16.

58 H. C. van de Hulst, draft of manuscript "Planning Space Science," dated 1983, provided by Professor van de Hulst to the STHP; Sir Harry Massey and M. O. Robins, *History of British Space Science* (Cambridge University Press, 1986), pp. 152–61. See also "A Report to the U.K. Science Research Council by the Astronomy, Space and Radio Board," *Spaceflight* 10(1968):423, and A. V. Cleaver, "The Crisis in European Space Affairs," *Spaceflight* 10(1968):370–1.

59 Massey and Robins, *History of British Space Science*, pp. 355–66; document entitled "Non-solar optical astronomy programs leading to LST," provided for the STHP by Jesse L. Mitchell, dated ca. 1971; A. Boggess et al., "The IUE Spacecraft and Instrumentation," *Nature* 275(1978):372–7; Y. Kondo (ed.), *Exploring the Universe with the IUE Satellite* (Dordrecht: Reidel, 1987), especially Chapter 1 by A. Boggess and R. Wilson, "The History of IUE."

60 ESRO report, "Focal Plane Instrumentation for the Large Space Telescope," December 16, 1974, MS(74)28, Appendix 1. See a note by C. R. O'Dell, February 22, 1974, project scientist papers (file "Reading File 1974"). This note was written after a discussion with Nancy Roman, who had attended the meeting.

61 ESRO, "Focal Plane Instrumentation for the Large Space Telescope," Appendix 1.

62 A. W. Schardt to R. J. H. Barnes, September 12, 1974, Aucremanne papers (file "LST–International/ESRO").

63 Minutes, LST Operations and Management Working Group, October 1–2, 1974, project scientist papers.

64 Minutes, LST Operations and Management Working Group, October 1–2, 1974, p. 3, project scientist papers.

65 "FY 1976 Budget Formulation: Action Items from OSS Review – September 11, 1974," Aucremanne papers (file "LST–International/ESRO").

66 Memorandum by M. Aucremanne, on "Report on Trip to London, England," October 22, 1974, and memorandum by C. R. O'Dell, on "Discus-

sions with UK and ESRO Officials," October 16, 1974, both in Aucremanne papers (file "LST–International/ESRO"). LST program manager Marc Aucremanne attended the London meeting too. The ESRO engineer was Robin Laurance, later to play an important role in the LST project.

67 John M. Logsdon, "U.S.–European Cooperation in Space Science," p. 13.

68 J. Naugle to J. Fletcher, August 7, 1974, NASA History Office, Naugle papers, box 1 (chronology file 3/19–12/31/74). See also J. A. Simpson to J. Fletcher, June 26, 1974, Naugle papers, box 1 (chronology file 3/19–12/31/74).

69 OHI, N. Hinners, with RWS and JT, November 16, 1984, p. 65.

70 Memorandum, C. R. O'Dell to M. Aucremanne, January 28, 1975, Aucremanne papers (file "LST–International/ESRO"). O'Dell had been asked to conduct the poll by Nancy Roman.

71 OHI, J. Bahcall, with PH, March 28, 1984, p. 7.

72 NASA document, "Summary of Current Status of European Interest in Participation in the Large Space Telescope (LST)," February 13, 1975, Aucremanne papers (file "LST–International/ESRO").

73 House Committee on Appropriations, *Department of Housing and Urban Development–Independent Agencies Appropriations for 1976: Hearings Before the Subcommittee on HUD–Independent Agencies, pt. 2, NASA*, 94th Cong., 1st sess., March 5, 1975, pp. 424–34.

74 The ministers of the governments representing ESRO and ELDO agreed on April 15, 1975, on the text of a convention that formed the basis of the European Space Agency at the European Space Conference in Brussels.

75 Memorandum, "Meeting between NASA and ESA Officials held at ESA HQ, Neuilly, 11 June 1975," Aucremanne papers (file "LST–International/ESRO").

5. SAVING THE SPACE TELESCOPE

1 This version of the text was employed in a briefing package prepared to help sell the Space Telescope to Congress. For the context of the quotation, see B. Jowett (trans.), *The Republic of Plato*, 3rd ed. (Oxford: Clarendon Press, 1888), pp. 230–1.

2 Minutes, LST Operations and Management Working Group, May 30–31, 1974, p. 1, project scientist papers.

3 J. Fletcher to J. Naugle, August 6, 1974, NASA History Office, Naugle papers (chronology file 3/19–12/31/74).

4 O'Dell daily notes, August 2, 1974, O'Dell papers.

5 L. Spitzer to C. R. O'Dell, August 7, 1974, with enclosures to minutes, LST Operations and Management Working Group, October 1–2, 1974, project scientist papers.

6 In current-year dollars, assuming 5 percent inflation each year, these figures become $350 million and $450 million: "Presentation to Dr. Hinners," September 1974, Marshall presentation files.

7 James A. Downey to distribution, "OSS Briefing to Dr. Hinners, Trip Report" (undated, but presumably September 1974), copy in project scientist papers.

8 There are some clear similarities here to the cost estimates for the space station NASA has been planning since the early 1980s.

9 O'Dell daily notes, August 28–29, 1974, O'Dell papers.

10 E. M. Burbidge to C. R. O'Dell, September 10, 1974, with enclosures to minutes, LST Operations and Management Working Group, October 1–2, 1974, project scientist papers. See also OHI, E. Margaret Burbidge, with RWS, October 29, 1984, pp. 5–6. The assistant was Jeffrey Rosendhal.

11 The scientific justification for the minimum LST was set out in C. R. O'Dell to M. Aucremanne, February 6, 1975, project scientist papers (file "Reading file 1975").

12 House Committee on Appropriations, *Department of Housing and Urban Development–Independent Agencies Appropriations for 1976: Hearings Before the Subcommittee on HUD–Independent Agencies, NASA*, 94th Cong., 1st sess., March 5, 1975, p. 444.

13 On August 13, O'Dell noted that it might be possible to study the three sizes, 3 meters, 2.5 meters, and 2 meters: O'Dell daily notes, August 13, 1974, O'Dell papers. The 2.5-meter and 2-meter options seem to have been quickly dropped.

14 The public literature on reconnaissance satellites, not surprisingly, is decidedly limited. These appear to be the most convincing examples of this material: John A. Adam, "Counting the Weapons," *IEEE Spectrum* 23(1986):46–56; James Bamford, "America's Supersecret Eyes in Space," *New York Times Magazine*, January 13, 1985, pp. 39, 50–5; William E. Burrows, *Deep Black: Space Espionage and National Security* (New York: Random House, 1986); Anthony Kenden, "U.S. Re-

connaissance Satellite Programs," *Spaceflight* 20(1978):243–62; Philip J. Klass, *Secret Sentries in Space* (New York: Random House, 1971); Curtis Peebles, "The Guardians: A History of the 'Big Bird' Reconnaissance Satellites," *Spaceflight* 20(1978):381–5, 400; Jeffrey Richelson, "The Keyhole Satellite Program," *Journal of Strategic Studies* 7(1984):121–53; Paul B. Stares, *The Militarization of Space: U.S. Policy, 1945–1984* (Ithaca: Cornell University Press, 1985).

15 John Pike, "The Future of American Verification Capabilities in the Wake of the *Challenger* Accident," in W. Thomas Wander, Richard A. Scribner, and Kenneth N. Luongo (eds.), *Science and Security: The Future of Arms Control* (Washington, D.C.: Committee on Science, Arms Control, and National Security of the American Association for the Advancement of Science, 1986), p. 10.

16 According to one calculation, if the Space Telescope and its existing complement of instruments were to be directed toward the earth, it would be unable to do better than record a fifty-yard-long blur: memorandum, John T. Clarke, on "Can the Hubble Space Telescope Be Used for Earth Imaging?" March 1985, project scientist papers (file "Reading File Clarke 1985").

17 Quoted in J. Kelly Beatty, "HST and the Military Edge," *Sky and Telescope,* April 1985, p. 302.

18 House Committee on Science and Astronautics, *1975 NASA Authorization: Hearings Before the Subcommittee on Manned Space Flight,* 93rd Cong., 2nd sess., March 2, 1974, pt. 2, p. 1002.

19 OMB papers, set of notes entitled "ST 4/6/76," record group no. 51, accession no. 81-8 (file "SSET-SPB-NASA-FY76").

20 On this last point, see OHI, J. Welch, with RWS and PH, August 20, 1984, pp. 10–11.

21 See also minutes, LST Operations and Management Working Group, October 1–2, 1974, project scientist papers.

22 Minutes, LST Operations and Management Working Group, October 1–2, 1974, p. 5, project scientist papers, and OHI, C. R. O'Dell, with RWS, American Institute of Physics, May 21–23, 1985, session I, pp. 11–12.

23 The 1.8-meter telescope became known as the Medium Aperture Optical Telescope.

24 Report on LST Operations and Management Working Group of December 13, 1974, by W. M. Burton, dated December 16, 1974, and included with enclosures to minutes of working group meeting, project scientist papers.

25 The plan was that it would also make considerable use of subsystems employed in the *HEAO-B* spacecraft that Marshall was already building as an orbiting x-ray astronomy observatory. See Marshall's study "Medium Aperture Optical Telescope Preliminary Study MSFC-LST-MAOT-12-74."

26 Minutes, LST Operations and Management Working Group, December 13, 1974, p. 4, project scientist papers. See also OHI, E. M. Burbidge, with RWS, November 29, 1984, pp. 5–6.

27 The interview was brought to the working group's attention by O'Dell: C. R. O'Dell to "Dear Colleague," December 23, 1974, project scientist papers (file "Reading File 1974").

28 A. L. H., "Fletcher Sees Major Thrust in Space Science in 1980s," *Science* 186(1974):1012.

29 C. R. O'Dell and N. Roman to "Dear Colleague," February 12, 1975, project scientist papers. Again there is a striking similarity between Schardt's suggestion and the story of NASA's space station plans in the 1980s. It is also worth noting that the scheme Schardt outlined had been suggested in 1972 by administrator Fletcher: See Chapter 2.

30 C. R. O'Dell to N. G. Roman, March 4, 1975, in enclosures to LST Operations and Management Working Group, April 3–4, 1975, project scientist papers.

31 OHI, R. C. Bless, with RWS, November 3, 1983, p. 20. See also OHI, J. Bahcall, with PH, November 3, 1983, p. 9.

32 Memorandum for the record, J. Naugle, on "Space Science Board Meeting on January 9–10, 1975," Naugle papers, NASA History Office, box 1 (chronology file 1975).

33 See, for example, OHI, W. C. Keathley, with RWS, December 5, 1984, p. 49, C. R. O'Dell, with RWS, American Institute of Physics, May 21–23, 1985, session II, pp. 61–2, and "Space Telescope FY-78 New Start Review," presented by W. Keathley, December 21, 1976, Marshall presentation files (file "Project Plan Presentation to Dr. Fletcher").

34 James A. Downey III, "Program Status," in *The Space Telescope,* NASA SP-392 (Washington, D.C.: NASA, 1976), p. 28. Downey's paper was delivered to the American Astronomical Society in August 1975.

35 Much of the material used in the briefing was taken directly from Downey's presentation of April 23.

36 For the April 23 and May 15 briefings, see file entitled "LST Presentation to Dr. Hinners (HQ) 4/23/75," Marshall presentation files.

37 Memorandum for the record, G. Low, on "Large Space Telescope," June 17, 1975, administrator's papers, 255-80-0608, box 4 (file "Astronomy 3-1 1975–1977 Space Telescope"), and personal notes, #146, June 14, 1975, box 66, folder 1, and personal notes, #151, September 7, 1975, box 65, folder 5, George M. Low papers, Rensselaer Polytechnic Institute Archives, Troy, New York.

38 Memorandum for the record, J. Naugle, on "LST," June 26, 1975, Naugle papers, NASA History Office, box 1 (chronology file 1975). On other pressures to cut costs, see memorandum for the record, J. Downey III, on "Record of Telephone Conversation," June 24, 1975, Aucremanne papers (file "LST 1975").

39 OMB, too, had been unhappy with the agency's cost performance on Viking; see, for example, OHI, M. Norman, with RWS, September 6, 1984, p. 10.

40 It was also the view in OMB that "Congress appears to want NASA to conduct more initiatives to apply NASA technology in areas such as energy, earth resources, etc.": "Background Material for the Director's Meeting with Dr. Fletcher, July 3, 1975, Domestic Council," Schleede papers (file "NASA, 1975: General, January–July"), Gerald R. Ford Library, Ann Arbor, Michigan.

41 House Committee on Appropriations, *Department of Housing and Urban Development–Independent Agencies Appropriations for 1976: Hearings Before the Subcommittee on HUD–Independent Agencies, pt. 2, NASA,* 94th Cong., 1st sess., March 4, 1975, p. 427.

42 It was, for example, a central argument of Vannevar Bush, one of the chief participants in the debate over federal funding for science after World War II. A clear statement of Bush's case was made in his *Science, The Endless Frontier: A Report to the President* (Washington, D.C.: GPO, 1945). For the context of Bush's report, see Daniel J. Kevles, "The National Science Foundation and the Debate Over Post War Research Policy," *Isis* 68(1977):5–26, J. Merton England, *A Patron for Pure Science: The National Science Foundation's Formative Years, 1945–57* (Washington, D.C.: National Science Foundation, 1982), and Nathan Reingold, "Vannevar Bush's New Deal for Research: Or The Triumph of the Old Order," *Historical Studies in the Physical and Biological Sciences* 17(1985):299–344.

43 David F. Noble, *Forces of Production: A Social History of Industrial Automation* (New York: Oxford University Press, 1986), p. 13. These themes are discussed in some detail in Michael Reagan, *Science and the Federal Patron* (New York: Oxford University Press, 1969), pp. 40–1. Tyndall's argument is presented in John Tyndall, *Lectures on Light. Delivered in the United States in 1872–73* (New York: D. Appleton & Company, 1873), pp. 174–83.

44 Derek de Solla Price, *Little Science, Big Science . . . And Beyond* (New York: Columbia University Press), p. 240. See, too, Price's *Science Since Babylon* (New Haven: Yale University Press, 1975), Chapter 6. For fine guides to the literature on the nature of technological knowledge and the interactions of science and technology, see Edwin T. Layton, Jr., "Through the Looking Glass, Or News From Lake Mirror Image," *Technology and Culture* 28(1987):594–607, and George Wise, "Science and Technology," *Osiris* 1(1985):229–46, and references cited therein.

45 This point was made, for example, by George Field in congressional testimony: House Committee on Appropriations, *Department of Housing and Urban Development–Independent Agencies Appropriations for 1978: Hearings Before the Subcommittee on HUD–Independent Agencies, pt. 7,* 95th Cong., 1st sess., March 31, 1977, p. 150.

46 Barry M. Blechman, Edward M. Graunlich, and Robert W. Hartman, *Setting National Priorities. The 1975 Budget* (Washington, D.C.: Brookings Institution, 1974), p. 133.

47 House Committee, *Testimony of W. Panofsky on the Stanford Accelerator Power Supply: Hearing Before the Joint Committee,* 88th Cong., 2nd sess., January 29, 1964, pp. 24–5.

48 OHI, J. Bahcall, with PH, December 20, 1983, p. 64.

49 OHI, J. Bahcall, with PH, March 22, 1984, p. 3.

50 OHI, J. Bahcall, with PH, November 3, 1983, p. 35.

51 Ibid.

52 G. Low to distribution, October 28, 1975, administrator's papers, 255-80-0608, box 4 (file "Astronomy 3-1 1975–77 Space Telescope"). See, too, memorandum, assistant administrator for public affairs to deputy administrator, November 5, 1975, NASA History Office (file "Space Telescope Documentation"). On the importance of image in the launching of a program, see W. Henry Lambright, *Governing Science and Technology* (New York: Oxford University Press, 1976), Chapter 2.

53 E. M. Burbidge to G. Low, November 13, 1975, included in enclosures to LST Operations and Management Working Group, February 6, 1976, project scientist papers.

54 G. Low to E. M. Burbidge, January 28, 1976, Aucremanne papers (file "LST 1975 [*sic*]"). But see also S. Sobieski to C. R. O'Dell, June 18, 1975, included in enclosures to LST Science Working Group minutes, July 16–17, 1975, and Ernst Stuhlinger's suggested change of name in 1974: O'Dell daily notes, August 29, 1974, O'Dell papers.

55 James T. Lynn to James C. Fletcher, July 25, 1975, Fletcher papers, NASA History Office (file "Correspondence 1976").

56 James Fletcher to James T. Lynn, September 1975, OMB active files (file "Director's Review Book FY 1977").

57 See, for example, "Ford Asks Slash in Tax, Matched by Spending Cut," *New York Times,* October 7, 1975, p. 1. A detailed analysis of the fiscal 1977 budget as it affected research and development is provided by Willis H. Shapley, *Research and Development in the Federal Budget FY 1977* (Washington, D.C.: American Association for the Advancement of Science, 1976).

58 James C. Fletcher to James T. Lynn, October 20, 1975, OMB papers (file "Director's Review Book FY 1977").

59 A study of the president's daily logs in the Gerald R. Ford Library does not provide any evidence for this recollection. By the time Fletcher did meet with Ford, the telescope had already been deleted, and Fletcher did not appeal it.

60 Shapley, *Research and Development in the Federal Budget FY 1977,* p. 55.

61 George B. Field to file, March 2, 1976, Field papers, Smithsonian Institution archives (file "Space Telescope Congressional Action 1974–1978"). See also John Bahcall to "Dear Colleague," February 19, 1976, Bahcall papers, James Fletcher to President Ford, November 26, 1975, White House central files, box 13 (file "FI4/FG 131–164"), and James Fletcher to James T. Lynn, November 20, 1975, White House central files, box 13 (file "FI4/FG 131–164"), Gerald R. Ford Library.

62 OMB active files (file "Director's Review Book FY 1978"), issue paper on "Earth Orbiting Space Telescope."

63 Memorandum, James C. Fletcher to Al Lovelace, May 4, 1977, Fletcher papers, NASA History Office (file "Correspondence 1977").

64 The shuttle's peak development years were expected to be FY 1977 and FY 1978.

65 Cost overruns in developing the shuttle meant that the "shuttle dividend" did not materialize, in fact, until fiscal 1983. See the comments of James Beggs in House Committee on Science and Technology, *1984 NASA Authorization: Hearings Before the Subcommittee on Space Science and Applications, vol. 2,* 98th Cong., 1st sess., February 3, 1983.

66 On the budget decisions, see, for example, briefing paper "FY 1977 Budget [NASA]," staff secretary, special files, box 10 (file "FY-1977–Fifty Issues [1]"), Gerald R. Ford Library. See also Craig Covault, "Aeronautic Stress, Shuttle Stretch Planned by NASA," *Aviation Week and Space Technology,* January 26, 1976, pp. 30–3. For Low's comments, see personal notes, #160, February 7, 1976, box 65, folder 4, George M. Low papers, Rensselaer Polytechnic Institute Archives, Troy, New York.

67 For the robust view of one critic, see James A. Van Allen's statement in "A Compilation of Papers Prepared for the Subcommittee on NASA Oversight of the Committee on Science and Astronautics, U.S. House of Representatives, Ninety-First Congress, Second Session" on "The National Space Program – Present and Future," pp. 90–2. See also OHI, J. Van Allen, with DHD and AN, August 6, 1981, pp. 355–9.

68 "Science on the Skids: A Conversation with Thomas Donahue," *Spaceworld,* November 1986, p. 18.

69 J. N. Bahcall to J. Fletcher, January 22, 1976, administrator's papers, 255-80-0608, box 4 (file "Astronomy 3-1 1975–77 Space Telescope"). For Fletcher's reply, see J. Fletcher to J. N. Bahcall, February 3, 1976, same file. See, in addition, John Bahcall to John Walsh, January 23, 1976, Bahcall papers, and John Walsh, "Astronomers Go Into Orbit," *Science* 191(1976):544–5. For letters in a vein similar to Bahcall's, see, for example, Ivan King to James Fletcher, January 30, 1976, Bahcall papers, and Robert Danielson to James Fletcher, February 3, 1976, Bahcall papers.

70 C. R. O'Dell to John Naugle, January 2, 1976, project scientist papers (file "Reading File 1976").

71 J. Fletcher to Michael Duval, October 22, 1974, Domestic Council, Schleede papers, box 20 (file "NASA, 1974: Budget for FY 1976 [2]"), Gerald R. Ford Library.

72 This was the "Positron Electron Project" (PEP).

73 Memorandum, J. Bahcall, "Regarding the Restoration of PEP Construction Funds by Congress to the ERDA Budget for FY76," January 8, 1976, Bahcall papers.

74 J. Bahcall to L. Spitzer, January 9, 1976, Bahcall papers. At that time, Spitzer was physically dis-

Notes to pp. 160–5

tant from the action as he was visiting the Institut d'Astrophysique in Paris.

75 See, for example, John Wolfe to David Williamson, February 11, 1976, Aucremanne papers (file "Large Space Telescope 1975 [sic]").

76 J. Bahcall to L. Spitzer, January 9, 1976, Bahcall papers.

77 See, for example, Harold L. Davis, "New Support for Basic Research," *Physics Today*, April 1976, p. 96.

78 OHI, J. Bahcall, with PH, November 3, 1983, p. 23. See also the numerous letters to congressmen, of which copies were often sent to NASA: administrator's papers, 255-80-0608, box 4 (file "Astronomy 3-1 1975–77 Space Telescope"); see, for example, Phillip Morrison to Edward P. Boland, February 20, 1976.

79 Hugh Heclo, "Issue Networks and the Executive Establishment," in Anthony King (ed.), *The New American Political System* (Washington, D.C.: American Enterprise Institute, 1979), p. 97.

80 See Allan J. Cigler and Burdett A. Loomis (eds.), *Interest Group Politics* (Washington, D.C.: CQ Press, 1983), particularly Chapter 1, "Introduction: The Changing Nature of Interest Group Politics," pp. 1–30. On lobbying, see, for example, Lewis Anthony Dexter, *How Organizations Are Represented in Washington* (Indianapolis: Bobbs-Merrill, 1969), and *The Washington Lobby*, 4th ed. (Washington, D.C.: Congressional Quarterly, Inc., 1982). See also the data presented in Warren E. Miller, Arthur H. Miller, and Edward J. Schneider, *American National Election Studies Data Sourcebook: 1952–1978* (Cambridge, Mass.: Harvard University Press, 1980).

81 OHI, J. Bahcall, with PH, November 3, 1983, p. 32.

82 Among others, see J. Leiper Freeman, *The Political Process: Executive Bureau–Legislative Committee Relations* (New York: Random House, 1964), on policy subsystems; Douglass Cater, *Power in Washington* (New York: Vintage, 1964), on subgovernments; Ernest S. Griffith, *The Impasse of Democracy* (New York: Harrison-Milton Books, 1939), on whirlpools; and Theodore J. Lowi, *The End of Liberalism* (New York: Norton, 1969), on iron triangles.

83 Heclo, "Issue Networks and the Executive Establishment." For an interesting analysis of Heclo's use of issue networks, see A. Grant Jordan, "Iron Triangles, Woolly Corporatism and Elastic Nets: Images of the Policy Process," *Journal of Public Policy* 1(1981):95–123.

84 RWS is most grateful to Henry Lambright for emphasizing the importance of ad-hocracy (Lambright's term).

85 OHI, N. Hinners, with RWS and JT, October 17, 1984, p. 24. In 1985, O'Dell, in recalling the omission of all Space Telescope funds, said that "it makes you really wonder if they're not going to scrap the whole damn thing, when they do something like that": OHI, C. R. O'Dell, with RWS, American Institute of Physics, May 21–23, 1985, session II, p. 27. That, of course, was exactly the response Hinners had hoped for.

86 For examples of Congress using the Washington Monument game, see William Greider, *The Education of David Stockman and Other Americans* (New York: New American Library, 1986), p. 50.

87 OHI, G. Field, with RWS, March 3, 1986, pp. 4–6.

88 *An International Discussion of Space Observatories: Report of a Conference held at Williamsburg, Virginia, January 26–29, 1976* (Washington, D.C.: National Academy of Sciences, European Science Foundation, 1976).

89 George Field to James Fletcher, February 12, 1976, Bahcall papers. See also J. Bahcall, E. M. Burbidge, A. Code, G. Field, and L. Spitzer to J. Fletcher, January 21, 1976, Bahcall papers.

90 George Field to John Naugle, March 2, 1976, Bahcall papers, and George Field to John Naugle, May 14, 1976, Bahcall papers. The agenda was worked out in advance by Naugle and Field.

91 J. Bahcall to George Field, May 20, 1976, Bahcall papers.

92 On the May 19 meeting, see OHI, E. Margaret Burbidge, with RWS, November 29, 1984, p. 7, John Bahcall, with PH, December 20, 1983, p. 78, C. R. O'Dell, with RWS, American Institute of Physics, May 21–23, 1985, session II, p. 28. See also the following documents: John Bahcall to George Field, May 20, 1976, Bahcall papers, George Field to James Fletcher, May 26, 1976, Field papers, Smithsonian Institution Archives, Washington, D.C., O'Dell daily notes, May 19, 1976, O'Dell papers, and the material in the file "Meeting with Fletcher Headquarters, May 19, 1976," project scientist papers. The latter contains the text of the four talks that were delivered: "Scientific Uses of the Space Telescope," by Lyman Spitzer, "Technological Challenges and Spin-offs of the Space Telescope," by Arthur Code, "The Benefits of Astronomical Research," by George Field, and "Astronomy at Century's End: A Prophecy for the Space Tele-

scope Era," by Philip Morrison. Fletcher's notes of the meeting are in the Fletcher papers (file "Correspondence 1976"), headed "Meeting with Astronomers – 5/19/76," NASA History Office.

93 L. Spitzer note, "Chronology 1976–77 Campaign Re LST," dated February 17, 1982, Spitzer papers, and OHI, R. C. Henry, with RWS and PH, November 29, 1983, p. 45.

94 *Congressional Record,* 94th Cong., 2nd sess., April 1, 1976, S.9063-4.

95 James A. Fletcher to James L. Mitchell, April 12, 1976, OMB papers, 51-81-8 3/33:24–7 (file "SSET: SSPB: NASA FY1976 [*sic*] Space Telescope"). See also memorandum, Memphis A. Norman to SSET, Science and Space Programs Branch, on "NASA's Proposal for Early Selection of a Contractor for the Space Telescope," April 14, 1976, same record file.

96 Memorandum, H. Newell to J. Fletcher (undated, but December 1971), NASA Headquarters, Newell papers, 255-79-0649, box 25 (file "AA Reading File").

97 Minutes, Operations and Management Working Group, February 6, 1976, project scientist papers.

98 See also memorandum, George Low to James C. Fletcher, April 21, 1976, on "Additional Items of Interest," Fletcher papers, NASA History Office (file "Correspondence 1976").

99 Quoted in Richard J. Fenno, Jr., *The Power of the Purse* (Boston: Little, Brown, 1966), p. 168. See also *Congressional Record,* May 26, 1964, p. 11566.

100 "Edward Boland," in *Citizens Look at Congress,* 92nd Congress profiles, vol. IV (Washington, D.C.: Grossman Publishers, 1972), p. 9.

101 House Committee on Appropriations, *Department of Housing and Urban Development–Independent Agencies Appropriations for 1977: Hearings Before the Subcommittee on HUD–Independent Agencies, pt. 2, NASA,* 94th Cong., 2nd sess., February 18, 1976, pp. 146–7.

102 OHI, J. Bahcall, with PH, December 20, 1983, p. 68.

103 George Field to John Bahcall, March 14, 1977, and attachment of text of speech given by Boland, Field papers, Smithsonian Institution archives (file "Space Telescope Congressional Action 1974–1978").

104 George M. Low to James C. Fletcher, April 21, 1976, Fletcher papers, NASA History Office (file "Correspondence 1976").

105 Shapley, *Research and Development in the Federal Budget FY 1977,* p. 77. On NASA's technology and national security initiatives, see G. Low to administrator, May 25, 1971, box 35, folder 6, George M. Low papers, and G. Low to J. Fletcher, October 16, 1974, box 34, folder 5, George M. Low papers, Rensselaer Polytechnic Institute Archives, Troy, New York.

106 James Fletcher to President Ford, June 4, 1976, Domestic Council, Schleede papers, box 21 (file "NASA, 1976: Fletcher letter to the President, June 4"), Gerald R. Ford Library.

107 September 8, 1976, was the tenth anniversary of the first television showing of the regular "Star Trek" episodes (this excludes pilots).

108 Memorandum from Ron Konkel to the director (James T. Lynn), June 21, 1976, on "Dr. Fletcher's June 4 letter to the President on the future of the civilian space program," Domestic Council, Schleede papers, box 21 (file "NASA, 1976: Fletcher Letter to the President, June 4"), Gerald R. Ford Library.

109 Private communication, H. Guyford Stever, to RWS, January 27, 1986. This anecdote was well known in OMB.

110 On Ford and space, see Eugene M. Emme, "Gerald R. Ford," in *Between Sputnik and the Shuttle: New Perspectives on American Astronautics* (San Diego: American Astronautical Society, 1981), pp. 111–21.

111 President Ford's budget message in *Issues '78: Perspectives on Fiscal Year 1978 Budget* (Washington, D.C.: OMB, 1977), p. 3.

112 See Claude E. Barfield, *Science Policy from Ford to Reagan* (Washington, D.C.: American Enterprise Institute for Public Policy Research, 1982), David Dickson, *The New Politics of Science* (New York: Pantheon, 1984), James Everett Katz, *Presidential Politics and Science Policy* (New York: Praeger, 1978), pp. 180–239, and Margaret Rossiter, "Science and Public Policy Since World War II," *Osiris* 1(1985):273–94. RWS is most grateful to Memphis Norman for pointing out that a useful means of following executive branch thinking on the funding of science and technology is provided by Section K of the *Special Analyses, Budget of the United States Government,* issued each year by OMB.

113 Issue paper 2, "Earth Orbiting Space Telescope (1983 launch date)," in OMB active files (file "Director's Review Book FY 1978 for NASA"). This issue paper also goes into detail on the benefits and disadvantages of European cooperation.

114 See Bahcall papers for these reports, and, for ex-

ample, see materials in file "ST–New Start", project scientist papers.

115   Committee on Appropriations, *Department of Housing and Urban Development–Independent Agencies Appropriations for 1978: Hearings Before the Subcommittee on HUD–Independent Agencies, pt. 5, NASA,* 95th Cong., 1st sess., March 29, 1977, p. 49.

116   House Committee on Appropriations, *Department of Housing and Urban Development–Independent Agencies Appropriations for 1978: Hearings Before the Subcommittee on HUD–Independent Agencies, pt. 7,* March 31, 1977, p. 147.

117   C. R. O'Dell to J. F. Toohey, March 3, 1977, project scientist papers (file "Congressional and legislative"). As O'Dell told Toohey, a legislative aide to Congressman Flippo, "The potential impact on security matters and the DOD budget I consider absolutely sound."

118   William L. Putnam to Edward P. Boland, February 11, 1977; copy in Field papers, 83-025, box 5, Smithsonian Institution archives (file "Space Telescope Congressional Action 1974–1978"). See, too, William Baum to William L. Putnam, March 3, 1976, William L. Putnam to William Baum, March 8, 1976, and William Baum to William L. Putnam, March 12, 1976, copies in STHP files. RWS is most grateful to William Baum for copies of this correspondence.

119   On Boland's friendship with O'Neill, see Tip O'Neill (with William Novak), *Man of the House: The Life and Political Memoirs of Speaker Tip O'Neill* (New York: Random House, 1987), and Martin F. Nolan, "Capital Odd Couple Calls It Quits," *Boston Sunday Globe,* February 20, 1977, p. 1. The anecdote is told in G. Field to J. Bahcall, April 8, 1977, Bahcall papers. Whether or not this encounter did any good for the telescope is, of course, another matter.

120   Copy in STHP files. The text was largely the work of R. C. Henry, and the charts were prepared by T. Bland Norris.

121   OHI, T. Bland Norris, with PH, April 19, 1984, p. 47.

122   Senate, *Hearings Before the Committee on Commerce, Science, and Transportation,* 95th Cong., 1st sess., nominations, April 7, 1977, pp. 37–98.

123   Memorandum, Frank Press to Frank Moore, on "Support for NASA's Space Telescope," April 14, 1977, OMB papers, 51-81-7, box 3 (file "SET-SSPB-NASA FY 1977–78 Space Telescope").

124   Ibid.

125   OHI, J. Bahcall, with PH, December 20, 1983, p. 66.

126   See, for example, Boggs's comments: House Committee on Appropriations, *Department of Housing and Urban Development–Independent Agencies Appropriations for 1978: Hearings Before the Subcommittee on HUD–Independent Agencies, pt. 5, NASA,* 95th Cong., 1st sess., March 29, 1977, pp. 100–1.

127   *Congressional Record,* 94th Cong., 1st sess., June 24, 1975, H.6010, comments by Mr. Boland; "House Committee Votes Pioneer Venus Delay, Seeks Choice Between It and LST," *Aerospace Daily,* June 20, 1975, p. 282.

128   Allen L. Hammond, "Pioneer-Venus: Did Astronomers Undercut Planetary Science?" *Science* 189(1975):270.

129   NASA document, "Strategy for Outer Planets Exploration," June 11, 1975, p. E6. RWS is most grateful to Craig Waff for bringing this report to his attention.

130   OHI, W. Baum, with RWS, June 22, 1986, p. 27. See also OHI, C. Chapman, with JNT, October 10, 1984, pp. 24–5, W. Brunk, with JNT, May 25, 1984, pp. 26–32, and J. Westphal, with JNT, September 28, 1985, pp. 53–6. See G. Field to C. Chapman, February 22, 1977, D. Hunten to C. Chapman, February 18, 1977, both copies in STHP files (copies provided by Clark Chapman).

131   W. Baum to N. Roman, June 8, 1976, program scientist papers, NASA Headquarters (file "ST–Faint Object Camera"). See also C. Chapman to J. Fletcher, April 7, 1976, George M. Low papers, box 34, folder 1, Rensselaer Polytechnic Institute Archives, Troy, New York.

132   We shall consider the choice of scientific instruments further in Chapter 7.

133   OHI, N. Hinners, with RWS and JT, October 17, 1984, p. 29. See also OHI, C. Chapman, with JT, October 10, 1984, pp. 27–8.

134   OHI, C. Chapman, with JT, October 10, 1984, p. 27.

135   *Congressional Record,* 95th Cong., 1st sess., July 19, 1977, pp. 23668–78.

136   See the copy of the memorandum of understanding in, for example, "European Space Agency Space Telescope Working Group: Report on the Utilisation of the Space Telescope" [ESA document ST-WG(79)9], Paris, November 30, 1979.

137   For example, OHI, W. Keller, with RWS, December 9, 1985, p. 75.

138   OHI, John Bahcall, with PH, December 20, 1983, p. 59.

6. MAKING AN INSTITUTE

1 Homer E. Newell, *Beyond the Atmosphere: Early Years of Space Science,* NASA SP-4211 (Washington, D.C.: NASA, 1980), p. 205.

2 On the changing influence of the National Academy, see Rexmond C. Cochrane, *The National Academy of Sciences: The First Hundred Years 1863–1963* (Washington, D.C.: NAS, 1978), A. Hunter Dupree, *Science in the Federal Government: A History of Policies and Activities* (Baltimore: The Johns Hopkins University Press, 1986), and Daniel J. Kevles, *The Physicists: The History of a Scientific Community in Modern America* (New York: Random House, 1979).

3 Newell, *Beyond the Atmosphere,* p. 204.

4 Ibid.

5 N. S. Hetherington, "Winning the Initiative: NASA and the U.S. Space Science Program," *Prologue* 7(1975):99.

6 L. Spitzer, Jr., to N. Roman, March 18, 1960, Spitzer papers (file "OAO Program Correspondence 1960–1971").

7 Draft memorandum, A. D. Code, L. Goldberg, L. Spitzer, and F. L. Whipple to N. Roman, June 15, 1960, Spitzer papers (file "OAO Correspondence 1960–1971"). On control, my own thoughts have been sharpened by Peter Galison, Bruce Hevly, and Rebecca Lowen, "Controlling the Monster: Stanford and the Growth of Physics Research, 1935–1962," papers presented at the Workshop on the History of Big Science, Stanford University, 1988.

8 On the operations of this committee and the opposition faced initially from the Space Science Board, see Newell, *Beyond the Atmosphere,* pp. 214–17.

9 Newell, *Beyond the Atmosphere,* p. 216.

10 James E. Webb to Norman F. Ramsey, January 14, 1966, Appendix A(1) of "NASA Ad Hoc Science Advisory Committee Report to the Administrator 15 August 1966," copy in STHP files.

11 The Ramsey committee, composed of scientists from a range of disciplines, was hardly in a position, even if it had wanted to, to make judgments about scientific priorities in astronomy.

12 "NASA Ad Hoc Science Advisory Committee Report to the Administrator 15 August 1966."

13 On the origins of AUI and Brookhaven, see Allan A. Needell, "Nuclear Reactors and the Founding of the Brookhaven National Laboratory," *Historical Studies in the Physical Sciences* 14(1984):93–122. The number of member universities would grow in future years.

14 On the founding of NRAO, see Allan A. Needell, "Lloyd Berkner, Merle Tuve and the Federal Role in Radio Astronomy," *Osiris* 3(1987):261–88.

15 J. Merton England, *A Patron for Pure Science: The National Science Foundation's Formative Years, 1945–57* (Washington, D.C.: NSF, 1982), p. 290. On the founding of Kitt Peak, see England, pp. 290–2, and see England, pp. 280–90, for the establishment of the National Radio Astronomy Observatory.

16 H. Newell and H. Smith, "Interim Response to the Report of the Ad Hoc Science Advisory Committee," June 7, 1967, Newell papers, p. 10.

17 Private communication, Leo Goldberg to RWS, July 8, 1987.

18 Leo Goldberg to Thomas O. Paine, June 5, 1970, copy in Newell papers, 255-79-0649, box 25 (file "AA Reading Files"). Goldberg also recalled that he had earlier been prepared to resign as chairman of the Astronomy Missions Board in order to accept the chairmanship of what later became the Greenstein committee, discussed in Chapter 4. He was persuaded not to resign by John Naugle, on the plea that it would soon be astronomy's turn for a major increase in funding, and Goldberg would be needed to help sell the Astronomy Missions Board program to Congress. The funds never materialized. Private communication, Leo Goldberg to RWS, July 8, 1987.

19 Homer E. Newell, "Notes on Science in NASA," November 14, 1969, and for Dessler's comments, see Appendix A, "Discontent of Space Science Community," Newell papers, 255-79-0649, box 24 (file "AA Chron Files").

20 J. Allen Crocker to Dr. Newell, May 15, 1970, Newell papers, 255-79-0649, box 25 (file "AA Reading Files").

21 Four interdisciplinary committees were placed under the council: physical sciences, life sciences, applications, and space systems. On the new advisory structure, see Newell, *Beyond the Atmosphere,* p. 219.

22 Memorandum, Homer Newell to Dr. Fletcher, "Relations with the Scientific Community and the Space Science Board," (undated but December 1971), Newell papers, 255-79-0649, box 25 (file "AA Reading Files"). Naugle, Newell wrote, also agreed with him.

23 C. R. O'Dell, "The Science Management Program for the Large Space Telescope," November 3, 1972, project scientist papers (file "Science Community Involvement in LST Project").

24 Minutes, LST Operations and Management

Working Group, November 5–6, 1973, p. 4, project scientist papers, but see also the memorandum in that file, T. Kelsall to C. R. O'Dell, November 8, 1973, containing Kelsall's "raw" notes of the meeting.

25 The other members were John Bahcall, Gerry Neugebauer, and Robert Noyes.

26 "Ad Hoc Committee, April 16, 1974. Summary of Activities," included in enclosures with minutes of LST Operations and Management Working Group, April 16–17, 1974, project scientist papers. It included notes on the experience of principal investigators on several space flight programs, together with the views of NASA personnel, including a discussion between the committee members and Goddard's project manager George Levin, O'Dell, and astronaut-scientist R. A. R. Parker.

27 Lyman Spitzer and George Field had been added to the uncommittee.

28 "LST Organization," report by the DATA IDT, September 1974, with enclosures to Operations and Management Working Group meeting, October 1–2, 1974, project scientist papers. Copies were sent to each member of the working group. See also OHI, R. C. Bless, with RWS, November 3, 1983, pp. 75–6.

29 "Large Space Telescope Post Launch Operations: NASA Laboratory for Space Astrophysics," September 1974, Goddard active files (file "Post-Launch Operations").

30 Such a scheme is essentially that eventually employed on the International Ultraviolet Explorer, launched in 1978.

31 In June, Pieper had told Naugle that Goddard did not support the idea of an institute. However, Goddard suggested "the creation of Regional Science Centers that would satisfy many of the same objectives of the single Science Institute": George F. Pieper to J. Naugle, June 18, 1974, Naugle papers, NASA History Office, box 1 (chronology file 3/19–12/31/74).

32 OHI, W. Keller, with RWS, November 28, 1984, pp. 41–2, and N. Hinners, with RWS and JT, December 4, 1984, p. 91.

33 OHI, R. Bless, with RWS, November 3, 1983, p. 19. This opinion has also been expressed to RWS in numerous private communications. Let us note here that Goddard's operation of the International Ultraviolet Explorer satellite, launched in 1979, has been regarded by many astronomers as a successful way to run the operations of a scientific satellite. The link between the International Ultraviolet Explorer and the Space Telescope is, of course, far from perfect because the IUE is a much simpler spacecraft than the Space Telescope, and the IUE, which is in geosynchronous orbit, is relatively straightforward to operate. Even though IUE is in many respects Big Science, Goddard has succeeded in operating it in the spirit of little science.

34 This was Gerry Neugebauer.

35 Minutes, LST Operations and Management Working Group, October 1–2, 1974, pp. 5–7, project scientist papers. See also memorandum, George M. Levin for the record, on "Results of Discussion Method [sic] of Operation of LST Observatory," included in enclosures to O'Dell's copy of minutes.

36 Memorandum, James A. Downey to G. Levin, "LST Science Institute," January 21, 1975, project scientist papers (file "Reading File 1975").

37 James A. Downey to George M. Levin, on "LST Mission and Data Operations Study," March 24, 1975, and James A. Downey to George M. Levin, on "LST Mission and Data Operations Study Guidelines," March 25, 1975. These are enclosures 3 and 4 to the minutes of LST Operations and Management Working Group meeting, April 3–4, 1975, project scientist papers.

38 O'Dell daily notes, February 28, 1975, O'Dell papers.

39 Minutes, LST Operations and Management Working Group meeting, April 3–4, 1975, pp. 5–6, project scientist papers.

40 J. Bahcall to C. R. O'Dell, April 14, 1975, project scientist papers (file "Reading File 1975").

41 C. R. O'Dell to J. Bahcall, April 18, 1975, project scientist papers (file "Reading File 1975").

42 See, for example, OHI, W. Keller, with RWS, November 28, 1984, p. 67, and N. Hinners, with RWS and JT, December 4, 1984, p. 95.

43 C. R. O'Dell to R. Giacconi, June 2, 1975, project scientist papers (file "Reading File 1975").

44 C. R. O'Dell to J. Naugle, May 30, 1975, project scientist papers (file "STSCI [1975] #1").

45 See also C. R. O'Dell to R. P. Kraft, June 5, 1975, project scientist papers (file "Reading File 1975"), and J. E. Naugle to C. R. O'Dell, August 8, 1975, administrator's papers, 255-80-0608, box 4 (file "Astronomy 3-1 1975–77 Space Telescope").

46 OHI, N. Hinners, with RWS and JT, November 6, 1984, p. 76.

47 N. Hinners, "NASA In-House Astronomers –

Verse and Converse," in M. S. Longair and J. W. Warner (eds.), *Scientific Research with the Space Telescope,* NASA CP-2111 (Washington, D.C.: NASA, 1979), p. 322.

48 OHI, N. Hinners, with RWS and JT, December 4, 1984, p. 46.

49 OHI, N. Hinners, with RWS and JT, December 4, 1984, p. 92.

50 See "Space Telescope (ST) Science Institute Study," April 27, 1976, enclosures 1 and 2 in minutes of ST Operations and Management Working Group, May 5–6, 1976, project scientist papers.

51 "A Program for High Energy Astrophysics (1977–1988) by the Ad-Hoc Planning Group of the High Energy Astrophysics Management Operations Working Group," Field papers, Smithsonian archives, 83-025, box 5 (file "Space Telescope Science Institute 1976–June 1977").

52 H. Friedman and R. Giacconi to J. Fletcher, November 13, 1975, Field papers, Smithsonian archives, 83-025, box 5 (file "Space Telescope Science Institute 1976–June 1977"). See also memorandum for the record, J. Naugle, on "Telephone Call from Gerald F. Tape (President, Associated Universities, Inc.) on September 3, 1975," September 5, 1975, Naugle papers, NASA History Office, box 1 (file "Chron File 1975"), and R. Giacconi to N. Hinners, July 1, 1975, Roman papers (file "ST"), NASA Headquarters, Washington, D.C.

53 "Draft of a Proposal for a National Institute for X-Ray Observatories," November 13, 1975, p. 9, Field papers, Smithsonian archives, 83-025, box 5 (file "Space Telescope Science Institute 1976–June 1977").

54 "9-5-75 Notes – O'Dell," project scientist papers (file "Reading File 1975").

55 R. P. Kraft to J. Naugle, September 3, 1975, Naugle papers, NASA History Office (file "Chron File 1975"). See also J. E. Naugle to R. P. Kraft, September 29, 1975, administrator's papers, 255-80-0608, box 4 (file "Astronomy 3-1 1975–77 Space Telescope").

56 G. Field to L. Spitzer, September 9, 1975, Field papers, Smithsonian archives, 83-025, box 7 (file "PSC Correspondence July 75 to Dec. 75"). Field's judgment was supported the following year by astronomer Halton Arp in a letter to N. Roman, June 17, 1976, project scientist papers (file "Shuttle Astronomy Working Group GSFC, 14–15 July 1976").

57 See "URA Proposal for the Space Telescope Science Institute . . . March 3, 1980," section D-4, business management proposal, 3-12, URA active files.

58 OHI, Noel W. Hinners, with RWS and JT, November 16, 1984, p. 76.

59 Newell, *Beyond the Atmosphere,* p. 241. See also personal notes, #93, May 16, 1973, box 68, folder 4, George M. Low papers, Rensselaer Polytechnic Institute Archives, Troy, New York.

60 Newell, *Beyond the Atmosphere,* p. 242.

61 That idea may have been due to a suggestion by Richard Goody of Harvard in the wake of the 1976 Williamsburg conference on international space observatories. The astronomers at the conference had empowered Goody to approach NASA to offer help in establishing a "Space Astronomy Institute," and Goody had told Naugle in February 1976 that a group could be organized to represent the National Academy of Sciences, the American Astronomical Society, and the European Science Foundation. Naugle had then informed Goody that there were indeed precedents for the academy assisting in the foundation of such facilities, including the Lunar Science Institute: J. Naugle to J. Fletcher, February 4, 1976, Naugle papers, NASA History Office (file "AA Chron Jan 1–April 30 1976").

62 *Institutional Arrangements for the Space Telescope* (Washington, D.C.: National Academy of Sciences, 1976), p. 36.

63 Ibid., p. 9.

64 Ibid., p. 35.

65 See, for example, "Summary Minutes, Management and Operations Working Group in Shuttle Astronomy," June 7, 1976, copy in Field papers, Smithsonian archives, 83-025, box 5 (file "AUI Meeting July 7 La Jolla"). Particularly illuminating are the notes on a telephone call Jay Bergstralh received from Goddard's study manager George Levin in late 1975. At that time, Bergstralh was executive secretary of a group called the Preston committee. It was composed of several leading astronomers and had been convened by NASA to examine the telescope's scientific usefulness. As Levin told Bergstralh, there were opposing viewpoints on the nature of operating the telescope, but George Pieper, Goddard's director of earth and planetary sciences, "found overwhelming sentiment in favor of institute, from astronomical community": Notebook "LST Committee J. T. Bergstralh," copy provided to STHP by Bergstralh.

66 OHI, J. Bahcall, with PH, March 22, 1984, p. 17.

67 OHI, N. Hinners, with RWS and JT, December 4, 1984, p. 93.

68 O'Dell and Roman were members, and others who sat on it were the ST mission operations manager and the director of mission control and data processing in NASA Headquarters. There were two ex officio members as well as four advisors, including a representative from the National Science Foundation.

69 Bahcall, for example, came to regard him as a "thoughtful, perceptive administrator" whose views counted "with everyone": OHI, J. Bahcall, with PAH, March 22, 1984, p. 11.

70 Memorandum, N. Hinners to distribution, on "Space Telescope Science Operations Management Working Group," April 5, 1977, being attachment 4 in memorandum from J. Warren Keller to "Space Telescope Science Operations Management Working Group Members," on "Report of the First Session of the Space Telescope Science Operations Management Working Group," May 3, 1977, project scientist papers (file "ST Sci. Opns. & Mgt. W.G.").

71 "Briefing to Hinners by Keller and O'Dell on STSc[I] Operation," July 21, 1977, Marshall presentation files.

72 A. G. W. Cameron to N. W. Hinners, July 27, 1977, project scientist papers (file "ST Sci. Opns. & Mgt. W.G.").

73 T. B. Owen to Philip M. Smith, August 4, 1977, GSFC papers, 255-82-296, box 9 (file "ST 1977–75").

74 Memorandum, M. Norman to H. Loweth, on "National Academy of Sciences Report Regarding Institutional Arrangements for the Space Telescope," April 6, 1977, OMB papers, 51-81-7, box 3 (file "SET-SSPB-NASA FY 1977–78 Space Science").

75 "Final Report of the Space Telescope Science Operations Management Working Group," November 1977, copy in project scientist papers (file "Keller Committee").

76 Memorandum, Bob (Cooper) to Ed Smylie, December 22, 1977, GSFC papers, 255-82-296, box 9 (file "ST 1977–75"). See also the attachment to this memorandum, "Presentation on Space Telescope Activities in Laboratory for Astronomy and Solar Physics," December 22, 1977.

77 House Committee on Science and Technology, *1979 NASA Authorization: Hearings Before the Subcommittee on Space Science and Applications, vol. 1, pt. 2,* 95th Cong., 2nd sess., February 2, 1978, p. 1576.

78 On the NASA Headquarters view of Cooper's po-

sition, see OHI, N. Hinners, with RWS and JT, December 4, 1984, p. 95.

79 L. Spitzer, Jr., to Robert A. Frosch, January 19, 1978, project scientist papers (file "STScI [1978]").

80 O'Dell daily notes, April 21, 1978, O'Dell papers.

81 See W. Keathley to W. Lucas, April 24, 1978, copy in O'Dell daily notes, April 21, 1978, O'Dell papers, and meeting record by Robert J. McCormick, "Space Telescope Institute (STI)," April 21, 1978, 3:30 p.m., copy in project scientist papers (file "STScI [1978]"), for the two meetings. The briefing charts for the second meeting are in GSFC papers, *Institutional Approach for Space Telescope Science Operations*. Briefing for NASA Administrator April 21, 1978," GSFC papers, 255-82-296, box 9 (file "ST 1979").

82 O'Dell daily notes, July 5, 1978, O'Dell papers. The briefing charts are in Goddard papers, *"Space Telescope Institute Planning*. Briefing for NASA Deputy Administrator July 5, 1978," GSFC papers, 255-82-296, box 9 (file "ST 1978"). Goddard again proposed a significant role for the center's Laboratory for Astronomy and Solar Physics, this time by offering to provide an office to act as a link between the institute and NASA. That proposal got nowhere. See "An Approach to Scientific Operations for the Space Telescope," May 1978, concurrence by G. F. Pieper and R. S. Cooper, in the same file.

83 House Committee on Science and Technology, *Space Telescope Program Review: Hearings Before the Subcommittee on Space Science and Applications,* 95th Cong., 2nd sess., July 13, 1978, pp. 6–7. O'Dell witnessed the hearing and thought Hinners's presentation "excellent": O'Dell daily notes, July 13, 1978, O'Dell papers.

84 Pamela E. Mack, "The Politics of Technological Change: A History of LANDSAT" (Ph.D. dissertation, University of Pennsylvania, 1983), p. 174. For an account of a dispute on the siting of a physics accelerator, see Anton G. Jachim, *Science Policy Making in the United States and the Batavia Accelerator* (Carbondale: Southern Illinois University Press, 1975).

85 On AURA's objections to picking a site, for example, see OHI, A. Code with RWS, February 22, 1984, pp. 34–5.

86 For a brief history of AURA, see *AURA: The First Twenty-Five Years 1975–1982* (Tucson, Ariz.: Association of Universities for Research in Astronomy, 1983).

87 Memorandum, C. Swanson to distribution, "AURA-ABMA meeting, August 5, 1959 SAT-

ELLITE TELESCOPE," copy in F. K. Edmondson's active files. RWS is grateful to Professor Edmondson for bringing this meeting to his attention.

88 Memorandum, M. Belton to distribution, on "STI Working Paper," June 16, 1976, and J. Greenstein to "Members, Ad Hoc Committee on the Space Telescope Institute," August 4, 1976, copies provided to STHP by J. Teem. See also OHI, J. Teem, with RWS, September 26, 1984, p. 6.

89 *AURA Issues*, 1(1978), 1.

90 AURA document, "Minutes of Meeting of Ad-Hoc Committee on Space Telescope Science Institute," Tucson, Arizona, December 5, 1978, copy provided by J. Teem to STHP (file "STScI Items of Historical Interest").

91 OHI, J. Teem, with RWS, September 26, 1984, p. 23.

92 OHI, J. Teem, with RWS, September 26, 1984, p. 23.

93 That was not done, and as we shall see in Chapter 9, it was soon widely regarded as a major mistake.

94 N. Hinners to A. Cameron, August 29, 1978, and D. Hornig to A. Stofan, April 13, 1978, AURA records, drawer STScI (file "STScI Hornig").

95 "AURA, Inc., Annual Meeting of Board of Directors, March 7–8, 1979. Subject: *Report of ad-hoc Committee on STSI*," copy in file provided by J. Teem to STHP (file "STScI Items of Historical Interest").

96 Memorandum from "Committee on the Space Telescope Science Institute (STScI)" to "URA Board of Trustees," on "Recommendations relative to the involvement of URA in the STScI competition and to the selection of a site," July 18, 1979, URA active files.

97 See minutes of the URA board of trustees, July 20, 1979, and September 20–21, 1979, URA active files.

98 I. King to N. Hinners, February 21, 1979, project scientist papers (file "Reading File 1979").

99 OHI, A. Davidsen, with RWS, January 29, 1984, p. 14.

100 See B. E. J. Pagel, "Ultraviolet Observation of a Quasar Spectrum," *Nature* 269(1977):195.

101 OHI, R. Henry, with RWS and PH, November 29, 1983, pp. 68–70.

102 OHI, W. Fastie, with RWS, June 6, 1986, p. 82.

103 OHI, A. Davidsen, with RMS, January 29, 1984,

p. 18, and J. Teem, with RWS, September 26, 1984, p. 17.

104 OHI, R. Henry, with RWS and PH, November 29, 1983, p. 70.

105 AURA document, "Minutes of Meeting of Ad Hoc Committee on Space Telescope Science Institute," Chicago, Illinois, September 8, 1979, copy provided by J. Teem to STHP (file "STScI Items of Historical Interest"). See also OHI, J. Teem, with RWS, September 26, 1984, p. 16, and A. Code, with RWS, February 22, 1984, p. 39.

106 "Minutes of Meeting of Ad-Hoc Committee on Space Telescope Science Institute," Tucson, Arizona, May 22, 1979, copy provided by John Teem to STHP.

107 The Johns Hopkins people also included members of the university's Applied Physics Laboratory.

108 OHI, J. Teem, with RWS, September 26, 1984, p. 50.

109 OHI, B. Lasker, with RWS, December 8, 1983, p. 15. See also OHI, J. Teem, with RWS, September 26, 1984, p. 36, and memorandum, J. Teem to A. Code, December 29, 1979, on "My Role in STScI Proposal Development," AURA records, drawer "ST Proposal" (file "Contract Negotiations").

110 OHI, B. Lasker, with RWS, December 8, 1983, p. 23.

111 J. Bahcall to J. Teem, January 16, 1981, AURA records, drawer "STScI" (file "STScI Miscellaneous").

112 OHI, J. Bahcall, with PH, March 22, 1984, p. 28. For an amusing account of how Bahcall sounded out James Westphal as a potential deputy director, see OHI, J. Westphal, with DHD, September 15, 1982, pp. 247–8.

113 OHI, J. Teem, with RWS, September 26, 1984, p. 23.

114 OHI, J. Bahcall, with PH, March 22, 1984, p. 23.

115 OHI, J. Bahcall, with PH, March 22, 1984, p. 23.

116 Memorandum, "Joseph Garcia, Director Program Operations Division, to 1200/Procurement Officer: Selection Statement for Space Telescope Science Institute," Goddard active files (file "STScI").

7 · UP AND RUNNING

1 OHI, D. Tenerelli, with RWS, December 23, 1985, pp. 154–5.

2 See the NASA documents "Space Telescope Project. Optical Telescope Assembly. Request for Proposal NAS8-1-6-PP-00597Q, January 28, 1977" and "Space Telescope Project. Support Systems Module. Request for Proposal NAS8-1-6-PP-00596Q, January 28, 1977."

3 See NASA, "Space Telescope Project. Optical Telescope Assembly. Request for Proposal NAS8-1-6-PP-00597Q, January 28, 1977": "Contract End Item Specification," p. 7.

4 See also OHI, J. Olivier, with RWS, January 18, 1984, p. 19.

5 OHI, W. C. Keathley, with RWS, December 5, 1984, pp. 45–6.

6 See, for example, OHI, W. Keller, with RWS, December 9, 1985, p. 47.

7 For a discussion of NASA's procurement practices in the 1960s and early 1970s, see Arnold S. Levine, *Managing NASA in the Apollo Era,* NASA SP-4012 (Washington, D.C.: NASA, 1982), pp. 84–105 and Chapter Two.

8 Douglas C. Harvey to James C. Fletcher, June 4, 1976, Aucremanne papers (file "Keller [Space Telescope]").

9 Cited in J. Kelly Beatty, "HST and the Military Edge," *Sky and Telescope* 69(1985):302.

10 Ron Ondrejka, former member of Itek's advance planning group, quoted in J. Kelly Beatty, "HST and the Military Edge," p. 302.

11 A. M. Lovelace to Juan J. Amodel, September 13, 1977, NASA History Office (file "Space Telescope Documentation"). See also OHI, C. R. O'Dell, with RWS, American Institute of Physics, May 21–23, 1985, session II, p. 53, J. Olivier, with RWS, February 27, 1985, pp. 15–16, and memorandum, C. Hartmann to R. Konkel, on "ST Project Review Feb. 9–10, 1983," S. Keller papers, NASA Headquarters, Washington, D.C.

12 Paul B. Stares, *The Militarization of Space: U.S. Policy, 1945–1984* (Ithaca: Cornell University Press, 1985), p. 31. See also, for example, William E. Burrows, *Deep Black: Space Espionage and National Security* (New York: Random House, 1986), pp. 84–90, 236–43.

13 See Chapter 5 (section entitled "Back to the Drawing Board").

14 C. R. O'Dell to J. Bahcall, July 27, 1977, project scientist papers (file "Reading File 1977").

15 "A Report to the Committee on Appropriations, U.S. House of Representatives, on the NASA Space Telescope Program," House Appropriations Committee, surveys and investigations staff, March 16, 1983, pp. 20, 23.

16 OHI, R. C. Bless, with RWS, November 3, 1983, p. 47. See also OHI, C. R. O'Dell, with RWS, May 21–23, 1985, session II, p. 57.

17 OHI, D. Leckrone, with RWS, August 14, 1984, p. 72. See also OHI, W. Keller, with RWS, December 9, 1985, p. 47.

18 See "Meeting with LMSC 10-12-77," notes on "ST Management Strategy (Cost Control)," Marshall presentation files.

19 OHI, J. Olivier, with RWS, February 27, 1985, p. 11.

20 The *Copernicus* satellite, the Einstein Observatory, and many of the space astronomy activities at the Naval Research Laboratory are examples.

21 Memorandum, Nancy Roman to associate administrator for space science, "Space Telescope Management," January 28, 1976, Aucremanne papers (file "Space Telescope 1976").

22 See GSFC active files (file "GSFC/MSFC Intercenter Agreements").

23 Memorandum, C. R. O'Dell to W. R. Lucas, October 26, 1976, and C. R. O'Dell and W. Keathley to Dr. Lucas, December 3, 1976, both in project scientist papers (file "Reading File 1976"). See also O'Dell daily notes, October 21, 1976, October 22, 1976, December 2, 1976, and December 7, 1976, O'Dell papers.

24 Private communication, T. B. Norris to P. Hanle, December 18, 1986, and OHI, W. J. Keller, with RWS, November 28, 1984, p. 15. On the MSFC/GSFC relationship, see also OHI, D. Fordyce, with RWS and PH, October 31, 1983, pp. 2–5.

25 "Project Plan for Space Telescope Design and Development Phase," Marshall Space Flight Center, July 1977, p. 47.

26 "A Report to the Committee on Appropriations, U.S. House of Representatives, on the NASA Space Telescope Program," House Appropriations Committee, surveys and investigations staff, March 16, 1983, p. 9.

27 Alex Roland, "Science and War," *Osiris* 1(1985):269.

28 Leslie R. Groves, *Now It Can Be Told: The Story of the Manhattan Project* (New York: Harper & Brothers, 1962), p. 169. On compartmentalization on the telescope program, see "A Report," House Appropriations Committee, surveys and investigations staff, March 16, 1983, p. 10.

29 "A Report," House Appropriations Committee, surveys and investigations staff, March 16, 1983, p. 9. On the Marshall Space Flight Center's reputation in NASA Headquarters for overloading contractors with requests for reports, see also OHI,

T. B. Norris, with PH, April 19, 1984, pp. 34–5.

30 OHI, J. Olivier, with RWS, February 27, 1985, p. 14. See also personal notes, #151, September 10, 1975, box 60, folder 8, George M. Low papers, Rensselaer Polytechnic Institute Archives, Troy, New York.

31 OHI, W. C. Keathley, with RWS, December 5, 1984, pp. 27–8.

32 In NASA terminology, this would be termed "co-located." It refers to members of one NASA center being placed at another center. So Goddard staff resident at Marshall are "co-located" at Marshall.

33 OHI, W. C. Keathley, with RWS, December 5, 1984, p. 30.

34 These were for interfaces, software, assembly and verification, and mission operations.

35 For a description of the working groups, see House Committee on Science and Technology, *Space Telescope Program Review: Hearings Before the Subcommittee on Space Science and Applications,* 95th Cong., 2nd sess., July 13, 1978, pp. 68–9.

36 Private communication, W. C. Keathley to RWS, November 20, 1986.

37 See Harvey M. Sapolsky, *The Polaris System Development: Bureaucratical Programmatic Success in Government* (Cambridge, Mass.: Harvard University Press, 1972), pp. 94–130.

38 For a brief discussion of PERT, see John Noble Wilford, *We Reach the Moon* (New York: Bantam Books, 1969), p. 54.

39 See "Project Plan Presentation by Marshall to Dr. Fletcher," December 21, 1976, Marshall presentation files, and "Briefing on ST Program Management to House Committee Staff Member, 12/7/77," Marshall presentation files.

40 Domenick J. Tenerelli, "Thermostructural Design Considerations to Achieve the Space Telescope Line of Sight Requirements," in *The Space Telescope,* NASA SP-392 (Washington, D.C.: NASA, 1976), pp. 166–8. See also OHI, D. Tenerelli, with RWS, January 11, 1985, pp. 7–8, and D. Tenerelli, with RWS and JT, December 27, 1985, pp. 51–2.

41 Perkin-Elmer news release, "Perkin-Elmer Space Telescope Fact Sheet No. 1," November 10, 1978.

42 House Committee on Science and Technology, *Space Telescope Program Review: Hearings Before the Subcommittee on Space Science and Applications,* 95th Cong., 2nd sess., July 13, 1978, p. 124.

43 Gary Watson, "Building the Space Telescope's Optical System," *Astronomy,* January 1986, p. 17.

See also Daniel J. McCarthy and Terence A. Facey, "The 2.4 m Space Telescope Program," *Perkin-Elmer Technical News* 9(1981):12–22.

44 House Committee on Science and Technology, *Space Telescope Program Review: Hearings Before the Subcommittee on Space Science and Applications,* 95th Cong., 2nd sess., July 13, 1978, p. 38.

45 See the Perkin-Elmer presentations in House Committee on Science and Technology, *Space Telescope Program Review: Hearings Before the Subcommittee on Space Science and Applications,* 95th Cong., 2nd sess., July 13, 1978, and L. Montagnino, R. Arnold, D. Chadwick, L. Grey, and G. Rogers, "Test and Evaluation of a 60-inch Test Mirror," *Proceedings of the Society of Photo-Optical Instrumentation Engineers* 183(1979):109–13.

46 Marshall, however, was familiar with it, to some degree, as it had been employed in the Einstein x-ray observatory, to be launched in 1978.

47 See viewgraphs, "Space Telescope Review for OSS, June 25, 1979," N. Roman papers, NASA Headquarters, drawer 1 (file "ST"). See also OHI, A. Reetz, with RWS, June 27, 1983, p. 13, and J. Olivier, with RWS, February 27, 1985, p. 8.

48 The provision of a catalog of guide stars would, as we shall see in Chapter 9, prove to be an enormous task and would be one of the most important activities of the Space Telescope Science Institute.

49 See OHI, D. J. Schroeder, with RWS, October 7, 1985, p. 19, as well as numerous comments in minutes of the Operations and Management Working Group between 1973 and 1976.

50 See, for example, OHI, N. Hinners, with RWS and JT, November 16, 1984, p. 46, and C. R. O'Dell, with RWS, American Institute of Physics, May 21–23, 1985, session I, p. 30. See also Chapter 5 (section entitled "Space Telescope and Galileo").

51 See remarks by L. Spitzer in *Scientific Research with the Space Telescope,* NASA CP-2111 (Washington, D.C.: NASA, 1979), p. 3. John Bahcall recalls the day of Danielson's death as "one of the worst days of my life. It was really a tragedy. I loved him. He was a courageous, magnanimous, very intelligent, wonderful companion, a lot of fun": OHI, J. Bahcall, with PH, March 23, 1984, p. 18.

52 See, for example, minutes, LST Operations and Management Working Group, April 3–4, 1975, project scientist papers.

53 OHI, N. Hinners, with RWS and JT, November 16, 1984, p. 39; see also pp. 39–42.

54 See *Report on Space Science* (Washington, D.C.: National Academy of Sciences, 1976), pp. 18–58.

55 *Aviation Week and Space Technology,* August 16, 1976, pp. 40–1.

56 Memorandum, N. Roman to "The File," on "Meeting on Planetary Uses of the ST," August 19, 1976, project scientist papers (file "Meeting to Review Instruments Proposed for ST from Standpoint of Planetary Use"). For O'Dell's rough notes on this meeting, see daily notes, August 5, 1976, O'Dell papers.

57 See B. A. Smith, letter to *Nature,* "Uranus Rings: An Optical Search," *Nature* 268(1977):32.

58 Memorandum, D. Leckrone to G. Pieper, August 12, 1976, on "Space Telescope Instrument Status Report, July–August 1976," Goddard papers, 255-81-305, box 3 of 4 (file "May–September 1976"). The planetary astronomers in attendance were M. Belton, R. Brown, B. Smith, and J. Westphal. On Roman's lobbying for a planetary camera, see O'Dell daily notes, September 13, 1976, O'Dell papers.

59 The early ideas on a Faint Object Spectrograph and a High Resolution Camera are briefly described in "Instrumentation for the Large Space Telescope (LST): Extract of a Report by the Mission Definition Study," November 20, 1974, European Space Research Organization, no. MS(74)28. Further descriptions of the two proposed instruments are contained in "Focal Plane Instrumentation for the Large Space Telescope (LST): Report on the Mission Definition Study," December 16, 1974, European Space Research Organization, no. MS(74)28.

60 "Focal Plane Instrumentation for the Large Space Telescope," Appendix 1. The "University College Imaging Photon Counting System" had been developed by a group under Alec Boksenberg, and the first astronomical observations with the system had been secured in 1973. See "Alec Boksenberg: King of the Castle," *Sky and Telescope,* April 1984, pp. 312–15.

61 The Phase A designs for the Faint Object Camera are described in detail in *"NASA-ESA Cooperation on ST.* The Faint Object Camera and Other Elements. Report on Phase A Study," May 31, 1976, European Space Agency, document DP/PS(76)19.

62 OHI, R. Laurance, with RWS, May 26, 1983, p. 69.

63 "Report of the NASA/USRA Faint Object Camera Capability Survey Team," July 1976, copy in STHP files. On the review team, see also OHI, R. Laurance, with RWS, May 26, 1983, pp. 67–70, D. Macchetto, with RWS, December 23,

1985, pp. 23–5, and J. Rosendhal, with JT and RWS, December 23, 1985, pp. 47–62.

64 OHI, D. Leckrone, with RWS, August 14, 1984, pp. 64–71. Memorandum, C. R. O'Dell to W. Keathley, August 26, 1976, project scientist papers (file "Announcement of Opportunity for Space Telescope"). See also the memoranda, D. Leckrone to N. Roman, on "Space Telescope Scientific Instruments Announcement of Opportunity," September 1, 1976, GSFC papers, 255-81-305, box 3 of 4 (file "May–September 1976"), and D. Leckrone to G. Pieper, on "Space Telescope Instrument Status Report August 15–September 15, 1976," September 15, 1976, GSFC papers, 255-81-305, box 3 of 4 (file "May–September 1976").

65 That was certainly the view of those at Goddard who had been intensively involved in planning the scientific instruments; see the numerous memoranda on the SEC Vidicon in GSFC papers, 255-81-305, box 3 of 4 (file "May–September 1976"). Goddard, it should also be noted, was managing Princeton's development of the SEC Vidicon.

66 OHI, J. Bahcall, with PH, December 20, 1983, p. 94.

67 See minutes, Space Telescope Operations and Management Working Group, October 18–19, 1976, project scientist papers. J. A. Westphal attended sections of that meeting, and for his recollections, see OHI, J. Westphal, with DHD, September 14, 1982, especially pp. 186–90. See also OHI, G. Field, with RWS, March 10, 1986, pp. 49–50.

68 OHI, G. Field, with RWS, March 10, 1986, pp. 49–50.

69 "Announcement of Opportunity for Space Telescope," AO no. OSS-1-77, NASA Headquarters.

70 Private communication, C. R. O'Dell to RWS, May 19, 1987. John Bahcall, for example, has also described Westphal as "one of the most original and creative, as well as non-conformist, people anywhere in science": OHI, J. Bahcall, with PH, March 28, 1984, p. 20.

71 And, it would turn out, *by* building the Wide Field Camera, they would not be doing space astronomy with that instrument ten years from then.

72 OHI, J. Westphal, with DHD, August 12, 1982, pp. 182–3. Gunn regards Westphal's account as basically correct, "if a little embellished": private communication, J. Gunn to RWS, March 6, 1986.

73 OHI, J. Westphal, with DHD, August 12, 1982,

p. 180. The papers presented at the meeting are included in *Advanced Electro-Optical Imaging Techniques,* NASA SP-338 (Washington, D.C.: NASA, 1972).

74 See OHI, S. Sobieski, with RWS, February 2, 1985. See also J. T. Williams, "The Intensified Charge Coupled Device as a Photon Counting Imager," in *The Space Telescope,* NASA SP-392 (Washington, D.C.: NASA, 1976), pp. 88–9.

75 James Samson, *Techniques of Vacuum Ultraviolet Spectroscopy* (New York: Wiley, 1967), pp. 220–2.

76 OHI, J. Westphal, with DHD, August 12, 1982, pp. 197–8.

77 See Robert W. Smith and J. N. Tatarewicz, "Replacing a Technology," *Proceedings of the IEEE* 73(1985):1221–35.

78 OHI, R. C. Bless, with RWS, November 3, 1986, pp. 3–4.

79 OHI, R. C. Bless, with RWS, November 3, 1983, pp. 34–5.

80 See active papers of J. D. Rosendhal, NASA Headquarters (file "Status of Detectors Proposed for the Space Telescope Wide Field Camera and Faint Object Spectrograph 9/77," and "Space Telescope Detectors – Working Papers").

81 This was Duccio Macchetto.

82 See "Space Telescope Announcement of Opportunity IDT Proposal Evaluation Procedures," project scientist papers (file "Scientific Instrument Proposal Evaluation"), and minutes of the Synthesis Section of the Peer Review for ST Science Selection, O'Dell papers (file "Selection of Space Telescope SIs [1977]").

83 OHI, D. S. Leckrone, with RWS, August 14, 1984, p. 72. As it turned out, the *Galileo* and Space Telescope programs would use two different kinds of CCDs, and so that commonality never transpired. Certainly the *Galileo* mission did not solve the development problems of the CCDs for the Space Telescope.

84 On the proposal led by Harms, see OHI, R. Harms, with RWS, May 26, 1987, pp. 4–14.

85 In 1987, the name of the High Resolution Spectrograph was changed to the Goddard High Resolution Spectrograph.

86 See, for example, E. A. Beaver and Carl E. McIlwain, "A Digital Multichannel Photometer," *Review of Scientific Instruments* 42(1971):1321–4, and John Meaburn, *Detection and Spectrometry of Faint Light* (Dordrecht: Reidel, 1976), p. 35.

87 OHI, D. S. Leckrone, with RWS, August 14, 1984, p. 75. See also the material in project scientist papers (file "Selection of Space Telescope Scientific Instruments [1977]").

88 Bless and his group had deliberately included only University of Wisconsin people on the proposal as a cost-saving measure: private communication, R. C. Bless to RWS.

89 See summary minutes of the Space Science Steering Committee, "Subject: Space Telescope Science Selection," October 25, 1977, project scientist papers (file "Selection of Space Telescope Scientific Instruments").

90 Spitzer would later become a special member for a time.

8. PROBLEMS ARISE

1 For a brief description of cost-plus-fee contracts, see Surveys and Investigations Staff, House Appropriations Committee, "A Report to the Committee on Appropriations, U.S. House of Representatives, on the NASA Space Telescope Program," March 16, 1983, p. 20.

2 Marshall document, "Project Plan for Space Telescope Design and Development Phase," July 1977, p. 40.

3 OHI, J. Olivier, with RWS, February 27, 1985, p. 3.

4 William C. Keathley to Warren Keller, March 14, 1978, Goddard chronology files.

5 See also R. Harms to G. Levin, March 15, 1978, R. Harms to G. Levin, March 16, 1978, memorandum for the record, William Keathley, on "FOS Proposal Increase," March 22, 1978, R. Harms to G. Levin, March 27, 1978, and R. Harms to G. Levin, May 8, 1978, all in Goddard chronology files. Part of the original increase was due to new requirements placed on the scientific instruments by NASA.

6 See U.S. General Accounting Office, *Report to the Congress of the United States by the Comptroller General: NASA Should Provide the Congress Complete Cost Information on the Space Telescope Program,* PSAD-80-15, January 3, 1980. For a cost history of the project's early years, see "Presentation to Associate Administrator for Space Science. Space Telescope Cost/Schedule Review. December 20, 1979," in GSFC papers, 255-82-296, box 9 (file "ST 1980").

7 See briefing charts for "Space Telescope POP 80-2 Review July 26, 1980," project scientist papers (file "ST Cost Cutting Activity").

8 "Notes 8-20-79 Keathley," Marshall project manager's files (file "Notes to Dr. Lucas [Gen Info 1978 to 1980]"); W. Keathley to W. Lucas, Au-

gust 27, 1979, Marshall project manager's files (file "Notes to Dr. Lucas [Gen Info 1978 to 1980]"). See also memorandum, Thomas A. Mutch to R. E. Smylie, on "Space Telescope Cost Review," January 7, 1980, GSFC papers, 255-82-296, box 9 (file "LST 1980").

9 "A Report to the Committee on Appropriations, U.S. House of Representatives, on the NASA Space Telescope Program," p. 23, and cost figures for the Support Systems Module in "Presentation to Associate Administrator for Space Science. Space Telescope Cost/Schedule Review. December 20, 1979," GSFC papers, 255-82-296, box 9 (file "ST 1980").

10 OHI, A. Reetz, with RWS, June 27, 1983, p. 17.

11 A. F. White to Dave Swenson, May 1978, LMSC-D663613, Lockheed chronology files.

12 R. H. Jones to A. F. White, May 23, 1978, LMSC-D619269, Lockheed chronology files. See also Telecon report, January 27, 1978, in Lockheed Space Telescope papers, box 31.

13 OHI, T. Bland Norris, with PH, April 19, 1984, pp. 81–2.

14 Material on the Norris cost review is contained in a file marked "79 Norris Cost Review," Goddard active files. However, the chief reference is "Presentation to Associate Administrator for Space Science. Space Telescope Cost/Schedule Review. December 20, 1979," copy in GSFC papers, 255-82-296, box 9 (file "ST 1980").

15 See also OHI, T. B. Norris, with PH, April 19, 1984, p. 81. There was only one dissenting view to that opinion.

16 On the High Energy Astronomy Observatories, see Wallace A. Tucker, *The Star Splitters,* NASA SP-446 (Washington, D.C.: NASA, 1984), and Wallace Tucker and Riccardo Giacconi, *The X-Ray Universe* (Cambridge, Mass.: Harvard University Press, 1985).

17 O'Dell daily notes, February 21, 1980, O'Dell papers.

18 O'Dell daily notes, March 20, 1980, O'Dell papers.

19 House Committee on Appropriations, *Department of Housing and Urban Development–Independent Agencies Appropriations for 1981: Hearings Before the Subcommittee on HUD–Independent Agencies, pt. 4, NASA,* 96th Cong. 2nd sess., April 18, 1980, p. 155.

20 This point is discussed in OHI, W. Keller, with RWS, December 9, 1985, p. 49.

21 O'Dell daily notes, April 1, 1980, O'Dell papers, and F. Speer to W. Lucas, April 4, 1980, Mar-shall project manager's files (file "Notes to Dr. Lucas [Gen Info 1978 to 1980]").

22 Note, F. Speer to W. Lucas, on "ST OSS Budget Review," April 25, 1980, Marshall project manager's files (file "Notes to Dr. Lucas [Gen Info 1978 to 1980]"). See also "Outline of Presentation Material" in same file, and O'Dell daily notes, April 23, 1980, O'Dell papers.

23 O'Dell daily notes, April 23, 1980, O'Dell papers.

24 Minutes to Science Working Group meeting, May 5–6, 1980, project scientist papers. See also R. C. Bless diary entries for May 5–6, 1980, copy provided to STHP.

25 Science Working Group minutes and enclosures, May 5–6, 1980, project scientist papers, O'Dell daily notes, May 6, 1980, O'Dell papers, and memorandum, J. C. Brandt to director (GSFC), on "Space Telescope as Viewed by a Goddard Principal Investigator – A Case History and Perceptions," October 2, 1980, GSFC papers, 255-82-296, box 9 (file "ST 1980").

26 F. Speer to D. McCarthy, June 27, 1980, and F. Speer to W. Wright, June 27, 1980, both in Marshall project manager's files (file "Project Manager's Reading File 1979–1981").

27 Briefing charts for "Space Telescope Budget Situation," July 18, 1980, Marshall project manager's files (file "ST Cost Assessment Team").

28 C. R. O'Dell to F. Speer, June 13, 1980, copy in GSFC chronology files.

29 Memorandum, W. Wright to F. Speer, July 8, 1980, copy in project scientist papers (file "ST Cost Cutting Activity").

30 Memorandum, "Project Management Directorate, Space Telescope Science and Operations Project [Goddard], to F. Speer," on "Impact of Sunshade Deletion of ST SIs," July 11, 1980, GSFC chronology files. The memorandum was written by David S. Leckrone.

31 Memorandum, C. R. O'Dell to F. Speer, June 27, 1980, project scientist papers (file "Reading File 1980"). See also E. Groth to C. R. O'Dell, July 28, 1980, GSFC active files (file "Chronology File July 1980").

32 Memorandum, C. R. O'Dell to F. Speer, June 27, 1980, project scientist papers (file "Reading File 1980"). The rest of the $580,000 would be spent in future fiscal years.

33 See notes by Goddard's director, Tom Young, regarding his meeting with the U.S. principal investigators in October 1980: notes "From the Desk of A. Thomas Young," GSFC papers, 255-82-296, box 9 (file "ST 1980"); the date is clear from

a memorandum, J. S. Brandt to director (Goddard), October 15, 1980, on "Costs of Science Instruments on the Space Telescope" (file "ST 1980").

34 Mutch had replaced Noel Hinners, who had left NASA in 1979 to become director of the National Air and Space Museum in Washington, D.C.

35 OHI, C. R. O'Dell, with DHD, April 12, 1984, p. 48.

36 F. A. Speer to C. R. O'Dell, June 30, 1980, Marshall project manager's files (file "Reading File 1980").

37 Memorandum, Thomas A. Mutch to R. E. Smylie, "Space Telescope Cost Review," January 7, 1980, GSFC papers, 255-82-296, box 9 (file "ST 1980").

38 Memorandum, Robert S. Kraemer to director, Goddard, on "Independent Assessment of Space Telescope Payload Status," July 31, 1980, copy in Marshall project manager's files (file "ST Cost Assessment Team"). See also handwritten notes, *"ST Payload Assessment* Kraemer 7-25-80," GSFC papers, 255-82-296, box 9 (file "ST 1980").

39 See O'Dell daily notes, July 17, 1980, O'Dell papers.

40 OHI, C. R. O'Dell, with RWS, American Institute of Physics, May 21–23, 1985, p. 112. O'Dell's scheme bears a striking resemblance to NASA's recent plans to build a space station "by the yard." See also p. 151 and note 8, p. 442.

41 O'Dell daily notes, January 24, 1980, O'Dell papers.

42 O'Dell daily notes, July 17, 1980, O'Dell papers.

43 Viewgraph entitled "Additional descoping," POP 80-2, July 26, 1980, Marshall project manager's files (file "ST Cost Assessment Team").

44 Private communication, C. R. O'Dell to RWS.

45 OHI, W. Keller, with RWS, December 9, 1985, p. 53. See also OHI, C. R. O'Dell, with DHD, April 12, 1982, p. 49, and OHI, A. Reetz, with RWS, September 21, 1983, p. 14.

46 O'Dell daily notes, July 28, 1980, O'Dell papers.

47 Memorandum, W. R. Lucas to Thomas A. Mutch (undated, but early August 1980), Marshall project manager's files (file "ST Cost Assessment Team").

48 OHI, R. C. Bless, with RWS, November 3, 1983, p. 70, and memorandum, J. C. Brandt to director (GSFC), on "Space Telescope as Viewed by a Goddard Principal Investigator – A Case History

and Perceptions," October 2, 1980, GSFC papers, 255-82-296, box 9 (file "ST 1980").

49 OHI, C. R. O'Dell, with RWS, American Institute of Physics, May 22–23, 1985, session II, p. 59, and OHI, C. R. O'Dell, with DHD, April 12, 1982, p. 54.

50 OHI, C. R. O'Dell, with DHD, April 12, 1982, p. 54. See also OHI, J. Bahcall, with PH, December 20, 1983, p. 87. RWS has linked Bahcall's recollection with that meeting of the Science Working Group. See also memorandum, J. C. Brandt to director (GSFC), on "Space Telescope as Viewed by a Goddard Principal Investigator – A Case Study and Perceptions," October 2, 1980, GSFC papers, 255-82-296, box 9 (file "ST 1980").

51 OHI, R. C. Bless, with RWS, November 3, 1983, p. 72.

52 See "Space Shuttle 1980: Status Report for the Committee on Science and Technology, U.S. House of Representatives, Ninety-Sixth Congress, Second Session" (Washington, D.C.: U.S. Government Printing Office 1980), p. ix.

53 That is not always the case. See, for example, Aaron Wildavsky, *The Politics of the Budgetary Process,* 3rd ed. (Boston: Little, Brown, 1979), pp. 21–3.

54 House Committee on Appropriations, *Department of Housing and Urban Development–Independent Agencies Appropriations for 1981: Hearings Before the Subcommittee on HUD–Independent Agencies, pt. 4, NASA,* 96th Cong., 2nd sess., April 18, 1980, p. 155.

55 C. R. O'Dell to R. C. Bless, September 16, 1980, project scientist papers (file "Reading File 1980").

56 Memorandum, "S/Associate Administrator for Space Science, B/Associate Administrator/Comptroller, to Distribution," on "Space Telescope Cost Review Team," October 6, 1980, Marshall project manager's files (file "ST Cost Assessment Team").

57 O'Dell daily notes, October 22–3, 1980, O'Dell papers.

58 Earl A. Reese, meeting record on Space Telescope executive session, November 5, 1980, copy in Marshall project manager's files (file "Notes to Dr. Lucas [Gen Info 1978 to 1980]"). See, too, Fred A. Speer, "Presentation of Space Telescope Rebaselining Status Report to NASA Administrator," November 5, 1980, copy in GSFC papers, 255-82-296, box 9 (file "ST 1980").

59 On the original plan for the use of orbital replacement units, see H. T. Fisher, "Space Telescope – Design for Orbital Maintenance," in George E.

Corrick, Eric C. Haseltine, and Robert T. Durst, Jr. (eds.), *Proceedings of the Human Factors Society: 24th Annual Meeting, October 1980, Los Angeles* (Santa Monica: HFS, 1980), pp. 214–18.

60 Marshall project manager's files (file "ST Replanning NASA Headquarters 11/3–5 [1980]").

61 Memorandum, C. R. O'Dell to F. A. Speer, December 8, 1980, included in enclosures to Space Telescope Science Working Group meeting, December 2, 1980, project scientist papers.

62 "CSAA Resolution to the SSB for Adoption and Transmission to Dr. Frosch," copy in Marshall project manager's files (file "ST Replanning NASA Headquarters, 11/3–5 [1980]").

63 On Speer's presentation, see "Space Telescope Rebaseline Review to the Administrator. December 9, 1980," copy in GSFC papers, 255-82-296, box 9 (file "ST 1980").

64 Administrator's meeting record on "Space Telescope Review," December 9, 1980, Marshall project manager's files (file "Briefing to Administrator December 9, 1980"), and Warren Keller's briefing charts for *"Space Telescope Cost Review,* Presentation to NASA Administrator, December 9, 1980," copy provided to STHP by Warren Keller.

65 OHI, J. Rosendhal, with JT, September 12, 1984, p. 46.

66 House Committee on Science and Technology, *1982 NASA Authorization: Hearings Before the House Subcommittee on Space Science and Applications,* 97th Cong., 1st sess., March 23, 1981, p. 3064. See also House Committee on Science and Technology, *1982 NASA Authorization (Program Review): Hearings Before the House Subcommittee on Space Science and Applications,* 96th Cong., 2nd sess., September 18, 1980, pp. 437–9, 496.

67 House Committee on Science and Technology, *1982 NASA Authorization: Hearings Before the Subcommittee on Space Science and Applications,* 97th Cong., 1st sess., March 23, 1981, p. 3065. See also OHI, W. Keller, with RWS, December 9, 1985, p. 47, and D. Schroeder, with RWS, October 7, 1985, p. 49.

68 F. Speer to J. Rehnberg, April 13, 1981, Marshall project manager's files (file "Project Manager's Reading File 1979–1981").

69 F. Speer to J. Rehnberg, April 29, 1981, Marshall project manager's files (file "Project Manager's Reading File 1979–1981").

70 OHI, M. Bensimon, with RWS, December 5, 1983, p. 12.

71 Briefing charts, "Preparation for Meeting with E. Ronan and J. Rehnberg, Perkin-Elmer," June

28, 1982, Marshall project manager's files (file "Information to Ron Crawford thru Art Reetz/ Prep for Meeting with Ronan, Rehnberg 6-28-82"), Marshall project manager's files (file "Memos & Notes FM & Staff DA01-Lucas 1982"). See also OHI, Marc Bensimon, with RWS, December 5, 1983, pp. 12–13.

72 F. Speer to W. Lucas, August 1980, Marshall project manager's files (file "Notes to Dr. Lucas [Gen. Info 1978–1980]").

73 F. Speer to W. Lucas, September 15, 1980, "ST Primary Mirrors," Marshall project manager's files (file "Reading File 1980").

74 "Perkin-Elmer Space Telescope Fact Sheet Number 11," April 30, 1981.

75 OHI, D. J. Schroeder, with RWS, October 7, 1985, p. 31.

76 F. A. Speer to C. R. O'Dell, August 25, 1981, project scientist papers (file "ST Removal Primary Mirror Control Actuators").

77 W. Fastie to C. R. O'Dell, September 17, 1981, project scientist papers (file "Reading File 1981").

78 C. R. O'Dell to F. A. Speer, "Impact of Removal of Primary Mirror Control Actuators," September 11, 1981, project scientist papers (file "Reading File 1981").

79 House Committee on Science and Technology, *Oversight Hearing Before the House Subcommittee on Space Science and Applications,* 97th Cong., 2nd sess., May 25, 1982, p. 49.

80 W. Keathley to W. Lucas, April 16, 1979, Marshall project manager's files (file "Notes to Dr. Lucas [Gen Info 1978 to 1980]").

81 F. Speer to W. Lucas, July 6, 1982, Marshall project manager's files (file "Weekly Notes").

82 F. Speer to K. Meserve, July 6, 1982, Marshall project manager's files (file "July/August 1982 Reading File ").

83 OHI, M. Bensimon, with RWS, October 5, 1983, p. 7. See also OHI, C. Pellerin, with RWS, August 1, 1983, p. 10, and J. Rehnberg, with RWS and PH, November 1, 1983, p. 35.

84 OHI, A. Reetz, with RWS, September 21, 1983, p. 20.

85 OHI, J. Rehnberg, with RWS and PH, November 1, 1983, p. 36.

86 F. Speer to M. Rosenthal, August 11, 1982, Marshall project manager's files (file "July/August 1982 Reading File").

87 Rapifax, July 29, 1982, F. Speer to M. Bensimon, Marshall project manager's files (file "July/August 1982 Reading File"). On the discussions over contamination and the problem of loss of lock, see the numerous discussions in the minutes of

the Science Working Group and the briefing packages for the Space Telescope quarterly meetings at Marshall in the active files of the Marshall project manager, with special reference to 1981 to 1983 for contamination and 1980 to 1983 for loss of lock. On loss of lock, see also OHI, D. Tenerelli, with RWS, December 31, 1985, pp. 107–21. The loss-of-lock problem prompted Marshall to search seriously for other concepts for the Fine Guidance Sensors. For a time, an alternative proposed by James Westphal and Jim Gunn, and promoted by Lockheed, based on a quadrant system came into the reckoning to replace the existing design.

88 OHI, M. Bensimon, with RWS, October 5, 1983, p. 14. See also F. Speer to W. Lucas, September 23, 1982, Marshall project manager's files (file "Notes to Lucas from Speer"), and viewgraphs of "ST Briefing to Mr. Beggs: 10/27," Marshall project manager's files.

89 OHI, C. Pellerin, with RWS, August 1, 1983, p. 9.

90 OHI, M. Bensimon, with RWS, October 5, 1983, p. 6.

91 OHI, M. Bensimon, with RWS, October 2, 1983, p. 13.

92 "ST Briefing to Mr. Beggs, 3 November 1982," Marshall project manager's files (file "25/1[2]").

93 OHI, S. Keller, with RWS and PH, December 15, 1983, p. 15. See also OHI, M. Bensimon, with RWS, October 5, 1983, pp. 12–13, and December 5, 1983, p. 31, E. J. Weiler, with RWS, October 20, 1983, p. 17, and viewgraphs of "ST Briefing to Mr. Beggs, 3 November 1982," Marshall project manager's files (file "25/1[2] Presentation to Mr. Beggs. 3 November 1982").

94 F. Speer to W. Lucas, November 10, 1982, Marshall project manager's files (file "From P-E 1982").

95 F. Speer to John D. Rehnberg, November 22, 1982, Marshall project manager's files (file "Reading File – 1982/1983").

96 OHI, A. Reetz, with RWS, June 27, 1983, p. 19. The same point had been made a few days earlier when Bensimon had visited Perkin-Elmer and there spoken with Perkin-Elmer's new project manager, Don Fordyce: OHI, M. Bensimon, with RWS, October 5, 1983, p. 15.

97 Memorandum, J. McCulloch to F. Speer, on "Perkin-Elmer Resources for FY 1983," December 13, 1982, Marshall Project manager's files (file "Reading File – 1982/1983").

98 Memorandum, C. Hartmann to R. Konkel, "ST Project Review Feb. 9–10, 1983," February 1983, Keller papers.

99 F. Speer to W. Lucas, December 20, 1982, Marshall project manager's files.

100 OHI, S. Keller, with RWS and PH, December 15, 1983, p. 16.

101 Sam Keller is not to be confused with the former Space Telescope program manager, Warren Keller.

102 OHI, S. Keller, with RWS and PH, December 15, 1983, p. 7.

103 OHI, S. Keller, with RWS and PH, December 15, 1983, p. 13.

104 OHI, R. C. Bless, with RWS, November 3, 1986, p. 31.

105 OHI, C. R. O'Dell, with RWS, American Institute of Physics, May 21–23, 1985, session II, p. 56. On the sanitized nature of quarterly meetings, see also R. Bless, "Space Science: What's Wrong at NASA," *Issues in Science and Technology* 5(1988–9):67–74; see also OHI, R. C. Bless, with RWS, November 11, 1986, p. 31, and R. Harms, with RWS, May 26, 1987, pp. 54–6.

106 OHI, S. Keller, with RWS and PH, December 15, 1983, p. 16. See also OHI, C. Pellerin, with RWS, August 1, 1983, p. 12.

107 OHI, M. Bensimon, with RWS, October 5, 1983, p. 16.

108 See "Notes on Conclusions and Recommendations to Accompany the Briefing Charts on the NASA Project Management Study," January 1981, copy provided to STHP by D. Hearth. See also Senate Committee on Commerce, Science, and Transportation, *NASA Authorization for Fiscal Year 1982: Hearings Before the Subcommittee on Science, Technology, and Space,* 97th Cong., 1st sess., April 27, 1981, pp. 639–707.

109 OHI, S. Keller, with RWS and PH, December 15, 1983, p. 17.

110 S. Keller to B. Edelson, February 4, 1983, Keller active files (file "[ST] Reports #2").

111 OHI, J. Rosendhal, with JT, September 12, 1984, p. 91. See also, for example, memorandum, Jeff Struthers to Fred Khedouri, on "Additional Funding for Space Telescope," April 26, 1983, copy in M. Norman workbook, OMB active files. This memorandum, in addition to examining the technical and management problems and NASA's actions in response to them, also details the ways the Space Telescope cost increases would affect other NASA space science and applications projects.

112 See "Marshall Interviews," February 9, 1983, Keller active files.

113 House Committee on Science and Technology,

461   *Notes to pp. 297–306*

1984 NASA Authorization: Hearings Before the House Subcommittee on Space Science and Applications, vol. II, 98th Cong., 1st sess., February 28, 1983, p. 1061.

114 House committee on Science and Technology, *1984 NASA Authorization: Hearings Before the House Subcommittee on Space Science and Applications, vol. II,* 98th Cong., 1st sess., February 28, 1983, p. 1102.

115 Draft of F. Speer's testimony for submission to the House Subcommittee on Space Science and Applications, February 28, 1983, copy in Keller active files.

116 Space Telescope Development Review Staff, "Briefing to the Administrator . . . By Mr. James C. Welch," March 16, 1983, copy in STHP files.

117 Space Telescope Development Review Staff, "Briefing to the Administrator," pp. 9, 13.

118 "A Report to the Committee on Appropriations, U.S. House of Representatives, on the NASA Space Telescope Program," p. ii.

119 OHI, E. J. Weiler, with RWS, October 20, 1983, p. 22. That Weiler did not know of Welch's existence helps to underscore how marginal the scientists were.

120 Quoted in "Space Telescope Scientists' View of Overall Space Telescope Program," March 1983, copy in STHP files.

121 House Committee on Appropriations, *Department of Housing and Urban Development–Independent Agencies Appropriations for 1984: Hearings Before the House Subcommittee on HUD–Independent Agencies,* part 6, *NASA,* 98th Cong., 1st sess., March 22, 1983, pp. 96–7.

122 House, *Hearings on Department of HUD–Independent Agencies Appropriations for 1984,* p. 86.

123 House, *Hearings on Department of HUD–Independent Agencies Appropriations for 1984,* p. 88.

124 House, *Hearings on Department of HUD–Independent Agencies Appropriations for 1984,* p. 90.

125 OHI, C. Pellerin, with RWS, August 1, 1983, p. 20.

126 Quoted in Craig Covault, "Problems Spur Shift in Telescope Program," *Aviation Week and Space Technology,* April 4, 1983, p. 14.

127 W. R. Lucas to James M. Beggs, April 8, 1983, Marshall project manager's files (file "Reading File 1982–1983").

128 Memorandum, J. M. Beggs to W. Lucas, "Large Space Telescope," March 24, 1983, S. Keller active files.

129 See, for example, OHI, N. Hinners, with RWS and JT, December 14, 1984, p. 115. There was bloodletting elsewhere in the program. The se-

nior vice-president of the Perkin-Elmer Optical Group was summarily reassigned.

130 OHI, S. Keller, with RWS and PH, December 15, 1983, p. 29.

9 · CLOSING IN

1 Norman R. Augustine, *Augustine's Laws* (New York: Penguin Books, 1987), p. 416.

2 NASA document, "Hubble Space Telescope. *Level I Requirements.* Office of Space Science and Applications, NASA Headquarters, December 23, 1983," copy in STHP files.

3 Some decisions were also designated "critical decision issues." Those were issues that might affect schedule, performance, or cost, and such were then to be discussed between Welch and project manager Odom, who would then select a plan for resolving the issue. Once it had been solved to Welch's satisfaction, it would be closed. On the "critical decision issues," see OHI, J. Odom, with RWS, February 26, 1985, p. 36, and J. Welch, with RWS and PH, August 20, 1984, p. 12.

4 R. C. Bless diary, March 18, 1983, copy in STHP files, and C. R. O'Dell daily notes, March 18, 1983, O'Dell papers.

5 OHI, J. Welch, with RWS and PH, August 20, 1984, p. 23.

6 R. C. Bless diary, May 5, 1983, and C. R. O'Dell daily notes, May 5, 1983, O'Dell papers. On the opposition of the scientists to reporting to Marshall, see also R. C. Bless diary entries and details of telephone conversations, May 19, 1983, and May 20, 1983, as well as O'Dell daily notes, May 24, 1983.

7 Minutes, ST Science Coordinating Committee, May 26, 1983, and R. C. Bless diary, May 26, 1983, and O'Dell daily notes, May 26, 1983, O'Dell papers.

8 R. C. Bless diary, May 31, 1983.

9 OHI, J. Odom, with RWS, February 26, 1985, p. 26.

10 OHI, C. R. O'Dell, with DHD, April 12, 1982, pp. 64–5.

11 OHI, C. R. O'Dell, with DHD, April 12, 1982, p. 45.

12 OHI, J. Welch, with RWS and PH, August 20, 1984, p. 18.

13 OHI, J. Welch, with RWS and PH, August 20, 1984, p. 21.

14 On this debate, see the minutes of the Science Working Group, February 1–2, 1984, project scientist papers, and notes taken by JT and RWS at that meeting, STHP files.

15 RWS notes on Science Working Group meeting, February 1, 1984, STHP files, and minutes to same meeting.

16 House Committee on Science and Technology, *Space Telescope 1984: Hearings Before the Subcommittee on Space Science and Applications*, 98th Cong., 2nd sess., May 22 and 24, 1984, p. 126.

17 J. Welch to L. Allen, November 6, 1984, copy in Marshall project manager's files (file "Reading File November 1984").

18 With various delays in the telescope's overall schedule, work on the clone had to be stretched out; the cost rose substantially, and design changes were also made.

19 Memorandum, A. Reetz to distribution, on "Minutes, ST Ad Hoc Project Designation Committee, November 12, 1982," January 11, 1983, O'Dell active files.

20 RWS notes on Science Working Group meeting, June 13–14, 1986, p. 4, STHP files, and minutes to same meeting.

21 Each of the four axial scientific instruments had three latches, the Wide Field/Planetary Camera had three, and each of the Fine Guidance Sensors had four. Taking account of mirror-image designs, there was a total of twenty different latch designs. See Jan D. Dozier and Everett Kaelber, "Latch Fittings for the Scientific Instruments on the Space Telescope," in *17th Aerospace Mechanisms Symposium*, NASA CP-2273 (Washington, D.C.: NASA, 1983), pp. 253–65.

22 For the latches, the specifications were that they would have to position the scientific instruments so that during ten hours in orbit they would not be out of alignment with respect to the length of the telescope tube by more than plus or minus 0.0039 inch, would not be out of alignment at right angles to the telescope tube by more than plus or minus 0.019 inch, and would not have rotated by more than plus or minus 10 arc seconds: "Space Telescope Project. Optical Telescope Assembly Request for Proposal NAS8-1-6-PP-00597Q January 28, 1977," exhibit C, p. 10.

23 OHI, J. Odom, with RWS, February 26, 1985, p. 12.

24 OHI, J. Odom, with RWS, February 26, 1985, p. 13. See also OHI, J. Rehnberg, with RWS and PH, November 1, 1983, p. 27.

25 Fred G. Sanders, "Space Telescope Neutral Buoyancy Simulations – The First Two Years," NASA technical memorandum 82485.

26 Report entitled "Space Telescope Scientists' View of Overall Space Telescope Program," prepared by E. Weiler, C. R. O'Dell, A. Boggess, and D. Leckrone, copy in STHP files. See also "Space Telescope Development Review Staff. Briefing to the Administrator of the National Aeronautics and Space Administration March 16, 1983. By Mr. James C. Welch," copy in STHP files.

27 Emphasis in original, "A Report to the Committee on Appropriations, U.S. House of Representatives, on the NASA Space Telescope Program," survey and investigations staff, March 16, 1983, p. 12.

28 R. C. Bless diary, March 18, 1983.

29 Quoted from Randy Quarles, "Space Telescope Firm Got Bonuses Despite Deficiencies," *Huntsville Times*, March 29, 1983, p. 11.

30 OHI, J. Odom, with RWS, February 26, 1984, p. 12.

31 OHI, S. Keller, with RWS, December 15, 1983, p. 13.

32 House Committee on Science and Technology, *Space Telescope 1984: Hearings Before the Subcommittee on Space Science and Applications*, 98th Cong., 2nd sess., May 22 and 24, 1984, p. 120.

33 Report, "WF/PC Science Team *Technical Report – July/August 1984*, J. A. Westphal, September 28, 1984," copy in Marshall project manager's files (file "GSFC Correspondence").

34 OHI, D. Tenerelli, with RWS, January 16, 1986, p. 20.

35 OHI, E. Weiler, with RWS, February 11, 1986, p. 23.

36 In fact, as would become clearer a year or two later, an alteration to the chips looked attractive as a fix for QEH for the WF/PC clone.

37 Presentation by D. Rodgers on "Quantum Efficiency Hysteresis," Marshall project manager's files (file "ST Quarterly Reviews 10–11 October 1984").

38 OHI, D. Tenerelli, with RWS, January 16, 1986, p. 208.

39 Memorandum, James Moore to distribution, "Minutes of Meeting Held at GSFC on Friday, October 26, 1984," Goddard active files.

40 Memorandum, Robert A. Brown to distribution, "WFPC QEH Science Impact Assessment Plan," November 1, 1984, Marshall project scientist papers (file "Dr. Brown 1984 Reading File").

41 See R. A. Brown's presentation charts on "The Impact of WFPC QEH on the HST Science Program," December 5, 1984, enclosed with minutes of Science Working Group meeting, January 8–9, 1985, copy in STHP files. On the WF/PC design as it stood in 1982 and some of the scientific programs it was planned to pursue, see James A. Westphal and the WF/PC Investigation Defi-

nition Team, "The Wide Field/Planetary Camera," in D. Hall (ed.), *The Space Telescope Observatory,* NASA CP-2244 (Washington, D.C.: NASA, 1982), pp. 28–39. Many scientific problems that astronomers in 1979 thought the Wide Field/Planetary Camera might tackle are described by M. S. Longair and J. W. Warner (eds.), *Scientific Research with the Space Telescope,* NASA CP-211 (Washington, D.C.: NASA, 1979).

42 W. Fastie to F. Carr, December 12, 1984, Goddard active files. See also the attached "Review of LMSC Meeting (Wm. G. Fastie, JHU)."

43 OHI, D. Tenerelli, with RWS, January 16, 1986, p. 213. For an account of QEH, as it was understood in November 1984, see J. Westphal and J. Gunn, "The Quantum Efficiency Hysterisis (QEH) Problem in the WF/PC," Goddard active files (file "WF/PC QEH Problem: Meeting of 12/5/84 at LMSC").

44 OHI, D. Tenerelli, with RWS, January 16, 1986, p. 214.

45 Memorandum, James B. Odom to distribution, "Action Items from the WF/PC Quantum Efficiency Hysteresis Meeting of December 19, 1984, at NASA HQ," Marshall project manager's files (file "Reading File December 1984").

46 RWS notes on Science Working Group meeting, January 8–9, 1985, STHP files, and minutes to same meeting, copy in STHP files.

47 See briefing charts "Wide Field/Planetary Camera WF/PC QEH Topics" and "Critical Design Issue AN-1792 WF/PC Quantum Efficiency Hysteresis Concern," Marshall project manager's files (file "WF/PC QEH Topics Presentation to Odom, Wojtalik, Mitchell"). As a backup, Lockheed was asked to fabricate and deliver the light baffle that would fit on the aperture door if sun pointing were indeed to be adopted later.

48 Wallace Tucker, "The Space Telescope Science Institute," *Sky and Telescope,* April 1985, p. 295.

49 That would lead institute staff to measure the brightnesses and positions of over twenty million stars to positions better than 1.5 arc seconds. NASA's conception of the Guide Star Selection System at the time that bids were tendered for the institute is contained in a NASA document, "Space Telescope Science Institute Request for Proposals," December 14, 1979, pp. 102–32. The institute's reports on subsequent changes to the Guide Star Selection System are contained in the institute's monthly progress reports and the series of institute documents tagged SO-04.

50 OHI, J. Russell, with RWS, December 18, 1984, p. 8.

51 That point was made by Peter Galison in his fascinating account of "Bubble Chambers and the Experimental Workplace," in Peter Achenstein and Owen Hannaway (eds.), *Observation, Experiment, and Hypothesis in Modern Physical Science* (Cambridge, Mass.: M.I.T. Press, 1985), p. 347.

52 Wallace Tucker and Riccardo Giacconi, *The X-Ray Universe* (Cambridge, Mass.: Harvard University Press, 1985), p. 72.

53 On Schreier's career trajectory, see OHI, E. Schreier, with RWS, July 6, 1987, and numerous references in Tucker and Giacconi, *The X-Ray Universe.* Rodger Doxsey was another x-ray astronomer who joined the institute in 1981 and became a key figure in planning for the telescope's scientific operations. Although his early career had entailed working on a variety of x-ray payloads for rocket flights, he had also performed a considerable amount of aspect analysis and data analysis for later x-ray spacecraft. Following his move to the institute, he would focus entirely on planning for scientific operations. See OHI, R. Doxsey, with RWS, July 22, 1987.

54 On Giacconi and his place in x-ray astronomy, see Wallace Tucker and Karen Tucker, *The Cosmic Inquirers: Modern Telescopes and Their Makers* (Cambridge, Mass.: Harvard University Press, 1986), pp. 45–96, 199–203, Richard F. Hirsh, *Glimpsing an Invisible Universe: The Emergence of X-Ray Astronomy* (Cambridge University Press, 1983), and Tucker and Giacconi, *The X-Ray Universe.*

55 Tucker and Giacconi, *The X-Ray Universe,* p. 41.

56 See John Teem's notes of conversations among J. Teem, A. Stofan and N. Hinners, May 1981, Teem papers, donated to STHP (file "STScI Items of Historical Interest").

57 Quoted in Tom Lachman, "The Universe and Dr. Giacconi," *Baltimore Magazine,* April 1983, p. 123.

58 OHI, S. Keller, with RWS and PH, February 11, 1985, p. 17.

59 On the selection of the institute's director, see "Report of the Space Telescope Institute Council to AURA Board of Directors . . . ," May 28, 1981, copy in file donated to STHP by J. Teem (file "STScI Items of Historical Interest").

60 That included the leader of the Source Evaluation Board that had reported in 1981 to the NASA administrator on the choice of the consortium for the institute.

61 "Consultant Team Report on Space Telescope Science Institute Program Management Review," enclosed with minutes of STOPAT meeting, July 13, 1983, copy in STHP files.

62 R. Giacconi, quoted in M. Waldrop, "Space

Telescope (II): A Science Institute," *Science* 221(1983):536.

63 OHI, J. Teem, with RWS, October 4, 1984, p. 76.

64 Notes to meeting of January 27, 1984, Marshall project manager's files (file "Presentation to Administrator 27 January 1984").

65 See " 'The STScI Staffing Space Matter' Presentation to Mr. Keller, February 27, 1984," by Goddard, and "Space Telescope Science Institute, Mr. Keller, NASA Headquarters, Feb. 27, 1984," by the Space Telescope Science Institute, both NASA Headquarters papers (file "6240").

66 OHI, J. Teem, with RWS, October 4, 1984, pp. 81–2.

67 The only one not to write was James Westphal.

68 C. R. O'Dell to J. Beggs, April 18, 1984, S. Keller active files. See also other letters on the topic in those papers.

69 George Clark to J. Rosendahl, May 7, 1984, S. Keller papers.

70 OHI, J. Teem, with RWS, October 4, 1984, p. 106.

71 J. Teem to B. Edelson, March 29, 1984, S. Keller active files.

72 S. M. Faber (chair), *Report of the Institute Visiting Committee on Its Visit to the Space Telescope Science Institute, May 21–23, 1984,* July 7, 1984, copy in STHP files.

73 Faber, *Visiting Committee,* 1984, p. 6.

74 Ibid.

75 S. Strom to F. Martin, November 7, 1980, Nancy Roman papers (file "ST"). See also E. J. Groth to J. Baniszewski, August 18, 1980, GSFC chronology files (file "August 1980"). Groth was the leader of the Space Telescope Data and Operations Team.

76 The issue was nevertheless studied at Goddard. See memorandum, John B. Martin to F. Speer, August 19, 1980, on "Proposed Addition of SOGS to the STScI Contract," GSFC chronology files (file "August 1980"). On the timing of the release of the SOGS "Request for Proposals," see F. Speer to W. Lucas, August 5, 1980, Marshall project manager's files (file "Notes to Dr. Lucas [Gen Info 1978–1980]").

77 OHI, E. Schreier, with RWS, July 6, 1987, p. 16.

78 On this review and a later one in September, see OHI, R. Doxsey, with RWS, July 22, 1987, pp. 35–6, and E. Schreier, with RWS, July 6, 1987, p. 28, and the enclosures to the minutes of the Science Working Group meeting, October 26–27, 1982, project scientist papers.

79 NASA document, "Announcement of Opportunity for Space Telescope," March 1977, AO no. OSS-1-77, p. 5. Much had also been made during the selling of the telescope of the chances of detecting planets around other stars, by direct observations and by use of the telescope's guidance system. On the use of the Fine Guidance Sensors for astrometry, including the detection of planets around other stars, see William H. Jeffreys, "Astrometry with the Space Telescope," *Celestial Mechanics* 22(1980):175–81. On the possible use of the Space Telescope to attempt to detect planets around stars astrometrically, see Jane Russell, "The Space Telescope's Search for Planets Around Other Stars," *Planetary Report,* September–October 1984, pp. 18–20, and "Prospects for Space Telescope in the Search for Other Planetary Systems," in M. D. Papagiannis (ed.), *The Search for Extraterrestrial Life: Recent Developments* (Dordrecht: Reidel, 1985), pp. 75–84.

On the telescope's possible use for solar system astronomy in general, see, for example, M. J. S. Belton, "Planetary Astronomy with the Space Telescope," and D. Morrison, "Investigation of Small Solar System Objects with the Space Telescope," in Longair and Warner, *Scientific Research with the Space Telescope,* pp. 47–75, 77–97, J. Caldwell's "Planetary Science with Space Telescope," *Advances in Space Research* 1(1981):199–213, and "Uranus Science with the Space Telescope," in G. Hunt (ed.), *Uranus and the Outer Planets* (Cambridge University Press, 1982), pp. 259–74.

80 OHI, C. R. O'Dell, with RWS, American Institute of Physics, May 21–23, 1985, pp. 137–8.

81 E. J. Groth, "Space Telescope Target Acquisition Presentation to Science Working Group," enclosure 3, minutes of Science Working Group meeting, January 10, 1980, project scientist papers.

82 OHI, R. Brown, with JT, April 3, 1984, pp. 31–2.

83 The story of the debate on the telescope's capability to perform planetary tracking, and when it should be added to SOGS, is extremely complicated and unfortunately too long for inclusion here. But on that topic, see, in particular, OHI, R. Brown, with JT, April 3, 1984, and June 7, 1984, T. Sherrill, with JT, September 28, 1985, the minutes and enclosures of the Science Working Group between 1983 and 1986, particularly those for September 12, 1984, May 15, 1985, January 14, 1986, and the documents produced by the committees formed after 1983 specifically to examine planetary tracking: the Plan-

etary Target Implementation Team and the Moving Targets Advisory Group.

84 A. Code to J. Beckers, July 30, 1984, copy in STHP files.

85 *Institutional Arrangements for the Space Telescope: A Mid-Term Review* (Washington, D.C.: National Academy Press, 1985), pp. 8–9.

86 Ivan R. King (chair), "Report of the Institute Visiting Committee Space Telescope Science Institute," June 10, 1985, p. 2, copy in STHP files.

87 NASA Report, "The Role of the Space Telescope Science Institute in the Hubble Space Telescope Project," prepared for House Subcommittee on Space Science and Applications, 1986, p. 4, copy in STHP files.

88 The tests and their results are detailed in a NASA report, "Space Telescope Project – Goddard. Verification and Acceptance Program (VAP) Test Program," STP-G-VAP-1000, November 1984.

89 F. Carr, presentation to Science Working Group, February 1, 1984, attachment to Science Working Group minutes for meeting, February 1–2, 1984, copy in STHP files.

90 JT notes on Science Working Group meeting, February 1, 1984, copy in STHP files.

91 Evan Richards, trip report, July 20–21, 1982, R. C. Bless papers (file "A&V Hassle"). Fred Speer was still project manager in that period; hence the reference to Speer's baseline.

92 Science Working Group minutes, October 26–27, 1982, Marshall project scientist papers, p. 4. See also Bless's enclosure, "Summary for SWG Minutes."

93 OHI, R. C. Bless, with RWS, November 11, 1986.

94 Memorandum, A. Guha to F. Speer, on "ST Assembly and Verification" (undated, but November 1982), R. C. Bless papers (file "A&V Hassle").

95 OHI, E. Richards, with RWS, February 20, 1984, p. 33. See also the GSFC chronology files of this period.

96 See memorandum, Gerald L. Burdett to distribution, on "Space Telescope Quarterly Review," October 8, 1982, Goddard active files.

97 OHI, E. Richards, with RWS, February 20, 1984, p. 34.

98 Memorandum, F. A. Speer to F. Carr, on "SI's Concern with Current A&V Planning," January 21, 1983, R. C. Bless papers (file "A&V Hassle").

99 OHI, N. Hinners, with RWS and JT, December 4, 1984, p. 106.

100 Memorandum, R. C. Bless to F. Carr, on "HSP Concerns with Current A&V Planning," January 31, 1983, and F. Carr to R. C. Bless, February 4, 1983, both in R. C. Bless papers (file "A&V Hassle").

101 OHI, F. Carr, with RWS, March 14, 1984, p. 50.

102 Memorandum, F. Carr to J. Odom, on "Conclusions and Recommendations from the April 15, 1983, 'A&V Meeting' at Goddard," May 11, 1983, R. C. Bless papers (file "A&V Hassle").

103 Notes of meeting on January 27, 1984, in Marshall project manager's files (file "Presentation to Administrator 27 February 1984"). On that meeting, see also memorandum, J. Welch to S. Keller, February 3, 1984, S. Keller active files. For a leading participant's view of NASA's attempts to win approval for the space station, see Hans Mark, *The Space Station: A Personal Journey* (Durham, N.C.: Duke University Press, 1987), pp. 126–209.

104 Memorandum, J. Odom to W. F. Wright, on "Assessment of LMSC Space Telescope Performance," April 12, 1984, Marshall project manager's files (file "SSM"). Other material on Marshall's formal interactions with Lockheed is contained in the same file.

105 See, for example, memorandum, F. Carr to J. Odom, on "Pre A&V Assessment," October 4, 1984, Marshall project manager's files (file "GSFC Correspondence").

106 OHI, F. Carr, with RWS, March 14, 1984, p. 55.

107 House Committee on Science and Technology, *Space Telescope 1984: Hearings Before the Subcommittee on Space Science and Applications,* 98th Cong., 2nd sess., May 22, 24, 1984, p. 38.

108 See Ariane Sains, "Perkin-Elmer Ships Telescope for Space," *The News-Times,* October 30, 1984, p. 1, GBL, "Perkin-Elmer Ships 2.3m Optical Space-Telescope Assembly," *Physics Today,* November 1984, pp. 17–19, NASA release no. 84-89, "Hubble Space Telescope Optics Arrive in California."

109 Memorandum, F. Carr to J. Odom, on "Pre A&V Assessment," October 4, 1984, Marshall project manager's files (file "GSFC Correspondence").

110 On Lockheed's test approach for the Space Telescope, see Clifford Gardner, "Space Telescope Verification Program," "Design and Verification for Space Telescope Maintenance (1)," and William F. Wright, "Space Telescope Perfor-

mance and Verification," all reprints in STHP files.

111 See, for example, OHI, J. Welch, with RWS, August 20, 1984, p. 20.

112 Memorandum, B. Edelson to D. Tellep, November 13, 1984, Marshall project manager's files (file "SSM"). See also memorandum, J. Odom to B. Bulkin, November 13, 1984, on "A&V Organizational/Operations Problems Areas," Marshall project manager's files (file "Reading Files November 1984").

113 Memorandum, J. Odom to Dr. Lucas, February 13, 1985, Marshall project manager's files (file "Reading File February 1985").

114 Senate Subcommittee on Science, Technology, and Space. *NASA Authorization for Fiscal Year 1986: Hearings*, 99th Cong., 1st sess., February 26, March 27 and 28, April 3 and 4, 1985, p. 279.

115 Memorandum, D. Pine to comptroller, on "ST POP Review and Quarterly Status Report," March 22, 1985, Marshall project manager's files (file "Correspondence from Headquarters").

116 Memorandum for the record, Bill Sneed, on "Space Telescope," December 1985, Marshall project manager's files (file "Notes to & from Dr. Lucas & Staff CY 1985"). See also attached memorandum on "Assessment of LMSC Performance Since October 12, 1984 Meeting with Val Peline."

117 David Friend, "Seeing Beyond the Stars," *Life*, December 1985, p. 29.

118 George Will, December 29, 1985. See also his column "Galaxies," *The [Baltimore] Sun*, January 5, 1986, p. 9M.

119 *U. S. Presidential Commission on the Space Shuttle Challenger Accident* (Washington, D.C.: U.S. Government Printing Office, 1986), vol. I. See also House of Representatives, *Report of the Committee on Science and Technology. . . . Investigation of the Challenger Accident*, 99th Cong., 2nd sess., 1986, House report 99-1016.

120 OHI, F. Carr, with RWS, May 29, 1986, p. 1.

121 David Leckrone, quoted in Glennda Chui, "NASA Space 'Eye' Still Shut," *San Jose Mercury News*, March 4, 1986, p. C1.

122 Burt Bulkin, quoted in Glennda Chui, "NASA Space 'Eye' Still Shut," *San Jose Mercury News*, March 4, 1986, p. C2.

123 "MSFC Project Manager's Monthly Summary to the Administrator Hubble Space Telescope (HST) Project April 1986," April 29, 1986, Marshall project manager's files (file "Reading File 1986").

124 "MSFC Project Manager's Summary to the Administrator . . . HST Project May 1986," May 28, 1986, Marshall project manager's files (file "Reading File 1986").

125 On the initial test results, see the Lockheed document "Review of Preliminary Results from HST Thermal-Vacuum Test 17 July 1986," briefing by Lockheed, LMSC/F130815.

126 See Marshall project manager's files, "Hubble Space Telescope Project Overview Presentation to J. R. Thompson, October 6, 1986," Marshall presentation files.

127 RWS notes on Science Working Group meeting, November 12, 1986, STHP files, and minutes to the meeting, copy in STHP files.

10. REFLECTIONS

1 OHI, A. Code, with RWS, February 21, 1984, p. 6.

2 OHI, M. Bensimon, with RWS, October 5, 1983, p. 20.

3 OHI, J. Bahcall, with PH, December 20, 1983, p. 65.

4 OHI, M. Disney, with RWS, May 1, 1984, p. 22.

5 OHI, R. Harms, with RWS, May 26, 1987, p. 3, and C. R. O'Dell, quoted in "Perkin-Elmer Ships 2.3-m Optical Space Telescope Assembly," *Physics Today*, November 1984, p. 19.

6 OHI, G. Emanuel, with RWS, December 2, 1985, p. 68.

7 OHI, J. Westphal, with JT, September 28, 1985, p. 1.

8 OHI, J. Westphal, with JT, September 28, 1985, pp. 15–16.

9 OHI, R. C. Bless, with RWS, November 11, 1986, p. 15, and E. Weiler to RWS, private communication.

10 This point was made by Robert Friedel in "The Wizard and the Expert," seminar paper presented at Johns Hopkins University, March 11, 1987, p. 22.

11 James E. Webb, *Space Age Management: The Large-Scale Approach* (New York: McGraw-Hill, 1969), p. 89.

12 See, for example, OHI, B. Bulkin, with RWS, January 11, 1985, p. 4.

13 See, for example, Charles Lindblom, *The Policy Making Process* (Englewood Cliffs, N.J.: Prentice-Hall, 1968). Lindblom, however, contends that although the process might seem disordered, it is not. For him, the pluralistic policy-making pro-

cess does in fact produce outcomes that resemble those that would be reached by a more "rational" decision-making process. A policy study of the Space Telescope, based on published sources only, forms a chapter of Richard Barke's *Science, Technology, and Public Policy* (Washington, D.C.: CQ Press, 1986).

14 *Institutional Arrangements for the Space Telescope* (Washington, D.C.: National Academy of Science, 1976), p. 9.

15 W. Henry Lambright, *Governing Science and Technology* (New York: Oxford University Press, 1976), p. 202. See, too, Ronald C. Tobey, "Big Science in the National Security State," in *Reviews in American History* 12(1984):138, as well as Michael D. Reagan, *Science and the Federal Patron* (New York: Oxford University Press, 1969).

16 OHI, J. Welch, with RWS and PH, August 20, 1984, p. 17.

17 See *Aerospace Daily*, May 24, 1978, pp. 133–4, *Aviation Week and Space Technology*, May 29, 1978, p. 13, July 31, 1978, p. 21, *Science* 200(1978):1132–3, *Nature* 92(1978):92, and House report, *Department of Housing and Urban Development–Independent Agencies Appropriation Bill, 1979*, 95th Cong., 2nd sess., May 25, 1978. I am grateful to Craig Waff for bringing this material to my attention.

18 OHI, D. Fordyce, with RWS and PH, October 31, 1983, p. 9.

19 John M. Logsdon, "The Space Shuttle Program: A Policy Failure," *Science*, May 30, 1986, p. 1104.

20 W. Henry Lambright, *Governing Science and Technology* (New York: Oxford University Press, 1976), p. 54.

21 OHI, D. Fordyce, with RWS and PH, October 31, 1983, p. 4.

22 OHI, J. Olivier, with RWS, January 18, 1984, p. 17. As James Welch notes, NASA typically makes "assessments of a proposal, budget figure, and says 'that's too low' or 'that's too high' and so forth and so on, but frequently, and I've seen this happen more than once, just in the short time that I've been with NASA, the center will be asked to go back and say, 'Well, can't you do this for X number of dollars less?' and the answer invariably is 'Yes, we can.'" This, as Welch underscores, is the wrong question to ask, as the answer is bound to be yes: OHI, J. Welch, with RWS, December 10, 1984, p. 24. In 1974, George Low visited Goddard and spoke with senior staff. "The more thoughtful people of the ones I talked with," Low noted, "voiced considerable concern that NASA's people weren't being honest with management and management, in turn, was not being honest with the outside world. Honesty here was used in the sense of buying-in on programs and projects." See personal notes, #125, July 20, 1974, box 67, folder 1, George M. Low papers, Rensselaer Polytechnic Institute Archives, Troy, New York.

23 OHI, C. R. O'Dell, with RWS, American Institute of Physics, session II, May 23, 1985, p. 57.

24 Senate Committee on Appropriations, *Senate Hearings Before the Committee on Appropriations. Department of Housing and Urban Development, and Certain Independent Agencies Appropriations. Fiscal Year 1978*, 95th Cong., 1st sess., March 21, 1977, p. 834.

25 See the comments of the chairman of the Space Science Board, Thomas Donahue, in "Space Science," in Theodore R. Simpson, *The Space Station: An Idea Whose Time Has Come* (New York: IEEE Press, 1985), p. 177.

26 See Riccardo Giacconi, "The Future of Space Astronomy," paper read at the AAAS annual meeting, Chicago, February 15, 1987, copy in STHP files, and Robert A. Brown and Riccardo Giacconi, "New Directions for Space Astronomy," *Science* 238(1987):617–19.

27 Freeman J. Dyson, "Science and Space," in Allan A. Needell (ed.), *The First Twenty-Five Years in Space: A Symposium* (Washington, D.C.: Smithsonian Institution Press, 1983), p. 98.

28 *Space Science Research in the United States: A Technical Memorandum* (Washington, D.C.: Office of Technology Assessment, 1982), p. 15. See, too, Jeffrey D. Rosenthal, "Space Science at NASA: Retrospect and Prospect," *Journal of the British Interplanetary Society* 41(1988):3–9. These questions are also discussed by, among many others, Mark Crawford, "Supercollider Faces Budget Barrier," *Science* 236(1987):246–8, Kim McDonald, "Big Ticket Science Research Is Pinching U.S. Projects, Harming Smaller Efforts," *Chronicle of Higher Education*, February 18, 1987, p. 1, and Eliot Marshall, "Big Versus Little Science in the Federal Budget," *Science* 236(1987):249. See, too, Freeman J. Dyson's "Science and Space." Similar issues permeate the recent debate on the proposed "Human Genome Project." See, for example, the spring 1987 issue of *Issues in Science and Technology*, which includes articles by Leroy Hood and Lloyd Smith, "Genome Sequencing: How to Proceed," and David Baltimore, "Genome Sequencing: A Small-Science Approach."

29 On Mohole, see Daniel S. Greenberg, *The Politics*

of Pure Science (New York: New American Library, 1967), pp. 171–208. On Isabelle, first known as the Intersecting Storage Accelerator and later the Colliding Beam Accelerator, see, for example, William J. Broad, "Perils of Isabelle," *Science 81*, December 1981, pp. 42–5, and H. Hahn, M. Month, and R. R. Rau, "Proton-proton intersecting storage accelerator facility IS-ABELLE at the Brookhaven National Labora-tory," *Reviews of Modern Physics* 49(1977):625–80. I am grateful to Allan Needell for the references to Isabelle.

30 See Allan Needell, "Nuclear Reactors and the Founding of the Brookhaven National Labora-tory," *Historical Studies in the Physical Sciences* 14(1983):93–122, and "Lloyd Berkner, Merle Tuve, and the Federal Role in Radio Astron-omy," *Osiris* 3(1987):261–88.

Index

The acronymns ST and LST refer to the Space Telescope and Large Space Telescope respectively.

Emanuel, Garvin, 378
energy crisis, 158–9
Enterprise, 176
ERDA (Energy Research and Development Agency), 163, 165
ERTS, *see LANDSAT*
ESA (European Space Agency), 141, 179, 226, 245–6, 255, 383, 407
ESRO (European Space Research Organization), 135–41, 245
Explorer I, 92
Explorer 12, 357

Faint Object Camera, 13, 179, 241, 245–6, 255, 377, 407, 414–15
Faint Object Camera Instrument Science Team, 255, 407
Faint Object Camera review team, 245–6
Faint Object Spectrograph, 14, 245, 253, 254, 255, 263, 280, 323, 353, 415; as core instrument, 241, 246
Faint Object Spectrograph Investigation Definition Team, 407–9
Fastie, William, 215, 217, 266, 293, 335
Federation of Atomic Scientists, 130
Fermilab, 206, 214
Field, George, 170, 199, 213; and Greenstein's committee, 131; and institute, 196–7, 206, 210, 213; and lobbying for (L)ST, 166, 170, 181, 386–7; and planetary science, 183; on Riccardo Giacconi, 342
Fine Guidance Sensors, 112–13, 231, 327, 338, 389, 414, 415–16; and astrometry, 240, 241, 257; development of, 239–40, 289, 297, 305, 315, 360
Fletcher, James, 134, 170, 445n59; and cost of (L)ST, 89, 99, 124, 135, 144, 151, 155, 161, 162, 163, 394; and justifications of LST, 143, 157; and lead center for LST, 84; and LST as international program,

140–1; and mirror size, 154; and OMB, 161–3, 171–2, 176; and Space Shuttle, 164–5, 171; and state of NASA, 175; and ST contractors, 223; and support for ST, 169–71, 171–2; and x-ray institute, 205, 206
flexure, 292
Flippo, Ronnie G., 180, 259
fluorescence, 251
Ford, Gerald, 135, 161, 162, 175, 176–7, 387
Fordyce, Don, 300, 301, 303, 306, 391
Fredrick, Laurence, 18, 53–4, 377
Friedman, Herbert, 341
Frosch, Robert, 210, 218, 224, 277, 285–7
Fuqua, Don, 171

galaxies, 19–20, 146, 334
Galileo, 7, 157
Galileo mission, 184, 255–6; *see also* Jupiter Orbiter Probe
Galison, Peter, 421, 426n17
Gamma Ray Observatory, 298
gang of four, 282
Gemini, 52
General Dynamics/Convair, 82
Giacconi, Riccardo, 202, 205, 334, 340; as director of institute, 343–7; and early career, 340–2
Glennan, T. Keith, 37
Goddard High Resolution Spectrograph, 14, 377, 415; *see also* High Resolution Spectrograph
Goddard Space Flight Center, 39, 192, 407, 417; and aperture door, 275; and assembly and verification, 345–6, 359, 361, 365; and design of LST, 61–2, 75; and ICCDs, 108–9, 250; and institute, 199–201, 207, 209–10, 343, 345–6, 452n82; and lead center for LST, 75–6, 82, 83–5; and (L)ST instruments, 82, 95, 246, 250, 353, 354–6; and (L)ST studies, 81–2; and management of (L)ST, 83, 317, 354–7, 382; and OAOs, 39,

42, 61–2, 71, 75; and quantum efficiency hysteresis, 333; and revision of ST program, 284; and ST costs, 273, 277–8; and ST operations, 196–7, 337; and ST quarterly meetings, 302; and "Verification and Acceptance" program, 353–61; *see also* Carr, Frank; Marshall Space Flight Center, and relationship with Goddard
"GOD" ("Great Optical Device"), 56–7
Goldberg, Leo, 30, 38, 44, 46–7, 194, 253, 449n18
Goody, Richard, 451n61
Greenstein, Jesse L., 131, 134, 181, 441n50
Greenstein's committee, 131–4, 144
Groth, Edward, 257, 349
Groves, Leslie, 230
Grumman, 121; and LST plans, 62, 74–5; and OAOs, 39, 60–1, 75, 224
Guide Star Selection System, 338, 464n49
Gunn, James, 221, 249–50, 334, 456n72

Hacking, Ian, 426n17
Hale, George Ellery, 11, 22, 58, 78, 387
Hale Telescope, 11–12, 30, 78, 190, 251
Harms, Richard, 256, 378
Harvey, Tom, 333
Heclo, Hugh, 168
Henry, Richard C., 171
Herschel, William, 292
high-dispersion spectrograph, 102
High Energy Astronomy Observatories, 84, 85, 131, 270, 436n10, 443n25; *see also* Einstein Observatory
High Resolution Spectrograph, 255, 256, 279–80, 323, 353, 457n85; *see also* Goddard High Resolution Spectrograph
High Resolution Spectrograph Investigation Definition Team, 409
High Speed Photometer, 13, 21,

multi-mirror telescopes, 12
Mundt, Karl E., 211
Mutch, Thomas, 271, 276, 277

National Academy of Sciences, 34, 37, 51, 64, 125, 188, 189, 208; *see also* Space Science Board
National Aeronautics and Space Act, 135
National Aeronautics and Space Administration, 36; and acquisition policy, 60; and advisory system, 191, 195–6; and agency budget, 175, 178; and agreement with Department of Defense on ST, 230, 283; and choice of projects, 117–19; and competition with Soviet Union, 37, 38, 41; and control of projects, 40, 75, 187–8, 189, 191, 193–4, 228; and Greenstein's committee, 131–2; and international projects, 135, 139, 140; and organizational structure, 59–60; and post-Apollo plans, 72, 175; and review of OAO program, 41; *see also entries under individual field centers*
National Aeronautics and Space Administration Headquarters, 59; and Black Saturday proposals, 281; and institute, 345–7, 352; and management of ST, 75–6, 315–17, 382, 390; and ST difficulties, 268, 271, 278, 297–9, 303–7, 310–13, 389–90; *see also entries under names of individual managers and individual offices*
National Astronomical Space Observatory, 51–2
National Radio Astronomy Observatory, 212, 398
National Science Foundation, 34–5, 44–5, 132, 178, 189, 193, 210, 337
Naugle, John, 80, 85, 121, 139, 152, 172, 203; and LST hearings, 120, 124–5
Naval Research Laboratory, 32–3, 39, 253, 341

Neugebauer, Otto, 419
neutral buoyancy tank, 325, 328
Newell, Homer, 54, 172, 187, 189, 195–6, 206
Newton, Isaac, 1, 5–7, 11, 12–13, 21, 425n4
Nixon, Richard, 115, 124, 135, 177, 195
Nordhausen, 28
Norman, Memphis, 209, 447n112
Norris, T. Bland, 167, 181, 229, 266
Norris cost review, 266–8, 277
Notre Dame Cathedral, 24
nuclear fusion, 158

OAOs (Orbiting Astronomical Observatories), 48–51, 60, 136, 144, 152, 190–1, 202, 224–5; and design, 38–40; and follow-on program, 48, 61–2, 71, 75, 90; and guest observers, 66–7; and loss of Goddard spacecraft, 42–3; and *OAO-I*, 42–2; and *OAO-II*, 42, 44, 199; and *OAO-III, see Copernicus;* and pointing and control systems, 39–40, 43–4; and requests for proposals, 39; scientific goals of, 38–9; as source of prestige, 38; and working group on, 37, 38–9
Oberth, Hermann, 28f
O'Dell, C. R., 92–5, 126, 144, 199, 377, 378, 446n85; and aperture door, 274, 275–6; and Black Saturday, 280–1, 282; and cost of ST, 225, 392; and detectors, 109, 247–8, 249; and institute, 199, 200–3, 205, 346; and lobbying for (L)ST, 129–30, 164, 386–7, 448n117; and LST as international program, 140; and LST briefings, 96; and minimum LST, 145, 146, 151–2; and project scientist, 93, 196, 320–1; and quarterly meetings, 303; and relationship with Goddard, 94–5; and revision of ST program, 284, 286, 293; on space astronomy, 27,

92–3; and ST instruments, 270–1, 272, 278–81
Odom, James B., 312, 314, 320, 356, 372, 382; and assembly and verification, 358; and latch mechanisms, 327, 329; and STOPAT, 319
Office of Advanced Research and Technology, 59, 68, 70–1, 72
Office of Management and Budget, 60, 117–19, 165, 383, 444n39; and institute, 209; and (L)ST, 160–3, 176–9, 185, 305
Office of Manned Space Flight, 59, 62, 73, 105, 120
Office of Science and Technology Policy, 117, 119, 177, 181, 209, 305
Office of Space Science (and Applications), 59, 68, 71, 72, 75–8, 82, 154, 185, 269, 301, 406–7
Office of Technology Assessment, 397
Olivier, Jean, 79–80, 226, 263, 391–2
O'Neill, Thomas P., 181
Optical Space Astronomy panel, 65
Optical Telescope Assembly, 81–2, 262–3, 360, 363, 407, 412, 413–14; and central baffle, 413; contracts for, 110, 236; and equipment section, 413; and focal plane structure, 413, 414; and metering truss, 239, 413; and secondary mirror, 413; selecting contractors for, 223–24; *see also* Eastman Kodak; Itek; Perkin-Elmer
Optical Telescope Technology Workshop, 68–9
orbiting solar observations, 39, 44
Owen, Tom, 209

Panama Canal, 367
PDR ("Preliminary Design Review"), 238–9
Peline, Val, 365
Perkin-Elmer, 64, 73, 97, 126, 224, 237, 407; and commu-

nication with Marshall, 310–11; and costs, 264, 290, 300, 304; and Fine Guidance System, 224, 231, 239–40, 289, *see also* Fine Guidance Sensors; and latch mechanisms, 295–6, 329; and management performance, 294–5, 299–300, 301, 304, 306, 359–60, 389–90; and mirror size, 146; and Optical Telescope Assembly, 237–40, 260–1, 273, 290, 293, 359; and Optical Telescope Assembly contracts, 110, 224, 225, 260–1, 290; and pointing and control system, 112–13; and primary mirror, 237–9, 290–3, 295, 300; and systems engineering, 232; and testing, 299, 300

"PERT" ("Program Evaluation and Review Technique"), 233

"Phased Project Planning," 67–8

photographic film, 13, 105–7

Physical Sciences Committee, 170, 184

Pieper, George, 94–5, 199, 205, 451n65

Pioneer Venus, 161, 162, 164, 183

Planck, Max, 47

Polaris, 391

policymaking: and "ad hoc" model, 26, 168, 385; and leadership model, 25, 167; for science in U.S., 25–6, 177–9, 383–4, 387

policy subsystem, 167

policy "whirlpool," 167

President's Scientific Advisory Committee, 387

Press, Frank, 181–2, 210

Price, Derek, 158

prime contractors, 88–9, 155

"Princeton Advanced Satellite," 107

Princeton–Huntsville axis, 129

Princeton University, 213, 216, 219

Princeton University Observatory, 23, 34, 107, 241

program design, 225–30, 231–6, 287, 381–3, 384, 388

Program Operating Plan, 264, 271

protoflight concept, 91, 365, 376, 384–5, 436n10

prototype concept, 91

Proxmire, William, 132, 134, 185

pulsar, 21

Putnam, William L., 180–1

quantum efficiency hysteresis, 324, 327, 330–1, 334, 335–6, 380, 414

Ramsey, Norman F., 192

Ramsey committee, 192, 194, 205, 449n11

RAND, 30–1, 34, 427n7; and "Preliminary Design of an Experimental World-Orbiting Spaceship," 31

rate gyroscopes, 297

reaction wheels, 112, 150

reconnaissance satellites, 34, 90, 106, 147–8, 224, 383

Rees, Eberhard, 85, 86

Reetz, Arthur, 264–5, 295, 299, 300

Rehnberg, John D., 289, 300

requests for proposals: for institute, 216–17; OAOs, 39; for Phase B, 100, 114–15; for ST, 221, 222–3

Richards, Evan, 353–4, 355–6

Richardson, R. S., 28

"rockoon," 428n22

Rodgers, David, 332–3

Rogers Commission, 302–3

Roman, Nancy, 55, 308, 377, 428n38; and detectors, 106, 246–8; and institute, 202; and LST planning, 55, 65, 69–70, 73, 77, 80, 103, 106; and minimum LST, 146, 151–2; and ST and planetary science, 183, 210, 242–4, 248

Rosendhal, Jeffrey, 254

Russell, H. N., 2, 23

"saddle bonding," 295, 300

Samson, James, 250–1

satellite control section, 97–8

SATS (Spacecraft Automated Test System), 359, 361

Saturn V, 70

Schardt, Alois, 151, 201

Schreier, Ethan, 339–40, 349

Schroeder, Daniel, 257, 291

Schwarzschild, Martin, 128, 194

Science Management Program, 93–5, 196

Science Research Council, 135, 139

Science Working Group, 127, 172; and assembly and verification, 354; and Black Saturday proposals, 282; composition of, 257; and detectors, 107, 108–9, 241, 242; influence of, 276, 317–18, 379; and institute, 196–202; and international partners, 138–40; and (L)ST descoping, 145, 150, 272; and (L)ST instruments, 102–4, 138, 139, 240–1, 242, 246–8, 253, 279, 324–5; and members as (L)ST advocates, 100–2, 130, 170, 386; and minimum LST, 144, 145, 149, 150, 151–2

scientific instruments: and arrangement inside LST, 103–4; and inorbit replacement, 375; *see also* Space Shuttle, and (L)ST; *entries under individual instruments*

Scientific Uses of the Large Space Telescope, see Spitzer, Lyman and Ad Hoc Committee on LST

Scowcroft, Brent, 175

SEC ("Secondary Electron Conduction") Vidicon, 107–9, 241, 242–4, 246, 247–8, 252

"seeing cells," 15

selling process, 117–19

Senate Appropriations Subcommittee (on HUD, Space Science, Veterans and Certain Other Independent Agencies), 126, 134–5, 174, 185

Senate Authorizations Subcommittee (on Science and Space), 171

serendipity mode, 104

Shane, C. D., 45

Shipley, George E., 125, 133, 155–6, 162, 180, 259, 289

shuttle dividend, 163, 283, 445n65